Werner W. Müller

HDPE Geomembranes in Geotechnics

Werner W. Müller

# HDPE Geomembranes in Geotechnics

With 124 Figures and 56 Tables

Springer

Dr. Werner W. Müller
Federal Institute for Materials Research and Testing (BAM)
Division IV.3
Unter den Eichen 87
12205 Berlin, Germany
werner.mueller@bam.de

Library of Congress Control Number: 2006932413

ISBN-10 3-540-37286-5 Springer Berlin Heidelberg New York
ISBN-13 978-3-540-37286-8 Springer Berlin Heidelberg New York



Springer is a part of Springer Science+Business Media

springer.com



Typesetting: by the Author
Production: LE-TEX Jelonek, Schmidt & Vöckler GbR, Leipzig
Cover: medionet AG, Berlin

Printed on acid-free paper 68/3100YL - 5 4 3 2 1 0

ET IN ARCADIA EGO

Christine Berger

# Preface

The reasons for writing this book are described in the following preface translated from the original German edition. They are still valid. The German edition has found widespread distribution and friendly reception in the German-speaking countries. Right from the beginning the Author was encouraged to publish an English edition. Time had to pass until the technical conditions were right and spare time was found to work on an English translation.

Such an accurate English translation of the German text was provided by Nigel Pye (npservices4u@gmail.com) in cooperation with Tamás Meggyes and the translation benefited from their technical and scientific understanding. The translation was updated by the author to include new developments and to adapt it to the requirements of an international audience. The final English text was then proofread by the translators. Omissions and errors have to be attributed to the Author. Naturally, I would be glad not only to hear positive comments but also to get any information on shortcomings and deficiencies.

Nowadays HDPE geomembranes are widely used for large-area liners and for creating seals in geotechnical engineering. Lining of water ponds, dams and dykes, landfill liners and capping (cover) systems, remediation of contaminated sites, waterproof liners for tunnels, under highways and for various other civil engineering purposes are just a few examples of their application. The book covers all aspects of HDPE geomembranes: materials, manufacture, textured geomembranes, properties, long-term performance and testing, installation and welding, quality assurance and control, protective layers, leak detection, standards, recommendations and regulations. Various important topics are dealt with in detail. The basic physical and chemical facts necessary to fully understand HDPE geomembrane properties and performance are thoroughly analysed and explained. The book may therefore serve as a practical handbook for manufacturers, designers, testing and inspection engineers and consultants and representatives of responsible administrative bodies providing all relevant facts for design, manufacture and installation, testing and performance assessment. Parts of it may also be used as a textbook for lectures or seminars on geosynthetics and geotechnical or geoenvironmental engineering in faculties of materials science and civil engineering.

It is a difficult task to translate a text which deals with such a specialist technical field because differences in national technical traditions and approaches find their way into the different languages. Therefore, Nigel Pye and Tamá Meggyes's professional contribution and commitment in producing the English edition is greatly appreciated. Furthermore, I express my gratitude to all the members of the German association of geomembrane manufacturers and installers (AK GWS).

Of course, I renew and strongly affirm the thanks which I expressed to many persons in the preface of the German edition and it is my deep desire to include Jenny, Uta and Hedwig Berger and Irmela Schröter.

Berlin, June 2006

Werner Müller

# Preface of the German Edition

The conception of large-area High Density Polyethylene (HDPE) geomembrane liners occurred in Germany in the early 1970s (Garling landfill, SCHLEGEL sheet). The Author defines HDPE geomembrane as a planar polymeric sheet at least one and a half millimetres thick, several metres wide and a few dozen metres long made of medium to high density polyethylene[1]. HDPE geomembranes are normally pitch black due to fine carbon black added to protect them from UV radiation. The use of HDPE geomembranes for minor lining tasks in hydraulic engineering goes back well into the 1960s[2]. In 1977 F. W. Knipschild et al. first reported the use of HDPE geomembranes in landfill liners in the journal[3] "Kunststoffe im Bau (Plastics in Buildings)" and later in a special edition of the journal[4] "Müll und Abfall (Waste and Refuse)". Another important milestone was the commissioning of a large flat die extrusion line for 5 m wide HDPE geomembranes by A. Gruber in Linz, Austria (AGRU geomembrane) in 1984, followed by similar equipment constructed by A. Schlütter in Kempen-Tönisberg (Carbofol geomembrane) some time later.

Beginning in Germany, HDPE geomembranes started a world-wide "journey of triumph" achieving many successes in the USA and South Africa during the 1980s. HDPE geomembranes are currently being used for all kinds of large area lining purposes: dams, dykes, reservoirs, all kinds of treatment basins – such as tailings ponds or leaching ponds in mineral and ore processing –, landfill basal liners, landfill capping, sealing large-areas for the containment and remediation of contaminated land, tunnel construction, canal construction, large-area contiguous liners in industrial plants and road construction.

---

[1] It is hoped that the term "HDPE foil" used previously has meanwhile died out, as foils are planar, very flexible sheets with a thickness of up to 0.5 mm.

[2] Zitscher F-F (1971) Kunststoffe für den Wasserbau, Bauingenieur-Praxis, Nr. 125. Verlag Ernst & Sohn, Berlin.

[3] (1977) Kunststoffe im Bau 12(4) pp 154–160 and (1979) 14(3) pp 130–134.

[4] Stief K (ed) (1979) Müll und Abfall, Beiheft 15, Deponiebasisabdichtung, Erfahrungen, Stand der Technik, Forschung. Erich Schmidt Verlag, Berlin.

Landfills were lined using plastic geomembranes of various materials in the early 1970s. Gradually standards for long-term resistance of geomembranes to the high impact of landfills were raised. Due to the influence of high profile contaminated land cases (e.g. the Georgswerder hazardous and municipal waste deposit in Hamburg, Germany) the focus was on durability against attack by a large variety of aggressive chemicals. It was essential that welding should be safe from the process engineering point of view using simple and reliable techniques and should be easy to control. The concept behind large-area geomembranes was to reduce the extent of welding and seam length. Over the years such requirements for increased performance and the favourable price/performance ratio have resulted in the clear success of HDPE thermoplastic geomembranes over geomembranes made of other materials. Appreciation of HDPE materials received another boost when issues of ageing and increased demands for service lifetime moved to the forefront as opposed to resistance to chemical attack. In Germany only geomembranes of selected HDPE materials have been certified for the lining of landfills and contaminated land since the end of 1980s. This development in landfill engineering has also influenced other fields of application. HDPE geomembranes have thus largely replaced other materials (e.g. soft PVC, bituminous geomembranes etc.) not only in landfill liners, but in other fields of civil engineering too, where long-lasting large-area liners are needed.

In Germany 2 to 4 million square metres of HDPE geomembrane are installed annually. This geomembrane remains cost-effective even when manufactured from high-quality plastic material, although the per square metre price of installed geomembrane does exhibit considerable variation. For BAM-certified products on the German market a per square metre price of 3 to 4 € per millimetre thickness of geomembrane can be assumed as a rule of thumb. About a dozen major suppliers of HDPE geomembranes compete world-wide offering an annual total of at least 100 million square metres.

In view of the wide-scale applications and large amounts of HDPE geomembranes used in foundation, construction and hydraulic engineering, or in more general terms, in geotechnical engineering[5], it appears appropri-

[5] The Brockhaus Encyklopaedia, 19th Edition, provides the following definition: Geotechnics is an umbrella term for those branches of civil engineering which deal with the construction of underground structures or those on the earth surface in which soils and rocks are key components of the structure... Environmental protection measures such as containment of landfills and contaminated land are included ... Examples are foundations of all kind, slopes, tunnels, dams and dykes, landfills, canals. Theoretical fundamentals are soil mechanics and fluid mechanics

ate to examine this building product in detail within the framework of a special monograph.

Since the 1980s experts of the Federal Institute for Materials Research and Testing (BAM) in Berlin have been dealing intensively with scientific and engineering issues regarding the use of plastic geomembranes and geotextiles in landfill engineering and containment of contaminated sites. Work was initiated by H. August, then head of the Laboratory for Plastics Physics and Technology, and later carried on headed by the Author in the successor Laboratory of Landfill Engineering since 1991.

The early study concentrated on the long-term behaviour of geomembranes and contaminant transport within liners. Starting in 1989, suitability tests were performed for plastic geomembranes in landfill basal liners and caps and certifications issued for the German state of Lower Saxony. Initially, use was made of own scientific results and the "Guidelines for Geomembrane Landfill Basal Liners (Deponiebasisabdichtungen aus Dichtungsbahnen)" published by the State Office for Water and Waste of North Rhine-Westphalia (Landesamt für Wasser und Abfall des Landes Nordrhein-Westfalen). However, since the requirement for the certification of plastic geomembranes has been incorporated in the Technical Instructions Hazardous Waste (TA Abfall)[6] and the Technical Instructions Municipal Waste (TA Siedlungsabfall)[7] "BAM certification" has been used throughout Germany as a proof of suitability of plastic geomembranes in landfill engineering.

---

and, currently, chemistry and microbiology in the field of landfill engineering. Plastics engineering can also be added due to an increasing use of geosynthetics.

Hydraulic Engineering is the construction activity with objectives in the field of water management, for the protection from natural catastrophes, to minimise loss of land, to avoid water shortages, to control soil water management, to reduce or avoid water pollution, to protect land and environment, to generate power, to build/maintain waterways and to serve fishery and recreation. Geosynthetics play a steadily increasing role in this field as well. Obviously there is a considerable overlap between geotechnical and hydraulic engineering.

Geotechnics is therefore an umbrella term which is somewhat too wide but still best suited to identify the field of application for HDPE geomembranes which are the objectives of our considerations.

[6] Second General Administrative Provision to the Waste Avoidance and Waste Management Act (Abfallgesetz (AbfG)), Technical Instructions Hazardous Waste, Part 1 (TI Hazardous Waste).

[7] Third General Administrative Provision to the Waste Avoidance and Waste Management Act (Abfallgesetz (AbfG)), Technical Instructions on Recycling, Treatment and Storage of Municipal Waste (TI Municipal Waste).

Opinions and experiences regarding certification have been exchanged with institutes and companies who are operating in the field of geomembrane liners in the USA where liners have undergone similar developments, although various characteristics have been emphasised in quite different ways. Developments in both Germany and the USA have significantly influenced the international course of events. At the peak of this development was the "First Germany/USA Geomembrane Workshop" in 1996, which offered a good opportunity to discuss and reflect critically on the "state of the art" achieved and on unresolved problems. Experience gained through these activities and the knowledge acquired provides the basis for this book.

Primarily this book deals with issues of materials science and materials testing, but aspects of engineering technology will also be dealt with. The specialist scientific background outlined above explains the selection of the topics: materials selection, manufacturing, testing, contaminant transport and in particular, long-term behaviour of HDPE geomembranes are all examined in detail. The fundamentals of chemical and physical characterisation of ageing processes in polyolefine plastics will also be discussed. The latter topic, i.e. long-term behaviour, is of fundamental importance to the development potential of plastic products in civil engineering. The scientific-technical level reached through our understanding of the long-term performance of HDPE geomembranes sets the standards to be achieved for other geosynthetic products.

Geomembrane welding is discussed in the light of the latest research results into welding properties of HDPE materials and characterisation of weld seam quality. A special chapter is devoted purely to the civil engineering topic of installation. Although the only example used is large-area landfill liner construction, experience and techniques developed with landfill liners are of great significance to all fields of application.

The topic of protective layers for geomembranes will be examined in somewhat greater detail, since properly designed protective layers are a prerequisite for a geomembrane's efficient functioning. Leak monitoring systems to locate leakage in installed geomembranes provide an additional interesting topic given that an increasing number of such systems are being offered and used.

An expert committee participated in developing BAM's certification procedures from the very beginning. The expert committee is a working group headed by a representative of the German Federal Environmental Agency (Umweltbundesamt (UBA)), its members being representatives of the responsible German State Authorities, consulting firms, third-party inspectors, testing institutes, resin manufacturers, geomembrane manufacturers, installers and BAM. The state of the art with respect to landfill liners has been extensively discussed and further developed within this expert committee. Not only has co-operation with the expert committee made a decisive contribution to forming the certification guidelines for geomembranes and protective layers, but has also produced various recommendations on requirements for installation companies, third-party inspectors and designing temporary landfill caps. Based on internal discussions within the expert committee, papers have, for example, been published on relevant technical contract conditions for the installation of geomembranes and geotextile protective layers in landfill liners. Thus in the last few years the joint activity of BAM and the expert committee has covered the entire field of geomembrane application in landfill engineering and contaminated land containment. This work has been referred to on numerous occasions.

The requirement tables of the Guidelines for the Certification of Geomembranes have been included in the appendix. Relevant standards and guidelines, plus an extensive list of the most appropriate references are also included. The book may therefore serve as a real handbook for manufacturers, designers, testing and inspection engineers and consultants and representatives of responsible administrative bodies providing all relevant facts for design, manufacture and installation, testing and assessment of performance.

Participating in meetings of the expert committee, either as members or guests over the years, many experts have indirectly contributed to this book: Dipl.-Ing. K.-H. Albers, Prof. Dr. H. August, Prof. Dr. H.-P. Barbey, Ing. K. Bohaty, Dipl.-Ing. W. Bräcker, Prof. Dr. E. Dahms, Dr. B. Engelmann, B.S. Ch. Eng. D. Etter, Dipl.-Ing. L. Glück, Dipl.-Ing. Ph. Frank, Mr. R. Hartmann, Prof. Dr. G. Heerten, Dipl.-Ing. G. Heimer, Dipl.-Ing. A. Hutten, Dipl.-Ing. W. Karczmarzyk (†), Dr. H. Klingenfuß, Dr. F. W. Knipschild, Dr. G. Koch, Dipl.-Ing. B. Kopp, Dr. G. Lüders, Dipl.-Ing. V. Olischläger, Dipl.-Ing. R. Preuschmann, Dipl.-Ing. W. Quack, Dr. F. Sänger, Dipl.-Ing. R. Schicketanz, Dr. S. Seeger, Dipl.-Ing. A. Schlütter, Dipl.-Ing. E. Spitz and Dipl.-Ing. K. Stief.

When such a book first sees the light of day, the Author would like to express his gratitude to numerous people. The book has been "nurtured" by the teamwork of the colleagues of the Laboratory of Landfill Engineering: Mr. H. Böhm, Mrs. B. Büttgenbach, Ms. I. Jakob, Dr. G. Lüders, Dipl.-Ing. R. Preuschmann, Dr. S. Seeger, G. Söhring and Mrs. Dipl.-Ing. R. Tatzky-Gerth. The role of Ms. I. Jakob, Dr. S. Seeger and Dr. G. Lüders should especially be emphasised since their specialist contribution was fundamental to Sections 5.4 and 7.2 (Ms. Jakob), Chapters 8 and 11 (Dr. Seeger) and Section 10.3 (Dr. Lüders). Prof. Dr. H. August has introduced the Author into this field and shared his treasure trove of knowledge and experience over many years. The Author's wife Mrs. Ch. Berger, Mr. C. Gerloff, Dr. G.

Lüders, Dr. F. W. Knipschild, Dipl.-Ing. R. Schicketanz, Prof. Dr. E. Schmachtenberg, Dr. S. Seeger, Dipl.-Ing. P. Trubiroha and Dipl.-Ing. N. Vissing have proofread various chapters of the German edition and given valuable advice. Mr. E. Klementz has also managed the book project on behalf of Birkhäuser Verlag in a very professional way. The Author gratefully acknowledges Dr. M. Bahner's special contribution to this book.

Responsibility for the content of the text and appendices and in particular, for any errors and omissions that may occur, lies solely with the Author. It is thus hoped that readers will provide him a wealth of criticism, additions and improvements.

Berlin, February 2001

Werner Müller

# Contents

# 1 Technical Regulations

Special technical regulations exist for geomembrane application primarily in the geotechnical fields of landfill, hydraulic and tunnel engineering. These regulations often explicitly refer to HDPE geomembranes. Also of relevance are the state-of-the-art rules for building sealing technology, which are described in standards issued by national and international organisations for standardization as well as in guidelines and recommendations of specialist technical organisations. In the following only a few hints can be given centring on regulations in Germany, USA and UK. In many other countries relevant regulations exist. All these regulations are in many ways similar. However, there are typical and important differences, such as in the requirements for minimum thickness, installation procedure and durability. All these topics will be discussed in the respective chapters.

A rich and useful source of information is the proceedings of the conferences of the International Geosynthetics Society and its national chapters (IGS, www.geosyntheticssociety.org) and the society's journals "Geotextiles and Geomembranes" and "Geosynthetics International" as well as the proceedings of the GRI-conferences published by the Geosynthetic Institute (GSI, www.geosynthetic-institute.org) in Philadelphia. An overview of the state of the art in the field of geomembranes from the middle of the 1980s might be taken from the proceedings of the 1984 International Conference on Geomembranes (N.N. 1984) and in the early 1990s from the Rilem Report 4, "Geomembranes, Identification and Performance Testing" (Rollin and Rigo 1991). On 10th and 11th June 1996 the first German-American Workshop on the use of geomembranes in landfill liners took place in the Federal Institute for Materials Research and Testing (Bundesanstalt für Materialforschung und –prüfung (BAM), www.bam.de) in Berlin. An intensive discussion took place in the workshop about experiences, state of the art and future development trends. The minutes were published as a special issue of "Geotextiles and Geomembrane" providing an overview of the state of the art in both countries (Corbet and Peters 1996).

The regulations for landfill basal liners and landfill capping systems, a field of special interest for HDPE geomembrane application and of exemplary significance for the whole geomembrane business, will therefore be discussed in greater detail (see also Chap. 9).

**Table 1.1.** Countries with regulations in which plastic geomembrane components in landfill liner systems are required or as alternative option recommended. Thickness (mm) and plastic material are indicated (Holzlöhner et al. 1995; Koerner and Koerner 1999)

| Country | Hazardous solid waste | | Municipal waste | |
|---|---|---|---|---|
| | Basal liner | Capping system | Basal liner | Capping system |
| Australia (NSW) | n/d, HDPE | | | |
| Austria | ≥ 2.5, HDPE | | ≥ 2.5, HDPE | |
| Belgium | n/d | | n/d | |
| Botswana | ≥ 2 | | | |
| Brazil | n/d | n/d | | |
| Canada (Alberta) | n/d | | | |
| Canada (New Brunswick) | | | ≥ 2.5, HDPE | |
| Canada (Novia Scotia) | | | 1.5, HDPE | 1.5, HDPE |
| Canada (Ontario) | 2.0, HDPE | | 1.5, HDPE | |
| Canada (Prince Edward Island) | | | ≥ 2.0, HDPE | ≥ 1.0, LDPE |
| Canada (Quebec) | | | n/d | ≥ 1.0, n/d |
| China (Hong Kong) | n/d, HDPE | | n/d, HDPE | n/d |
| Denmark | | | n/d | n/d |
| France | n/d | n/d | n/d | n/d |
| Germany | ≥ 2.5, HDPE | ≥ 2.5, HDPE | ≥ 2.5, HDPE | ≥ 2.5, HDPE |
| Hungary | ≥ 2.0, HDPE | | ≥ 2.0, HDPE | |
| Israel | | | ≥ 1.5, n/d | |
| Italy | ≥ 2.5, HDPE | | n/d | |
| New Zealand | | | ≥ 1.5, n/d | |
| Poland | | | n/d | |
| Portugal | n/d | | n/d | |
| Russia | n/d, HDPE | | n/d | |
| South Africa | 1.5–2.0, n/d | | | |
| Sweden | | | ≥ 1.0, n/d | |
| Swiss | | | ≥ 2.5, HDPE | n/d |
| China (Taiwan) | | | ≥ 1.5, HDPE | |
| Thailand | ≥ 1.5, HDPE | ≥ 1.0, HDPE | ≥ 1.5, HDPE | ≥ 1.5, n/d |
| United Kingdom | ≥ 2.0, n/d | | ≥ 2.0, n/d | |
| USA | ≥ 1.5, HDPE | ≥ 1.5, HDPE | ≥ 1.5, HDPE | ≥ 0.5, n/d |

n/d: thickness and/or plastic material (polymer) not designated

Since the middle of the 1980s considerable developments and changes have taken place in landfill engineering as well as in other fields of waste management in Germany and in a similar way in other industrial countries. The ideas of the multi-barrier concept (Stief 1986) stimulated extensive research into the characteristics and interactions of barriers (geological and hydro-geological landfill site conditions, landfill liners and capping systems, waste body, landfill use and aftercare). In this research, the issues of further development of landfill liner systems were dealt with and in particular the performance, durability and installation of geomembranes and protective layers were investigated (August et al. 1997; Landreth 1984). The overall goal of all these research efforts was to ensure reliable protection against environmentally hazardous emissions from landfills and, most importantly, guarantee long-term groundwater protection. Meanwhile, geomembranes are stipulated as an integral part of landfill basal liner and capping systems in the regulations of many countries (Table 1.1) and in most cases HDPE geomembranes are explicitly recommended or, if not, the plastic material is not explicitly specified (Holzlöhner et al. 1995; Koerner and Koerner 1999). Still in most countries a compacted clay liner or even a simple soil layer is recommended as capping (Table 1.1).

In the 1980s and early 1990s it was thought that a plastic geomembrane had a service lifetime of less than 100 years, which was (and is) true for various plastic materials, and that a soil layer, especially a compacted clay liner, was considered as the long-term component in the landfill basal liner and capping system emulating a naturally watertight geological formation. Nowadays, however, the view has changed completely: it has been shown that the time for the functional engineering properties of certified and properly installed HDPE geomembranes to become significantly affected by aging processes is so long (many centuries) that aging is not relevant in design considerations for landfill capping systems using such properly selected HDPE geomembranes. On the other hand it has been revealed that a single conventional compacted clay liner on top of the waste body and underlying a drainage and restoration layer, which is typically 1 m thick, can be destroyed by root penetration and burrowing animals and, above all, by crack formation due to desiccation processes (Melchior et al. 2001). In the long term, processes of soil formation will transform the conventional compacted clay liner unprotected by a geomembrane into a part of the restoration layer (Suter et al. 1993). The time scale over which this will happen will depend on local conditions (weather, vegetation, restoration layer etc.). Therefore today, the HDPE geomembrane is considered as the part of the composite liner having equal if not greater importance than the compacted clay liner and there is considerable concern about the compacted clay liner as a single liner in a capping system (Simon and Müller 2004).

HDPE geomembranes can be combined with various other components (leak monitoring system, geosynthetic clay liners of high durability, polymer amended sand-bentonite-mixtures, capillary barriers) to form alternative composite liners as reliable and cost effective capping systems (Simon and Müller 2004). Such alternative capping systems are tested and increasingly applied for landfills and for the containment of contaminated sites.

The European Council Directive on the landfill of waste, which entered into force in July 1999 and had to be implemented by Member States by July 2001, divided landfills into three classes and established requirements on the geological barrier as well as on the basal liners and capping systems. The Member States must ensure that existing landfill sites cannot continue to operate unless they comply with the provisions of the Directive as soon as possible. Since many landfills in Europe have neither an acceptable geological barrier nor a basal liner, a capping will be of great importance to achieve the required protection against potential hazards of landfills to ground water, soil and air. Therefore, the performance and cost of capping systems will become an important issue in many European countries.

The European Council Directive on the landfill of waste recommends an impermeable mineral layer for the capping system of non-hazardous landfills and a composition of an artificial sealing layer and an impermeable mineral layer for hazardous landfills. However, the regulations of the Member States and the competent local authorities are allowed to take thoroughly justified technical developments into account. In the light of the above mentioned concerns the recommendations should be definitely modified to the extent that HDPE geomembranes might also be used as artificial sealing layers for surface lining of non-hazardous landfills.

Based on the research achievements and with support from an expert committee, BAM's own Laboratory of Landfill Engineering developed the technical specifications for plastic geomembranes and protective layers used to line landfills and contaminated sites, which have been published in certification guidelines[1] (Müller 1995; Müller 2001), where recommendations and guidelines of specialist associations have been referenced as far as possible. In accordance with these specifications only HDPE geomembranes have been certified for landfill applications.

The Geosynthetic Research Institute (GRI) has published a detailed listing of HDPE geomembrane requirements[2], GRI Test Method GM13, including quality assurance measures, which, in similar fashion to the BAM certification guidelines, have been developed in close co-operation with

---

[1] Available via www.bam.de

[2] Available via www.geosynthetic-institute.org

experts in manufacture and other institutions in relevant fields (GRI 2006). It also publishes special test methodologies for geosynthetics as GRI test methods and standards.

In the Netherlands, the quango KIWA (Keurinsinstitut for Waaterkeiding, www.kiwa.nl), which offers certification and research, training and consultancy in various fields, has published requirements for the certification of HDPE geomembranes, namely the BRL-K358 certification guideline. In France, the certification organisation ASQUAL (association des Centres Techniques pour assurer la promotion de la qualité et la certification, www.asqual.com) has established a quality assurance system for HDPE geomembrane manufacturing. However, the overall level of requirements on index and performance properties is not nearly as extensive and strict as those of BAM certification or GRI GM 13 specification, especially concerning long-term behaviour.

In a similar way to BAM granting its certification for geomembranes in landfill engineering, the German Institute for Construction Engineering (Deutsches Institut für Bautechnik (DIBt), www.dibt.de), in Berlin, acting as common state authority in the field of construction, certifies (under the German Groundwater Protection Act) geomembranes for the use in all kinds of plants for storage, filling and trans-shipment of substances dangerous to water. These specifications have been compiled in the "Geomembrane Certification Principles" (Zulassungsgrundsätze (ZG) Kunststoffbahnen für LAU-Anlagen) produced by an expert committee "Coatings and Geomembranes" under the auspices of DIBt (DIBt 2000).

The internet platform www.geosynthetica.net provides comprehensive construction documents and technical support information for engineers, regulators, contractors, installers, and facility owners involved with civil and geotechnical works containing geosynthetics and especially plastic geomembranes.

The Working Group 5.1 of the chapter "Geosynthetics in Geotechnics" (Kunststoffe in der Geotechnik) of the German Society for Geotechnical Engineering (Deutsche Gesellschaft für Geotechnik e. V. (DGGt), www.dggt.de) also deals with issues of geomembrane application in geotechnics, mainly in relation to landfill and hydraulic engineering. Their results have been published as the so called GLR-recommendations (GDA-Empfehlungen) jointly with the Working Group 6.1 of the chapter "Geotechnics of Sanitary Landfill and Waste Disposal" (Geotechnik der Deponiebauwerke) of DGGt (DGGt 1997b), see Appendix 2, Table A2.4. Current drafts and amendments of GLR-recommendations are to be found in the September issue of the German journal "Bautechnik" (Construction Technology), Ernst & Sohn Publishers, Berlin.

The technical regulations for tunnel engineering should also be mentioned here, as medium and high-density PE geomembranes are steadily gaining importance due to their long-term performance. A sub-group of Working Group 5.1 of the DGGt is working on developing recommendations for geomembranes used in tunnel liner applications. Extensive recommendations have already been published for waterproof liners in traffic tunnel structures (DGGt 1997a). With regard to the fundamentals of engineering design for tunnel liners using geomembranes the guidelines of Deutsche Bahn AG are of importance and in particular, Guideline 853:99-03 *Design, Construction and Maintenance of Railway Tunnels, Part 10 Liners and Drainage* (Eisenbahntunnel planen, bauen und instand halten, Teil 10, Abdichtung und Entwässerung). This guideline[3] expects geomembranes to exhibit "long-term durability"; however, no special plastics engineering tests or requirements have been stipulated for their long-term performance. As to the properties of PE geomembranes, the SIA standard V280:1996 *Plastic Geomembranes (Polymer Geomembranes) – Threshold Values and Materials Testing* (Kunststoff-Dichtungsbahnen (Polymer-Dichtungsbahnen) – Anforderungswerte und Materialprüfung), of the Swiss Society of Engineers and Architects (Schweizerischer Ingenieur- und Architektenvereins (SIA), www.sia.ch) is sometimes referred to in tunnel engineering.

Obviously, for the identification, performance and suitability testing of HDPE geomembranes one should use standard test methods as far as possible, see Appendix 2, Table A2.1 and A2.2. The International Geosynthetics Society (IGS) published an "Inventory of Geomembrane Standards" which contains the standards of national and international standardization bodies[4]. National standards of European states have now mostly been replaced by European standards of the European Committee for Standardization (CEN, www.cenorm.be). The standardization work for geomembranes takes place in the Technical Committee TC 154 "Geosynthetics" and TC 254 "Flexible Sheets for Waterproofing", sometimes forming joint working groups.

Relevant standards can be researched, ordered and, more recently, viewed at the expense of the user via the websites of standards organizations such as German Institute for Standardization (Deutsches Institut für Normung (DIN), www.din.de or www.beuth.de), Austrian Standards Institute (Österreichisches Normungsinstitut, www.on-norm.at), Swiss Stan-

[3] To be obtained via: DB Netz, Zentrale NEF 1, Theodor-Heuss-Allee 7, D-60486 Frankfurt am Main.

[4] To be obtained via IGS Secretariat, P.O. Box 347, Easley, South Carolina 29641-0347, USA, email: IGSsec@aol.com.

dards Association (Schweizerische Normen-Vereinigung, www.snv.ch), British Standards Institute (BSI, www.bsi-global.com), AFNOR (Association Française de Normalisation, www.afnor.fr), UNI (Ente Nationale Italiano di Unificazione, www.unicei.it), AENOR, (Asociación Española de Normalización y Certificación, www.aenor.es), and so on.

The current edition of the DIN pocket book 150 (DIN 1998) contains a compilation of DIN, CEN or ISO standards that are used for geomembranes. The Austrian Institute for Standardization published two special standards for geomembrane application in landfill engineering: ÖNORM S 2073:1998-03 *Landfills – Plastics Geomembranes – Requirements and Testing* (Deponien – Dichtungsbahnen aus Kunststoff – Anforderungen und Prüfungen) and ÖNORM S 2076-1:1999-10 *Landfills – Plastics Geomembranes – Installation* (Deponien – Dichtungsbahnen aus Kunststoff – Verlegung), which, however, are out of date, since the minimum service lifetime assumed is only 25 years and the testing of various important properties is missing.

All the standards relevant for geosynthetics from the American Society for Testing and Materials (ASTM, www.astm.org) are regularly published in its annual book of ASTM Standards, Volume 04.13, Geosynthetics (ASTM 2003), see Appendix 2, Table A2.2. Further information might be taken from the website of the Committee D35 "Geosynthetics" and its subcommittee D35.10 on geomembranes.

Geomembranes are dealt with in the Technical Committee T 221 of the International Organization for Standardization (ISO, www.iso.org).

The German Society for Welding Technology and Associated Methods (Deutschen Verband für Schweißen und verwandte Verfahren e. V. (DVS), www.dvs-ev.de) publishes guidelines and standards on geomembrane welding (DVS 1998), see Appendix 2, Table A2.4. Within the society the Working Group W4 "Geomembrane Welding" (Fügen von Kunststoffen), and in particular its subgroup W4.7, deals, among others, with the use of HDPE geomembranes in landfill engineering. Some of the standards are available in English (www.dvs-verlag.de/en/). On the basis of DVS standards and recommendations for the qualification of geomembrane welders the German association of geomembrane manufacturers and installers (Arbeitskreis Grundwasserschutz (AK GWS), www.akgws.de), has organized a certification procedure for geomembrane installers.

Comparable to the German DVS, in the United Kingdom, The Welding Institute (TWI, www.twi.co.uk) offers information and advice on welding plastic geomembranes and, in cooperation with the British Geomembrane Association (www.bga.uk.net), a certification scheme for geomembrane welders was established. The UK Environment Agency has specified that from April 2004 on all landfill sites in the UK at least two site operators must be accredited to the full level (based on BS EN 13067:2003 *Plastic Welding Personnel – Qualification Testing of Welders – Thermoplastics Welded Assemblies*) of the certification scheme and all other site operators (except for a maximum of one trainee) to the entry level of the scheme.

The International Association of Geosynthetic Installers (IAGI, www.iagi.org) has published a HDPE geomembrane installation specification and offers a HDPE welding certification program.

Finally, the plastics database CAMPUS should be mentioned here. About forty plastics manufacturers have joined forces to compile a database for their customers. This provides information on the most important physical-chemical parameters and processing properties of all plastic materials on the market and in particular of the HDPE resins according to a unified scheme. The website www.campusplastics.com provides information on the database and access to the various data sets of the companies involved.

## References

ASTM (2003) Annual Book of ASTM Standards, Volume 04.13, Geosynthetics. American Society for Testing and Materials (ASTM), West Conshohocken

August H et al. (eds) (1997) Advanced landfill liner systems. Thomas Telford, London

Corbet SP and Peters M (1996) First Germany/USA Geomembrane Workshop. Geotextiles and Geomembranes 14: 647–726

DGGt (1997a) Empfehlung Doppeldichtung Tunnel-EDT. Verlag Ernst & Sohn, Berlin

DGGt (1997b) GDA-Empfehlungen. Verlag Ernst & Sohn, Berlin

DIBt (2000) Zulassungsgrundsätze für Kunststoffbahnen als Abdichtungsmittel von Auffangwannen, Auffangräumen, Auffangvorrichtungen und Flächen für die Lagerung und das Abfüllen und das Umschlagen wassergefährdender Stoffe (ZG Kunststoffbahnen in LAU-Anlagen). Deutsches Institut für Bautechnik (DIBt), Berlin

DIN (1998) DIN-Taschenbuch 150, Kunststoff-Dachbahnen, Kunststoff-Dichtungsbahnen, Kunststoff-Folien und kunststoffbeschichtete Flächengebilde (Kunstleder). Beuth Verlag, Berlin

DVS (1998) Taschenbuch DVS-Merkblätter und -Richtlinien, Fügen von Kunststoffen. DVS-Verlag, Düsseldorf

GRI (2006) GRI Standard GM13: Test Properties, Testing Frequency and Recommended Warrant for High Density Polyethylene (HDPE) Smooth and Textured Geomembranes. Geosynthetic Institute (GSI), Folsom, USA

Holzlöhner U et al. (eds) (1995) Landfill liner systems, a state of the art report. Penshaw Press, Cleadon, UK

Koerner JR and Koerner RM (1999) A survey of solid waste landfill liner and cover regulations: Part II – worldwide status, GRI report #23. Geosynthetic Research Institute, Folsom, USA

Landreth RE (1984) The Role of Flexible Membrane Liners in Support of RCRA Regulations. In: Proceedings of the International Conference on Geomembranes. Industrial Fabrics Association International (IFAI), St. Paul, Minnesota (USA), pp 21–23

Melchior S et al. (2001) A comparison of traditional clay barriers and the polymer-modified material trisoplast in landfill covers. In: Cossu R et al. (eds) Proceedings Sardinia 2001, Eighth International waste management and landfill symposium. Environmental Sanitary Engineering Center (CISA), Cagliari (Italy)

Müller WW (ed) (1995) Anforderungen an die Schutzschicht für die Dichtungsbahnen in der Kombinationsdichtung, Zulassungsrichtlinie für Schutzschichten. BAM, Berlin

Müller WW (ed) (2001) Certification Guidelines for Plastic Geomembranes Used to Line Landfills and Contaminated Sites. BAM, Berlin

N.N. (1984) Proceedings of the International Conference on Geomembranes. Industrial Fabrics Association International (IFAI), St. Paul, Minnesota (USA)

Panofsky E (1983) Et in Arcadia Ego: Poussin and the Elegiac Tradition. In: The Meaning of the Visual Arts. The University of Chicago Press, Chicago, pp 295–320

Rollin A and Rigo J-M (eds) (1991) Geomembranes, Identification and Performance Testing. Chapman and Hall, London

Simon FG and Müller WW (2004) Standard and alternative landfill capping design in Germany. Environmental Science & Policy 7: 277–290

Stief K (1986) Das Multibarrierensystem als Grundlage von Planung, Bau, Betrieb und Nachsorge von Deponien. Müll und Abfall 18: 15

Suter GW et al. (1993) Compacted Soil Barriers at Abandoned Landfill Sites are Likely to Fail in the Long Term. Journal of Environmental Quality 22: 217–226

# 2 HDPE Materials and Geomembrane Manufacture

## 2.1 Materials

The plastic produced by the polymerisation of ethylene molecules $H_2C = CH_2$ is called polyethylene (PE). As a result of the polymerisation process, polyethylene molecules of different molecular structure are produced, depending on the physical conditions, the type of polymerisation reaction and the chemical environment. This leads to materials with various densities and differences in the morphology, thus to very different material properties. The plastic essentially consists of the polymeric material. However, during manufacturing or processing a package of various chemicals is usually added which may significantly influence the materials behaviour. PE is therefore an umbrella term for a wide range of plastic materials whose only common feature is that $-CH_2-$ (methylene group) is their basic component within the polymer chain.

The oldest manufacturing process dating back to the 1930s is the so-called high-pressure polymerisation in which oxygen or an other free-radical generator triggers a polymerisation reaction in ethylene at high temperature (up to 275 °C) and high pressure (up to 280 MPa) (Ehrlich and Mortimer 1970). The polymer produced exhibits a highly branched structure with a small number of long side chains (1–5 per 1000 C atoms) and a large number of small side chains (20–30 per 1000 C atoms) (Fig. 2.1). Accordingly, the polymeric material has a low density and low crystallinity.

However, using transition metal catalysts (so called Ziegler-Natta type catalysts (Ziegler 1965)), ethylene can be polymerised at even considerably lower pressures (< 10 MPa) and temperatures (< 200 °C) in what is termed the low-pressure polymerisation (Table 2.1 and Fig. 2.2) (Nowlin 1985). Depending on the chemical environment, the physical conditions and characteristics of the catalyst, different manufacturing processes can be distinguished. In the solution phase polymerisation the polymers emerge in a dissolved state in a liquid hydrocarbon (Standard Oil process), in the suspension polymerisation (Slurry or Phillips process) they are produced as solid particles suspended in a liquid hydrocarbon and in the gas phase polymerisation (also called Unipol process) ethylene gas (and co-monomers, see below), hydrogen gas and inert gas carriers are fed into a bed of powdered catalyst particles. As a result of all three processes, linear polyethylene polymers are produced almost completely lacking a branching structure, i.e. they contain only 1–2 ethyl side chains per 1000 C atoms (Fig. 2.1) and the polymeric material has a high density and high crystallinity.

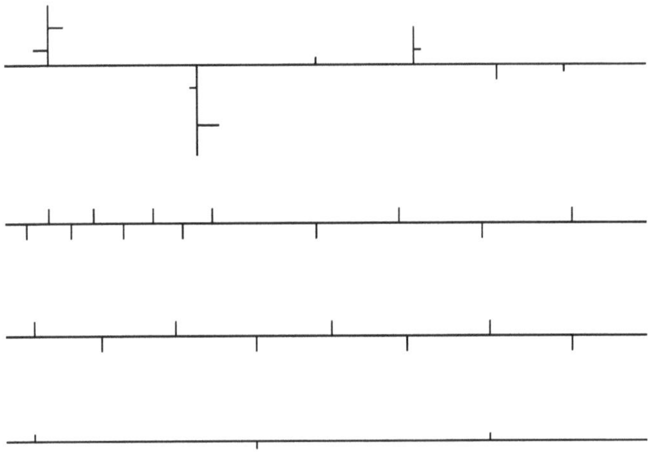

**Fig. 2.1.** Schematic illustration of polyethylene molecular structure of various density ranges (Elias 1992). Top: LDPE, radical polymerisation yields a number of (long) side chains. Bottom: HDPE, catalytic polymerisation gives rise to linear chains with a small number of short branches. Both drawings in the middle illustrate LLDPEs produced by catalytic polymerisation with α-olefines. Small amounts of butene-1, hexene-1 or octene-1 co-monomers lead to ethyl, butyl or hexyl side chains. Polymerisation in the gaseous phase produces chains arranged in a block-shaped fashion and distributed at various frequencies along the chain. Solution phase polymerisation provides a statistical random distribution along the whole chain

A density-based method has been introduced for polyethylene classification relying on ASTM-standards (Table 2.2). The density-based classification coincides to some extent with that based on the polymerisation processes. Consequently, polyethylene with low density (LDPE) is usually a polyethylene material produced by high-pressure polymerisation and polyethylene with high density (HDPE) is a polyethylene material manufactured by low-pressure polymerisation. Since the end of the 1970s copolymers of ethylene and α-olefines (butene-1, hexene-1, octene-1) have been produced in large quantities mainly using low-pressure polymerisation processes (Fig. 2.2). This results in a polyethylene polymer with ethyl groups ($-C_2H_5$), butyl groups ($-C_4H_9$) and hexyl groups ($-C_6H_{13}$) branching off along the linear polyethylene chain (Fig. 2.1).

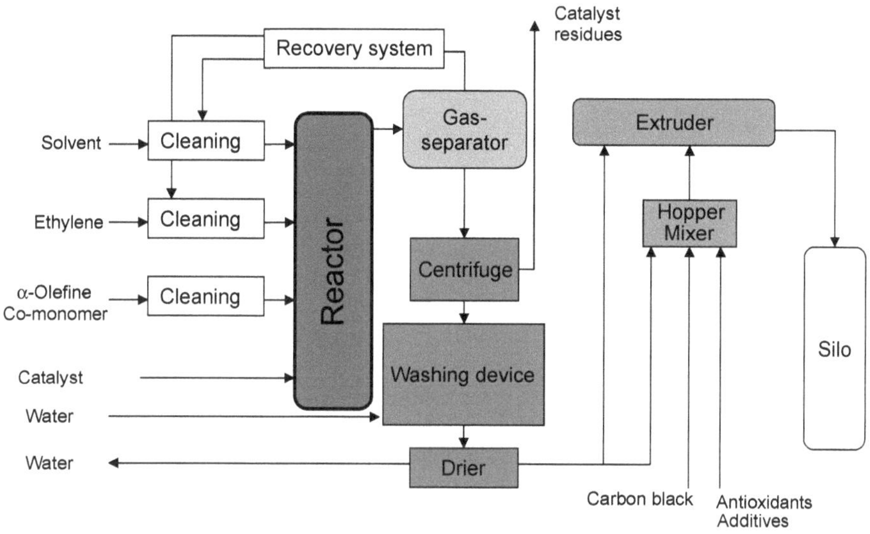


**Fig. 2.2.** Simplified flow chart of the Phillips process (solution phase polymerisation) (Elias 1992). The solvent (for example isobutane), ethylene, co-monomers and the catalyst are fed into the reactor. The polymer solution is degassed (flashed) from the reactor through a gas separator after a certain reaction time. In a cleaning unit (centrifuge, washing device, drier) the polymer is separated out and ethylene and solvent residues are processed. The polymer emerges in the drier in the form of snow white flakes. Flakes, carbon black, stabiliser and further additives, for example Ca-stearate, are mixed in a mixer and this mix is fed into an extruder at an adequate mix ratio with the polymer flakes. Here the mix is melted, homogenised and finally granulated. The black pellets are then transported to the storage facilities

This polymeric material exhibits many advantageous features similar to those of HDPE (for example resistance to chemicals), it does not, however, show the disadvantages of HDPE, which is initially a tendency towards stress crack formation. As the density of this material is relatively low, it is difficult to put it into any of the above categories and it has been therefore termed Linear Low Density Polyethylene (LLDPE) (Bork 1984). It should also be noted that the terms MDPE and LMDPE are hardly used any longer these days: the materials being included in the LDPE or HDPE and LLDPE groups.

**Table 2.1.** Examples of process parameters for low-pressure polymerisation of polyethylene (Elias 1992; Miles and Bristion 1979; Vieweg et al. 1969)

| Process | Ziegler | Phillips | Standard Oil | Unipol |
|---|---|---|---|---|
| Physical reaction conditions | Polymer-aggregates suspended in hexane | Suspended in butane or dissolved in cyclo-hexane | Dissolved in xylol | Imbedded in gas flow |
| Catalyst | $TiCl_4(C_2H_5)_2AlCl$ on Mg-carrier | Partially reduced $Cr_2O_3$ on $Al_2O_3$ | Partially reduced MoO on $Al_2O_3$ | Cr/Ti-Mg on carrier |
| Pressure (MPa) | 0.8–3.5 | 2.8–5.0 | About 7 | 0.7–2.1 |
| Temperature (°C) | 50–90 | 85–175 | < 200 | 85-100 |
| Reactor | CSTR | Pipe loop reactor | | Fluidised bed reactor |
| Residence time (h) | 2–3 | 1.5 | | 3–5 |
| Control of molecular mass via | $H_2$ | Temperature | | $H_2$ |
| Catalyst-productivity (g PE/g Catalyst) | 3,000 | 3,000–10,000 | | 9,000 |

CSTR: continuous stirred tank reactor

Geomembranes suitable for civil engineering purposes are manufactured from polyethylenes produced using low-pressure polymerisation with a few per cent by weight of butene or hexene or octene co-monomers. Densities of the pure polymeric materials are around 0.932 to 0.942 g/cm³. There is a great variety of materials starting from those with a very narrow molecular mass distribution (polydispersity $M_w/M_n \approx 4$, see Sect. 3.3) up to those with a broad distribution width (polydispersity $M_w/M_n \approx 15$). The typical properties of these resins are displayed in Table 2.3. However, since the density of geomembranes coloured with carbon black and made of LLDPE resins is usually above 0.941 g/m³, the term HDPE geomembranes has also become common usage for these geomembranes.

Data such as density, melt flow rate, melting temperature, crystallinity, number-average or mass-average of molecular mass distribution and polydispersity are not yet satisfactory to fully characterise the properties of polymeric materials. The specific manufacturing process and process-related parameters define further properties such as type and extent of impurities, for example catalyst residues, low-molecular fractions and co-monomer distribution within and between polymer chains, which are difficult to quantify.

**Table 2.2.** Polyethylene classification and PE resin nomenclature as per ASTM D1248-84 *Standard Specification for Polyethylene Plastics Moulding and Extrusion Materials* (formerly) and ASTM D883-96 *Standard Terminology Relating to Plastics* (currently)

| Density* ($g/cm^3$) | Manufacturing process | Old terms as per ASTM D1248 | New terms as per ASTM D883 |
|---|---|---|---|
| 0.910–0.925 | Radical polymerisation | Low Density (LD) | Low Density (LD) |
| 0.919–0.925 | Catalytic polymerisation | | Linear Low Density (LLD) |
| 0.926–0.940 | Radical polymerisation | Medium Density (MD) | Medium Density (MD) |
| 0.926–0.940 | Catalytic polymerisation | | Linear Medium Density (LMD) |
| 0.941 and above | Catalytic polymerisation | High Density (HD) | High Density (HD) |

*) Measured on pure, non-pigmented materials

**Table 2.3.** HDPE resin properties for geomembranes in geotechnical engineering

| Characteristic | Property |
|---|---|
| Co-monomer | Butene-1, hexane-1, octene-1 |
| Co-monomer fraction | < 10 % by weight |
| Density | 0.932–0.942 $g/cm^3$ |
| Melt mass-flow rate (190/5) | 0.3–3 g/10min |
| Melting temperature | ≈ 130 °C |
| Crystallinity* | 50–55 % |
| Number-average molecular mass, $M_n$ | 15,000–50,000 |
| Polydispersity, $M_w/M_n$ | 4–15 |

*) Heat of melting relating to 293 J/g for crystalline HDPE

Even this fails to fully characterise the final plastic material, since additives (e.g. antioxidants and light stabilisers) are mixed into the polymer resin, from which the geomembrane is extruded, either early by the resin manufacturer or during the course of geomembrane manufacture (Gugumus 1990; Zweifel 2001). Such additives are of paramount importance for the use of plastics, since they help to guarantee an appropriate service lifetime of the plastic products.

Oxygen can oxidise polyethylene as well as other organic molecules. Radicals act as trigger agents for the complex chemical reactions. A radical emerges when paired electrons of a chemical bond (for example those between hydrogen and carbon atoms along the polymer chain, denoted by *P*) get separated and the molecule fragments with the unpaired electrons cannot find each other immediately. An unpaired electron looking for a new bond is denoted with a point (˙) in the chemical formulae. Important examples are the alkyl radical $\dot{P}$ and the peroxy radical $PO\dot{O}$, the latter formed with oxygen. The radicals, together with oxygen, initiate a chain of reactions, which means that a reaction cycle is repeated a number of times before it stops, see Sect. 5.2.1. Each cycle produces hydroperoxide molecules $POOH$. This molecule disintegrates into radicals again (for example into the peroxy and the alkyl radical) which in turn initiate new oxidising reaction cycles. In the course of the reaction cycle and due to the termination reaction, polymer chains are decomposed or cross-linked. The polymeric material becomes embrittled and the product from the material can lose its capability to function properly.

Radicals can be created by impurities (catalyst residues). Thermal fluctuations, stresses at high temperatures and the attack by oxygen molecules itself can all result in radical formation. UV radiation has a considerable impact. Radicals are created without UV radiation only extremely infrequently under normal conditions. Oxidation therefore progresses very slowly. Only at high temperatures does radical formation and oxidative reaction mechanisms get considerably accelerated, for example under conditions prevailing during polyethylene processing.

To inhibit the oxidation process, certain chemical compounds (antioxidants or stabilisers) are added to the polyethylene, see Sect. 5.2.2. The so-called chain-breaking-donors or primary antioxidants, also called inhibitors, react with the chemical radicals. They thus interrupt the reaction chain. The so-called hydroperoxide decomposer or secondary antioxidants react with the hydroperoxide before it can disintegrate into radicals. Thus they prevent the start of new reaction chains.

Compounds of the following groups belong to the primary antioxidants: steric hindered phenols, secondary aromatic amines and steric hindered amines. To the secondary antioxidants belong compounds of the following groups: phosphites and phosphonites, organic sulphides or thioether (thiosynergists). Sometimes a special nomenclature is applied to the antioxidants: compounds in the phenol and amine groups are numbered with symbols "AO-*n*" and the hindered amines with symbols "HALS-*n*", while phosphites and phosphonites are denoted with "PS-*n*" and thioether with "S-*n*". Table 2.4 displays the antioxidant groups, together with the commercial names used to market the products and temperature ranges, in which they are effective. A detailed and comprehensive overview is given in (Zweifel 2001).

**Table 2.4.** Overview of antioxidants (based on (Hsuan and Koerner 1998; Zweifel 2001))

| Substance name | Examples of trade names[2] (Zweifel 2001) | Effective temperature range (Fay and King III 1994) |
|---|---|---|
| Primary antioxidants | | |
| Phenols | Irganox® 1076 (AO-3), Irganox 1330 (AO-13), Irganox 1010 (AO-18) | up to 300 °C |
| Steric hindered amines (HALS)[1] | Tinuvin® 770 (HALS-1), Tinuvin 622 LD (HALS-2), Chimassorb® 944 (HALS-3), Hostavin® , N 30 (HALS-31), Uvinul® 4050H (HALS-21), 4049H (HALS-20), 5050H (HALS-22) | up to 150 °C |
| Secondary antioxidants | | |
| Phosphites | Irgafos® 168 (PS-2) | 150–300 °C |
| Organic sulphides (thioether) | Dilauryl thiodipropionate (DLTDP), Distearyl thiodipropionate (DSTDP) | up to 200 °C |
| Stabiliser packet | | |
| Phosphite and phenol, 1:1 | Irganox B225 | |
| Phosphite and phenol, 2:1 | Irganox B215 | |

[1]) Hindered Amine Light Stabiliser (HALS)
[2]) Irganox, Chimassorb, Tinuvin, Irgafos are trade names of Ciba Speciality Chemicals, Hostavin is a trade name of Clariant and Uvinul is a trade name of BASF

An additive package is, as a rule, added to polyethylene, which includes compounds effective as process stabilisers at high process temperatures and other antioxidants also effective as long-term stabilisers at application temperatures. A combination of various antioxidants can enhance the efficacy of the individual components by synergetic effects. The amount of a component added is about a few hundred ppm, thus approx. 0.5 to 2 % by weight of antioxidants is present in the material. The "conventional" stabiliser package consists of a high-molecule phenolic antioxidant (long-term stabiliser) and a phosphite (process stabiliser). Behaviour and efficacy of this package have been extensively investigated.

In addition to the antioxidants, other co-stabilisers, such as metal soaps as acid acceptors, are also added. Metal soaps are used to neutralize residues of acids formed from catalyst carriers and to prevent corrosion of the extruder. To some extent they act as a lubricant to improve flow properties. Transition metal ions effectively catalyse peroxide decomposition. Therefore metal deactivators are sometimes added. These substances form stable complexes with metal ions. The oxidative stability of HDPE in contact with metals is thereby largely improved.

The polymer manufacturers keep the type and amount of antioxidants secret. In spite of the small amounts of antioxidants, their addition is of great importance for the long-term behaviour of the material[1] - not only from a technical point of view. The chemicals are very expensive and, in spite of their small amounts, make up a considerable part of the material price.

The damaging effect of the sunlight's ultraviolet (UV) fraction on plastics has already been mentioned; this applies in particular to polyethylene. Absorption of UV photons on impurities and structural irregularities, such as hydroperoxides, carbonyl groups, double bonds formed during processing, on catalyst residues and on charge-transfer complexes from the polyethylene and molecular oxygen can give rise to production of free radicals at high quantum yield, which initiate oxidative chain reactions. This is called photo-oxidation and will be discussed in greater detail in Sect. 3.2.14. Light stabilisers have the task of protecting the materials from UV radiation. They are substances that absorb the damaging UV light or "quench" the excited states preceding the formation of free radicals. Antioxidants that "neutralize" the emerging free radicals also exert a light protection effect in this sense. There are a large number of chemical compounds that can be used as light stabilisers, their description is however beyond the scope of our discussion (Zweifel 2001).

---

[1] Ageing and old-age diseases (for example diabetes) are also caused in humans by radicals and their oxidising effect. "Fortunately our bodies are not completely defenseless against the attacks of free radicals. Antioxidants such as vitamin E and C can soak up free radicals and thereby slow the propagation of reaction chains. Beside these vitamins, a number of other enzymes break up free radical reactions by removing molecules formed in the middle of reaction sequence. Nevertheless, free radicals are thought to be an important continuing source of chemical damage to vital macromolecules, particularly DNA, proteins and fatty lipids of cell membranes." (Ricklefs RE and Finch CE (1995) Aging, A Natural History, Scientific American Library. Freeman and Company, New York, p 24). As the stabiliser package determines the operational life of the plastic material, a diet, rich in vitamins and ballast materials, is of paramount importance for the life expectancy of the materials scientist.

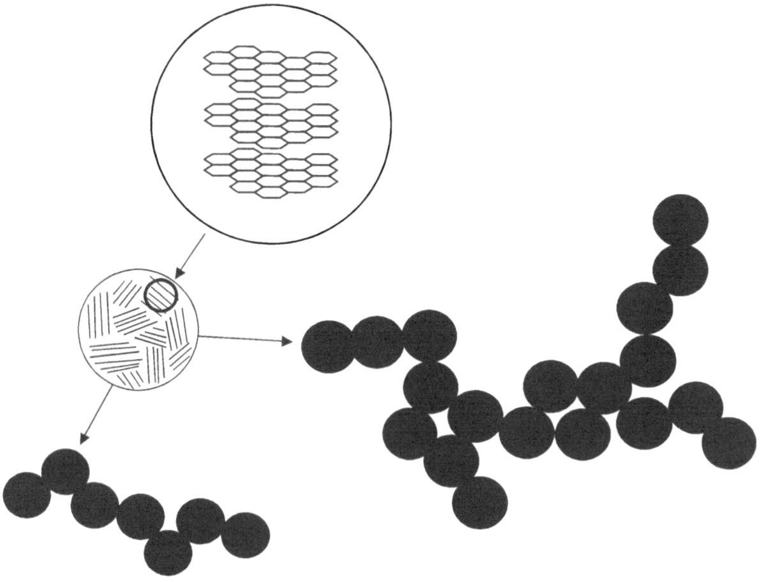

**Fig. 2.3.** Schematic illustration of carbon black morphology. Primary particles (black circles) are built of stacks of layers in which, similar to graphite, the carbon atoms are arranged in a hexagonal lattice. The distance of the layers is about 3.5 Å, the size of the stacks approx. 2 nm, while the size of the primary particles is about 10 to 100 nm. The primary particles agglomerate to primary aggregates, i.e. the actual carbon black particles, which in turn can stick together. Depending on the size and complexity of the agglomerates they are called low-structured carbon black (left) or high-structured carbon black (right). The size of the agglomerates for low-structured carbon black is about 1 micrometer. See also Figs. 3.1a and b

Finely dispersed carbon black is a light stabiliser, which absorbs UV radiation and has proven to be the most effective (and relatively cheap) light screen for polyolefins (Accorsi and Romero 1995; Bode 1969), with the only disadvantage that it dyes the plastic deep black. Although aesthetic aspects must be considered in connection with many consumer goods, fortunately they are of lesser importance for the geomembranes in civil engineering. However, black geomembranes exposed to the sun in the open become more strongly heated, and this has a great influence on wave formation due to thermal expansion. Nevertheless, experienced installers have mastered this problem, so HDPE geomembranes are almost always stabilised with carbon black.

Carbon black is, as a rule, produced by imperfect combustion of hydrocarbons (mineral oil, bituminous coal, tar oil, natural gas etc.). Depending on the process (channel process, gas carbon black process, furnace process[2] and flame carbon black process), carbon black of various properties can be produced (Bode 1969; Gilg 1989). The following basic properties are used to characterise carbon black (Fig. 2.3): medium size of primary particles (spherical structures made up from graphite layer stacks), size and structure of the aggregates (i.e. extent of agglomeration of primary particles forming bigger grape-like structures, which are the primary carbon black aggregates), area and chemical properties of the carbon black surface (oxygen-containing groups on the surface and physically adsorbed substances) and moisture adsorption capacity. The size of primary carbon black particles used for pigmentation, UV stabilisation or conductivity-enhancement is around 10 to 100 nm and their surface around 25 to 1500 $m^2/g$. Fine-grained carbon black is best suited for UV stabilisation because absorption of UV quanta increases with reducing primary particle size, reaching a maximum at 20 nm (Accorsi and Romero 1995). On the other hand, carbon black must be as homogeneously dispersed as possible in the polyethylene to achieve good absorption and the agglomerates of carbon black powder must be broken up into the primary aggregates. For this purpose a rather highly structured carbon black with a not too high surface is best suited. All aspects considered, a fine-grained but highly structured furnace carbon black is used for UV stabilisation. Homogeneity of carbon black distribution is usually assessed by microscopic examination, see Sect. 3.2.3 and Figs 3.1a and b.

Carbon black and further additives can be added to the polymer powder (flakes) as the last process step of HDPE resin production itself, from which black resin pellets with homogeneously dispersed carbon black can then be extruded. However, the granulated, naturally coloured resin can also be mixed with carbon black during geomembrane production. For this purpose a special additive, the so-called carbon black batch, is produced from a carrier resin, usually LDPE, with about 40 % by weight of carbon black. During geomembrane extrusion (see Sect. 2.3) the pellets of the carbon black batch and the pellets of the naturally coloured HDPE resin are mixed in a hopper preceding the extrusion process. Extrusion causes

---

[2] In the furnace process, mineral oil bottoms or bituminous coal tar oils are sprayed into and burnt in a closed refractory brick furnace in a fuel gas flow (natural gas or manufactured gas) at a specified oxygen deficiency. The oil fragments dehydrogenate and polymerize to carbon black particles. The residence time of the particles in the combustion chamber is controlled by water injection. Various carbon black types can be manufactured, depending on the process parameters.

the carbon black to be homogeneously dispersed. The batch is dosed in such a way that the final product contains approx. 2 % by weight of carbon black to which about 3 % by weight of LDPE carrier resin is also added.

The mixing of carbon black into polyethylene can drastically change its conductivity (Gilg 1989). Depending on the carbon black properties, a network of carbon black aggregates can form with a concentration of about 8 % by weight and above, which will enable electron transport. From the non-conducting polyethylene (specific resistance > $10^{15}$ Ω cm) an electrically conducting material is thus formed with a specific resistance of around $10^{0}$ Ω cm. Using co-extrusion it is also possible to manufacture geomembranes consisting of a conducting and an insulating layer.

All this indicates that polyethylene geomembranes are manufactured from a plastic material of fairly complex composition. The HDPE material of the geomembrane is characterised by the properties of the base polymer, the manufacturing process, the composition of the additive package and carbon black batch. Changes in the parameters of the polymer, of the manufacturing process and composition of the additives may considerably influence the properties of the geomembrane, especially its long-term behaviour. Extensive investigation into the long-term behaviour and the subsequent suitability tests make it a prerequisite that the material of a tested HDPE geomembrane should be characterised unambiguously. The suitability tests are only valid for those products whose material, usually within very narrow limits, corresponds to the material of the tested product.

## 2.2 Morphology

In addition to the material composition, the morphology, which forms during processing, determines the geomembrane properties and not only the melting behaviour, mechanical properties, resistance to chemicals etc., but also long-term behaviour, such as creep, oxidative degradation and stress crack formation (Barhem 1993; Kanig 1975; Strobl 1997).

Polyethylene is a thermoplastic material. Above the melting point it is transformed into a molten material of a completely irregular, amorphous tangle of polymer chains. A pseudo-order exists only in the range below 20 Å, similar to liquids. Crystalline structures are formed from this amorphous phase during cooling. The lamella is the microcrystalline building component of the crystalline structures: it is a planar formation to which polymer chain segments fold together perpendicular to the lamella surface (Fig. 2.4) while the overall polymer coils do not untangle. Certain parts of a polymer chain fold in a section of the lamella, the chain then exits the lamella and submerges into either the same lamella in a different place or into another adjacent lamella and folds once again. Therefore, amorphous areas of loops, chain ends and sections of the polymer chains bridging the lamellae (so-called tie molecules) are present between the lamellae (Fig. 5.16). The typical lamella thickness (c-axis) is between 100–300 Å. The planar extension (a-axis and b-axis) can be of up to a few micrometres.

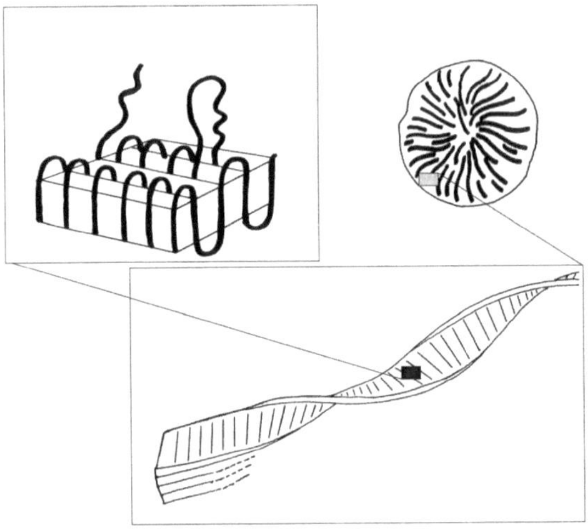

**Fig. 2.4.** Sketch of the basic structural components in HDPE geomembrane morphology – based on our understanding as of today. Folded polymer chains form extended lamellae (top left). Twisted stacks of lamellae create longitudinal fibrils (bottom). The sphere-shaped spherulite (top right) is composed of radial fibrils pointing outwards with amorphous areas in between. This structural model enables the interpretation of basic processes, such as stress crack formation, see Sect. 5.3.4

On the macrocrystalline level in linear polyethylene, typical, closely arranged, sphere-shaped crystalline areas, the so-called spherulites, can be observed in the crystalline phase with a size of about up to 10 μm (Fig. 2.4). The lamellae are arranged into longish lamella stacks, which are directed outwards from the centre of the spherulite. The b-axis along the greater longitudinal dimension of the lamellae points in the direction of the spherulite radius, while the a-axis of the lamellae in the stacks, being perpendicular to the radius, rotates. Thus the stacks create outward directed spirals (fibrils) (Fig. 2.4).

The lamella stacks, which grow during spherulite formation, may break and branch out, new lamella stacks may emerge or grow together with others. The ball-shaped spherulite is filled with such crystalline fragments.

The space between the microcrystalline areas within a spherulite and between the spherulites is filled with the amorphous mass of chain loops, chain ends and bridging tie-molecules emerging from the lamellae. Half of the PE geomembrane material typically consists of amorphous ranges and another half of crystalline ranges.

The carbon black particles are dispersed within the material composed of spherulites of various sizes, joined and connected by the “cement” of the amorphous areas. Low-molecular polymer fractions, antioxidants and other low-molecule additives are dissolved in the amorphous areas. This is the place where contaminants or oxygen may dissolved and diffuse and oxidative degradation takes place. Microcracks within the amorphous areas, caused by disentanglement of the tie-molecules between the crystalline areas while the material is exposed to stress, are the initiators of stress crack formation, see Sect. 5.3.4.

Apart from material properties, the manufacturing process has a considerable influence on shaping the morphology. The manufacturing process imposes a certain orientation on the complex structure of amorphous and crystalline components in the geomembrane. Therefore, the mechanical properties and sensitivity to stress crack formation are different across and along the extrusion direction. Warming-up of the geomembranes relaxes this orientation to some extent, see Sect. 3.2.5. If the orientation is too strong, warming-up may cause considerable elongation resulting in wave formation and stress development in the weld seam area during welding. Morphology of the geomembrane changes locally when hot polyethylene particles are impinged on the geomembrane surface, where they melt together with the base geomembrane forming a surface texturing, see Sect. 6.1. Such inhomogeneity in morphology might be the cause of a strong sensitivity in stress cracking behaviour to the process parameter and texturing material, which is observed in impinged textured geomembranes (Seeger et al. 2000).

## 2.3 Manufacture

Geomembranes for large-area liners are manufactured using the flat die extrusion (or cast sheeting) or the circular die extrusion process (or blown sheeting). Since installation must be performed quickly under the constraints of the construction schedule and weld seams as potential weak points should be avoided, geomembranes should be manufactured as wide as possible. The minimum width of state-of-the-art geomembranes used for examples in landfill engineering today is 5 m. Powerful machines, currently available are capable of manufacturing 9 m wide geomembranes using flat die extrusion and 7 m wide ones using the blown sheet extrusion process. However, geomembrane rolls must also enable easy handling during transport and installation. With a minimum thickness of 2.50 mm, as required in the technical instructions for landfill liners in Germany, and a typical length of 100 m, a 10 m wide geomembrane roll would weigh 2.5 tones. If geomembranes much wider than 10 m were applied, typical civil engineering projects with installation areas of a few hectares would require very expensive handling and transport logistics.

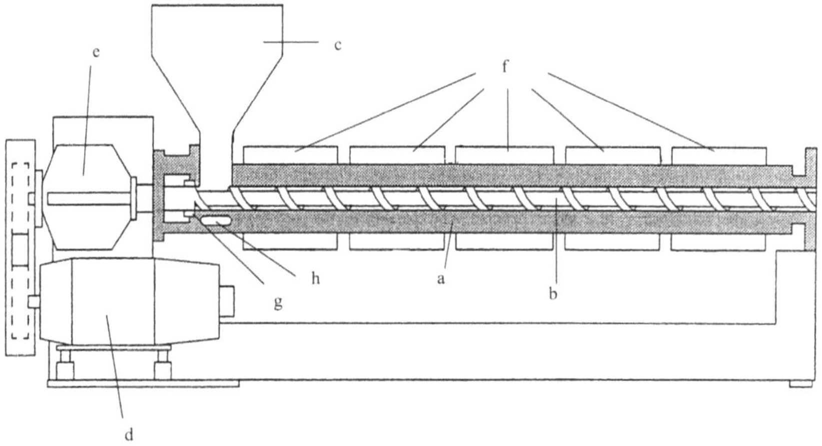


**Fig. 2.5.** The basic components of an extruder are the barrel (a), with a screw (b) running within it. The pellets are fed into the extruder via a mixer and a feed hopper (c). A motor (d) and a gearbox (e) drive the screw. The speed is about 100 $min^{-1}$. The barrel is fitted with adjustable heating and cooling devices (f). Connection points for screw heating (g) and cooling ducts (h) in the barrel are indicated in the figure. In the extruder the pellets are melted, homogenised and simultaneously transported. This should be performed in a steady-state process at the lowest possible thermal and mechanical impact. The barrel must be able to cope with a considerable pressure (a few tens MPa). The surface of the barrel and screw are made of special alloys selected with respect to the polymers to keep friction and adhesion of the molten material as low as possible. Source: (Struve 1994)

The common feature of all geomembrane manufacturing processes is that the resin pellets (and carbon black batch) are melted or plastisised and homogenised with respect to both temperature and materials distribution in an extruder, followed by the moulding of the molten material using suitable tools. This thermoplastic material processing technique is highly developed and specially adapted to various applications. The extrusion processes are described in more detail in Hensen et al. 1997, while a comprehensive summary of the basic aspects of geomembrane extrusion can be found in (Struve 1994). Figures. 2.5 and 2.6, courtesy of F. Struve, and a few photographs will be used to illustrate the processes.

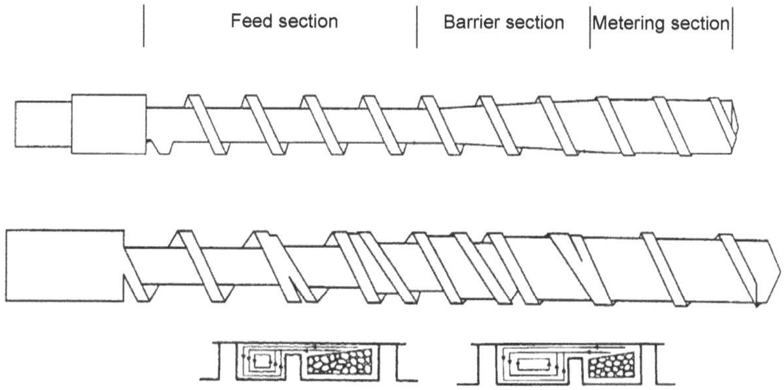


**Fig. 2.6.** The extruder screw (top) can be divided into three sections. The initial screw flight depth of the feed or transportation zone is quite deep, since here only slowly melting pellets of small density are transported. In the compression or transition section the pellets are converted into a homogeneous molten material with greater density, therefore the depth must steadily decrease. In the output or metering section the molten material is finally transported to the extrusion exit and pressure for the supplementary transport of the delivered melting strand is developed. The pellets melt on the surface of the barrel and the screw. A core of unmelted material initially remains in the melting strand of the barrel canal which makes constant and complete plastisising and homogenisation more difficult. Therefore so-called barrier screws (centre) with dual threads are used in the transition section. Depth and volume of the first barrel channel decrease, those of the second increase. The barrier screw flight is lower than the normal screw flight. The molten material formed near the wall therefore enters the second channel and accumulates there. The solid mass remaining in the first channel can be more easily melted (bottom). Many varied forms and geometry of extrusion screws are possible. The simplest form of such a screw, the so-called Maillefer screw is shown in this figure. Particularly good homogenisation and dispersion of pellet mixtures can be achieved when two or more screws are interlocked (double-screw or planetary screw extruder) (Limper and Stehr 1990; Müller and Dienst 1990). The design of extruders is a minor science in itself (Hensen et al. 1997)

A feed hopper and a mixer continuously feed resin pellets (and carbon black batch pellets if necessary) and a certain amount of granulated edge trimming from the running geomembrane production into the extruder (Fig. 2.5). The main part of the extruder is the extrusion screw, which rotates in a heated barrel (Fig 2.6). Typical process temperatures for HDPE resins are around 200 to 230 °C. The extruder is often followed by an independently driven gear pump, which ensures a steady mass flow at a constant pressure out of the extruder into the die. Before the stream of molten materials enters the die it is first pressed through the sieves of a screen pack exchanger where impurities (foreign bodies and non-disintegrated pellets) are retained. The screen pack usually consists of three coarse, medium and fine meshed woven stainless steel wire cloths which are backed up by a steel plate which has a dense pattern of holes, the so called breaker plate.

**Fig. 2.7.** Flat die of equipment for 5 m wide HDPE geomembranes. The die is fed by two extruders, connections far right. The arrangement of the countersink screws on the top indicates the line of the coathanger-shaped duct. The choker bar can be adjusted by the nuts of the protruding bolts, and the die lips can be adjusted by the hexagon socket screws on the far left. The carpet of molten material leaves the die lip horizontally and is pulled into the gap between the two bottom rollers of the polishing roll stack. This and the following two figures courtesy Naue GmbH & Co. KG

In the flat die process the melting strands flow into the flat die (Fig. 2.7) and spread across the flow direction into the breadth through the coathanger-shaped distribution duct of the die. The distribution duct leads to a slit-shaped canal across the whole width of the die, which the molten material enters and where the "carpet" of molten material is formed. The molten material is thereby directed above a choker bar, which can be used to locally adjust various slit heights. The carpet then leaves the die lips whose aperture width can also be adjusted. The choker bar and lips are adjusted in such a way that the carpet of molten material leaves the die at the same velocity and thickness across its whole width. Choker bar and lip adjustment allow to control the geomembrane thickness with high accuracy.

**Fig. 2.8.** Overall view of a 5 m equipment. The two extruders are in the foreground. Degassing units are installed in the rear third of the extruders. Both extruders are connected through a filter screen and a melt pump to the flat die, which is arranged at right angles. In the cooling and polishing roll stack there are three chromium-plated, water-cooled rollers arranged vertically one above the other. Instead of smooth rollers, rollers with engraved surface patterns can also be used which shape and form a structure onto the surface of the carpet of molten uncrystallized material. The rollers in the cooling and polishing roll stack can be arranged in completely different ways (horizontal, vertical or combinations). The different arrangements in each case have pros and cons (Groß 1997). The railings and stairs in the foreground give an indication of the dimensions of the equipment

This viscous carpet is then usually polished and cooled in a cooling and polishing roll stack by three or more cooled rolls (Figs. 2.8 and 2.9). Instead of smooth rolls with polished surfaces, engraved rolls can be used to form the first nip in the roll stack. The pattern engraved into the rolls is then embossed onto the still highly viscous surfaces of the carpet of molten material. Using this technique geomembranes with different surface structures can be calendered (see Sect. 6). After transporting over an ambient temperature cooling line the geomembrane moves to the wind-up station where it is wound on a take-up spool (Fig. 2.10).

**Fig. 2.9.** Cooling and polishing roll stack and flat die of a production unit for 9 m wide geomembranes. The size of the railing and the 120 l waste container give an impression of the giant dimensions of the equipment. Two rolls are horizontally arranged and the third roll is offset downwards in the roll stack. The carpet of molten material is being drawn diagonally at 45° from the die into the roller nip

Before it enters the winder, about 5 to 10 cm are cut off from both right and left margin of the geomembrane to obtain sharp and straight edges. Following this, about 10–15 cm wide, thin protective foil or edge tape is laminated or glued[3] onto the edges on both surfaces to protect this area for welding from dirt accumulated during storage, transport and installation. The parameters of the flat die extrusion process are automatically recorded and controlled. Thickness measurement is part of the process control, for which a β-radiation thickness gauge can be used. Geomembranes manufac-

[3] The protective foil or edge strip must allow its removal prior to welding without any adhesive residue remaining on the geomembrane.

tured in this way exhibit a uniform and faultless surface. Thickness can be adjusted within narrow limits. The flat die extrusion enables a relatively strong orientation to be introduced into the material. The orientation can be characterised by measuring the dimensional stability (Sect. 3.2.5). The process parameters of flat die extrusion can be chosen so as to keep this test parameter within acceptable limits.

**Fig. 2.10.** Cooling system line of a production unit for 7 m wide geomembranes. Behind the cooling and polishing roll stack the geomembrane first passes through an ambient temperature cooling line, which can be as long as 20 m depending upon the size of the equipment, and continues to be cooled down by the ambient air. At the end of the cooling line the geomembrane is rolled up on a take-up spool (foreground). Typically 100 to 150 m long geomembranes are manufactured. Roll changes take place every 1 to 2 hours. In each case a cross-stripe is cut off to be tested in the manufacture quality assurance laboratory. The roll is often wrapped. If the properties are satisfactory according to various test results, the roll is released and transported to the delivery store

In the blown sheet extrusion process, after leaving the extruder, gear pump and screen pack, the resin moves to a circular spiral mandrel die in which a tube of molten material is formed and inflated with air from the inside. The tube is then lifted in a haul-off tower up to 40 m high where it is cooled by an internal and external flow of cool air. It then proceeds to the gable-roof-shaped layflat boards and into the nip roll assembly, responsible for the lifting and transport of the tube. The flattening of the tube when passing the nip rolls will cause two creases, which are characteristic of blown sheet extrusion geomembranes (Fig. 2.11). Before winding, the flattened tube is longitudinally cut and unfolded to form the geomembrane. Finally, a protective foil or edge tape is attached to the edges.

**Fig. 2.11.** View of a blown sheet extrusion line. In the background, bottom left, the circular die can be seen of which an HDPE hose, supported by supporting air, exits. The cut tube runs down from the top and covers the view onto the closed tube running upward into the layflat boards. The cut hose enters the take-up mechanism (middle ground) and is finally wound on a spool in the winding mechanism (foreground). The plant tower may be as high as 40 m and is visible on the factory site from a great distance. Figs 2.10 and 2.11 courtesy Serrot International

As with other processes, fully automated process control is applied. Surface quality and thickness variation of blown extrusion geomembranes may cause problems. Besides, properties within the area of the two creases may potentially deviate from those of the base material. Modern equipment however is capable of mastering all these problems.

Friction forces between geomembrane and subgrade, for example a mineral liner, or between the geomembrane and the overlying protective layer can be increased by providing one or both sides of the geomembrane with a surface structure or texture. Types of surface structure or texture, manufacturing methods and testing of textured geomembranes will be dealt with in detail in Chap. 6.

Equipment, requirements on geomembrane thickness, surface appearance and dimensional stability and resin type determine the production speed of HDPE geomembranes. High-quality 2.5 or 3 mm thick geomembranes with an embossed surface structure can only be manufactured slowly: the production speed is around 1 m/min. Significantly higher speeds can be achieved in the production of thin and smooth geomembranes amounting to up to a few metres per minute. Equipment output for 5 m wide geomembranes is around 1–2 t/h.

Quality management plays an important role in the manufacture of geomembranes for geotechnical applications (see Sect. 9.3). Within the framework of production quality assurance, samples are taken from resin deliveries and manufactured geomembranes at regular intervals to be tested in the geomembrane manufacturer's own laboratory. For each roll delivery, an acceptance protocol is issued in accordance with EN 10204:2004 *Metallic Products – Types of Inspection Documents*, Section 3.1B. Type, extent and frequency of the tests of this so-called in-house monitoring for the manufacture of BAM-certified geomembranes are displayed in Appendix 1, Tables A1.5 and A1.6, Part 1 and 2. Table A1.7 of Appendix 1 describes the test procedures and frequencies for third-party control of resin, carbon black batch and geomembranes as part of the in-house QMS of the manufacturers of BAM certified HDPE geomembranes.

Table 2.5 gives an overview of HDPE geomembrane manufacturers and their products[4] certified by BAM for landfill liners in Germany. Other manufacturers, like JUTA a. s., Dukelská 417, CZ-54415 Dvúr Králové n. L., www.juta.cz or Atarfil, Central Office and Factory, Ctra. Córdoba Km 429, Complejo El Rey, E-18230 Atarfe (Granada), www.atarfil.com, which manufacture up to 3 mm thick and at least 5 m wide HDPE geomembrane, are present on the European and world-wide market. More in-

[4] Status as per 2005. The current list of certified geomembranes can be viewed at www.bam.de/de/service/amtl_mitteilungen/abfallrecht/index.htm.

formation might be obtained from the European Geomembrane Association (EGMA), www.eupc.org/egma/. There are, of course, a lot of manufacturers who do not manufacture geomembranes for landfill liners (with a 2.5 mm minimum thickness) but for various other purposes. Information about manufacturers, who possess a construction supervisory certification by the German Institute for Construction Technology (Deutsches Institut für Bautechnik – (DIBt)) in the field of sealing plants for storage, filling and trans-shipment of substances dangerous to water, can be obtained from DIBt. Further reference and source of information are given in Chap. 1.

**Table 2.5.** Manufacturers of HDPE geomembranes certified for landfill liners in Germany

| Manufacturer | Contact and marketing partner | Equipment |
|---|---|---|
| AGRU Kunststofftechnik GmbH<br>Kunststoffwerk<br>Ing.-Pesendorfer-Str. 31<br>A-4540 Bad Hall | Contact might be obtained via www.agru.at | flat die extruder, 5 m wide geomembranes, with embossed structures on one or both sides |
| GSE Lining Technology GmbH<br>Großmoorring 4<br>D-21079 Hamburg<br>Produktionsstätte:<br>D-17248 Rechlin | Contact might be obtained via www.german.gseworld.com | flat die extruder, 7 m wide geomembranes, with sprayed on (or impinged) textures on one or both sides |
| NAUE GmbH & Co. KG<br>D-32339 Espelkamp-Fiestel<br>Gewerbestraße 2 | Contact might be obtained via www.naue.com | flat die extruder, 5 m and 9 m wide geomembranes, with embossed structures on one or both sides |

## References

Accorsi J and Romero E (1995) Special Carbon Black for Plastics. Plastics Engineering: 29–32

Barhem PJ (1993) Crystallization and Morphology of Semicrystalline Polymers. In: Thomas EL (ed) Structure and Properties of Polymers. VCH Puplishers Inc., New York, pp 153–212

Bode R (1969) Ruß als Pigment für Kunststoffe. Kautschuk und Gummi - Kunststoffe 22: 167–174

Bork S (1984) Lineares Polyethylen niedriger Dichte (LLDPE) – Eigenschaften, Verarbeitung und Anwendung. Kunststoffe 74: 474–486

Ehrlich P and Mortimer GA (1970) Fundamentals of the Free-Radical Polymerization of Ethylene. Advances of Polymer Science 7: 386–448

Elias H-G (1992) Makromoleküle, Band 2, Technologie. Hüthig & Wepf Verlag, Basel, Heidelberg, New York

Fay JJ and King III RE (1994) Antioxidants for geosynthetic resins and applications. In: Hsuan G and Koerner RM (eds) Proceedings of the 8th GRI Conference, Geosynthetic Resins, Formulations and Manufacturing. Industrial Fabrics Association International (IFAI), St. Paul, USA, pp 74–92

Gilg R-G (1989) Ruß für leitfähige Kunststoffe. In: Mair HJ and Roth S (eds) Elektrisch leitende Kunststoffe. Carl Hanser Verlag, München, pp 55–76

Groß H (1997) Wie sieht die günstigste Walzenanordnung aus? Kunststoffe 87: 564–568

Gugumus F (1990) Antioxidantien. In: Gächter R and Müller H (eds) Taschenbuch der Kunststoff-Additive. Carl Hanser Verlag, Munich, pp 1–103

Hensen F and Berghaus U (1997) Plastics Extrusion Technology. Hanser Verlag, Munich

Hsuan YG and Koerner RM (1998) Antioxidant Depletion Lifetime in High Density Polyethylene Geomembranes. Journal of Geotechnical and Geoenvironmental Engineering 124: 532–541

Kanig G (1975) Neue elektronenmikroskopische Untersuchungen über die Morphologie von Polyethylen. Progr Colloid & Polymer Sci 57: 176–191

Limper A and Stehr R (1990) Der Planetwalzenextruder - ein vielseitiges Aufbereitungsaggregat. Kunststoffe 80: 26–30

Miles DC and Bristion JH (1979) Polymer Technology. Chemical Publisching Co., Inc., New York

Müller WW and Dienst M (1990) In-line-Herstellung von Platten und Folien. Kunststoffe 80: 21–25

Nowlin TE (1985) Low Pressure Manufacture of Polyethylene. Prog Polym Sci 11: 29–55

Seeger S et al. (2000) Long term testing of geomembranes and geotextiles under shear stress. In: Cancelli A et al. (eds) Proceedings of the Second European Geosynthetics Conference. Pàtron Editore, Bologna, pp 607–610

Strobl G (1997) The Physics of Polymers. Springer Verlag, Berlin, Heidelberg

Struve F (1994) Extrusion of Geomembranes. In: Hsuan G and Koerner RM (eds) Proceedings of the 8th GRI Conference, Geosynthetic Resins, Formulation and Manufacturing. Industrial Fabrics Association International (IFAI), St. Paul, USA, pp 94–112

Vieweg E et al. (eds) (1969) Kunststoff-Handbuch, Vol. IV, Polyolefine. Carl Hanser Verlag, München

Ziegler K (1965) Consequences and Developement of an Invention. Rubber Chem Technol 38: 22

Zweifel H (ed) (2001) Plastic Additives Handbook. Carl Hanser Verlag, Munich

# 3 Testing of HDPE Geomembrane Properties

## 3.1 Overview

Tests on geomembranes serve various objectives. First performance properties must be determined, then whether the quality standards were adhered to while manufacturing the geomembrane from the selected material and installing it on the construction site must be checked. Part of this test is the identification of the materials, i.e. to decide whether or not the agreed upon resins and batches were used. For geotechnical applications, it is not only the performance properties, but the long-term behaviour, which is of particular importance.

Accordingly, it is common to classify the tests in the following groups: performance tests to check the performance properties, index tests to determine various parameters which characterise the product and may serve to define quality (see Sect. 9.4) and durability tests to check the long-term properties. A test should be described as clearly, unambiguously and completely as possible in a test standard. In a performance test the function of the geomembrane and the impacts acting on the geomembrane under field conditions must be simulated as realistically as possible. However, this simulation will be usually imperfect. The more clearly and precisely the test method is specified, the more the test is converted into a pure index test. The prime objective of an index test is to describe a certain property as clearly, selectively and reproducibly as possible in terms of a well-defined test quantity. In the durability or long-term performance tests, the test conditions are selected in such a way that the repercussions of ageing and long-term effects of impacts occur accelerated and can be investigated within an experimentally available timeframe. Here too, the test conditions are rather far from the real situation. Tests of performance properties and long-term behaviour thus resemble more or less the index tests. On the other hand, the specialist must often deduce the performance properties and long-term behaviour from index tests through experience and comparisons. The transitions between the three groups of tests are therefore not sharp and the classification is ambiguous.

**Table 3.1.** Testing of HDPE geomembranes

| Property (test procedure) | Performance | Quality assurance | Identification | Durability |
|---|---|---|---|---|
| Surface condition, appearance | | • | | |
| Homogeneity of cross section | | • | | • |
| Carbon black content | • | • | | • |
| Carbon black dispersion | • | • | | |
| Skew | • | • | | |
| Waviness | • | • | | |
| Thickness | • | • | | |
| Density | | • | • | |
| Melt flow rate (MFR) and change in MFR | | • | • | |
| Dimensional stability | • | • | | |
| Permeability to hydrocarbons | • | | | |
| Melting enthalpy and point | | | • | |
| Oxidation stability (OIT) | | | • | |
| Tensile properties | | • | • | |
| Behaviour under planar deformation (burst test) | • | | | |
| Tear resistance | • | | | |
| Resistance to static puncture | • | | | |
| Resistance to dynamic puncture | • | | | |
| Flexibility in low temperature | • | | | |
| Relaxation behaviour | • | | | |
| Seam quality (peel, tensile test) | • | • | | |
| Resistance to chemicals | • | | | • |
| Stress crack resistance | • | | | • |
| Resistance to thermo-oxidative degradation | • | | | • |
| Long-term behaviour under combined stressing | • | | | • |
| Weathering resistance | • | | | • |
| Microbe resistance | • | | | • |
| Root penetration resistance | • | | | • |
| Resistance to rodent damage | • | | | • |
| Stress crack resistance of textured geomembranes (long-term tensile test) | • | | | • |
| Adhesion of texture particles (long-term shear strength test) | • | | | • |
| Friction (Shear box test) | • | | | |

**Table 3.2.** Testing procedures and standards for HDPE geomembranes within the BAM certification, see Appendix 1

| Property (test procedure) | Standard | Reference* |
|---|---|---|
| Surface condition, appearance | DIN 16726, Section 5.1 | Section 3.2.1 |
| Homogeneity of cross section | | |
| Skew | DIN 16726, Section 5.2 | |
| Waviness | | |
| Thickness | DIN 16726, Section 5.3 | Section 3.2.2 |
| Carbon black content | DIN EN ISO 11358 or ASTM 1603 | Section 3.2.3 |
| Carbon black dispersion | ASTM D5596 | |
| Density | DIN EN ISO 1183-1, -2 | Section 3.2.4 |
| Melt flow rate (MFR) and change in MFR during manufacturing | DIN EN ISO 1133 | |
| Dimensional stability | DIN 16726, Section 5.13.1, DIN 53377, BAM procedure B14 | Section 3.2.5 |
| Permeability to hydrocarbons | ASTM 5886 | Section 3.2.6 |
| Melting enthalpy and melting point | ISO 11357-3 | Section 3.2.7 |
| Oxidation stability (OIT) | DIN EN 728 | |
| Tensile properties | DIN EN ISO 527-3 | Section 3.2.8 |
| Behaviour under planar deformation (burst test) | DIN 53861 | Section 3.2.9 |
| Tear resistance | DIN 53356-A | Section 3.3 |
| Resistance to static puncture (CBR test) | DIN EN ISO 12236 | Section 3.3 |
| Resistance to dynamic puncture | DIN 16726, Section 5.12 | Section 3.3 |
| Flexibility at cold temperature (bending test) | DIN EN 1876-1 | Section 3.3 |
| Relaxation behaviour | DIN 53441 | Section 3.2.10 |
| Seam quality (peel test, tensile test)) | DVS 2226-2 DVS 2226-3 | Section 10.2 |

Reference to sections of this book, in which the details of the test procedure and the requirements (usually on 2.5 mm thick HDPE geomembranes) are discussed

For example, a performance property, e.g.. flexibility (expressed by Y-oung's modulus), taken here as the deformation behaviour at small forces, can be deduced from the tensile test. At the same time, test quanities determined by the tensile test, such as yield stress and yield strain as well as maximum tensile force (strength) and strain at break serve as parameters for quality assurance and materials identification.

**Table 3.2, Part 2.** Testing procedures and standards for HDPE geomembranes within the BAM certification, see Appendix 1

| Property (test procedure) | Standard | Reference* |
|---|---|---|
| Welding properties of resin | BAM procedure | Section 10.3 |
| Resistance to chemicals | DIN EN ISO 175 | Section 3.2.11 |
| Stress crack resistance | ASTM D5397, DIN EN 14576 | Section 3.2.12 |
| Resistance to thermo-oxidative degradation | BAM procedure following ASTM D5721 | Section 3.2.13 |
| Long-term behaviour under combined stressing | DIN 16887, ISO 9080 | Section 3.2.13 |
| Weathering resistance | DIN EN 12224 | Section 3.2.14 |
| Microbial resistance | DIN EN ISO 846, Procedure D | Section 3.2.15 |
| Root penetration resistance | FLL test procedure for root penetration resistance (www.f-l-l.de) | |
| Stress crack resistance of textured geomembranes (long-term tensile test) | DVS 2203-4 | Section 3.2.16 |
| Adhesion of texture particles (long-term shear strength test) | BAM procedure | Section 3.2.17 |
| Friction (Shear box test) | DIN EN ISO 12957-1 GDA E3-8 | Section 3.2.18 |

*Reference to sections of this book, in which the details of the testing procedure and the requirements (usually on 2.5 mm thick HDPE geomembranes) are discussed

Misunderstandings and misinterpretations can occur in the interpretation of tests and test results if this lack of clarity in the meaning of the tests is not taken into account. This can be illustrated by two examples. At first sight, the multi-axial tension test (burst test) seems to provide information about deformation behaviour in a plane state of stress – an important performance property of geomembranes. However, the strain at yield or at burst of the geomembrane in the multi-axial tension test may not be read as the generally permissible strain limit. Indeed, this parameter is irrelevant for the long-term application since only the limiting strain permissible over the long-term is significant (see Sect. 8.3.1 and Sect. 5.3.4) since long-term performance is strongly influenced by the stress crack resistance of the material. Some parameters, the strain at break of textured geomembranes in the tensile test, for instance, must be interpreted with care. The strain at break is sometimes strongly reduced by texturing the HDPE geomembrane surface (see Sect. 6.2). However, strain at break is certainly not a relevant performance property: permissible deformations in a building structure are much smaller. Strong reduction in strain at break is therefore irrelevant for the performance of textured geomembranes. However, the test of strain at break may be used for quality assurance and investigations of long-term behaviour, since the test quantity is very sensitive to changes in the materials properties. Strain at break in geomembranes with a textured surface depends highly on the type of specimen and the specimen preparation technique. The type and position of the surface structure components on the specimen influence the extent and range of the shear flow above the yield point where a rapid tear off can occur at notches or places of stress concentration due to the surface structure. Therefore, the strain at break of a textured geomembrane can only be used for quality assurance when the test conditions are precisely specified for each individual case. Strains at break of various textured geomembranes may then not be compared to one another.

A large number of tests are in principle possible on plastic geomembranes and geomembrane materials to qualify the products for the use in geotechnical structures. The scope of tests chosen for the characterisation and suitability approval of a geomembrane, however, will certainly not include all conceivable tests. The tests are selected depending on the expected short-term loads and deformations, the effective long-term impacts, the requirements for quality assurance, which must identify deficiencies rapidly and clearly, and on the need to identify the material as unambiguously as possible. Some tests, e.g. the creep behaviour of an HDPE geomembrane, are not directly relevant as HDPE geomembranes must always be installed in such a way that they do not carry loads over the long term. Other performance properties, such as robustness to mechanical loads, can be determined by many different tests. In this case one must decide upon a selection which is limited to only a few tests. Nevertheless, redundant tests cannot completely be avoided. Furthermore, the selection of the identification and durability tests depends greatly on the type of geomembrane material itself.

A detailed description of all tests and relevant test methods on plastic geomembranes is beyond the scope of this book. In the following, those procedures will be dealt with that are used within the BAM certification for plastic geomembranes. First of all those tests of particular importance will be examined for which no full and unambiguous description for HDPE geomembranes exists in any standard specification or no agreed upon standard specification is yet available. A detailed but to some extent outdated overview of test methods for geomembranes used in geotechnics is published in (Rollin and Rigo 1991).

Table 3.1 shows the tests on HDPE geomembranes that are generally, or in special cases, used within the BAM certification framework assigned to different test purposes. Special tests are necessary for geomembranes with a textured surface: testing the "core thickness", testing the repercussion of the surface texture on stress crack resistance, testing the long-term shear strength of applied structure particles and testing the friction properties. Testing of the properties of textured geomembranes is dealt with in greater detail in Sect. 6.2. The testing of weld seams is described in Sects. 10.2 and 10.3.

Table 3.2 lists the properties tested and the test procedures used in the BAM certification. The last column contains references about the respective section of this and other chapters in which the tests and requirements are described in detail. A compilation of requirement tables can be found in Appendix 1.

**Table 3.3.** Testing, standards and requirements for HDPE geomembranes according to test method GM 13 of the Geosynthetic Institute (GSI)

| Property, Test | Test Standard | Test Value (for 2.5 mm nom. thickness) |
|---|---|---|
| Thickness (lowest individual of 10 values) | ASTM D5199 | Nom. thickness ≥ ave. nom. thickness – 10% |
| Density | ASTM D1505<br>ASTM D792 | 0.940 g/ml (min.) |
| Tensile Properties<br>Yield stress<br>Break stress<br>Yield elongation<br>Break elongation | ASTM D638, Type IV | <br>37 kN/m (min. ave.)<br>67 kN/m (min. ave.)<br>12 % (min. ave.)<br>700 % (min. ave.) |
| Tear Resistance | ASTM D1004 | 311 N |
| Puncture Resistance | ASTM D4833 | 800 N |
| Stress Crack Resistance | ASTM D5397 | 200 h |
| Oxidative Induction Time (OIT)<br>Standard (Std) OIT<br>High Pressure (HP) OIT | <br>ASTM D3895<br>ASTM D5885 | <br>100 min (min. ave.)<br>400 min (min. ave.) |
| Oven Ageing at 85 °C | ASTM D5721 | 55 % Std-OIT retained after 90 days<br>80 % HP-OIT retained after 90 days |
| UV-Resistance | GM 11 | 50 % HP-OIT retained after 1600 hrs |

ave.: mean value; min. ave.: lower limit for mean value

As a comparison, Table 3.3 shows a list of the properties tested and test procedures specified by the Geosynthetic Institute, Philadelphia, for testing and certification of HDPE geomembranes. The requirements are also indicated in Table 3.3.

## 3.2 Test Methods

### 3.2.1 External Appearance, Homogeneity, Skew and Waviness

Testing of these test quantities is performed in accordance with DIN 16726:1986 *Plastic Roofing Felt and Waterproofing Sheet; Testing* which is commonly used for testing plastic sheets in Germany. In the last years the test procedures describe in DIN 16726 were step by step substituted by CEN standards, like EN 1848-2:2001 *Flexible Sheets for Waterproofing – Determination of Length, Width, Straightness and Flatness – Part 2: Plastic and Rubber Sheets for Roof Waterproofing* or EN 1850-2:2001 *Flexible Sheets for Waterproofing – Determination of Visible Defects – Part 2: Plastic and Rubber Sheets for Roof Waterproofing.*

The external appearance is inspected qualitatively only by a visual investigation for blisters, scratches, cracks and pores. No damage may be present. A carefully prepared and performed fabrication provides a smooth and nonporous geomembrane surface which is free of streaks. Occasionally, fine grooves or scratches and tiny imprints of dust and dirt particles may develop. Only where manufacture was unsatisfactory due to technical deficiencies can the problem arise to distinguish between non-acceptable and still acceptable pores and scratches or imprints that impair the external appearance. In cases of doubt, it should be checked whether there are any repercussion on stress-strain behaviour (tensile test, multi-axial tension test) and stress crack resistance (NCTL test). However, geomembranes with large numbers of scratches, streaks, pores and dirt imprints spread over the whole surface do not fulfil the requirements anyway. The foreman has to continuously inspect the external appearance during the manufacturing process.

Homogeneity is supervised within certification testing and third-party control of manufacture where sections are investigated at a 6-fold enlargement in accordance with DIN 16726, Sect. 5.1. The sections must be free of voids, pores or foreign inclusions. If the external appearance gives cause for doubts, homogeneity should always additionally be tested.

DIN 16726, Sect. 5.2 describes the test of skew and waviness in detail. The tests are usually performed as part of the manufacturer in-house quality control before each operation start-up. Any re-start of the extrusion line after a stoppage, a resin change or a change of thickness is considered as an operation start-up. To perform the test, a minimum 12 m long section of geomembrane is unrolled on an even base. The test is performed at ambient temperature (see Sect. 3.2.8, Footnote 1). The greatest distance of the geomembrane edge from the imaginary straight line connecting both end points (skew) and the highest crest of the emerging waves from the even base (waviness) is determined along a length of 10 m. The deviation must be smaller than 50 mm in each case. The fulfilment of this requirement, however, is not always sufficient to achieve an even and intimate contact with the foundation layer on the construction site as well (see Sect. 9.3). A large number of short waves in the centre or at the edges of the geomembranes, where each fulfils the requirement, may together create welding problems, because large differences in length may occur in certain parts of adjacent geomembranes.

### 3.2.2 Thickness

The thickness test on smooth geomembranes is performed in accordance with Sect. 5.3 of the DIN 16726 standard. The BAM certification guidelines refers to DIN 53370:1976 *Testing of Plastic Films; Determination of the Thickness by Mechanical Feeling.* The measurement is carried out using a vernier calliper or mechanical feeler. The necessary accuracy of the measuring tools depends on the requirements for minimum thickness and thickness tolerances. DIN 16726 basically refers to the old standard DIN 53353:1971 *Determination of Thickness using Mechanical Feeler.* This standard has been replaced by DIN EN ISO 2286-3:1998 *Rubber- or Plastic-coated Fabrics–Determination of Roll Characteristics– Part 3: Method for Determination of Thickness* and EN 1849-2:2001 *Flexible Sheets for Waterproofing–Determination of Thickness and Mass per Unit Area – Part 2: Plastic and Rubber Sheets for Roof Waterproofing.* The accuracy of the measuring instrument required is, however, often not sufficient. The mechanical feeler device should be able to measure the thickness with an accuracy of 0.005 mm. Thickness measurements on geomembranes and geotextiles are described in detail in the standard ASTM D5199 *Standard Test Method for Measuring Nominal Thickness of Geotextiles and Geomembranes.*

It has already been mentioned that the thickness during production can be measured and monitored with a β radiation thickness gauge. A radioactive β-ray emitter of known intensity is arranged above the geomembrane. The attenuation of radiation decreases exponentially with the thickness. After a calibration (determination of the coefficient of attenuation) the thickness can be calculated from the intensity measured below the geomembrane.

The thickness can also be determined by simple, portable ultrasonic devices. The delay time of the echo of an ultrasonic impulse from the back side of the geomembrane is measured which is sent by an ultrasonic measuring head placed on top of the geomembrane. This method is mainly used for the measurement of the thickness of weld seams (see Sect. 10.2).

Finally, thickness can also be measured optically. A smooth cut edge is observed by a handheld microscope. More expensive optical methods can be used for the determination of the thickness of geomembranes with a textured surface which are more difficult to investigate (see Sect. 6.2).

In geotechnical engineering, HDPE geomembranes with different thicknesses are used and there is an ongoing discussion about the minimum thickness necessary. One has to choose the optimum between mechanical robustness, stress crack resistance, performance during installation and above all, welding on the one hand and financial expenditure on the other. A short-term saving in purchase can easily be lost through repair and remediation costs required over the long-term. The HDPE geomembrane should be a minimum 2.0 mm thick (Knipschild and Tornow 1979), but under no circumstances thinner than 1.5 mm. However, there are reports about regular problems with extrusion fillet welding of 1.5 mm thick HDPE geomembranes (Hein et al. 2003). Usually, 2.0 to 2.5 mm thick HDPE geomembranes are recommended, as can be seen in Table 1.1. In Germany, the thickness is the only property of the plastic geomembrane in the landfill liner for which the TA Municipal Waste and the TA Hazardous Waste directly specify a requirement: it must be  2.5 mm. According to the BAM certification requirements for landfill liner geomembranes, each individual value measured at 0.2 m intervals along the width of the geomembrane must be  2.50 mm and the maximum permissible deviation of any individual value from the average is ± 0.15 mm. For other nominal thicknesses greater than 2.5 mm – for example, 3 mm thick HDPE geomembranes are sometimes used for basal liner systems – it is ± 0.20 mm.

### 3.2.3 Carbon Black Content and Distribution

Homogeneously distributed, fine-grained carbon black proved an effective and cost-efficient stabiliser against UV radiation. It is either added to the polyethylene by the raw material manufacturer or during geomembrane production. Properties, content and homogeneity of the distribution of carbon black determine the quality of UV stabilisation.

The carbon black content can be determined by thermogravimetry (TG measurement). TG measuring instruments with evaluation units are available from a number of instrument manufacturers. The acquisition is expensive, the execution of a carbon black content test, however, fast and easy. It is increasingly replacing the former common procedure in accordance with ASTM D1603-94 *Standard Test Method for Carbon Black in Olefin Plastics*, or ASTM D4218-96 *Standard Test Method for Determination of Carbon Black Content in Polyethylene Compounds by the Muffle-Oven Technique*. The basics of thermogravimetry for polymers are described in EN ISO 11358:1997, *Plastics – Thermogravimetry (TG) of Polymers – General Principles*. In the TG measurement of carbon black content, a tiny specimen from the geomembrane is heated up on a microbalance. The temperature and the mass loss are continuously measured by the microbalance. The heating first occurs in a nitrogen atmosphere where the polyethylene in the specimen is completely pyrolysed at a temperature of approx. 480 °C. Afterwards an oxygen-containing atmosphere (e.g. synthetic air) is introduced and the carbon black is burned. The mass fraction of carbon black can be determined by a standard evaluation of the specimen mass via temperature curve.

In BAM the following test conditions are used. A 10 to 50 mg piece of specimen punched out from the geomembrane is weighed into a 150 µl $Al_2O_3$ or platinum crucible. It is heated in an atmosphere of continuously flowing (200 ml/min) nitrogen ($N_2$ purity 5) up to 600 °C. At that temperature the gas flow is changed to 200 ml/min of synthetic air. The start temperature is 30 °C, the final temperature 900 °C and the heating rate is 20 K/min. This test method produces hardly any residue (< 0.05 % by mass).

The carbon black content is determined in accordance with the standard evaluation method of the actual device from the step in the specimen mass via temperature curve due to the burning of the carbon black. At least three individual measurements are carried out. A prerequisite for a reliable TG measurement is the calibration of the temperature measurement and a buoyancy correction. In so doing the equipment manufacturer's instructions have to be followed.

After the sharp drop in the mass via temperature curve due to polyethylene pyrolysis, the remaining mass does not usually remain constant with temperature. Rather, the curve continues to gradually decrease. The point of time or temperature for switching over to synthetic air gas flow can therefore affect the measurement result of the carbon black content. Such measurement conditions must be observed to achieve comparable results. Comparing the measurement results with measurements obtained by conventional methods or by previously used and checked TG measuring instruments is recommended when commissioning a TG measuring instrument.

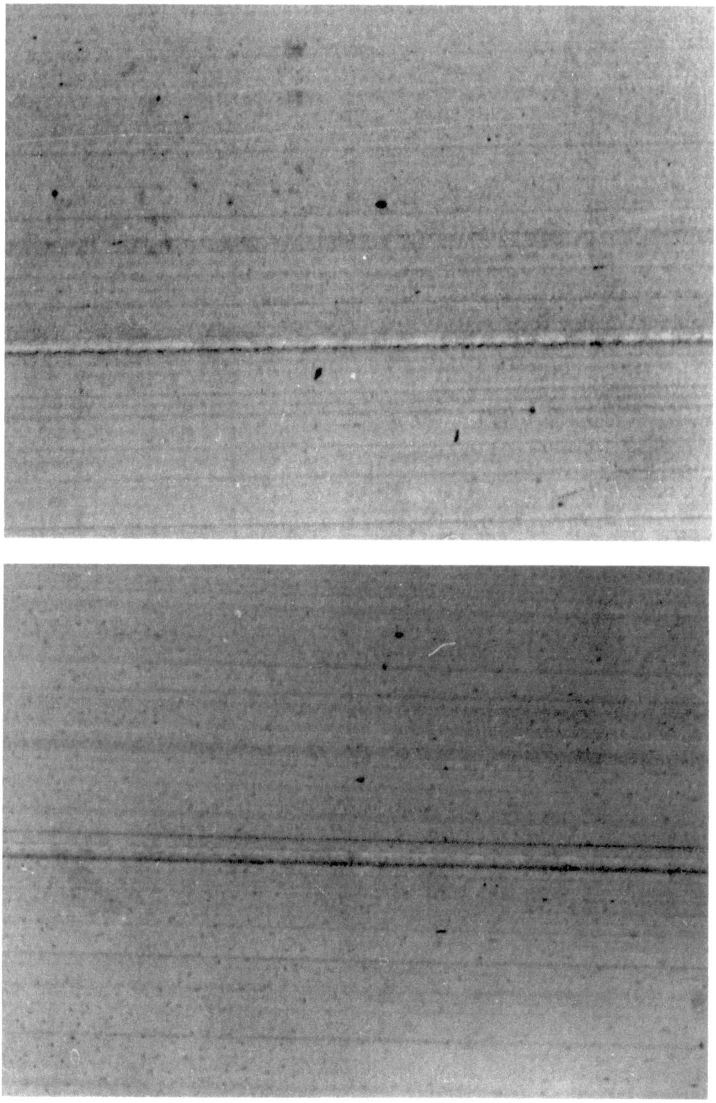

**Fig. 3.1a.** Examples of a sufficiently homogenous carbon black distribution. The photographs were taken of a microtome section from an HDPE geomembrane with 100fold enlargement. The width of the pictures corresponds to 1.20 mm and the height to 0.90 mm

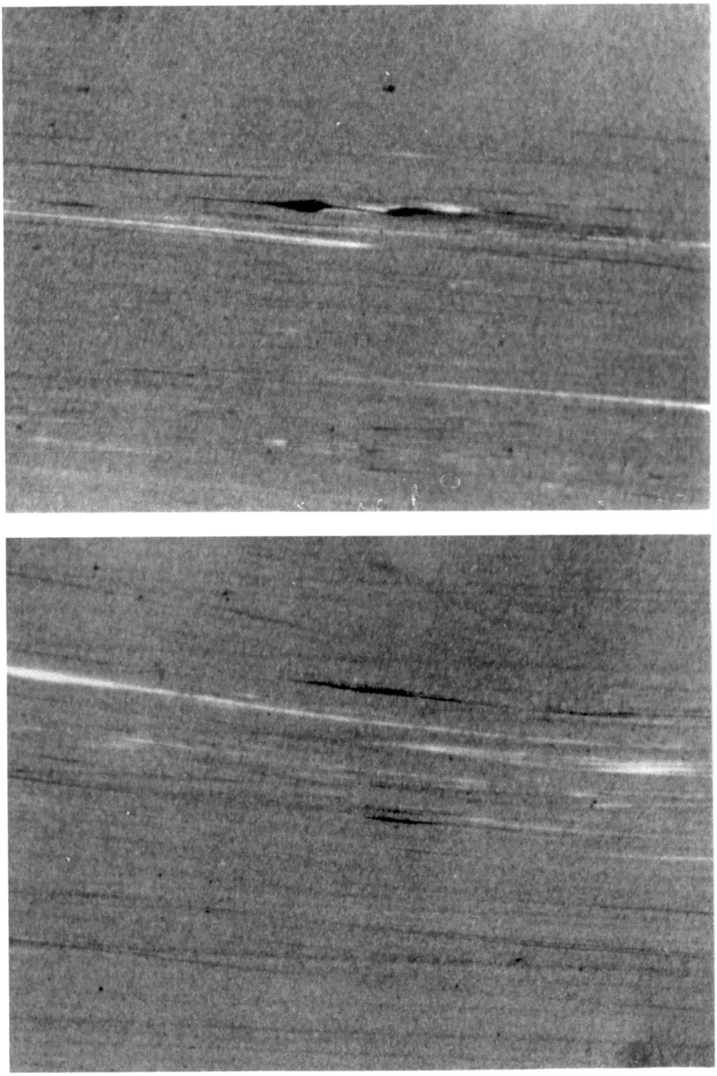

**Fig. 3.1b.** Examples for a poor carbon black distribution (impermissible large agglomerates and unsatisfactory mixing). The photographs were taken of a microtome section from an HDPE geomembrane with 100fold enlargement. The width of the pictures corresponds to 1.20 mm and the height to 0.90 mm

During the development of the meanwhile outdated draft DIN 16739:1994 *Geomembranes from Polyethylene (PE) for Landfill Liners; Requirements, Tests* an interlaboratory comparison was organised with the TG test method described (Süddeutsches Kunststoffzentrum (SKZ), the former Hüls AG and BAM). A very good agreement was obtained both comparing the results of the different testing laboratories and the results of the two methods, TG measurement and ASTM 1603.

The assessment of the homogeneity of carbon black distribution is described in the standards ISO 18553 DAM 1: 2006 *Method for the Assessment of the Degree of Pigment or Carbon Black Dispersion in Polyolefin Pipes, Fittings and Compounds* and ASTM D5596-94 *Standard Test Method for Microscopic Evaluation of the Dispersion of Carbon Black in Polyolefin Geosynthetics.* Sometimes the French standard NF T 51-142: Février 1992 *Plastiques – Compositions à base de polyéthylènes et de leurs copolymères – Détermination du degré de dispersion du noir de carbone*, or the British standard BS 2782-8 *Methods 823A and 823B. Methods for the Assessment of Carbon Black Dispersion in Polyethylene Using a Microscope* had been used.

The test of homogeneity of carbon black distribution within the framework of the BAM certification is performed in accordance with ASTM D5596-94. Approx. 10 μm thick microtome sections are taken from the geomembrane cross-section of 10 test samples which were cut from randomly chosen places in the geomembrane. The microtome sections are tested for the homogeneity of carbon black distribution and any faults with a 100-fold magnification in a microscope. A minimum of a 10 mm$^2$ surface is investigated.

The homogeneity is assessed according to the following criteria. No carbon black strips or naturally coloured material strips, streaks of poorly mixed areas, no carbon black concentrations larger than 30 μm or patches of naturally coloured material or faults such as bubbles or voids may be visible. A qualitatively superior homogeneity of the distribution shows only fine carbon black spots (up to approx. 30 μm diameter) speckled evenly over the grey-pigmented area. Figures 3.1a and 3.1b show examples of conforming and unsatisfactory carbon black distributions. According to ASTM D5596-94, homogeneity is classified by comparison with sample chart pictures (ASTM-Adjunct D35-Carbon Classification Chart for Geosynthetics). Homogeneity of distribution required by the BAM certification corresponds as a minimum to Category 1 in all pictures. According to ISO 18553 the frequency of the size of carbon black spots is assessed quantitatively and the appearance of the distribution qualitatively by the comparison with reference pictures.

### 3.2.4 Melt Mass-flow Rate and Density

Architecture and chemical composition of the polymer macromolecule determine the flow behaviour of the polymer melt. This flow behaviour, or more exactly the relation between shear stress and shear velocity (viscosity), is a key property for processing. The determination of the viscosity (see Sect. 3.3) is difficult for polyethylenes, since a test solution can only be produced at high temperatures due to their high resistance to chemicals and must be handled at that temperature. However, the melt mass-flow rate is an easy-to-determine parameter for the flow behaviour from which even the viscosity can be estimated using an appropriate extrapolation method (Menges 1990). The melt mass-flow rate measurement can be performed according to EN ISO 1133:2005 *Plastics – Determination of the Melt Mass-flow Rate (MFR) and the Melt Volume-flow Rate (MVR) of Thermoplastics.* ASTM D1238-99 *Standard Test Method for Flow Rates of Thermoplastics by Extrusion Plastometer* is most often used among US standards.

The device used for melt mass-flow rate measurement consists of a heated cylinder with a nozzle attached to the lower end. Cylinder and nozzle opening must be manufactured according to the exact specifications of the test standard within narrow tolerances. A matching piston to whose piston rod weights can be applied moves in the cylinder. The sample material (granules or geomembrane sample reduced to small pieces) is place in the cylinder, heated to a specified temperature above the melting temperature of the specimen and extruded through the nozzle by applying a specified weight on the piston. The emerging extrudate is cut off automatically at specified time intervals and the sections are weighed. The test quantity, i.e. the melt mass-flow rate, is the mass of polymer melt discharged at a certain test temperature and a certain mass of the test weight related to 10 minute extrusion time periods. Therefore the unit is g/10 min. The test temperature in °C and the mass of the test weight in kg must always be indicated. For HDPE materials a temperature of 190 °C is used and any of the three test weights of 2.16 kg, 5 kg or 21.6 kg applied.

In addition, the volume of the polymer melt discharged through the nozzle per time interval can also be calculated from the piston movement. This test quantity, with a unit of $cm^3$/10 min, is called melt volume-flow rate. Usually, the abbreviation MFR is used for the test quantity of melt mass-flow rate and MVR for the melt volume-flow rate. However, the old abbreviation, MFI (melt flow index) is also often used. Another parameter is the ratio of MFR measured with a large test weight to that measured with a small test weight. This quantity designated as melt flow ratio is unfortunately often also abbreviated as MFR. It may serve as an indicator for the width of the molecular mass distribution.

The density can be determined using methods common in other branches of material testing. The buoyancy method, the determination of density using a pyknometer, the floatation method and the density-gradient method are described in detail in EN ISO 1183-1:2004 *Plastics – Methods for Determining the Density of Non-cellular Plastics – Part 1: Immersion Method, Liquid Pyknometer Method and Titration Method* and EN ISO 1183-2:2004 *Plastics – Methods for Determining the Density of Non-cellular Plastics – Part 2: Density Gradient Column Method.* Within the GRI test method GM13 the ASTM D1505-98 *Standard Test Method for Density of Plastics by the Density-Gradient Technique* and ASTM D792-98 *Standard Test Methods for Density and Specific Gravity (Relative Density) of Plastics by Displacement* are used. All procedures are related to Archimedes' principle. A method based on another physical principle is described in ASTM D4883-99 *Standard Test Method for Density of Polyethylene by Ultrasound Technique.*

Small differences in crystallinity affect the density. For an exact and reproducible density determination the specimen must be conditioned and the history of the specimen preparation must be known. Density is often measured together with the melt mass-flow rate and both serve as characteristic material parameters. Therefore, preparation and conditioning method based on the ASTM D2839-87 *Standard Practice for Use of a Melt Index Strand for Determining Density of Polyethylene* has proved suitable: The density is determined on extrudates from the melt mass-flow rate measurement after boiling in water for one hour.

The measurement of melt mass-flow rate and density is usually the first step in the specification and identification of polyethylene materials. It should be noted that the density of the base polymer differs from the density of the finished geomembrane mixed with carbon black. The confusion, which can occur in the classification of the PE resin, has briefly been dealt with in Sect. 2.1. As a rule, only geomembranes coloured black by carbon black, and not their natural-coloured resin, reach a density which properly meets the classification of HDPE plastic material (see Sect. 2, Tables 2.2 and 2.3). However, one may not consider these classification limits as strict technical criteria: even if the density of the black geomembranes is just below 0.940 g/cm³, it may correspond to what is called HDPE geomembrane over the whole spectrum of its characteristics.

The outdated standard DIN 16776-1:84 *Polyethylene and Ethylene Copolymer Thermoplastics; Classification and Designation* and the replacing European standard EN ISO 1872-1:1999 *Plastics – Polyethylene (PE) Moulding and Extrusion Materials – Part 1: Designation System and Basis for Specifications* have introduced a designation system for polyethylene resins based on appropriate levels of the designatory properties density and mass melt-flow rate. The melt mass-flow ranges used in the specification data blocks, Table 3.4, are important for the assessment of the ability to weld two HDPE geomembranes from different resins as well, see Sect. 10.1. The HDPE geomembrane usually has melt mass-flow rates which are in the ranges T012 and T022.

**Table 3.4.** Code numbers of melt mass-flow rate ranges and test conditions according to EN ISO 1872-1

| Code number | Range of melt mass-flow rate (g/10 min) |
|---|---|
| 000 | ≤ 0.10 |
| 001 | 0.10 < ...≤ 0.20 |
| 003 | 0.20 < ...≤ 0.40 |
| 006 | 0.40 < ...≤ 0.80 |
| 012 | 0.80 < ...≤ 1.5 |
| 022 | 1.5 < ...≤ 3.0 |
| 045 | 3.0 < ...≤ 6.0 |
| 090 | 6.0 < ...≤ 12 |
| 200 | 12 < ...≤ 25 |
| 400 | 25 < ...≤ 50 |
| 700 | 50 < |

| Code letter | Test condition Temperature/nominal load |
|---|---|
| E | 190 °C/0.325 kg |
| D | 190 °C/2.16 kg |
| T | 190 °C/5 kg |
| G | 190 °C/21.6 kg |

Set of conditions T is used only for materials having an MFR less than 0.1 g/10min when tested under set of conditions D. Set of conditions G is used only for materials having an MFR less than 0.1 g/10min when tested under set of conditions T. Set of conditions E is used only for materials having an MFR greater than 100 g/10min when tested under set of conditions D

### 3.2.5 Dimensional Stability

Geomembrane production imposes an oriented morphological structure on the material and thus "freezes" certain internal stresses. These stresses and the orientation are relaxed by heating and the dimensions of a test specimen change. The changes are moderate in small test specimens. The dimensional change can be both positive and negative (Fig. 3.2). The geomembranes shrink in the extrusion direction (often called machine direction), while elongation and shrinkage may occur crosswise. The amount of dimensional change or the dimensional stability depends on the manufacture process and is not uniform along the width of the geomembrane.

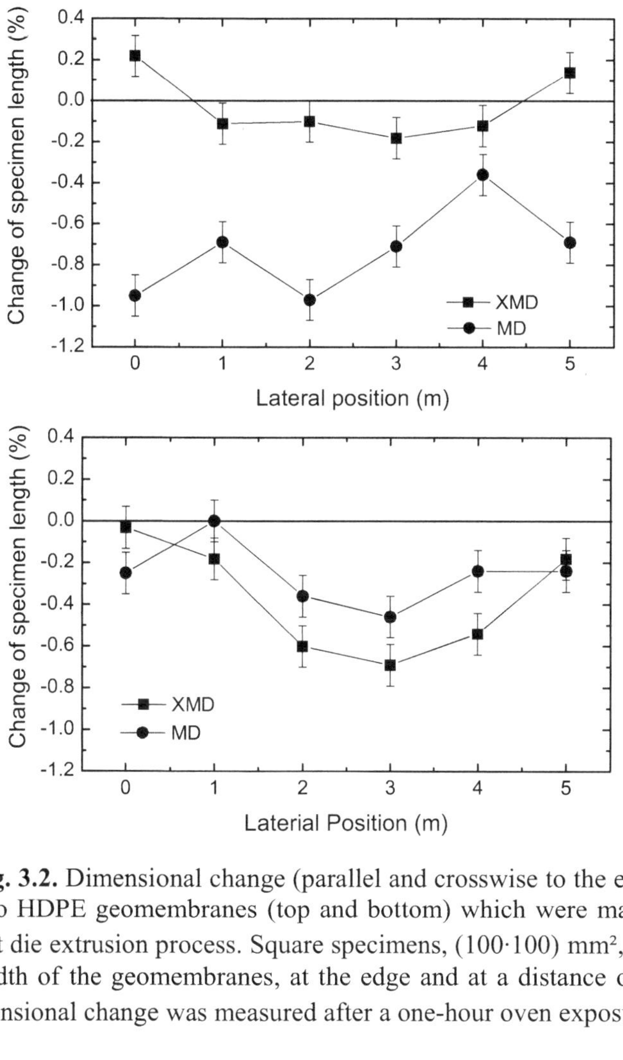


**Fig. 3.2.** Dimensional change (parallel and crosswise to the extrusion direction) of two HDPE geomembranes (top and bottom) which were manufactured using the flat die extrusion process. Square specimens, (100·100) mm$^2$, were taken along the width of the geomembranes, at the edge and at a distance of one metre. The dimensional change was measured after a one-hour oven exposure at 120 °C

The black geomembranes can reach temperatures up of 70 °C in the field when exposed to direct sun irradiation. The thermal expansion due to temperature differences up to 40 °C is reversible to a large extent. However, this does not apply to dimensional changes due to the orientation from the manufacturing process. Since the dimensional change varies along the width of the geomembranes, too large dimensional changes or too low dimensional stability cause permanent wavy distortions in the geomembrane and a smooth, wave-free installation will become impossible. Dimensional change emerges in a similar way due to heat input in the weld area: small waves are created along the seam and thus stresses in the weld seam can develop there as well. Orientation and internal stress have an effect on stress crack resistance. In the production of foils, sheets and geomembranes therefore, a small heat-induced dimensional change generally qualifies as an evaluation criterion for an efficient and material conformal processing.

BAM uses DIN 53377:69 *Testing of Plastic Films; Determination of Dimensional Stability* for the determination of dimensional change of HDPE geomembranes. Quadratic specimens (plates) with 100 mm edge lengths are cut from the geomembrane. The edges must be right-angled and the lateral faces must be even. The specimens are kept in an oven at 120 °C for an hour. Temperature fluctuation within the range of the specimen may not exceed ± 2 °C. The plates must be placed in the oven in such a way that shrinkage and expansion are not obstructed. For this purpose, specimens are placed on glass plates coated with talcum powder, aluminium foil or baking foil. The BAM certification guidelines require these test conditions, since as much as 50 % of melt heat is used up when heating up to 120 °C, so that smaller crystallites are melted, therefore, orientation and internal stresses should have been eliminated to a large extent.

The edge length ($l_{before}$ and $l_{after}$) of the plates before and after heating in the oven is measured in the extrusion direction and crosswise and the change of relative length

$$\delta l = \left( \frac{l_{after}}{l_{before}} - 1 \right) \cdot 100 \tag{3.1}$$

is calculated. The plates may not be bent during the edge length measurement and a completely even alignment must be achieved using a simple clamping device. The length measurement device must have a minimum measurement accuracy of 0.01 mm. The dimensional change is then defined as in Eq. 3.1, expressed in % and rounded to a decimal figure (i.e. ‰ values).

Another procedure for specimen preparation and length measurement is described in EN 1107-2:1999 *Flexible Sheets for Waterproofing – Determination of Dimensional Stability – Part 2: Plastic and Rubber Sheets for Roof Waterproofing*. Here quadratic specimens of 250 mm edge length are cut from the geomembrane. Two durable markings are applied on the two imaginary transversal and longitudinal median line of the specimen at a distance of 200 mm symmetrical to the centre. Their distance must be optically or mechanically measured before and after oven exposure with a minimum accuracy of 0.1 mm. The relative change in distances of the markings after oven exposure is calculated and similarly rounded to ‰-values. In analogy to this standard procedure the dimensional stability for HDPE geomembranes is often determined as follows: 120 mm quadratic specimens are used and four markings (e.g. cuts by a razor blade) or four holes are made (with a drill and reamer) at a distance of 100 mm. The type of the marking must enable a clear and reproducible identification of the measuring point. The inner surface of the hole must be smooth and sharp-edged. The measuring instruments must have a minimum measuring accuracy of 0.01 mm.

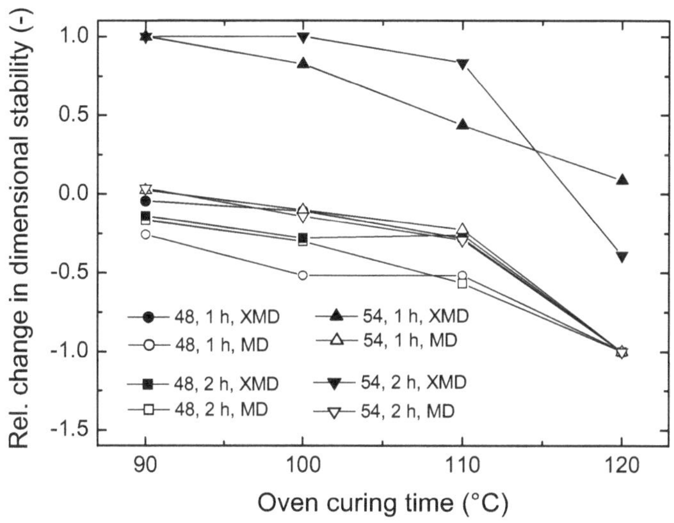


**Fig. 3.3.** Relative variation of the dimensional change (not the dimensional change itself (!)) related to the respective maximum dimensional change as a function of oven exposure temperatures. The measurements were performed after one- and two-hour oven exposure at 90 °C, 100 °C, 110 °C and 120 °C on two HDPE geomembranes (specimen 48 and 54) in machine direction (MD) and crosswise machine direction (XMD). In all cases, where a dimensional change can be observed, their absolute value clearly increases with increasing oven temperature. Differences between a one-hour and two-hour exposure cannot be recognised. A very long oven exposure over several months at 80 °C, however, will cause dimensional changes which are in the range of the changes to be measured at 120 °C after one hour

Oven temperature, oven exposure time and specimen size are either not specified at all in the available standards for the measurement of dimensional stability or in different ways. DIN 16726 for plastic roofing felts and waterproofing sheets requires an ageing at 80 °C over 6 hours for a specimen size of 100 mm: this requirement underestimates the residual dimensional change which may develop in an installed geomembrane heated during several days by sun irradiation, see Fig. 3.3. The oven exposure time is too long for a continuous manufacturing quality control anyway. The EN 1107-2 requires 250 mm specimens. Such large specimens are difficult to handle and therefore not practicable. A test in conformity with the standard would take at least 27 hours according to the test specimen preparation and test procedure (20 hours conditioning in the standard atmosphere of (23 ± 2) °C temperature and (50 ± 5) % relative humidity, 6 hours oven exposure at 80 °C and 1 hour conditioning at standard atmosphere after removal from the oven). Therefore this method is simply not practicable for quality assurance measures. The ASTM D1204-94 *Standard Test Method for Linear Dimensional Changes of Nonrigid Thermoplastic Sheeting or Film at Elevated Temperature* also requires this specimen size, stipulates the temperature at 100 °C and leaves the oven exposure time open.

The determination of the dimensional change is not part of the requirements for HDPE geomembranes of the Geosynthetic Institute (Table 3.3), although this test quantity is an important and easy-to-measure parameter for a high quality processing.

The permissible change in dimensions under specified test conditions should be as small as possible. However, the manufacturing technology, in particular the flat die extrusion method followed by a calendering process, inevitably causes a considerable orientation and an associated certain dimensional change. Textured geomembranes with embossed structures are especially affected. The BAM certification guideline requires that the absolute value of the dimensional change in machine as in cross machine direction must be ≤ 1.0 % for smooth geomembranes and ≤ 1.5 % for geomembranes with an embossed surface pattern. These values are not too strictly chosen since the thermal expansion, as a comparison, is only 1 % at a temperature difference of 40 °C (calculated with a coefficient of thermal expansion of $1.5–2.5 \cdot 10^{-4}\,K^{-1}$ for HDPE materials within the range of 20 °C to 70 °C). An appropriate processing technology can ensure adherence to these limiting values without problems.

### 3.2.6 Permeation

Permeation of pollutants through geomembranes plays an important role in the selection of suitable plastic materials because in many cases geomembranes are used not only to prevent water but also hazardous substances contained in water, liquid chemicals or gases from migrating into the environment (e.g. subgrade, soil or groundwater). The permeation behaviour of a plastic geomembrane must always be assessed in connection with the chemical resistance of the plastic material to a test liquid. Diffusion and solution are the physical processes that determine permeation in geomembranes. The values for the physical parameters of diffusion and solution processes in HDPE geomembranes, i.e. solubility $s$ or partition coefficient $\sigma$ and diffusion coefficient $D$ are known for a large number of organic substances and aqueous solutions of organic substance, see Chap. 7. The test to determine the parameters can be performed as a permeation test or an immersion test. An introduction to the methods for the investigation of diffusion in polymeric materials is presented in (Crank and Park 1968).

Permeation rate $J$ and induction time $t_{ind}$ can be determined directly in the permeation test (Fig. 3.4). Consider the following typical experimental situation: a liquid chemical or an aqueous solution of this chemical at concentration $c_0$ is accumulated on top of a geomembrane specimen of thickness $d$ in an appropriate test device. The concentration of the chemical is always kept at zero under the geomembrane by continuously removing the chemical. The geomembrane is then under the influence of a constant external concentration difference $\Delta c = c_0$. The chemical dissolves to a certain extent in the geomembrane surface and a constant internal concentration difference is formed which is not necessarily equivalent to the external concentration difference. It is the internal gradient which drives the diffusive mass transport. If the concentration equilibrium develops fast enough at the interface, then the internal concentration difference in the geomembrane is given for a steady state by $s\Delta c$ for a pure test liquid and by $\sigma\Delta c$ for an aqueous solution. However, it takes a certain time (induction time) for the molecules to diffuse through the geomembrane and for the steady-state internal concentration difference to develop. Afterwards a constant-rate diffusive mass transport takes place. The permeation rate and the induction time can then be expressed for the pure liquid as:

$$J = \frac{s\,D\,\Delta c}{d} \qquad t_{ind} = \frac{d^2}{6D}\,. \tag{3.2}$$

$s$ is replaced by $\sigma$ for an aqueous solution.

The imperviousness of a material is characterised by the permeability $P = Jd/\Delta c$ which is the ratio of the permeation rate to the concentration gradient as driving force generating diffusive mass transport. The reciprocal value $d/P$ is often called diffusion resistance. The permeability of a geomembrane for a chemical or its aqueous solution is thus given by the product of diffusion coefficient and solubility or partition coefficient.

An overview of measurement methods for the permeation of gaseous and liquid substances is presented in ASTM D5886-95 *Standard Guide for Selection of Test Methods to Determine Rate of Fluid Permeation Through Geomembranes for Specific Application.* Permeation through the geomembrane under impact by a pure liquid can in many cases be measured using a simple gravimetric method, as was for example described in the (outdated) standard DIN 53532:1989 *Testing of Elastomers, Determination of Permeability of Elastomer-Sheetings to Liquids.*

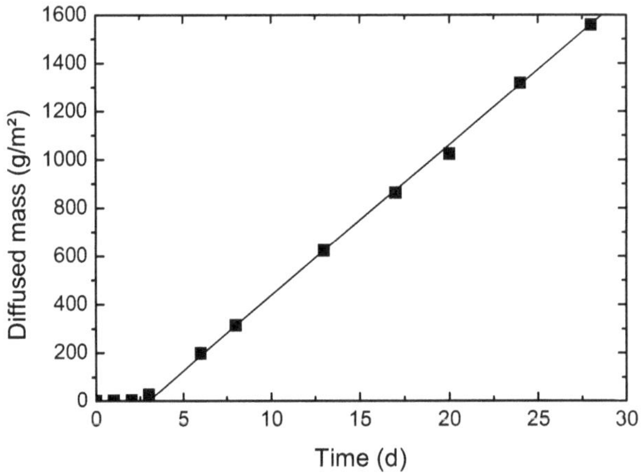


**Fig. 3.4.** Diffused mass of trichloroethylene related to unit specimen area which escapes from a test cell filled with liquid trichloroethylene in gravimetric permeation measurement as a function of time. The test specimen was cut from a 2.5 mm thick HDPE geomembrane. The linear regression fit to the data points in steady state yields the permeation rate (slope of the line defined as mass which diffuses through the geomembrane per unit time and area, here 30 g/(m²d)) and the induction time (time axis intercept of the regression line, here 3 days)

The tests are carried out in cylindrical test containers made of aluminium or another chemically resistant, impervious and light material. The container is open at one end. The open end is covered in a gas-tight fashion by the test specimen, i.e. a circular geomembrane segment. BAM uses an aluminium pot with a threaded cap. The flange of the pot is provided with three sealing ribs and the moveable flange in the cap with two sealing ribs offset in relation to the pot flange sealing ribs. The sealing flange in the cap is mounted on a ball-bearing. When screwing the pot and cap together, the ribs press into the geomembrane specimen and a gas tight contact is formed. The capacity of the pot is about 150 ml. Depth and filling level of the container must be chosen in such a way that swollen and thus wavy and deformed test specimens are still completely covered by the test liquid. The measurement is performed at a standard atmosphere of (23 ± 2) °C temperature and (50 ± 5) % relative humidity. Loss of weight of the test container due to permeation of the test liquid is measured as a function of time (Fig. 3.4). Induction time and permeation rate are determined from the linear regression of the data points in the steady-state phase.

The parameters $D$ and $s$ or $\sigma$ can also be determined experimentally by immersion tests. In an immersion test a geomembrane test specimen is immersed in a liquid chemical or its aqueous solution with a specified constant concentration $c_0$. Absorption of the chemical by the test specimen with a mass $G$ is measured by its gradual increase in mass $\Delta G(t)$. If the solubility of the chemical in the geomembrane material is limited, a saturation value for increase in mass $\Delta G(\infty)$ develops over the course of time. In a diffusion process obeying Fick's laws with a constant diffusion coefficient $D$, the slope of mass change as a function of time $t$ is proportional to $(Dt)^{½}$ up to about 2/3 of the saturation value. The proportionality factor depends on the shape of the test specimen. It holds for a plate with a large area compared to its thickness $d$ (Crank and Park 1968):

$$\Delta G(t) = \Delta G(\infty)\left(\frac{4}{d\sqrt{\pi}}\right)\sqrt{Dt}\,. \tag{3.3}$$

The solubility can be determined for the immersion in the liquid chemical from the saturation value as:

$$s = \frac{\Delta G(\infty)}{G}\,. \tag{3.4}$$

The partition coefficient for the immersion in an aqueous solution can be determined from the concentration $c(\infty) = \Delta G(\infty)/V$ of the chemical in the test specimen (with a volume $V$ and a density $\rho$) and the concentration of the chemical in the aqueous solution $c_0$, in which the specimen was immersed, according to:

$$\sigma = \frac{c(\infty)}{c_0} = \frac{\dfrac{\Delta G(\infty)}{G}\rho}{c_0}\,. \tag{3.5}$$

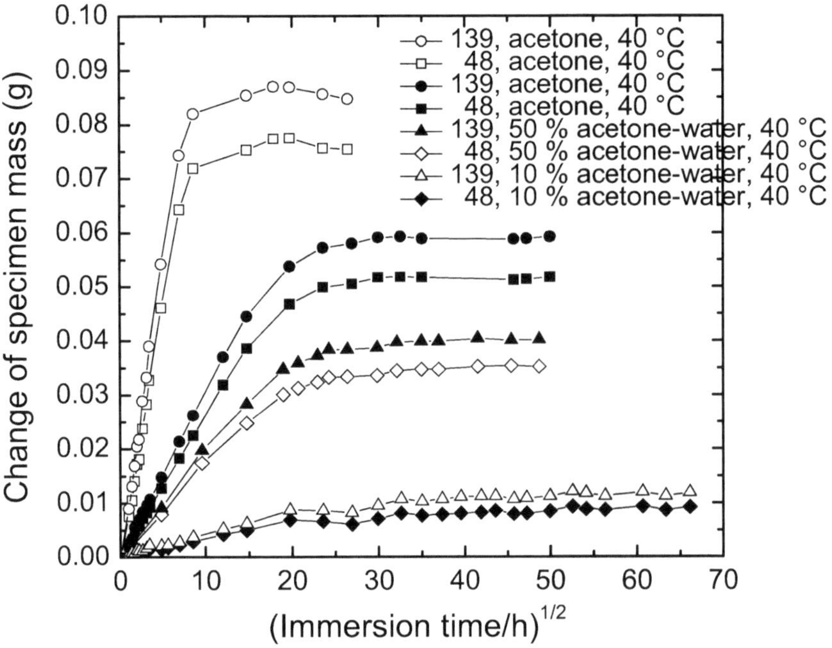


**Fig. 3.5.** The change of specimen mass (g) vs. square root of time *t* (*t* measured in h) of immersion test data on a 2.5 mm thick HDPE geomembrane. The immersion was performed in pure acetone and in an acetone-water solution at 23 °C and 40 °C. Circular geomembrane disks were used as test specimens (thickness: 2.5 mm, diameter: 58 mm)

The immersion test can be performed as follows (Müller et al. 1997; Müller et al. 1998). Circular disks with a diameter of approximately 5 to 6 cm are punched out from the geomembrane using a screw press punching device. The specimens are cleaned with water and then with ethanol. The cleaned specimens are dried in a vacuum-drier at 40 °C to a constant weight and cooled in a desiccator over silicagel as a drying medium. The specimens are stacked in glass dishes with glass needles as spacers and weighted with glass beads. Afterwards the test liquid is added and the glass container is covered with a glass cover and a plastic (parafilm©) foil. The glass containers are placed in an air-conditioned room at a temperature of 23 °C. The immersion tests can also be performed at an elevated temperature by placing them in an oven. If necessary, an oven suitable for flammable or explosive liquids has to be used. The immersed specimens are removed at regular intervals with tweezers, dabbed with an absorbent, fluff-free paper fleece, placed into a lockable weighing glass and weighed immediately using a precision balance with an accuracy of 0.1 mg. Subsequently, the specimens are again placed into the glass containers where the test liquid can be inspected and renewed if necessary. Fig. 3.5 shows an example of the mass change vs. time curves determined in this way for acetone used as test liquid.

### 3.2.7 Thermal Analysis and Measurement of Oxidation Stability

Thermoplastic materials have different states of aggregation: at elevated temperatures a liquid-like amorphous, highly viscous polymer melt develops. When the temperature drops below a certain limit during cooling, polymer chains starts to fold into crystals. However, polymers never crystallize entirely in practice. The persisting, more or less large, amorphous ranges are transformed into a glassy state at low temperatures. The transition into the glass state can be imagined in such a way that the randomly ordered polymer chains "stiffen" and contract. The mobility of the polymer chain segments is strongly reduced and the associated free volume drastically decreases. Melting of the partial crystalline plastic and crystallisation of the polymer melt, solidifying into and thawing from the glassy state take place in a narrow range of temperatures. Melting required input of heat (endothermal process), crystallisation is accompanied by the release of heat (exothermal process). The transitions between the states are, however, not as sharp and well defined as it is known from many inorganic materials. Also, there exists no gaseous polymeric state.

At very high temperatures irreversible chemical transformations take place. At a high oxygen partial pressure and high temperatures an oxidative degradation can take place within minutes. In an inert gas atmosphere the polymer decays by thermally driven decomposition – a procedure called pyrolysis. All these chemical reactions start at certain temperatures depending on the environment of the specimen. They are connected with the absorption and release of substantial amounts of heat. Pyrolysis is an endothermic process, while oxidation is an exothermic one. The temperatures, at which transitions and transformations take place, and the specific amount of heat, which is transferred – or more precisely the amount of transition enthalpy, since the transition usually takes place under constant pressure – are characteristic parameters for a thermoplastic material (Table 3.5). One of the tasks of thermal analysis is to measure such characteristic temperatures and transition enthalpies. In the following the so-called Differential Scanning Calorimetry (DSC) will be discussed in more details. Thermal analysis, however, whose terms are explained in DIN 51005:93 *Thermal Analysis (TA), Terms* or ASTM E473:2000 *Standard Terminology*

*Relating to Thermal Analysis* is an umbrella term for various analytical methods, such as thermogravimetry (TGA) described briefly in Sect. 3.2.3. An overview of the methods of thermal analysis is given in (Hemminger and Cammenga 1989).

A DSC apparatus consists in principle of two ovens. In one of the ovens a crucible filled with the specimen is installed, and an empty crucible as a reference specimen is placed in the other oven. The ovens are located in a specified gas atmosphere. Both ovens are heated, cooled or annealed at a constant temperature according to a controlled temperature programme. The difference in the heat flow of the two ovens, which is necessary to carry out the temperature programme, is measured as a function of time or oven temperature (thermal curve).

**Table 3.5.** Transition temperatures (in °C) of thermoplastic materials used for geosynthetics (geomembranes, nonwoven geotextiles, geocomposite drains, geogrids etc.). In addition to polyolefins PE and PP, polyester PET and polyamides PA 6.6 and PA 6 and polyvinyl chloride PVC are also used, the letter, however, only to a small extent

| Material | Glass transition temperature (°C) | Melting temperature (°C) | Temperature of thermal decomposition (°C) |
|---|---|---|---|
| PE-LD | -130 | 85 to 125 | ~ 420 |
| PE-HD | -125 | 130 to 140 | ~ 480 |
| PP | -20 to -5 | 165 to 175 | 328 to 410 |
| PET | 70 to 80 | 245 to 265 | 283 to 306 |
| PA | 40 to 60 | 210 to 265 | 310 to 380 |
| PVC | 65 to 85 | - | 180 |

For measuring the melting curve and enthalpy of melting, the temperature of both ovens is increased simultaneously at the same constant rate. When the specimen begins to melt and absorbs heat, the heat flow $W$ toward this oven must be considerably increased in order to maintain the constant rate of temperature increase. The curve of the difference in the heat flow $\Delta W$ between the specimen oven and reference oven is initially approximately constant as a function of oven temperature, it increases, however, to ever larger values as the specimen melts, goes through a maximum and returns to the initial constant value when the specimen is completely melted (Fig. 3.6). The temperature (peak temperature) in the maximum of the melting curve is usually called melting temperature. Also, the temperature at the toe of the increasing peak line is often indicated (onset temperature). The overall peak area can be used as a measure for the enthalpy of melting. EN ISO 11357-1:1997 *Plastics – Differential Scan-*

*ning Calorimetry (DSC) – Part 1: General Principles* and ISO 11357-3:1999 *Plastics – Differential Scanning Calorimetry (DSC) – Part 3: Determination of Temperature and Enthalpy of Melting and Crystallization* or ASTM D3418-99 *Standard Test Method for Transition Temperature of Polymers by Differential Scanning Calorimetry* should be consulted for the thermoanalytical measurement of melting behaviour. EN ISO 3146:1997 *Plastics –Determination of Melting Behaviour (Melting Temperature or Melting Range) of Semi-crystalline Polymers by Capillary Tube and Polarizing-microscope Methods* describes other methods for the determination of melting temperatures.

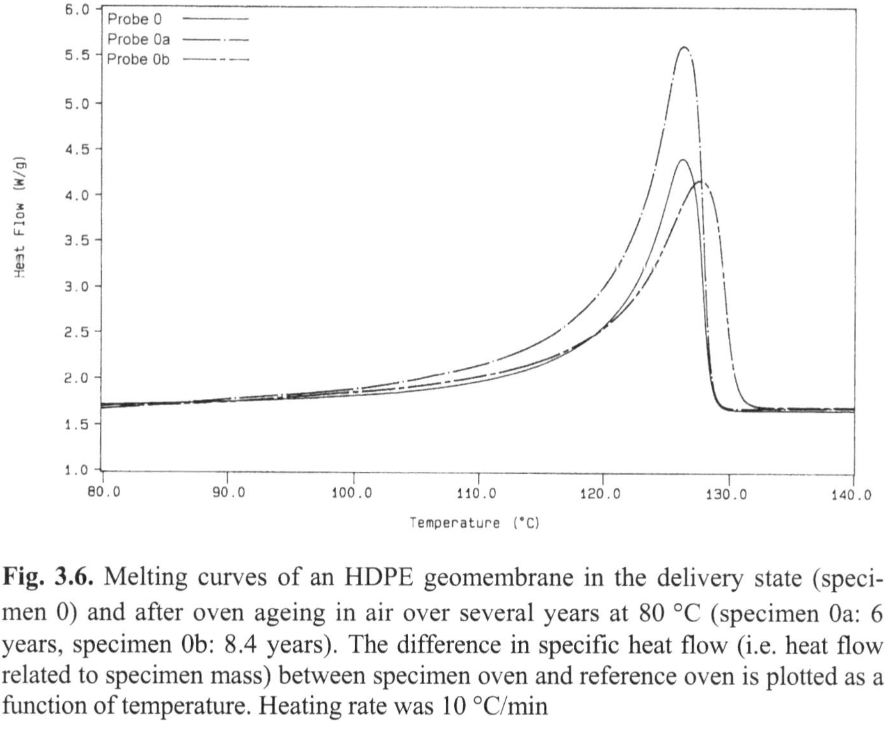


**Fig. 3.6.** Melting curves of an HDPE geomembrane in the delivery state (specimen 0) and after oven ageing in air over several years at 80 °C (specimen 0a: 6 years, specimen 0b: 8.4 years). The difference in specific heat flow (i.e. heat flow related to specimen mass) between specimen oven and reference oven is plotted as a function of temperature. Heating rate was 10 °C/min

The glass transition can be measured in an analogous way. Depending on thermal history and heating or cooling rate a step is formed in the thermal curve at the glass transition. The onset temperature or the midpoint temperature of the step may be designated as glass transition temperature. Details are described in ASTM D3418 or ISO 11357-2:1999 *Plastics – Differential Scanning Calorimetry (DSC) – Part 2: Determination of Glass Transition Temperature*. The ovens must be cooled for this purpose, however, by liquid nitrogen to achieve the very low temperatures of the glass transition in polyethylenes.

The oxidation stability of a thermoplastic material is characterised by the so-called oxidative induction time or OIT for short (Mueller and Jakob 2003). In the OIT measurement the ovens of the DSC apparatus are placed within an inert gas environment (nitrogen gas at a certain flow rate) and heated to a specified high measurement temperature. When the measurement temperature has been reached the ovens are exposed to an oxidizing atmosphere where, for example, pure oxygen gas replaces the inert gas flowing through the ovens. For a high-pressure OIT measurement the ovens are placed in a high-pressure measuring cell, in which a high oxygen partial pressure (typically up to 5 MPa) is established before or after reaching the measuring temperature. The OIT measurement starts with the initiation of the oxygen flow. The difference in heat flow $\Delta W$ between reference oven and specimen oven is measured as a function of time. A rapidly accelerating auto-oxidation of the specimen begins after a certain time. Because the reaction enthalpy is released, the heat flow in the specimen oven must be throttled back considerably to prevent a further heating of the oven. The heat flow difference curve, initially running horizontally as a function of time, will increase steeply (exotherm). The period of time from the beginning of the measurement until the onset of the exotherm in the thermal curve is the oxidative induction time. The point of onset in the curve can be chosen as the point where a threshold value of the deviation from the horizontal base line is exceeded (threshold method) or as the point of intersection of the tangent to the base line with the tangent of the steepest slope in the exotherm (tangent method) (Fig. 3.7). Obviously, the OIT value highly depends on measuring temperature and oxidising atmosphere. Statements about OIT value have to be always accompanied by data about measurement temperature TM and the oxygen atmosphere in which the measurement has been performed.

For measuring the oxidative induction time for HDPE geomembranes, the standards EN 728:1997 *Plastics Piping and Ducting Systems– Polyolefin Pipes and Fittings – Determination of Oxidation Induction Time* and ASTM D3895-95 *Standard Test Method for Oxidative-Induction Time of Polyolefins by Differential Scanning Calorimetry* can be consulted. The high-pressure OIT measurement on polyolefins is described in the standard ASTM D5885-95 *Standard Test Method for Oxidative Induction Time of Polyolefin Geosynthetics by High-Pressure Differential Scanning Calorimetry*.

Occasionally, the oxidative induction temperature (also abbreviated as OIT) is also measured. The specimen is exposed to an oxygen atmosphere from the start, heated at a constant rate and the heat flow difference measured as a function of temperature. Oxidation begins at certain high temperatures, which again leads to a sharp exothermal change in the curve of the heat flow difference. The temperature of the onset is then taken as the oxidative induction temperature.

The detailed practical implementation of the measurement method described here may be different in principle. Table 3.6 shows an example of the process steps, as they can be performed in a commercial dynamic power-compensation differential scanning calorimeter and in a corresponding pressure cell. The measurement of the melting curve on an HDPE specimen consists of the consecutive process steps 1, 2 and 5. The standard OIT measurement consists of process steps 1, 2, 3 and 5 and the high-pressure OIT measurements of process steps 1, 2, 4 and 5. However, the equipment manufacturers' instructions, in particular safety instructions for the use of high-pressure measurement cells, must always be followed.

**Table 3.6.** Process steps which can be combined in a commercial DSC apparatus. In the apparatus two open platinum pans are used as ovens which are fitted into a standard or high-pressure measuring cell. Liquid nitrogen cooling keeps the ovens' environment at a constant temperature (e.g. 0 °C). The cells can be purged and blown through by a gas

| No. | Process step | Process parameter |
|---|---|---|
| 1 | Isothermal gas pre-purge | Initial cell temperature; continuous $N_2$ gas flow (30 ml/min); duration about 1 min |
| 2 | Heating | Up to measurement temperature $T_M$ or maximum temperature $T_{max}$ of temperature programme; heating rate 10 °C/min; continuous $N_2$ gas flow (30 ml/min) |
| 3 | Isothermal oxidation (standard) | Measurement temperature $T_M$ 5 min equilibrium time for the test specimen; changeover from continuous $N_2$ to $O_2$ gas flow (30 ml/min); changeover point defines zero time of the measurement |
| 4 | Isothermal oxidation (high-pressure) | Changeover from continuous $N_2$ to $O_2$ gas flow (30 ml/min); time of $O_2$ purge about 5 min; closing of outlet valve; pressure increase up to measurement pressure; closing of inlet valve; start of heating up to measurement temperature; initiation of the temperature programme is zero measurement time |
| 5 | Cooling | Cooling rate 50 °C/min down to initial cell temperature; continuous $N_2$ gas flow (30 ml/min) |

To measure the melting curve, the specimens, weighing about 5 mg are placed in an aluminium pan (50 ml capacity) and usually covered with a perforated aluminium lid. The OIT measurements take place using an open pan. When a copper pan is used, the catalytic effect of the transition metal cupper has an influence on the oxidative degradation and the OIT values are considerably reduced. The interpretation of the measured times thus becomes more difficult. In the measurement of the melting curve, the temperature is usually increased up to $T_{max}$ = 180 °C. Standard OIT values of HDPE materials are measured at $T_M$ = 200 °C or 210 °C. The high-pressure OIT value is measured just above the melt range at $T_M$ = 150 °C and at an oxygen pressure of 3.4 MPa.

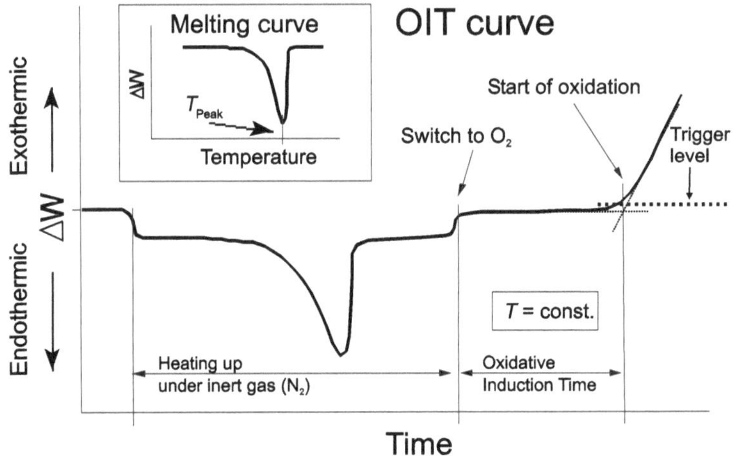


**Fig. 3.7.** Schematic illustration of OIT curve measurement (large diagram) and melting curve (small diagram). (Note that in this case the difference in heat flow between reference oven and specimen was plotted)

Typical melting temperatures and melting heat values of HDPE geomembranes or the associated degrees of crystallinity, calculated from the melting enthalpy with an extrapolated value of 293 J/g for a fully crystallised HDPE material, are displayed in Tables 2.2 and 3.5. The standard OIT values of well-stabilised HDPE resins are supposed to be ≥ 20 min at 210 °C (Müller 1999a) and ≥ 100 min at 200 °C (GRI 1998) and the high-pressure OIT values ≥ 400 min at 150 °C and 3.4 MPa (GRI 1998).

Misunderstandings occur frequently when OIT values are interpreted to obtain long-term oxidative stability characterisation of plastic geomembranes under actual field conditions. Some important aspects of OIT measurements are sometimes not taken into account (Mueller and Jakob 2003).

Type and quantity of antioxidants in essence determine the oxidative induction time of an HDPE resin. The OIT value of a non-stabilised HDPE resin is very small at elevated temperatures: a few minutes in the standard OIT measurement, a few tens of minutes in the high-pressure OIT measurement. It is, however, the chemical effectiveness and the migration behaviour of the stabiliser at the high measurement temperature $T_M$, not at the application temperature, which affect the OIT measurement. Both effectiveness and migration behaviour (e.g. volatibility) of the stabiliser may depend highly on temperature. In particular, phosphitic stabilisers are used as processing stabilisers for protection against oxidation at the high processing temperatures and contribute therefore substantially to OIT values. However, they make minor contribution to a long-term stabilisation at application temperatures in comparison. Conversely, steric hindered amines, the so-called HALS (Hindered Amine Light Stabilisers), have no influence on oxidative stability in measurements well above 150 °C. The OIT value in itself therefore says nothing about the quality of long-term stabilisation under application conditions. It cannot be used for the comparison of long-term oxidative stability under field conditions of different stabiliser packages in a resin or of different resins with the same stabiliser package.

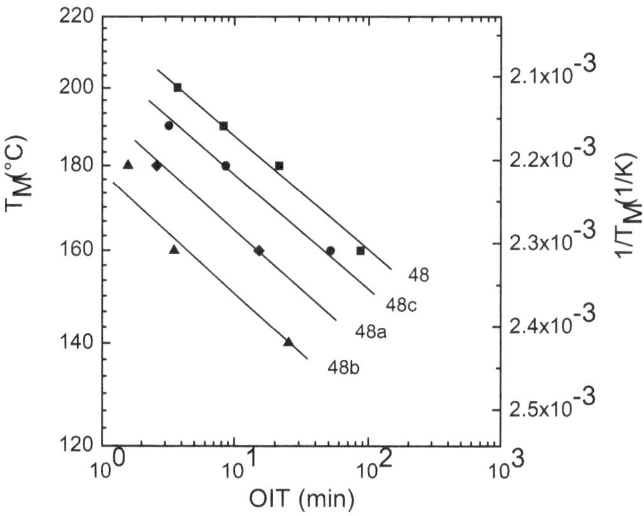


**Fig. 3.8.** OIT values of an HDPE geomembrane in the state of delivery (specimen 48), measured at different OIT measurement temperatures $T_M$ after oven ageing in air at 80 °C over 6 years (specimen 48a) and over 8 years (specimen 48b) (Mueller and Jakob 2003) and after an exposure to a mixture of liquid hydrocarbons over 2977 days at room temperature (specimen 48c) (Kalbe et al. 2000). The data show an approximately linear relationship between the logarithm of oxidative induction time and the inverse absolute measuring temperature ($1/T_M$)

One should always remember that in the DSC apparatus the specimens are oxidised in pure oxygen flow at high temperatures to measure the OIT value. In the following the example of a purely stabilised HDPE geomembrane will be discussed. Since the material is very sensitive to oxidation, its behaviour in the OIT test can be investigated more closely. Figure 3.8 shows an Arrhenius plot of the reciprocal absolute OIT testing temperature ($1/T_M$) versus the logarithm of the corresponding OIT values of sample 48 measured during air oven ageing and exposure to liquid hydrocarbons. Figure 5.27 (Mueller and Jakob 2003) shows additional data for immersion in water. OIT measurements were performed down to $T_M = 120$ °C below the melting point of the HDPE resin (onset of melting peak: 118 °C, peak maximum: 128 °C). In this case the variation is fairly large: minimum: 305 min, maximum: 580 min over 5 measurements. At each stage of ageing, the data follow an Arrhenius line with the same slope as the data from the initial sample. However, the lines are shifted parallel to each other to lower OIT values according to the severity of ageing. The high temperature oxidation of the specimens in an oxygen atmosphere is therefore characterised by a well defined activation energy of 140 kJ/mol, which is completely independent of the ageing conditions the samples have gone through. Only the intercept with the horizontal OIT axis for a given $T_M$ decreases with ageing or exposure time. This finding can be readily explained by assuming a continuous reduction of the antioxidant concentration during ageing, while the intrinsic properties of the polymer remain intact. Extrapolating the line in Fig. 5.27 further, an oxidative induction time of 148 days would be obtained at 80 °C for sample 48 when exposed to a flow of pure oxygen. On the other hand the oven ageing data show that the induction time in air at 80 °C is > 5000 days. Therefore, the errors in service lifetime prediction, that occur when OIT data are extrapolated to ambient temperatures, easily amount to many orders of magnitude (Howard 1973).

The typical stabiliser package, which is used in carbon black-stabilised HDPE materials but also in other polyolefin plastics, mainly consists of a mixture of phenolic and phosphitic antioxidants. HALS have been rarely used in HDPE geomembranes so far. For a given composition of this stabiliser package and a specific HDPE material, its OIT value is a monotonously increasing function of stabiliser quantity in the material. The OIT value is very small at small quantities and increases over a wide range with an increasing content. A calibration curve can be established which specifies the exact OIT values as a function of stabiliser quantity for the special stabiliser package and material (Gray 1990). A calibration curve is, however, only necessary if quantitative statements are needed about the stabiliser content from the OIT values. For the typical stabiliser package one can assume that, as a rough estimate, proportionality exists between the OIT value and stabiliser content without any specific knowledge about the calibration curve. If an HDPE geomembrane manufactured from such a material is subjected to long-term tests on oxidative stability, such as oven ageing in air and water, monitoring of the change in the relative OIT value can provide an indication of the change of the stabiliser content in the long-term test. OIT measurements only in this respect may provide the basis for service lifetime predictions. The approach is discussed in detail in Sect. 5.4.2.

The change of the OIT value in the long-term oxidative degradation test can thus characterise the long-term oxidation stability of a material. If the OIT value only changes a small amount in a long-term test and a considerable OIT value can still be measured at the end of the test, then high oxidation stability can be assumed under field conditions for the period of time which is covered by the long-term test, taking account of the accelerated test conditions. The BAM guideline for geomembrane certification (Müller 1999b) requires that the OIT value *OIT*(1y) after a one year's forced air oven ageing at 80 °C equals at least 70% of *OIT*(0.5y) after half a year:

$$\frac{OIT(1y)}{OIT(0.5y)} \geq 0.7\,. \tag{3.6}$$

Since the OIT value often drops more sharply at the beginning of the ageing test, the value achieved after a half year is considered a suitable reference value. However, this value, measured at 210 °C, should be $OIT(0.5y) \geq 10$ min. The requirements of the Geosynthetic Institute (GRI 1998) demand a residual value of 55 % of the standard OIT or a residual value of 80 % of the high-pressure OIT after a forced air oven ageing at 85 °C for 90 days:

$$\frac{OIT(90d)}{OIT(0)} \geq 0.55\,(\text{standard - OIT}) \qquad \geq 0.8\,(\text{HP - OIT})\,. \tag{3.7}$$

The OIT value should further be used as a parameter in geomembrane quality assurance. In their resin specifications, resin manufacturers indicate minimum values for the OIT value which are usually measured at 200 °C or 210 °C. The adherence to the minimum value is guaranteed in the acceptance test certificate. The incoming resin OIT value should be regularly checked as part of the geomembrane manufacturer's incoming quality control programme and within the third-party control of geomembrane production.

The determination of OIT values of excavated polyolefin geosynthetics and the comparison with the initial values should become a regular part in trial pits investigations and field studies dealing with long-term behaviour of polyolefin geosynthetics. This is the only easy way to obtain substantial information about the change of oxidation stability under field conditions, since the monitoring of the mechanical properties is not conclusive with respect to long-term behaviour, see Sects. 5.1 and 5.2.

When OIT values are compared within quality assurance or case history studies, the random distribution of the individual measurements must be taken into account and the significance of statements about changes must be proved using a statistical analysis. The relevant statistical method is described for example in (Hald 1952; Motulsky 1995).

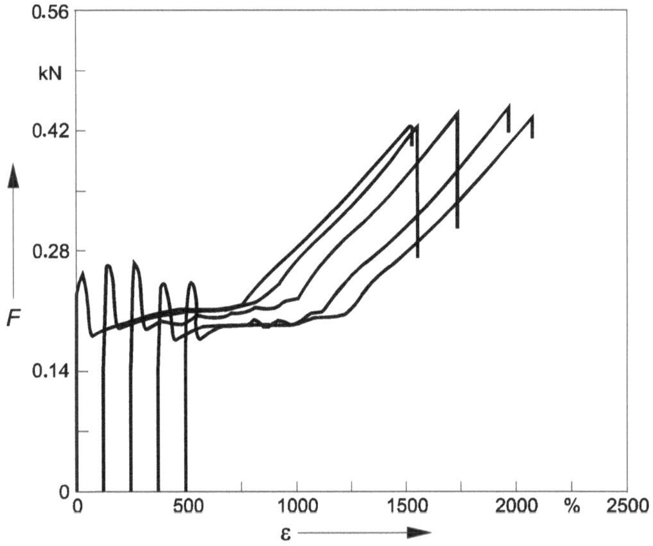


**Fig. 3.9.** Typical load-extension curve (or stress-strain curves) of tensile tests on five test specimens from an HDPE geomembrane. The test specimens (test specimen 5 relying on EN ISO 527-3) were taken crosswise to the extrusion direction. Tensile force is plotted against elongation. The test cross-section of the test specimen is approx. 2.3 · 6 mm². The individual curves were offset plotted by 100 % for better visibility. The test speed was 50 mm/min. The test was performed at a standard atmosphere

### 3.2.8 Tensile Test

The tensile test properties of geomembranes are tested according to DIN ISO 527-1:1996 *Determination of Tensile Properties, Part 1, General Principles*, and DIN ISO 527-3:2002 *Plastics – Determination of Tensile Properties – Part 3: Test Conditions for Films and Sheets*. ASTM D638-

99 *Standard Test Method for Tensile Properties of Plastic* should be mentioned among US American standards which largely correspond to DIN ISO 527. In addition, the standard ASTM D6693-01 *Standard Test Method for Determining Tensile Properties of Nonreinforced Polyethylene and Nonreinforced Flexible Polypropylene Geomembranes* is often referred to. Figure 3.9 shows typical stress-strain curves, or more precisely load-extension-curves, for HDPE geomembranes. The following characteristic test quantities can be read off from this curve: yield stress, i.e. the first stress value where strain increases without any increase in stress; stress at break, i.e. the stress where the specimen breaks; yield strain, i.e. the strain at yield stress; strain at break, i.e. the strain at the stress at break and finally the modulus of elasticity in tension (Young's modulus), i.e. the difference quotient of stress and strain difference between the strain values 0.0025 and 0.0005. Alternatively the 2 % secant modulus may be used as an approximation to the modulus of elasticity (ASTM D5323-92 *Standard Practice for Determination of 2 % Secant Modulus for Polyethylene Geomembranes*). Stress is indicated in units of MPa (= $N/mm^2$). Strain is defined as the change of the gauge lengths of the specimen referred to the initial gauge length and measured in the dimension of % or used as a dimensionless quantity. For the strain at break the nominal elongation is often used which is defined as the change in the distance between the clamping jaws (clamping length) compared to the initial distance of the clamps. When strain at break values are compared, this difference must be taken into account.

The tensile test results depend on the test atmosphere, i.e. temperature and moisture conditions, test speed, and type and conditioning of the specimen used. According to the standard, the test atmosphere[1] can be

---

[1] According to DIN 50014:1985 *Climates and their Technical Application; Standard Atmospheres* the standard atmosphere 50014-23/50-2 comprises an air temperature of 23 °C and a relative humidity of 50 %. The additional requirements for dew point temperature, air pressure and air velocity usually play no role in geomembrane testing. The digit 2 at the end marks an accuracy class (± 2 °C and ± 6 %). The term "room temperature" in this standard refers to the temperature of a room "in that air temperature is within a specified range, without taking into account relative humidity, air pressure and air velocity". Normally, a temperature range of 18 °C–23 °C is assumed. The EN ISO 291:1997 *Plastics–Standard Atmospheres for Conditioning and Testing* defines a laboratory atmosphere with an air temperature of 23 °C and a relative humidity of 50 % as standard atmosphere 23/50, Class 2. Class 2 refers to the accuracy (± 2 °C and ± 10 %). The standard introduces the term ambient temperature which is defined in a similar way to room temperature in DIN 50014. However, 18 °C–28 °C is now specified for the temperature range. The conditioning of plastic specimens is described in Appendix A of EN ISO 291.

agreed upon. Strictly speaking, the specimens must be equilibrated to the test atmosphere and tested in the standard atmosphere, (23 ± 2) °C temperature and (50 ± 5) % relative humidity. However, tests are often carried out at ambient atmosphere. The test speed is typically 100 mm/min. When tensile tests are performed as a part of the quality control of a running geomembrane production, the speed is changed during the test to 200 mm/min above the yield strain to reduce test time. With a gauge length of 50 mm and an elongation of 1000 % the test of one specimen would otherwise last 5 min. DIN ISO 527-3 formally refers only to films and sheets with a thickness up to 1 mm. Nevertheless, this part is also used to test geomembranes with thicknesses of 1.5 mm and greater. Type 5 is the specimen usually used which is illustrated in Fig. 3.10 together with the two other specimen types 1B and 2 occasionally used as well.

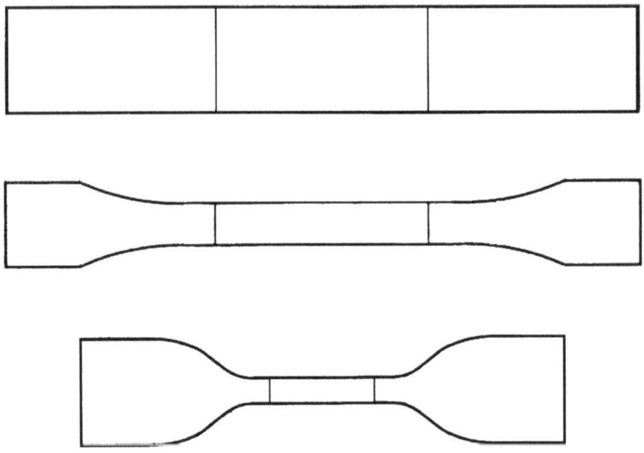

**Fig. 3.10.** Test specimen type 2 (top), type 1B (middle) and type 5 (bottom) which is used for testing HDPE geomembranes based on EN ISO 527-3. The test specimens have the following sizes

| Type of test specimen | | 5 | 1B | 2 |
|---|---|---|---|---|
| Width of narrow section | $b_1$ | (6 ± 0.4) mm | (10 ± 0.2) mm | - |
| Width overall | $b_2$ | (25 ± 1) mm | (20 ± 0.5) mm | 10 to 25 mm |
| Gauge length | $L_0$ | (25 ± 0.25) mm | (50 ± 0.5) mm | (50 ± 0.5) mm |
| Length of narrow section | $l_1$ | (33 ± 2) mm | (60 ± 0.5) mm | - |
| Distance between grips | $L$ | (80 ± 5) mm | (115 ± 5) mm | (100 ± 5) mm |
| Length overall | $l_3$ | ≥ 115 mm | ≥ 150 mm | ≥ 150 mm |

Test quantities derived from the tensile test can be used for the assessment of performance properties, as a quality criterion within the framework of quality assurance and for a rough material identification. Plastic geomembranes should be able to withstand forces at limited deformations, so that imposed deformations are distributed over a wider range. On the other hand geomembranes should not be too rigid to allow for medium forces to deform them. For instance F. W. Knipschild suggested a tensile force of a minimum of 400 N for each 50 mm of plastic geomembrane length at a strain of 5 % (Knipschild 1985). HDPE geomembranes have a Young's modulus of approximately 600 MPa. The yield stress exceeds 15 MPa at the test conditions specified above and the yield strain exceeds 10 %. Stress at break reaches values of approximately 30 MPa. The strain at break is usually above 1000 %. The strain at break reacts sensitively to material changes which are, for example, due to imperfect manufacturing or degradation processes. The change in strain at break can therefore be used in long-term investigations as an indicator for degradation. The visco-elastic deformation behaviour of thermoplastic materials is dealt with in detail in Chap. 4.

### 3.2.9 Multi-Axial Tension Test (Burst test)

Installed geomembranes are deformed by subsidence or settlement in the geotechnical structure. The geomembrane is usually forced into a planar state of stress accompanied with imposed multi-axial deformations (elongations along both axes in the tangential plane and a perpendicular thickness reduction). In addition to local indentations due to small objects such a large area deformation forms the main mechanical impact imposed on installed geomembranes. Therefore a test which enables direct testing of the multi-axial stress-strain behaviour and characterisation of it by appropriate parameters was sought. DIN 53861:1970 *Testing of Textiles, Vaulting Test and Bursting Test* describes an experimental setup for textile planar formations which was introduced in a modified form for geomembrane testing by the then working group AK 14 "Plastics in Geotechnics" of DGGt in 1984 (DVWK 1989). The test procedure rapidly spread world-wide. The test was also included in BAM's certification requirements and a test description can be found in the BAM certification guideline. Meanwhile, it was standardised by the American Society for Testing and Materials as ASTM D5617-94 *Standard Test Method for Multi-Axial Tension Test for Geosynthetics* and a draft standard was issued by the European Committee for Standardisation EN 14151:2001 *Geoynthetics – Determination of Burst Strength*. In a somewhat misleading way, the multi-axial tension test is often called bursting pressure test or burst test. However, only the strain values and the characteristics of the stress-strain behaviour are relevant and not the bursting pressures reached.

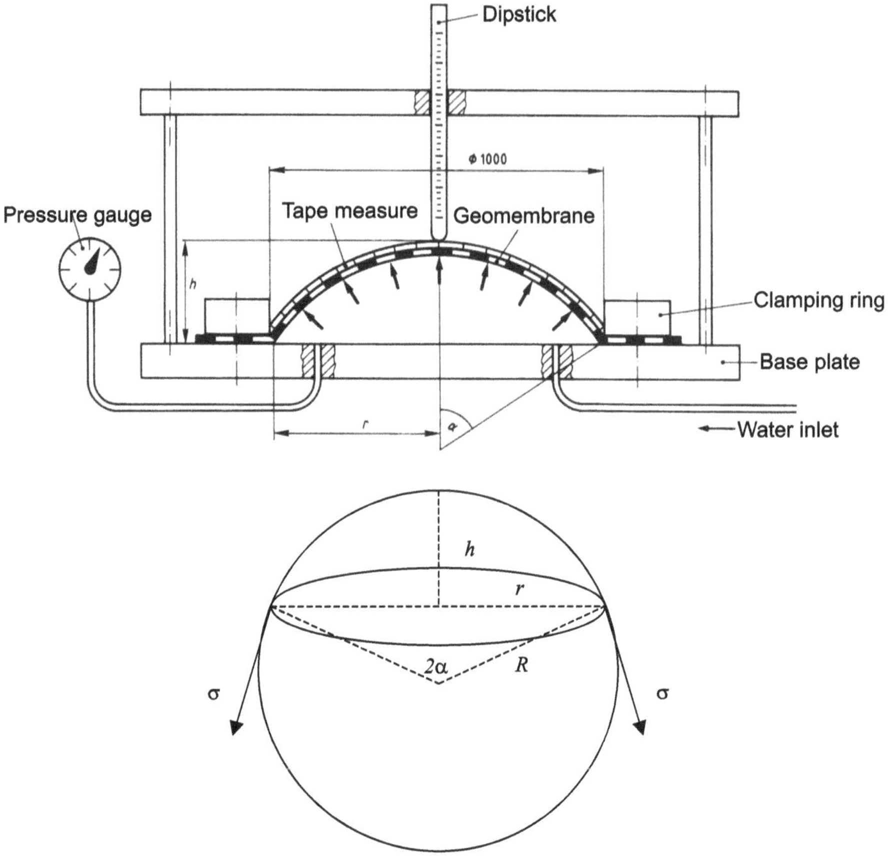


**Fig. 3.11.** Schematic of the device for a burst test (source: Schicketanz Consulting Engineers, State Institute for Materials Testing, Hanover) (Müller 1992). The attached sketch (bottom) explains the geometrical meaning of the quantities used to describe the test parameters

Dissimilar technical variants are available in which the experimental procedure is performed in practice. Figure 3.11 shows a schematic of a possible test device. The test is performed as follows: A disk of the geomembrane is clamped between the rings of the clamping devices in such a way that a free, circular geomembrane area of 800 mm to 1000 mm in diameter can be loaded. For sealing purposes elastomer rings can be applied in the clamped area between clamping ring and geomembrane disk and between base plate and geomembrane disk. Loading takes place using air or water as a pressure-transmitting medium. When the medium flows in, the geomembrane bulges and the pressure builds up. When using water, the formation of enclosed air cushions beneath the geomembrane must be avoided. The pressure build-up can be performed in steps or continuously by controlling the intake valve. The arc height (or the centrepoint) deflection of the deformed specimen from the base plate and the pressure are measured. The ASTM standard requires a rate of centrepoint deflection of 20 mm/min for the continuous pressure build-up.

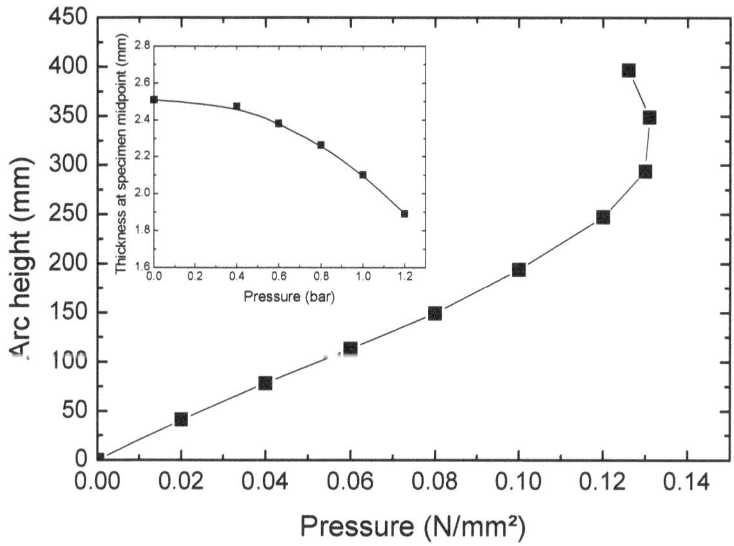


**Fig. 3.12.** Arc height (or centrepoint deflection) as a function of water pressure measured in a burst test on a 2.5 mm thick HDPE geomembrane (Kreiter and Hutten 1990). The clamp free diameter was 1050 mm. The pressure was stepwise increased by 0.2 bar (0.02 N/mm²) per two minutes. The material starts to yield at approximately 1.3 bar (0.13 N/mm²). When yielding has started, the rate of water flow is not sufficient to compensate for the volume increase under the geomembrane: the pressure drops. Finally, the yielded range tears: the test ends with a spectacular water fountain. The small diagram shows the decrease in the thickness of the geomembrane in the centre of the arched test specimen

According to the requirements of the BAM certification guideline (Müller 1992) and the above mentioned recommendation of the DGGt working group (DVWK 1989), the pressure is increased in steps of 20 kPa or 10 kPa and maintained at each step for 2 min. The arc height is measured at least every 10 seconds when continuous pressure build-up is used, and in the middle of the time intervals between pressure increases when step-wise pressure build-up is applied. The test is carried out until a usually clearly visible yielding (plastic flow) of the arched specimen is achieved. Figure 3.12 shows a typical curve obtained for the arc height as a function of water pressure.

This arc height-pressure diagram can be converted into a stress-strain diagram. It can be assumed for HDPE geomembranes that the arched geomembrane has approximately the form of a sphere calotte (Fig. 3.11). $R$ is the radius of the relevant sphere, $h$ is the height of the calotte (the arc height or centrepoint deflection). The free clamping radius $r$ is the radius of the basis circle of the calotte.

The so-called arc elongation $\varepsilon_b$ can be determined from the arc height $h$ and the free clamping radius $r$ as follows:

$$\varepsilon_L = \left(\frac{\mathrm{arc}\,\alpha}{\sin\alpha}\right) - 1 \quad \text{with} \quad \sin\alpha = \frac{2rh}{r^2 + h^2} \quad \text{and} \quad \mathrm{arc}\,\alpha = \frac{2\pi}{360}\alpha\,. \tag{3.8}$$

The arc elongation can also be determined by direct measurement of the change in the arc length $b$: $\varepsilon_L = \Delta b/b$, $\Delta b = b - 2r$. The relationship between the test pressure $p$ and tangential stress $\sigma$ is given as (see Fig. 3.11):

$$\sigma = \frac{pR}{2d}\,. \tag{3.9}$$

$d$ is the thickness of the geomembrane tested.

The deflection enforced by the test creates a plane state of stress in the geomembrane specimen, see Sect. 5.3.3. The accompanying deformation state is, however, three-dimensional: the thickness of the geomembranes also changes, see Fig. 3.12. The bursting pressure and the change of the arc elongations with pressure depend highly on the variant of the test procedure, for example on free clamping radius or deformation rate. Therefore, the test conditions must always be precisely recorded. The test is usually performed at room temperature (see Sect. 3.2.8, Footnote 1). In Germany the test was most often carried out using the step-wise pressure increase as described above. Test results of smaller scattering of data and a more unambiguous characterisation of the deformation behaviour of the geomembranes can be achieved if the pressure is continuously increased at a constant rate and the arc height is continuously measured. The arc height-pressure curve and the calculated arc elongation-tangential stress curve describe quantitatively the multi-axial deformation behaviour at a specified deformation rate. The test report should always include a qualitative description of the failure mode. HDPE geomembranes usually fail in the shape of "a hole in cat eye", which is described in ASTM 5617 as follows: "Circular or elliptical hole in an area where the material has significantly necked down and thinned. The large thinned area resembles a pupil of a cat eye".

The experimental set-up can also be used for the testing of stress relaxation in a plane state of stress. In this test a specified quantity of water is pumped rapidly into the device producing a certain arc height and associated multi-axial deformation. Care must be taken to prevent air cushion formation. The gradual decrease in pressure as a function of time can be obtained from the manometer reading. The test procedure and the evaluation of the relaxation test data can be performed in analogy to the stress relaxation test under an imposed uniaxial deformation. Figure 3.13 shows such a stress relaxation curve at an imposed multi-axial deformation.

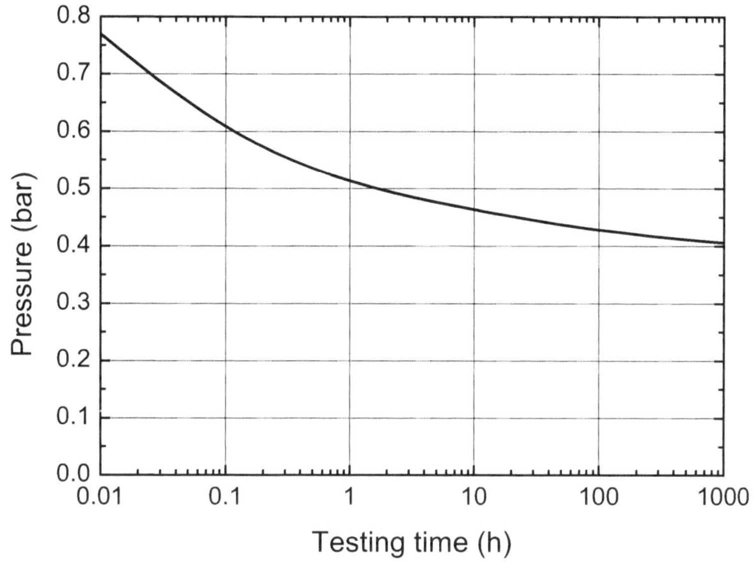


**Fig. 3.13.** A certain arc height is imposed on the clamped-in-place HDPE geomembrane specimen by the inflow of a specified amount of water, which rapidly enters the burst test device. Because of the incompressibility of the water body the imposed centrepoint deflection does not change with time. However, water pressure will drop as a function of time and a relaxation curve for the plane state of stress can be determined, which typically occurs in the case of subsidence of the geomembrane foundation layer (Kreiter and Hutten 1990)

### 3.2.10 Relaxation Test

The relaxation behaviour, i.e. the stress relief at imposed deformations, is included in the assessment of long-term behaviour of plastic geomembranes. However, creep behaviour, i.e. the continuous deformations due to an imposed load, has no relevance to practical use. The situation in the investigation of plastic pipes is just the opposite. The ring stiffness test, both as a short-term or long-term creep test, or the pipe pressure test provide characteristics with direct practical relevance for pipe application. The different approaches follows from the fact that geomembranes must be installed in such a way that they are not forced to carry load over the long term (Knipschild 1985). In a geomembrane anchored on the top of a berm, which has only little friction with the subgrade, even the dead weight may cause tensile stresses to form. However, this stress (approx. 1 $N/mm^2$) is not critical with respect to ductile or brittle failure even on long, steep slopes. The downhill slope forces or lateral spreading forces caused by other superstructures, however, would rapidly create a large tensile stress in the geomembrane because of the its thin cross-section. Such stresses become effective for instance when the friction force between the geomembrane and overlying layer is greater than that between the geomembrane and subgrade and the friction force in this boundary surface cannot support the downhill slope forces or the lateral spreading force. Under such conditions the geomembrane will rapidly creep, yield and finally break (ductile failure). Even if the stress remains below the level of ductile failure, stress cracks may develop over the long term, see Sect. 5.3.

In individual cases even a sudden brittle failure was observed in HDPE geomembranes manufactured from "true" high density PE-resins, when they were exposed to a stress over a large area (EPA 1992). This phenomenon, however, seems to occur solely in extremely cold weather. The brittle cracks extend and branch out and propagate rapidly over the whole surface. This kind of cracking is dealt with in the technical literature under the term "rapid crack propagation (RCP)" as an independent mechanism of stress cracking.

Various deformation, however, are imposed unavoidably on geomembranes during the installation and operational phase, even if they are handled and installed correctly. The imposed deformations also result in stresses which are partially relieved over the course of time by stress relaxation. Chapter 4 deals with creep, relaxation and deformation behaviour of HDPE geomembranes in more detail. Since imposed deformations represent a frequent and a likely event, not only particularly stress crack resistant resins should be chosen but the relaxation behaviour of the geomembranes should be checked as well.

The test of relaxation behaviour is described in (outdated) DIN 53441:1984 *Testing of Plastics, Stress Relaxation Test*. The principle of the test method is that an elongation is imposed on the test specimen by a uniaxial tensile force which is then kept constant. The force necessary to maintain constant elongation decreases as a function of time due to relaxation. The temporal reduction of the force and the stress, i.e. the ratio of the force to the original cross-section of the specimen, will be measured. The test result is therefore a stress versus time or a relaxation curve $\sigma(t)$ for a specified constant elongation $\varepsilon$. Straining the specimen requires a certain time $t_S$. After a short pre-relaxation the specimen will reach the specified elongation $\varepsilon$ at time $t_0$, the first stress measurement takes place at time $t_1$. The standard requires that the straining rate is chosen in such a way that $t_1 - t_0 \geq 10\, t_S$. The relaxation curve depends highly on temperature. The test must therefore take place in a well-defined atmosphere, such as the standard atmosphere (see Sect. 3.2.8, footnote 1). Parallel bar specimens are chosen as test specimens (Fig. 3.10, type 2).

According to the requirements of the BAM certification guidelines the first measured value has to be determined after one minute, thus $t_1 - t_0 = 1$ min or $t_S \approx 6$ s. The measurement is carried out over 1000 hours. Then the ratio of the stress in 1000 hours, $\sigma(1000\text{h})$, to the stress in one minute, $\sigma(1\text{min})$, is determined. The stress relaxation must be so great that the stress reduction is at least 50 %:

$$\frac{\sigma(1000\ \text{h})}{\sigma(1\ \text{min})} \leq 0.5 \,. \tag{3.10}$$

### 3.2.11 Resistance to Chemicals

Depending upon type and properties, chemicals can affect plastics in various ways. Molecules of a chemical can diffuse into the plastic and, depending upon their solubility, can become highly enriched. The physical properties of the plastic, such as volume, mass, tensile properties and deformation behaviour, may all change. The phenomenon is called swelling and it is accompanied with a swelling pressure. The swelling forces can cause stresses within the plastics (Menges 1990). Such swelling chemicals can eventually fully dissolve the plastic.

The actual chemical effects of chemicals have to be distinguished from such physical effects. The molecules of a chemical can enter into chemical reactions with the reactive groups of the polymer chain. For example, oxidation is possible in polyolefins, and hydrolysis in polyesters and polyamides. Chemical reactions can result in various detrimental effects from discoloration of the plastic to its complete destruction.

Apart from these "primary" effects secondary effects can also emerge. Chemicals can extract or modify additives and thus indirectly affect the properties of plastics. Detrimental repercussion of mechanical stress on plastics can be increased by chemicals. First of all stress crack formation can be drastically accelerated by chemicals. Conversely, mechanical stress may increase the resistance of a plastic against chemical degradation, oxidative or hydrolytic attack for instance.

Resistance against chemicals can be tested by immersion tests. The test specimens are kept in the chemical according to a specified procedure and specified conditions, removed after a certain period of immersion and tested for irreversible or, after conditioning, e.g. by oven drying, for reversible changes of the properties. Investigating the alteration of the following properties is reasonable in HDPE geomembranes: external appearance (surface properties), mass of the test specimens immediately after removal and after re-drying, mechanical parameters from the tensile test (in particular stress and strain at failure) immediately on removal and after re-drying, oxidation stability (oxidative induction time) after re-drying, stress crack resistance after re-drying (NCTL test). For weld seams it is the short time peel test and short time tensile shear test and in addition, the long-term creep peel test that provide information about detrimental effects of chemicals. The change of adhesion of the applied structural particles can be tested in textured geomembranes for which a planing-off test or an abrasion test is used (see Sect. 6.2). The results of immersion tests may only be applied to mechanically non stressed plastic components. The combined effect of chemical and mechanical stress can be tested in pipe pressure tests using particular chemicals as test liquids, see Sect. 3.2.13. Also the long-term shear test, Sect. 3.2.18, can be performed using a liquid chemical.

EN ISO 175:2000 *Plastics Methods of Test for the Determination of the Effects of Immersion in Liquid Chemicals* and DIN 53521:1987 *Testing of Rubber and Elastomers, Determination of the Resistance to Liquids, Vapour and Gases* provide instructions about the procedure of immersion tests. Based on these standards, BAM has developed a test method for the test of the resistance of geotextiles against liquid chemicals in the "Certification Guidelines for Protective Layers" (Müller 1996, 1995) which can also be applied in an analogous way to geomembranes. ASTM D5747-95a *Standard Practice for Tests to Evaluate the Chemical Resistance of Geomembranes to Liquids*, in connection with ASTM D5322-92 *Standard Practice for Immersion Procedures for Evaluating the Chemical Resistance of Geosynthetics to Liquids*, describes immersion tests for resistance against liquid chemicals (liquid wastes, industrial chemicals, leachate) of geomembranes which are used in geotechnical engineering.

Type and composition of chemicals, immersion temperature and immersion time are the key test conditions. A characterisation of the chemical resistance must always be seen in connection with these test conditions. There are various methods which can be illustrated by two examples from the field of landfill liner applications. The US Environmental Protection Agency (US EPA) has developed a test guideline EPA Method 9090:1986 *Compatibility Tests for Wastes and Membrane Liners* for testing landfill liner geomembranes (White and Verschoor 1990). The geomembranes are tested in those leachates with which they (are expected to) come into contact in the planned landfill cell. For this purpose a leachate is delivered by the landfill operator or a sample is taken from a comparable, existing landfill and used as a test liquid. The test is performed in closed containers so that as little as possible of the volatile components in the leachate escape. Immersion temperatures are 22 °C and 50 °C and the immersion time is 120 days. Then a number of characteristics are tested: hardness, properties in tensile test, Young's modulus, resistance to punctual load, resistance to tear propagation, burst test, volumetric change, change of mass, content of volatile components, content of extractable components. The test method is very close to a real application as far as the selection of the test liquids and test temperature is concerned. Initially, the test had to be performed for each individual construction project. The test method is rather expensive and "unpleasant" because original leachates are used as test liquids.

Such performance test methods with test liquids and test temperature apparently comparable to application conditions have various disadvantages: the immersion time is not even approximately sufficient even at slightly increased temperatures, e.g. 50 °C, to reflect the actual time of impact in a real landfill liner system. The test liquid cannot be identified precisely enough and may change during the test. Though a number of test parameters are evaluated, these "macroscopic" parameters express the chemical-physical changes at a "microscopic" molecular level only, when these have already progressed to a large extent. This condition, however, is seldom reached within the short immersion time. Geomembranes, which do not pass this test i.e. exhibit significant, irreversible changes in the test parameter, are certainly useless as landfill liners. However, the opposite case, i.e. even when the geomembrane passes the test, it fails to provide a sufficient proof of the product's real suitability in the long term.

Another approach for the requirements of resistance to chemicals is pursued in the BAM certification procedure. The initial consideration was that the service lifetime of a geomembrane in a landfill liner is extremely long. Therefore the type and extent of the chemical stress cannot be representatively described at all, particularly since elevated temperatures may also develop over longer times. Therefore only those geomembranes should be used that have extensive functional reserves against a broad spectrum of chemicals right from the beginning (see Sect. 5.1). The geomembranes are therefore exposed to an attack of highly concentrated chemicals from different material groups. Only geomembranes that have proved resistant to this attack possess these large functional reserves.

The details of the test method are described in the above mentioned BAM "Certification Guidelines for Protective Layers" which can also be applied to plastic geomembranes. The immersion time is at least 3 months. The parameters are sequentially determined after an immersion of one week, fourteen days and 3 months on specimens which were always immediately re-dried after removal (5 days at 50 °C followed by 2 days at 23 °C in a vacuum oven). The immersion temperature is 23 °C. The immersion may not cause any change in the external appearance. The tensile test parameters and the mass may not change more than 25 % during immersion and after re-drying. For specimens immersed in swelling media, which regain sufficient mechanical strength only after re-drying, OIT measurements have to clarify to what extent an impairment of stabilisation against thermal-oxidative degradation has occurred. The test liquids are listed in Table 3.7.

While the EPA Method 9090 differentiates very little between the different plastic geosynthetic materials, only geosynthetics made of a few special plastic materials, especially HDPE geomembranes, will pass the BAM test.

Since polyethylene consists of saturated C-C bonds and C-H bonds, which are structurally simple, have relatively high bonding energy and therefore only show a low affinity to chemical reactions, they belong to the plastic materials of high resistant to chemical effects. Medium to high density polyethylene cannot be dissolved by chemicals at normal temperatures. A thorough analysis of the effects of the material groups specified in Table 3.7 on medium to high density polyethylene will show, however, how different the detrimental effects of chemicals may be, even for this highly resistant plastic (Dolezel 1978). Several cases can be distinguished.

In an ideal case, no chemical or physical changes take place over a wide range of temperature (no oxidation and no swelling), also none of secondary nature (no acceleration of stabiliser depletion and stress cracking). This is the case for the material groups 10 and 11, inorganic caustic solutions and inorganic neutral salts have no detrimental effects on HDPE materials.

A second case is when no major physical or chemical effects can be observed, i.e. when no chemical reactions take place and solubility and thus swelling is very small over a wide range of temperature, but secondary effects can be observed, e.g. acceleration of stress cracking. This is the case for substance groups 3 (aromatic amines), 4 (alcohols), 6 (aromatic esters and ketones), 7 (aldehydes) and 8 (organic acids). These substance groups, in particular alcohols and organic acids, accelerate stress cracking. The so-called stress crack resistance factors can be used for the assessment of the effect of concentrated liquid media on stress cracking (Sect. 3.2.13). These factors have been measured for many HDPE materials and a large number of chemicals in pipe pressure tests. The influence of chemicals on stress cracking is described in more detail in Sect. 5.3.4.

**Table 3.7.** List of test liquids for the testing of the chemical resistance of plastic geomembranes for landfill liners

| No. | Material group | Composition of test liquids |
|---|---|---|
| 1 | Petroleum (Otto fuels) and aromatic hydrocarbons | 40 % by volume 2,2,4-trimethylpentane (isooctane)<br>30 % by volume methylbenzene (toluene)<br>20 % by volume dimethyl- benzene (xylene)<br>10 % by volume methylnaphthalene |
| 2 | Heating oils, diesel fuels, paraffin oils and lubricating oils | 35 % by volume diesel fuel<br>35 % by volume paraffin oil ($C_{10}$–$C_{20}$)<br>30 % by volume HD 30 lubricating oil |
| 3 | Amines | 40 % aqueous solution of dimethylamine |
| 4 | Alcohols | 30 % by volume methanol<br>30 % by volume propanol-(2) (isopropanol)<br>40 % by volume ethandiol-(1,2) (glycol) |
| 5 | Aliphatic hydrocarbons | 30 % by volume trichloroethene (trichloroethylene)<br>30 % by volume tetrachloroethene (tetrachloroethylene)<br>40 % by volume dichloromethane (methylene chloride) |
| 6 | Aliphatic esters and ketone | 50 % by volume ethane acid ethyl ester (ethyl acetate)<br>50 % by volume 4-methyl-pentanol-(2) (methyl isobutyl ketone) |
| 7 | Aliphatic aldehydes | 37 % aqueous solution of methanal (formaldehyde) |
| 8 | Organic acids | 50 % by volume ethanoic acid (acetic acid)<br>50 % by volume propanoic acid (propionic acid) |
| 9 | Inorganic mineral acids (oxidizing) | 50 % by volume sulphuric acid (95–97 %)<br>50 % by volume nitric acid (65 %) |
| 10 | Inorganic bases | 60 % caustic soda solution |
| 11 | Inorganic neutral salt solution | Saturated $NaCl/Na_2SO_4$ solution (ratio 1:1) |

The substance groups 2 (diesel fuels, paraffin oils and lubricating oils) and 5 aliphatic chlorinated hydrocarbons and aliphatic hydrocarbons (e.g. petrols) do not cause any chemical reactions either, but physical effects due to swelling, e.g. change in the mechanical properties occur during their influence. This may be considered as the third case. Solubility of the substance group 2 is small, that of aliphatic hydrocarbons and chlorinated hydrocarbons, however, fairly large. The diffusion and solubility of chemicals in HDPE geomembranes are dealt with in detail in Chap. 7. Because of the still limited solubility these effects are completely reversible even after very long immersion times. However, swelling is accompanied with a gradual change in stabilisation in this case. Figure 3.8 shows the measurement of the change of the OIT value of an HDPE geomembrane which was exposed to a mixture of aromatic and aliphatic hydrocarbons and chlorinated hydrocarbons in a test cell for permeation measurement through a composite liner over 12 years. After oven re-drying the removed specimens were identical in all properties to the untreated reference specimen, only OIT values showed noticeable changes which can be attributed to stabiliser extraction.

The behaviour at normal temperatures against aromatic hydrocarbons (toluene, xylene) corresponds to the case just discussed. At high temperatures (≥ 80 °C), however, the effect of swelling is irreversible: organic solvents can dissolve even a HDPE material under these conditions. This very poor solubility of polyethylene requires special high-temperature equipment in liquid chromatography for the determination of the molecule mass distribution or in viscosity measurement, since these measurements must be performed with a hot test solution.

Finally, there is the case when the chemicals enter into proper chemical reactions with polyethylene. This is only the case for substance group 9: oxidising acids (sulphuric acids, nitric acids etc.). Here too, noticeable effects can only be observed at high concentrations and temperatures. Oxidation by these acids differs from the special radical chain reaction of oxidation by oxygen. Sections. 5.2, 3.2.7 and 3.2.12 are dedicated to the oxidative degradation by oxygen.

This overview of the various cases of repercussions of chemicals on HDPE materials justifies the conclusion already addressed several times that, though HDPE geomembranes are not entirely inert, they are extraordinarily resistant to a broad spectrum of high-concentration chemicals at normal temperatures. This can also be seen by the assessment of chemical resistance as described in DIN 8075-Attachment 1:1984 *High Density Polyethylene (HDPE) Pipes; Chemical Resistance of Pipes and Fittings.* The standard classifies for a large number of chemicals whether HDPE materials are resistant, only conditionally resistant or are not resistant at all in the temperature range between 20 °C and 60 °C. The latter is only the case with a few chemicals (e.g. chromosulfuric acid, aqua regia, bromine, chlorine or fluorine gas and dichloroethylene). For the special resins selected for HDPE geomembranes, the test of the chemical resistance in the certification procedure is usually nowadays no longer necessary due to the readily available experience. Immersion tests are only preformed on other geosynthetic products from polyolefin materials (e.g. geotextiles and geocomposite drains) with the focus on resistance against swelling or oxidising chemicals and highly stress crack-initiating media. Tests with the substance groups 2, 3, 6, 7, 10 and 11 of Table 3.7 are therefore usually not necessary.

Finally, the different interpretation of the term “resistance to chemicals” has to be briefly dealt with as it is used in geotechnical engineering (Müller et al. 1997). The term plays an important role in the assessment of sealing materials and there are misunderstandings because of the different interpretations.

Within the framework of the BAM certification procedure for plastic geomembranes, resistance to chemicals is understood in such a way that the geomembranes withstand the impact of chemicals of high-concentrations over a large period of time without any serious changes. It is therefore interpreted in the sense of large functional reserves of the plastic geomembrane (see Sect. 5.1).

In geotechnical engineering the effect of high-concentration, aggressive chemicals occur only in very special cases, possibly in the case of hazardous waste storage or containment of contaminated land. It is usually gaseous chemicals in soil vapour or aqueous solutions of chemicals that have an effect on geomembranes or other liner materials. In accordance with Henry’s law, a certain concentration develops in the geomembrane at equilibrium with a specified concentration of the chemical in the aqueous solution or specified partial pressure in the soil vapour. When the concentration in the solution or the partial pressure at equilibrium is sufficient small, no unfavourable material changes develop in the geomembrane, or it will at least need a longer period of time for detrimental effects to occur, even if the plastic material is not resistant to the concentrated chemical. The material can be considered resistant to aqueous solutions and gas contamination in soil vapour up to a certain concentration. However, statements about the resistance must include the permissible concentrations in aqueous solutions or the permissible partial pressures. The EPA Method 9090 raises requirements for the resistance in this sense. The regulation itself does not speak about resistance, but only about compatibility of the plastic materials with chemicals of potential influences.

Finally, the term resistance is occasionally used even in the case of only a low-concentration and very short term impact by the chemical. If the amount of the chemical and the time period over which the chemical exerts its influence is small compared to the mass of the liner material and the time of the solution process or other material changes, and if the naturally stronger impairment of the initially near-surface range of the material does not have any unfavourable effect on the effectiveness of the whole bulk material, then even a non-resistant material can withstand the impact without unfavourable changes with regard to the performance. Statements about resistance, however, must then describe the respective restriction of quantity of the chemical and exposure time of the liner material, the special impact scenario and the material behaviour under these conditions in detail. In such cases one should only speak about conditional resistance or compatibility, in order to emphasize that the material remains functional only under special conditions (low concentration in the aqueous solution or soil vapour, temporal and quantitative limitation of the exposure).

### 3.2.12 Resistance to Thermal-Oxidative Degradation

Thermal-oxidative degradation is understood as the oxidative degradation at higher temperatures. The oxidative degradation of polyolefins usually proceeds very slowly at normal application temperatures. Noticeable effects within reasonable test periods can only be observed at higher temperatures. Therefore the expression thermal-oxidative degradation became a common term. The resistance is tested in oven ageing tests. The changes in geomembrane properties, which occur during the ageing time, are recorded by taking specimens in regular intervals from the oven and testing them. Sect. 3.2.7 already discussed how oven ageing tests and the measurement of the change in oxidation stability can generally characterise the resistance of HDPE geomembranes against oxidative ageing. It is discussed in Sect. 5.4 how estimates about service lifetime can be derived from such tests. The objective here is to describe the test method of oven ageing in greater detail.

The terms used in the description of oven tests and the requirements for the devices within the BAM certification guideline had originally been defined in the old outdated standard DIN 50011-1:1978 *Testing of Materials, Structural Components and Equipment, Warming Cabinet, Definitions, Requirements*, and in the standards replacing it: DIN 50011-11:1982 *Climates and their Technical Application; Controlled-Atmosphere Test Installations; General Terminology and Requirements* and DIN 50011-12:1987 *Artificial Climates in Technical Applications; Air Temperature as a Clima-*

*tological Quantity in Controlled-atmosphere Test Installations*. The DIN 50011-2:1960 *Testing of Materials, Structural Components and Equipment; Hot Cabinets; Directions for the Storage of Specimens* gives recommendation for oven ageing. ASTM E145-94 *Standard Specification for Gravity-Convection and Forced-Ventilation Ovens*, specifies requirements for ovens within the range of US standardisation. ASTM D5721-95 *Standard Practice for Air Oven Aging of Polyolefin Geomembranes* describes a test method of oven ageing of polyolefin geomembranes which are used in earth works and foundation engineering. Oven ageing is mentioned in DIN 16726. The Technical Committee (TC) 189 of the European Committee for Standardisation (CEN) compiled a manual ISO TR 13434:1998 *Guidelines on Durability of Geotextiles and Geotextile-related Products*, which makes a reference to a suggestion about a "screening test" for oxidation stability of geosynthetics: EN ISO 13438:2004 *Geotextiles and Geotextile-related Products – Screening Test Method for Determining the Resistance to Oxidation*.

There are different types of oven, i.e. devices in which specimens are exposed to elevated temperatures without humidity regulation or control. In the oven with natural ventilation, air exchange takes place gradually: air enters the lower part of the oven through an air inlet opening and rises upward by buoyancy due to density difference (gravity convection) and leaves the oven through an exit at the top. In the forced ventilation oven air is kept in circulation by a fan. The oven can be operated in two ways. In the fresh-air method the fan pumps the interior air outward through a vent and fresh air flows through an inlet. With closed vents, air is kept in circulation without renewal; here one speaks of forced air circulation.

In addition to various volatile components, stabilisers also escape during oven ageing of polyolefin geomembranes. This stabiliser evaporation and degradation is one of the key aspects of oxidative ageing. Oxygen is consumed during oxidative degradation to some extent depending on oxygen partial pressure. In order to have as uniform conditions in the interior atmosphere as possible, oven ageing in both forced ventilation and fresh air supply should be performed (ASTM D5721-95). When ovens with natural ventilation are used, at least an extensive air exchange should be ensured.

The actual usable volume of the oven, i.e. the volume, where the temperature is kept within the limits specified by the manufacturer or the limits necessary for the test, is smaller than the total internal oven volume. Its size also depends on the loading of the oven with specimens. The temperatures in various places of the oven differ at a certain time. The temperature at the same place also varies as a function of time. Depending upon the extent of these spatial and temporal variations in temperature in the usable volume of the empty oven, different accuracy classes are distinguished in

DIN 50011-12. In the usual tests the sum of spatial and temporal variation in temperature should not be greater than ± 2 °C and the temporal fluctuation alone no greater than ± 1 °C.

The specimens must be placed with sufficient distance between each other (at least 20 mm according to ASTM D5721) and with sufficient distance to the wall so that the air flow touches them as extensively as possible. The specimens are usually hung up on an insert lattice or shelve. Any contact with metallic surfaces must be avoided (ASTM D5721). A homogenous impact on the specimens is achieved when the specimens are regularly relocated from the front to the rear and from the top to the bottom (ASTM D5721).

Rather different test conditions are chosen in the standards and guidelines. DIN 16726 only contemplates an oven ageing test at (80 ± 2) °C for 7 days. The rough rule of thumb[2] that a temperature increase of about 10 °C accelerates the ageing processes by a factor of 2–3, can give some indication of relevant ageing times (Figs. 3.15). A 7 day ageing is not enough to ensure sufficient resistance for even a short service lifetime of 10 years. The BAM certification guideline requires an ageing of 1 year at (80 ± 2) °C, the GRI Standard GM 13 requires an ageing of 90 days at (85 ± 2) °C. In these tests the change of the OIT value is evaluated due to oven ageing, see Sect. 3.2.7. The test aims at normal to very long service lives (50 to > 100 years).

The screening test according to ISO TR 13434 has the aim of selecting geosynthetic products from polyethylene and polypropylene which are suitable for at least a medium service lifetime of 25 years. Nevertheless, oven ageing is performed at a high test temperature in order to achieve as short test times as possible. The test conditions and requirements are specified for two types of application: first, reinforcement or other applications where long-term strength is a significant parameter, second, all other applications. If the long-term strength is significant, the loss of strength may not exceed 50 % after an oven ageing at 110 °C and 28 days for polypropylene and 100 °C and 56 days for polyethylene. However, for other applications, the ageing time is halved: the loss of strength may not exceed 50 % after an oven ageing at 110 °C and 14 days for polypropylene and 100 °C and 28 days for polyethylene. Oven ageing tests at such high tem-

[2] The rule was established by the Dutch chemist Jacobus van't Hoff for the speed of chemical reactions in 1884. It is often also called van't Hoff law or RRT (reaction rate temperature) rule. This rule is based on the fact that the reaction rate depends exponentially on the inverse absolute temperature (see Sect. 5.1) and that the activation energy of many chemical procedures is within the range of approximately 50–100 kJ/mol.

peratures must neither be extrapolated using the Arrhenius law to application temperatures, nor service lifetimes be estimated with the rough rule of thumb (see Footnote 2). Overestimations for the service lifetimes of at least one order of magnitude can occur for polypropylenes (see Sect. 5.2.2). This method can therefore only serve as a rough screening with respect to the specifically indicated service lifetime of 25 years. It is based on the experience that polyolefin geosynthetic products, which have proved suitable over many years in practice, usually pass this test as well.

Oven ageing not too close to the melting temperature requires a long test time. It was therefore suggested to test at high oxygen pressure. However, the results have to be interpreted with caution. Under a high oxygen pressure the stabilisers are rapidly consumed by oxidative reactions and the associated hydroperoxid formation and oxidative degradation are greatly accelerated. On the other hand, high pressure has minor influence on the stabiliser migration processes. Therefore the half-time of auto-oxidative degradation of mechanical properties and the induction time of the auto-oxidation process are considerable reduced, while the depletion time due to migration remains constant or is possibly increased. Within the common interpretation scheme, described in Sects. 5.2.2 and 5.4, the failure time in an ageing test at high oxygen pressure is therefore essentially determined by the short high pressure values of the half-life of the auto-oxidative degradation and to some extent of its induction time (including rapid antioxidant consumption under the specific test conditions). The high pressure test ignores completely antioxidant migration effects because the stabilisers are consumed by oxidative reactions within a short time period compared with antioxidant depletion time under normal oxygen partial pressure. Brittleness of polyolefine geosynthetics and polyolefine geomembrane specimens is indeed achieved in high pressure tests at 80 °C within a few weeks (Schröder et al. 2000). Whereas antioxidant depletion by migration and by antioxidant degradation at that temperature and at normal oxygen partial pressure requires a few months testing time for geotextiles and some years testing time for HDPE geomembranes. For HDPE geomembranes the service lifetime is determined by an extremely large value of the antioxidant depletion time and not by induction time of auto-oxidation and the half life of auto-oxidative degradation of mechanical properties (see Sect. 5.4.2). As shortly described in Sect. 8.2.3, the antioxidant depletion time is generally very small for polyolefin geotextiles stabilised with a package of phenolic and phosphite antioxidants. In the high pressure test a sample of a poorly chemically stabilized but highly stretched and therefore structurally stabilised monofilament fabric or slit film fabric might therefore perform as well as or even better than a chemically well stabilised geomembrane sample, while in normal pressure oven ageing and therefore in "real life" performance the service lifetime of the geomembrane might possibly be an order of magnitude larger than that of the fabric. It is therefore not possible to compare the oxidative resistance of different geosynthetics under field conditions by doing a ranking according to the failure time in high pressure tests as it is not possible to qualify oxidative resistance by high temperature initial OIT values (Mueller and Jakob 2003). Therefore, high pressure tests might only be applied to compare oxidative resistance of products of a certain type of geosynthetics group, see Chap. 8, footnote 1.

### 3.2.13 Stress Crack Test: Pipe Pressure Test and NCTL Test

Geomembranes from partially crystalline thermoplastics can fail due to the so-called stress cracking which is understood to be a gradual brittle cracking under the long-term effect of a low-level stress. Stress cracking is highly accelerated by the influence of surface tension reducing substances dissolved in water. In contrast to stress crack corrosion, it is a purely physical process. Based on a very simple model concept, the long-term effect of stress causes a gradual de-anchoring and disentanglement of the tie-molecule chains that connect the crystalline regions. Thereby a microcrack is formed within the intermediate amorphous region. It grows gradually to a large crack leading to eventual failure. Sect. 5.3 deals with the phenomenon of stress crack formation in detail.

The really high-density polyethylenes (i.e. with a density > 0.960 g/cm³) are usually highly sensitive to stress cracking. However, HDPE geomembranes are made of so called LLDPE polyethylenes resins, which contain a few % by mass of other α-olefins (butene-1, hexene-1, octene-1) (see Chap. 2). In these polymers the α-olefins form side branches at the long ethylene chain, which increases the anchoring strength. Therefore, such HDPE materials have a high resistance to stress crack formation. The test of stress crack resistance is nevertheless an important element of plastic geomembrane selection and very high requirements must be demanded.

In the usual tests on stress crack resistance (see Sect. 5.3.2), e.g. the pin impression test EN ISO 4600:1998 *Plastics – Determination of Environmental Stress Cracking (ESC) – Ball or Pin Impression Method*, or in the so-called "bend strip test" or "Bell-test" as per ASTM D1693-97a *Standard Test Method for Environmental Stress-Cracking of Ethylene Plastics*, no failure of HDPE geomembranes is observed under the test conditions required by the standards. These index tests therefore cannot be used for the assessment of stress crack resistance of plastic geomembranes. The requirements are simply too low.

A concept based on the hoop stress versus time-to-failure curve of the pipe pressure test was developed for the assessment of the stress crack resistance and long-term behaviour of HDPE pipes (Gebler 1989). This concept can also be applied to HDPE geomembranes when such stress-rupture characteristics are available for pipes made of the HDPE material from which the geomembranes were manufactured (Koch et al. 1988). Service lifetime predictions based on this concept will be dealt with in Sect. 5.4 in detail. In the following a short description of pipe pressure testing is given (ASTM D1598-02 *Standard Test Method for Time-to-Failure of Plastic Pipe Under Constant Internal Pressure*).

In the pipe pressure test a pipe section is exposed to a specified internal test pressure. For this an end closure is attached to the section and the specimen is filled with water. Then the open end is closed and attached to a pressurising device and the whole section is immersed in a water bath having a specified test temperature. After some conditioning a pressure is applied and the time until the pipe section fails, the so-called time-to-failure, is measured. Failure is defined as loss of internal pressure.

Pressurising the pipe section causes a three-axial state of stress in the pipe wall. There is a radial outward directed stress component $\sigma_r$, a stress component in circumferential direction $\sigma_\varphi$ and a stress component in the direction of the axis of the pipe section $\sigma_z$. Relevant for failure are the tensile stresses within the pipe wall, i.e. the stress components $\sigma_\varphi$ and $\sigma_z$. The ratio of $\sigma_\varphi$ to $\sigma_z$ to is approximately 2:1. The largest tensile stress component $\sigma_\varphi$ is defined as hoop stress. The hoop stress can be calculated as:

$$\sigma_\varphi = p\frac{d-s}{s}, \tag{3.11}$$

where $p$ is the internal pressure, $d$ is the measured average outside diameter of the pipe and $s$ is the measured minimum pipe wall thickness. Most significant are two failure modes. At high internal pressure a necking down of the pipe wall will start at some weak surface point. The thinned area grows under continuous yielding along the pipe axis perpendicular to the hoop stress. Finally a balloon occurs, grows and ruptures. This failure is considered as ductile failure due to creep under high load. At lower pressure a brittle failure mode is observed. A brittle crack surface starts to open and to grow inward into the pipe wall and in the direction of the pipe axis. The thickness of the pipe wall is thereby continuously reduced and finally ductile failure occurs. This failure mode is typical for stress crack formation. Time-to-failure increases with decreasing internal pressure. However, even at very low pressures brittle failure will occur at some time due to oxidative degradation of the pipe material.

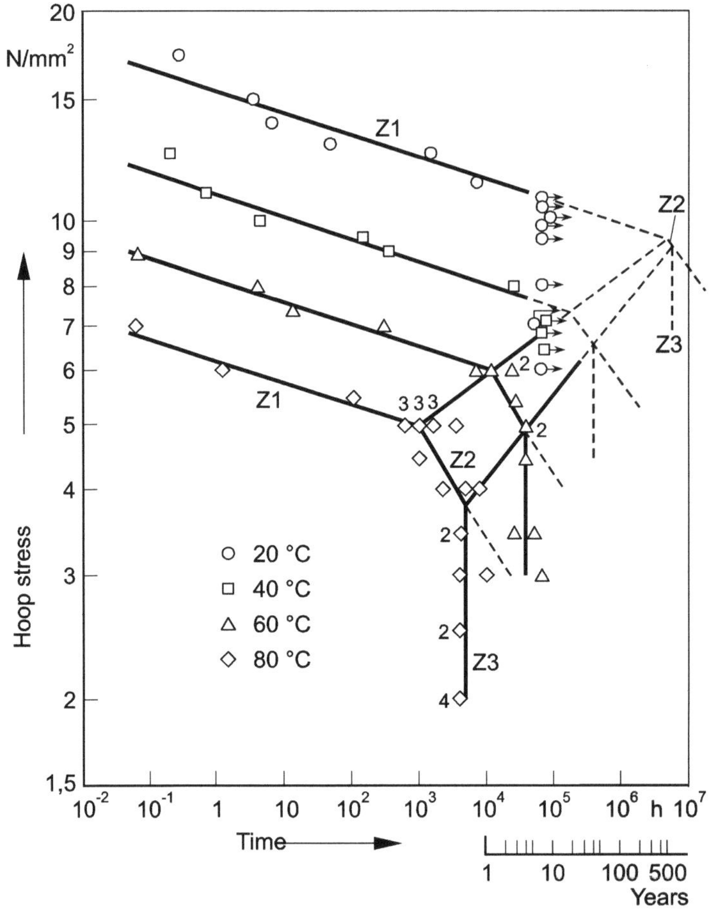


**Fig. 3.14a.** Stress-rupture characteristic of HDPE pipes made of Hostalen GM 5010 T2. Compared to modern materials this outdated resin had very low resistance to stress crack and oxidative degradation. Therefore, the various failure modes could be observed within manageable testing times even at 60 °C. Z1, Z2, Z3 designate the sections of the hoop stress versus time-to-failure curve, which result from different failure modes: Ductile failure with excessive deformation (Z1), stress crack formation and brittle failure (Z2) and oxidative degradation (Z3). The figure has been taken from reference (Koch et al. 1988)

For a specified internal pressure and test temperature the measured times-to-failure exhibit a logarithmic normal distribution. To attribute a certain time-to-failure to the applied test condition (hoop stress and temperature) a fractile of this distribution is chosen, for example, the 5 % fractile, i.e. 95 % of the pipes will have time-to-failure at the specified test conditions larger than the 5 % fractile. Plotting the hoop stress and associated time-to-failure data of thermoplastic pipes for a specified test temperature in a diagram with log stress versus log time-to-failure coordinates, a typical straight-line behaviour is found (Figs. 3. 14a and b).

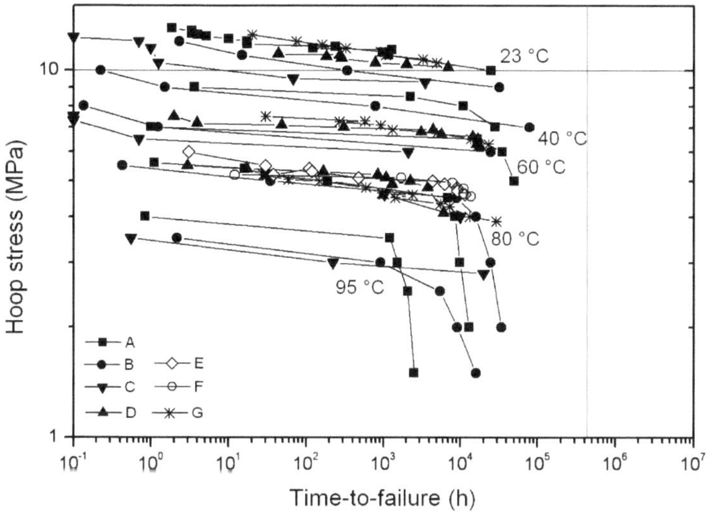


**Fig. 3.14b.** Stress-rupture data from pipe pressure testing of various HDPE resins used for geomembranes

The stress-rupture curve may be best fitted by sections of straight lines, where each section corresponds to one of the above mentioned failure modes. The first section corresponds to a failure accompanied with excessive ductile deformation. With the transformation of the failure mechanism from ductile failure to brittle failure by stress crack initiation and growth, a significant downward shift of the slope occurs and second straight line section of deeper slope results characterising brittle failure. After very long testing time oxidative degradation of the material starts, and brittle failure of the oxidised materials occurs even at very low stress level, quite independent of the hoop stress. With decreasing test temperature the stress-rupture line is shifted to the right in the diagram, i.e. to higher stresses and longer times-to-failure in a systematic way determined by the temperature dependence of the rate of creep, stress crack formation and degradation processes (Fig. 3.15). Stress-rupture behaviour at a higher temperature may therefore be used as a predictor of longer-time behaviour at a lower temperature. Details of the methods for the forecasting of the longer-term hydrostatic strength of pipes are given in the following standards: ISO 9080:2003 *Plastic Piping and Ducting Systems–Determination of the long-*

*term Hydrostatic Strength of Thermoplastics Materials in Pipe by Extrapolation,* ASTMD2837-01a *Standard Test Method for Obtaining Hydrostatic Design Basis for Thermoplastic Pipe Materials* and DIN 16887:1990 *Determination of the Long-term Hydrostatic Pressure Resistance of Thermoplastics Pipes.*

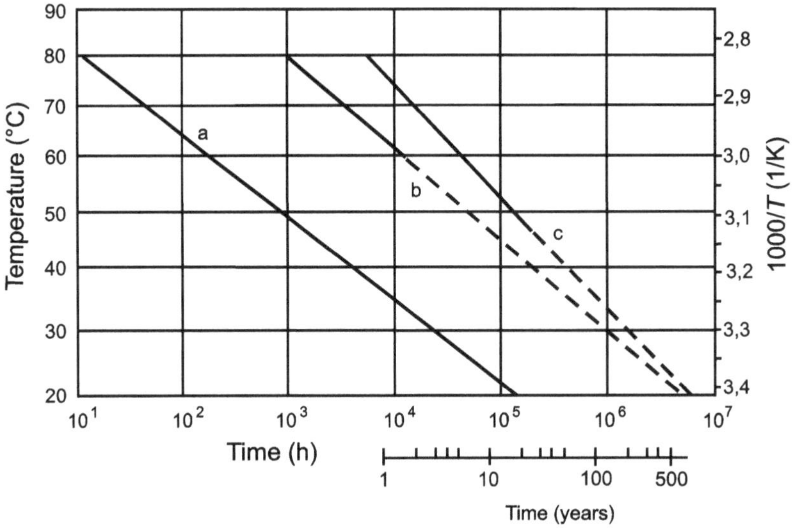


**Fig. 3.15.** Arrhenius-Diagram of the temperature dependence of the ductile-brittle transition and stress crack-oxidative degradation transition in the stress rupture curve of HDPE pipes. a: 1. ductile-brittle transition of Hostalen 5010, b: ductile-brittle transition of Hostalen GM 5010 T2 (see Fig. 3.14a), c: start of oxidative degradation of resins GM 5010 and GM 5010 T2 (see Fig. 3.14a). The figure has been taken from reference (Koch et al. 1988)

**Table 3.8.** Extrapolation factors $K_e$(according to DIN 16887) which determine the time limits for which extrapolation of stress-rupture data are allowed as a function of the difference between maximum test temperature and service temperature $\Delta T$

| $\Delta T$ (°C) | $K_e$ |
|---|---|
| ≤ 10 | 1 |
| > 10 to ≤ 15 | 3 |
| > 15 to ≤ 20 | 5 |
| > 20 to ≤ 25 | 9 |
| > 25 to ≤ 30 | 16 |
| > 30 to ≤ 35 | 28 |
| > 35 to ≤ 40 | 50 |

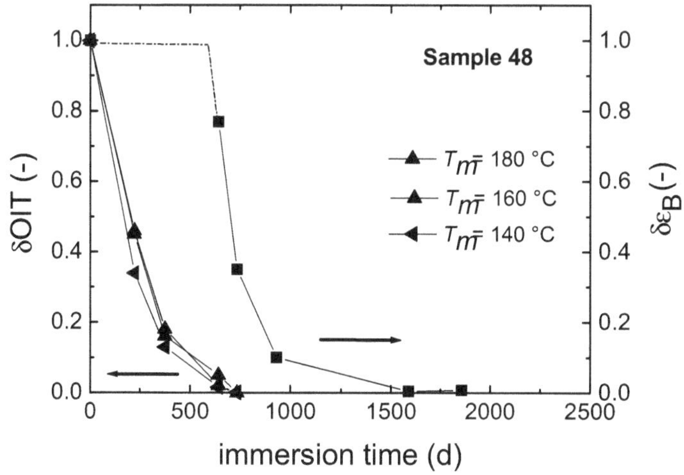


**Fig. 3.16.** Change of the relative OIT value *δOIT* and of relative strain at break $\delta\varepsilon_B$ of a geomembrane made of the outdated resin Hostalen 5040 T12 during immersion in hot water at 80 °C. The OIT-value was measured at different OIT measuring temperatures $T_M$ using Cu pans. At all measuring temperatures the decay rate of the OIT during immersion is identical: It may serve as an indictor of antioxidant depletion. After about 2 years OIT is zero and auto-oxidation starts, as can be seen in the reduction of strain at break, with a slow oxidation rate. Additional 2 years have to pass until strain at break has reduced to 6 %, the permissible limiting strain in the long-run (see Sect. 8.3.1). According to the RRT-rule (see Chap. 3, footnote 2 and Sect. 5.1) 4 years testing time at 80 °C, will correspond to roughly 1000 years of service lifetime at 20 °C. This result corresponds to the extrapolation of stress-rupture data from pipe pressure testing on pipe sections from comparable HDPE resins (see curve c in Fig. 3.15)

Sufficient data for HDPE geomembrane materials are not usually available from pipe pressure tests which would enable a direct extrapolation of the stress-rupture curve at 23 °C or 40 °C according to the instructions of ISO 9080 for small stresses and longer times. The emphasis is on the brittle branch of the stress-rupture curve which results from brittle failures and determines the failure behaviour at small stresses and extremely long service lifetimes. However, one can use the experience accumulated with HDPE materials over many decades (Schulte 1997; Krishnaswamy 2005) and fall back on the extrapolation factors of DIN 16887 or ISO 9080 which were checked by measured stress-rupture data of polyethylene pipes. These factors $K_e$ give the time limits of a permissible extrapolation of the stress-rupture curves in the following sense. Let us assume that hoop stress versus times-to-failure data are measured at several higher test temperatures with $T_{max}$ as the maximum test temperature and that the entire stress-rupture curve at a certain low application temperature $T_S$ is extrapolated, for example by multi-linear regression, from these data. Let $t_{max}$ now be the average value of the five longest times-to-failure at $T_{max}$. The calculated overall creep curve at $T_S$ may then be used up to a time $t_e = K_e \cdot t_{max}$ for assessment of long-term behaviour. $K_e$ is a function of $T_{max} - T_S$, i.e. the higher the temperatures are at which stress-rupture data are obtained and entered into the extrapolation and the longer these times-to-failure are, the longer the permissible time limits for extrapolation at $T_S$ are, see Table 3.8. For example if $T_{max} = 80$ °C and $T_S = 40$–45 °C then $K_e = 50$ (or $T_S \leq 30$ °C then $K_e = 100$ according to ISO 9080).

The specification of the permissible extrapolation limits implies that the ratios of the times of the transition from the ductile to the brittle branch at different testing temperatures is greater than the extrapolation factors in any case and that the times-to-failure due to oxidative degradation of well stabilised HDPE pipes is at least as large as the ductile to brittle failure transition time. Furthermore, the sections of the creep curves at different test temperatures for the HDPE materials can be assumed to be roughly parallel to each other. One can therefore use the extrapolation factors themselves for the estimation of a lower limit of the service lifetimes by applying them to the curve measured at $T_{max}$. If the times-to-failure at 80 °C and 4 N/mm² are longer than 8760 h (1 year) then the times-to-failure at 4 N/mm² and 40 °C (20 °C) will certainly be longer than 50 years (100 years)[3]. The HDPE materials can be thus assessed only based on the stress-rupture data measured on pipes at 80 °C in accordance with DIN 16887. The measurement must take longer than $10^4$ h (417 days).

The requirement for stress crack resistance of the HDPE material used for geomembranes formulated in this way has been included in the BAM certification guideline for geomembranes in landfill liners (Müller 1992, 1999b) since it is based on genuine long-term tests. This evaluation procedure is very conservative since it assumes that the long-term available residual stress at an imposed deformation initiates a stress crack, in the same way as an imposed durable external stress does (see Sect. 5.4.1).

It is, however, desirable to be able to test the stress crack resistance on the geomembranes themselves using a laboratory method. M. Fleißner suggested an index test method in 1987 for thermoplastic pipes which could also be transferred to geomembranes (Fleißner 1987). In Fleißner's method a perimeter notch was cut with a razor-blade on a quadratic paral-

[3] According to the comparison curves of the new DIN 8075:1999 *Polyethylene (PE) Pipes - PE 63, PE 80, PE 100, PE-HD - General Quality Requirements, Testing* even an internal pressure strength of 5.8 N/mm²might be assumed.

lel bar specimen. The tensile stress-rupture curves were then measured in a long-term tensile test at 80 °C in ethylene glycol. The typical transition from the ductile to the brittle fracture behaviour was observed in the log tensile stress versus log time-to-failure diagram whose coordinates can be used as a measure of stress crack resistance.

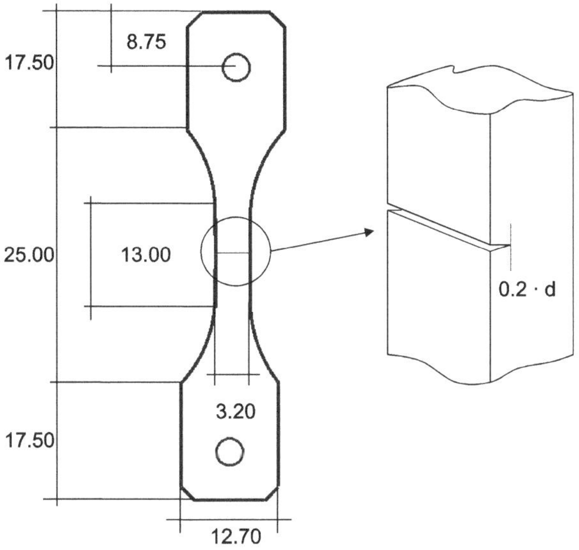


**Fig. 3.17.** Dimensions of the test specimen for the Notched Constant Tensile Load test (NCTL test). The test specimen is punched from the geomembrane by a suitable punching device. In a notching device (notcher) the specimen is notched with a razor blade in such a way that the notch depth is 20 % of the medium thickness of the geomembranes

At the beginning of the 90's, R. M. Koerner and staff members of the Geosynthetic Research Institute (GRI) checked and finally standardised a similar procedure for geomembranes in interlaboratory tests as ASTM D5397-95 *Standard Test Method for Evaluation of Stress Crack Resistance of Polyolefin Geomembranes Using Notched Constant Tensile Load Test*, as the so-called NCTL test (Hsuan 2000). The reasons for developing this method were cases of damage by stress crack formation within the weld seam area in HDPE geomembranes observed in the USA. The field investigation into the occurrence of stress cracks and the method for testing was described in a detailed report to the US EPA (EPA 1992; Hsuan 2000). Meanwhile the procedure has been accepted world-wide as the relevant test for the stress-crack resistance of plastic geomembranes. It was standardised in Europe as EN 14576:2005 *Geosynthetics – Test Method for*

*Determining the Resistance of Polymeric Geosynthetic Barriers to Environmental Stress Cracking.*

The NCTL test is a long-term tensile test on notched test specimens taken from a geomembrane sample. Small dumbbell specimens (Fig. 3.17) are punched from the geomembrane sample (along or crosswise to the direction of manufacture). Notching is carried out with a razor-blade. The depth of the notch is 20 % of specimen thickness. The stress-rupture curve is then determined in a long-term tensile test at (50  1) °C in a bath of 90 % by volume water and 10 % by volume surfactant (e.g. Igepal CO-630®). Only careful attention to the details of the test method, in particular the procedure of notching the specimens, can provide reproducible and comparable test results in the NCTL test. In Germany, the test is occasionally performed using specimen type 5 or type 1B of the EN ISO 527-3 tensile test standard (Sect. 3.2.8).

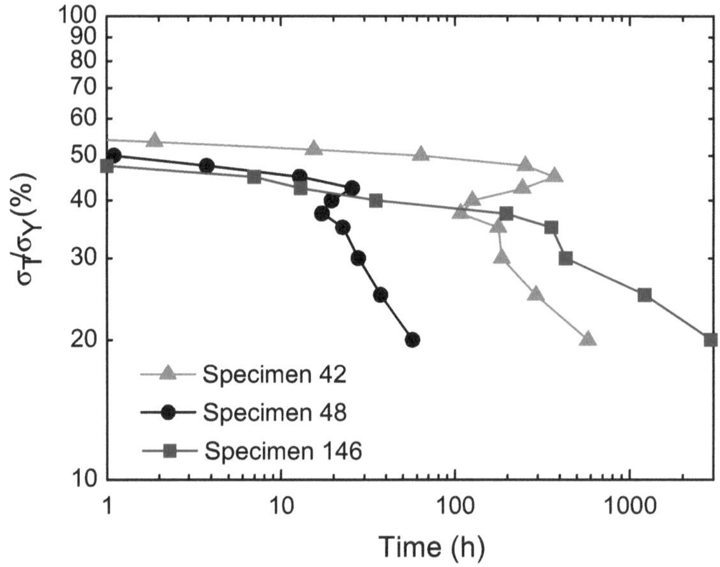


**Fig. 3.18.** Stress-rupture curves of HDPE geomembranes from different resins (specimen 42, 48 and 146) in the NCTL test. The relative stress, i.e. the test stress $\sigma_T$ related to the yield stress $\sigma_Y$ is plotted against the time-to-failure (average value from three measurements). Various shapes can be formed in the transition from the ductile to the brittle branch of the stress-rupture curve. Sometimes a "nose" develops in the curve (specimen 42 and, less pronounced, specimen 48), see Sect. 5.3.4

In the evaluation of the NCTL test the logarithm of relative stress – i.e. ratio of test stress to yield stress of the material at room temperature – is plotted against logarithm of time-to-failure. Relative stress is defined as applied force divided by $w \cdot t_L$, where $w$ is the width of the neck of the test specimen (3.2 mm) and $t_L$ is the ligament thickness of the notched specimen (80 % of nominal thickness). Typically, test stresses are applied in the range between 20 % and 65 % of relative stress in 5 % steps. The average value of time-to-failure is calculated from at least three individual measurements for each test stress. Similarly to the evaluation of the pipe pressure test, a stress-rupture curve is obtained composed of two sections of straight lines (Fig. 3.18). The coordinates (relative stress and time-to-failure) of the transition from the ductile failure section to the brittle failure section of the curve can be used for the quantitative assessment of stress crack resistance. This transition may occur after 100 h at the earliest and should be above a relative stress of 30 % of the room temperature yield stress. Currently, mainly so-called single-point measurements are only performed (NCTL SP (single point) test). The average time-to-failure is determined at a one relative stress of 30 %. This must exceed 200 h. These requirements were derived from the results of the field tests cited above: the HDPE materials which showed stress cracking in the field, just did not meet this criterion.

The NCLT test has the advantage of only requiring a short test time and can directly be performed on specimens from the geomembrane. This test is highly sensitive to the stress crack resistance of the HDPE materials. Chapter 2 discussed the term HDPE for encompassing a broad range of materials. The top of this range, the real high-density HDPE materials, are not suitable for geomembranes because of their stress crack sensitivity. The NCTL test can safely expunge such materials. Based on the available experience, NCTL testing is therefore sufficient to ensure high stress crack resistance of an HDPE geomembrane when the relevant requirements are met. Compared to NCTL testing pipe pressure testing is very complicated and time consuming. However, the pipe pressure test is a genuine long-term test. It has the advantage that it is also an implicit test of oxidation stability. In the case of insufficient stabilisation, a third, vertical branch appears in the stress rupture curve which indicates the embrittlement of the material due to oxidative degradation. The assessment of HDPE materials based on a pipe pressure test should therefore only be omitted if the NCTL test is accompanied with appropriate long-term testing of oxidation stability. Within the framework of the BAM certification procedure for landfill liner geomembranes, the stress crack resistance is tested as per ASTM D5397-95 alternatively to the pipe pressure test. However, immersion tests (80 °C, $10^4$ h, air and water) are also performed for the determination of oxidation stability.

The effect of chemicals on stress crack formation or, in general, the resistance to the combined effect of chemical and mechanical impact can also be determined by the pipe pressure test (Kempe 1984). A specific liquid chemical or a gas is used as a test medium instead of water. So-called resistance factors can be derived from the stress-rupture curve. The time-related resistance factor indicates how the time-to-failure is reduced in the liquid (or gaseous) chemical at a given hoop stress and test temperature in comparison to the relevant time-to-failure in water (or air). The stress-related resistance factor indicates how the test hoop stress must be reduced in order to reach the same time-to-failure in the liquid or gaseous chemical as in water or air. The resistance factors usually depend highly on the magnitude of the hoop stress and the test temperature. The resistance factors for HDPE, PP and PVC materials have already been measured for numerous substances: the DVS 2205-1:1987 *Calculation of Thermoplastics Tanks and Apparatuses from Thermoplastics, Parameters* lists the stress reduction coefficients (reciprocal value of the stress-related resistance factor). A resistance factor of 1, i.e. no reduction in comparison to the behaviour in water, was obtained at a test stress of 4 N/mm$^2$ and temperatures up to 80 °C in the following substances: sea water, uric acid, caustic potash solution (50 %), caustic soda solution (50 %), saline solutions up to saturation, spirits of all kinds, natural gas and many other chemicals. Waste water, however, exhibits a certain stress cracking promoting potential although this is very low. Thus, at 4 N/mm$^2$ and 80 °C, waste water from a cellulose factory exhibited a stress-related resistance factor of 0.95, waste water from a chemical fibre factory a factor of 0.75, waste water from whey utilisation a factor of 0.73 and a natural gas condensate showed a resistance factor of 0.78 (Kempe 1984).

### 3.2.14 Weathering Resistance

In civil engineering, geomembranes are exposed to bad weather phenomena, in particular to intensive direct and diffuse sun radiation. Normally, geomembranes are only exposed to weather influences during short installation and operation periods, usually they are soon covered. In some structures, however, e.g. water ponds and reservoirs, geomembranes remain unprotected over many years. Experience gathered particularly in hot regions with long sun exposure in particular, shows that carbon black-stabilised HDPE geomembranes exhibit an extraordinarily high weathering resistance. Such experience was confirmed in weathering tests. When sufficient quantity and homogeneous distribution of high quality carbon black additive was used for UV stabilisation, weathering resistance test of the

HDPE geomembrane is unnecessary. However, natural-coloured HDPE geomembranes dyed with white pigments are also used which are protected by the admixture of special additives as UV stabilisers. Geosynthetic products from only poorly or unstabilised polyolefins can be destroyed by the weather within weeks or months. Therefore, it might be interesting to shortly discuss weather influence and the weathering resistance test (Gugumus 1990; Hamid et al. 1992; Hufenus et al. 1998; Zweifel 2001).

The actual cause of ageing due to weather is the UV radiation of the sun which lies within the wavelength range of 290 nm to 400 nm. Visible light and infrared radiation, which are next in the spectrum of the electromagnetic radiation of the sun, only have an indirect effect inasmuch as they convey thermal energy to the geomembranes or, more generally, to the geosynthetics and in so doing heat them up and accelerate all chemical and physical degradation processes. By encourageing extraction, diffusion or degradation of stabilisers humidity also has an indirect effect on weather resistance of polyolefin materials.

Saturated carbon-carbon bonds in polyethylenes and polypropylenes do not directly absorb UV quanta. Structural anomalies with unsaturated bonds, impurities, such as catalyst residues from the polymerisation process or oxidation products already formed, which can develop in particular during processing, absorb parts of the radiation within the ultraviolet range. Such components of polymers that absorb the UV radiation are called chromophores. The UV quanta energy ($h\nu$) is converted into chemical excitation energy of the polymer molecules. Excitation finally results in the breaking of chemical bonds and in the emergence of free radicals. Together with the generally pervasive oxygen, free radicals set oxidative reaction chain and the degradation in motion. UV radiation thus acts as a free-radical generator or as a "fuel" for auto-oxidation. One also speaks of photo-oxidation in this context. Chemical processes of auto-oxidation will be dealt with in Sect. 5.2. All chemical equations of the oxidative chain reaction discussed there are valid. In photo-oxidation, however, in addition the reaction:

$$\text{Chromophore} + PH + h\nu \rightarrow \text{free radicals}\,(P^{\bullet}, PO^{\bullet}, HO^{\bullet}, HO_2^{\bullet}) \qquad (3.12)$$

emerges as the dominant chain initiation. Since the oxidation products, for example hydroperoxide, absorb UV radiation, the radiation interferes with the chain propagation reactions as well.

If the auto-oxidation process has been set in motion by the UV radiation then it will develop further even in darkness. In unsatisfactorily stabilised polyolefin materials therefore even a short weathering effect may cause detrimental degradation processes in the long run.

Plastics can be protected against UV radiation in three ways. One possibility is the addition of molecules or pigments, which extensively absorb UV radiation and convert the radiation energy directly into heat energy. The particular effectiveness of fine carbon black has already been discussed. Carbon black not only absorbs UV radiation but also binds free radicals through a number of organic compounds adsorbed on the carbon black surface. Absorbent light-protective substances, such as carbon black, display their full effectiveness only when a certain thickness of the exposed layer is available.

So-called quenchers were used in the past because of their layer thickness independent effectiveness, particularly for the stabilization of fibres and thin films. Quenchers are organometallic compounds which absorb the excitation energy of molecules excited by UV radiation and convert it into chemically ineffective heat energy.

Finally, antioxidants (radical inhibitors, hydroperoxide decomposer) can also be added to obstruct photo-oxidation. The so-called hindered amine light stabilisers (HALS) have proved particularly effective (Klingert 1995; Schrijver-Rzymelka 1999). HALS compounds are the derivatives of 2,2,6,6-tetramethylpiperidin. The compounds can act both as radical scavengers and peroxyradical and hydroperoxide decomposers and may to some extent even become regenerated. Light-stabilising agents based on HALS compounds have already conquered half of the light-stabilising agent market (Schrijver-Rzymelka 1999).

Weather resistance can be tested by open-air or artificial weathering. The effects of solar radiation on the geomembranes or more generally on geosynthetics, in connection with heat, humidity, rain, dryness, oxygen and other components of air, can be tested naturally and directly by open-air weathering and measuring the accompanied changes in the properties. The procedure for open-air weathering is described in EN ISO 877:1997-05 *Plastics – Procedure for Natural Weathering, Irradiation behind Window Glass and Accelerated Weathering by Solar Irradiation using Fresnel Mirrors*. The weather is, however, very different at different places and during its seasonal changes. In addition, climatic fluctuations can be observed over periods of many years, e.g. in the overall solar irradiation or precipitation at a certain place. Therefore, results obtained in open-air weathering tests in one place cannot be applied to other places with different climatic conditions. The test results are most often not reproducible. The test periods of open-air weathering must be at least as long as the planned service lifetime. Thus usually very lengthy tests are needed.

Therefore a very early attempt was made to develop laboratory test methods which artificially imitate and accelerate the impacts by the weather, based on unambiguously specified test conditions, and are thus capable of ensuring reproducible and comparable test results (Kamal and Huang 1992). It was intended to establish criteria as to how the test conditions must be chosen to guarantee a certain service lifetime under certain weather conditions. The source of light of an artificial weathering test facility must reflect the spectral distribution of the irradiation intensity of the solar global irradiation, i.e. the direct solar irradiation and the scattered radiation from the atmosphere in the relevant UV range. The thermal loading of the test specimen is another key test condition. The black panel temperature serves as a measure for this. It is the temperature which develops during artificial weathering in a black-lacquered metal plate with a thermally insulated back in the same position as the specimen. Other important test conditions are the period of water spraying and dryness and the prevailing test temperatures and test humidity.

Test equipment choice, conditions to be adjusted and necessary irradiation test periods are still subject to discussion. In Europe, a state of the art for the test of weather resistance is described by DIN EN 12224:2000 *Geotextiles and Geotextile-related Products – Determination of the Resistance to Weathering* (Greenwood et al. 1996; Trubiroha and Schröder 1997). This standard specifies a single UV fluorescent lamp or UV fluorescent lamp combinations as a source of light. The spectral distribution of the irradiation intensity to be maintained by the lamps is precisely specified. Weathering is carried out in wet/dry cycles. Each cycle consists of a 5 hour dry period with a black standard temperature of (50 ± 3) °C at a humidity of (10 ± 5) % and a one-hour spraying with water at a black panel temperature of (25 ± 3) °C. The standard requires, however indirectly, temperature fluctuations to be kept within certain limits (e.g. ± 1 °C) as far as possible because an increase in temperature of 1 °C in polyethylene for instance, increases the speed of photo-oxidation by about 8 %, thus can also cause a fairly large change in the results. The testing of geotextiles and related products should be carried out up to applied irradiation energies of 50 MJ/m². Depending upon the type of lamp an irradiation period of 320 hours to 430 hours should be applied. A UV irradiation of 100 MJ/m², for instance, is about equal to the maximum irradiation during two summer months in Central Europe or the medium irradiation during three summer months (Trubiroha and Schröder 1997).

Test standard ASTM 4355-02 *Standard Test Method for Deterioration of Geotextiles by Exposure to Light, Moisture and Heat in a Xenon Arc Type Apparatus* describes artificial weathering using a xenon arc light source. According to this standard, five specimens of a geotextile taken in machine direction and for the crosswise machine direction are exposed in a xenon arc device for 150, 300 and 500 h. The exposure consists of 120 min cycles composed of 90 min of light only at (65 ± 3) °C un-insulated black panel temperature and (50 ± 5) % relative humidity, followed by 30 min of light plus water spray. The average breaking strength of the exposed specimens is determined and the degradation curve is plotted as percent strength retained versus exposure time.

The guidelines on durability, published by the European Committee for Standardisation (CEN) as ISO TR 13434, attributes the permissible exposure times to the test results in accordance with DIN EN 12224 as illustrated in Table 3.8.

**Table 3.8.** Maximum exposure times according to the chance of mechanical strength found in accelerated weathering tests (EN 12224) and specific application

| Application | Retained strength after weathering test | Maximum time of exposure during installation |
|---|---|---|
| Reinforcement or other | > 80 % | 1–4 months[1)] |
| applications where long-tern | 60–80 % | 2 weeks |
| strength is a significant parameter | < 60 % | Cover on day of installation |
| | > 60 % | 1–4 months[1)] |
| Other applications | 20–60 % | 2 weeks |
| | < 20 % | Cover on day of installation |

[1)] depending on the season and the location in Europe

The much shorter test times of artificial weathering in comparison to open-air weathering or field exposure might fail to properly consider stabiliser migration and extraction or stabiliser degradation processes which may considerably influence the long-term behaviour when weathering has a long-term influence. It was suggested therefore that an immersion of the specimens over 100 hours in hot water of 60 °C with a two-hour water exchange should precede the actual weathering test. This method is included in the weathering test requirements in German technical regulations in the field of road construction. According to these regulations high weather resistance is certified for the product with a residual strength > 80 % after artificial weathering, medium weather resistance at 60–80 % and low weather resistance at < 60 %. Materials with low weather resistance must be covered or protected within a week, those with medium weather resistance within two weeks and those with high weather resistance in two months at the latest.

The permissible exposure times in ISO TR 13434 are particularly generous. The auto-oxidation process of polyolefin materials is already in a rather advanced state when noticeable strength losses can be measured. One has to assume that degradation will continue even after the geosynthetic is covered. In principle, geosynthetics should therefore be exposed to the weather only so long as no changes in the mechanical properties can be observed in the artificial weathering test over the testing time which corresponds to the field exposure time.

The admixture of carbon black to the HDPE geomembrane ensures a very high weather resistance. The proof that chemical UV stabilisation ensures a similar weather resistance, therefore requires very long irradiation periods. BAM certification requires that in an artificial weathering test according to DIN EN 12224 chemically UV-stabilised geomembranes must be able to cope with an irradiation period of at least 1500 hours without significant changes in the mechanical properties. This irradiation period corresponds to open-air weathering over a period of approximately 2 years. In the GRI test method GM13 of the Geosynthetic Institute the internal GRI standard GM 11:1997 *Accelerated Weathering of Geomembranes using Fluorescent UVA-Condensation Exposure Devices* is used to test the weathering resistance of HDPE geomembranes. The high pressure OIT should retain 50 % of its original value after 1600 hours testing time.

### 3.2.15 Resistance to Biological Effects

Biological impacts, i.e. attacks by living organisms, can be distinguished according to the organisms and the damage caused by them in plastic geomembranes or more generally in geosynthetics: damage due to bites by vertebrate animals and insects, damage by roots and material modification or degradation by micro-organisms (bacteria and fungi).

However, biological impacts can be classified according to damage mechanisms (Albertson 1992) as well:

1. Biophysical effects where mechanical impacts cause damage like cracks and holes, e.g. bites by animals or the bursting and swelling pressure of growing cell agglomerations of plant roots.
2. Secondary biochemical effects when damage results from chemicals excreted by the growth or metabolism of organisms. These chemicals might directly attack the plastic material (Sect. 3.2.11) or might change the environment thereby intensifying other degradation processes. Thus, for instance, acid is set free at root points which accompany the biophysical effect of root penetration.

3. Primary biochemical effects (or direct enzymatic actions) where the materials are directly included into the metabolism of the organisms by enzymatic degradation.

Damage by rodents and insects can occur if the biting organs of the animals can find points of attack. The smooth, nonporous and very firm surface of thick HDPE geomembranes which is very difficult to deform locally exhibits great resistance to the bites of animals and insects. The "behaviour of geomembranes to rodents" was investigated in detail by R. Rumberg and collaborators in a research project (Rumberg et al. 1982). The research report contains an overview of the state of the art. The test method described below was developed within the project. HDPE geomembranes with a sufficient thickness (> 1.5 mm) provide an "optimum protection" according to these test results compared with other geomembranes. It is hardly possible for small rodents to bite through a thickness of 2.5 mm, even at the edge of a geomembrane. In addition, polyethylene is not accepted or used as a food. However, animals, which gnaw and bite particularly aggressively, can damage even HDPE geomembranes. Thus bite edges were found in trial pits in certain cases.

In the old version of the certification guidelines of the German Institute for Building Technology (DIBt) of 1982 for plastic geomembranes used to line all kinds of storage or handling facilities for substances dangerous to water, a test method of the resistance to rodents was described: a trough is welded from geomembrane test samples, constructed in the soil of an enclosure and filled with earth. In addition, loose pieces of geomembrane specimen are laid out in the feeding site. The same structure with a reference material is constructed for comparison. Large voles (water vole, arvicola terrestris) are kept in the enclosure. The population must be renewed regularly with newly caught wild living animals. The geomembrane specimens are kept in the enclosure for a minimum of six months, or until the reference specimens are gnawed through in at least five places. 2.5 mm thick HDPE geomembranes had regularly passed this test, so currently this rather expensive test method is omitted in HDPE geomembrane certification for landfill liners[4]. HDPE geomembranes certified by DIBt as a build-

[4]The test method is not entirely harmless. Grzimek's "Animal Life" (Tierleben) says: "As carrier and transmitter of tularemy, a rodent plague, which can also infect humans, water voles play a role in spreading epidemics. ... Anyone who catches water voles on cropland or in a cellar should beware of being bitten by animals which are still alive in the trap. In order to exclude every danger of infection, dead water voles should be buried a half metre deep in the earth and a thorough hand wash is highly recommended". Grzimek B (1979) Tierleben, Vol. 2, Säugetiere (Mammals). Deutscher Taschenbuchverlag, Munich, p 331.

ing authority (DIBt 2000) and certified by BAM for landfill liners (Müller 1999b) are rodent-resistant according to the requirements of this test.

BAM has carried out tests on termite attack (coptotermes formosanus, zootermopsis angusticollis, neotermes jouteli) against HDPE geomembranes relying on EN 117:2005 *Wood Preservatives – Determination of Toxic Values against Reticulitermes Species (European Termites) (Laboratory Method)* (Hertel 1999). Especially cut edges were attacked with vigour by the termites. HDPE geomembranes are therefore not termite-resistant, just like more or less all normally unprotected plastic materials (Dolezel 1978).

The following procedure (container test) is in principle common to all tests of resistance against root penetration. Three layers are placed in a container: a humidity soil layer, a geomembrane and a soil reclamation layer into which seeds are sown or young shoots are implanted. The test procedure consists of the actual test lot in which several containers are filled with geomembrane specimens and, in addition, of a reference test lot of several containers with specimen from a reference material, for example bitumen layer or a bitumen sheet. The test is carried out for as long as strong root-penetration of the bitumen layer had occurred in the reference test, but at least until the end of a specified test period. The geomembrane is considered as root-resistant if no root-penetration of the geomembranes has occurred by the end of the test.

In the certification guideline of DIBt and DIN 16726 the resistance to root penetration is tested by relying on the following test procedure. Unglazed clay pots are used as containers. The humidity soil layer and the reclamation soil layer consist of field soils. Lupins of the lupinus albus sort are sown in the reclamation soil layer. A sheet of bitumen 85/40 serves as a reference specimen. The test usually takes between 6 to 8 weeks.

However, today the *Method for Investigation of Root Penetration Strength for Roof Greening* (1995) developed by the Landscape Research, Development & Construction Society (FLL, www.f-l-l.de) is regarded as state of the art for testing the resistance of geomembranes against root penetration. A large-volume, cuboid-shaped container with minimum edge lengths of at least $800 \cdot 800 \cdot 350$ mm³ is used. A layer of ground slate/blue clay is used as a humidity soil layer, overlain with a thin geotextile as a compensating and protective layer. A box is welded together from individual pieces of geomembrane which is placed into the container on the geotextile. The sides of the geomembrane box reach to the upper edge of the container. Thus there are bottom corner, wall corner and bottom seams in the test specimen. In the bottom seam there is a T-junction. A mixture of high moor land peat and ground slate/blue clay is used as a reclamation soil layer. White alder (alnus incana) and European aspen (populus tremula) are planted and seeds of couch grass (agropyron repens) are sown in the reclamation layer. Several simultaneous tests are carried out. The reference specimen is made of 20 mm thick bitumen 85/25 sheets. The bottom of the test containers is transparent, so that the extent of root-penetration into the humidity layer test can be inspected in regular intervals. The test period takes at least 4 years. In the opinion of the research society this is "the shortest period of time for vegetation tests with wood and herbs under open land conditions to obtain dependable results and to be able to make reliable statements about root resistance".

Working group 4 of the CEN Technical committee 254 has issued a draft standard EN 13948:2000 *Flexible Sheets for Waterproofing – Bitumen, Plastic and Rubber Sheets for Roof Waterproofing – Determination of Resistance to Root Penetration* in analogy to the test procedure of FLL. However, modifications were introduced which found strong criticism by the experts of FLL.

The plant roots do not find any points of attack for a mechanical effect in the surface of HDPE geomembranes and along perfectly manufactured extrusion fillet seams or dual hot wedge seams. Secondary biochemical effects do not play a role because of the very high resistance of the HDPE geomembrane to chemicals. Root strength of HDPE geomembranes was therefore confirmed in these tests.

The most interesting topic under the keyword "biological resistance" is the primary biological effect, however, thus the degradability of polymer materials by micro-organisms. Extensive research has been carried out in this field for many decades (Albertson 1992; Dolezel 1978). The objective has not been to improve the resistance of the plastic materials, rather, to find polymer materials which are biodegradable in natural environments as far as possible and to characterise this property by tests. Both most widespread polyolefin materials (PE and PP) have been tested in various ways in this regard and always proved very resistant in the tests. The rule of thumb derived from the biodegradation tests says that polyolefins only up to a degree of polymerisation of about 800 can be degraded and used as a carbon source for micro-organisms (bacteria and fungi) (Albertson 1992). Low-molecular mass fractions, the so-called polyethylene waxes, may be contained to a small extent in polyethylene materials. These waxes diffuse to the surface and form a thin film. Therefore in micro-organism resistance tests, e.g. in the soil burial test, it can be observed occasionally that fungi spread on the surface of PE geosynthetics. This only changes the appearance; the performance properties of the product are not affected.

Plastic garbage made of polyethylene is biologically degraded in the environment only if the photo-oxidative degradation initiated by UV radiation and oxygen has progressed to a large extent already. Unexposed to

UV radiation the time period that plastic garbage can survive in the environment can be very long. Trial pits in landfill bodies have shown that plastic packaging, plastic foils and especially polyethylene bags remain unchanged to a large extent for decades[5], even in this very rich microbiological environment.

The micro-organism resistance test of plastics is regulated in EN ISO 846:1997 *Plastics – Evaluation of the Action of Microorganisms*. The methods described in the standard deal with the issue of whether a plastic contains food for fungi (method A), for bacteria (method C) or whether antimycotic components are contained (method B) or how the plastic behaves in a microbiologically active soils. The European Committee for Standardisation submitted a draft for a standard on soil burial tests for geotextiles: EN 12225:1997 *Geotextiles and Geotextile-related Products – Method for Determining the Microbiological Resistance by a Soil Burial Test*. In this test, specimens are buried in microbially active soil. Microbial activity is controlled by reference test specimens: bleached and untreated cotton fabrics must be completely destroyed by the end of the test. The test containers are kept at (95 ± 5) % relative humidity and (26 ± 1) °C under a fresh air supply for at least 16 weeks followed by sample characterisation by visual inspection and testing the mechanical properties. According to the manual ISO TR 13434, mentioned above, no tests on resistance to microbiological effects are necessarily for geosynthetics from virgin (non-recycled) polyethylene, polypropylene, polyester (polyethylene terephthalate PET) and polyamide 6 or 6.6.

---

[5] I recall one of the landfill conferences of the State Institute for Trade and Industry (Landesgewerbeanstalt) Nuremberg. A paper was presented about investigations on the landfill body of the central landfill in Hanover. Photographs were shown of decades-old plastic garbage which had been excavated completely unchanged. A little later, another paper dealt with pollutant emissions through composite liners. In the calculations it was assumed that the HDPE geomembrane would completely disappear in 50 years. In his inimitable manner, K. Stief suggested in the following discussion that in the future short-lived household items be manufactured from high-quality polyethylene resins of the geomembranes since these obviously rapidly decay, and landfill liner geomembranes be made from as cheap plastics as possible which are used for household items, since these seem to last for very long periods of time in a landfill. He only managed to make half of the audience laugh. The other half was bitterly silent: the reservation against plastic geomembranes was so great in that initial period of landfill engineering. Nevertheless, ageing behaviour of plastic garbage is an interesting topic. It is worth noting in connection with the assessment of long-term shear strength of landfill bodies that the very high shear parameter values are a result of a "reinforcement" of the waste by plastic bags, plastic packaging and similar plastic items.

### 3.2.16 Long-Term Tensile Test

The long-term tensile test is used to test stress crack resistance of geomembrane specimens. Manufacture of geomembranes with textured surface or welding of geomembranes (extrusion fillet seams or dual hot wedge seams) produces characteristic possible weak points for stress crack initiation, such as sharp edges, points of stress concentrations due to geometrical shape of the structure or seam geometry or anomalies in the morphology, which might lead to a considerable reduction in stress crack resistance. Long-term tensile tests can reveal such weak points. When specimens of smooth HDPE geomembrane are tested, stress cracks nearly always develop from the specimen edge or from the clamping device where ghost lines such a machining defects from the specimen preparation or traces from the clamping initiate the cracking: punched smooth specimens fail relatively rapidly, sawn specimens with smoothed cut edges can resist for a very long time. The long-term tensile test on smooth geomembrane specimens would only reflect the quality of specimen preparation and clamping technique and not the stress crack resistance characteristic of the geomembrane (see also Sect. 10.2). Therefore, stress crack resistance of smooth geomembranes has to be tested in a long-term tensile test with notched specimens (NCTL test), see Sect. 3.2.13. However, stress crack resistance of textured HDPE geomembranes can be tested by the long-term tensile test described in the following (see Fig. 5.6). The long-term tensile test is an index test. By specifying the minimum time-to-failure, the extent to which the textured surface is allowed to impair stress crack resistance of the geomembrane is restricted.

The test is performed according to EN ISO 6252:1998 *Plastics – Determination of Environmental Stress Cracking (ESC) – Constant-Tensile-Stress Method* and the guideline DVS 2203-4:1997 *Testing Welded Joints on Thermoplastic Plates and Pipes – Long-term Tensile Test.*

In the long-term tensile test a test specimen is exposed to uniaxial tensile force in the long term and immersed in a stress crack-encouraging liquid at an elevated temperature. The time-to-failure is measured. This principle can be put into practice using different test apparatus. Figure 3.19 shows a schematic of possible test devices in which a well-defined and constant force is applied. The weight of the lever mechanism for force transmission must be correctly taken into account and the effect of the friction in the lever bearing should be checked. The specimen temperature must be equal throughout all its part and kept constant within small limits during the test period. A mixture of 2–6 % by mass of surfactant, e.g. Arcopal N 150®, Marlophen 812® or Laventin W®, and water is usually used as a test liquid. Depending upon the solution behaviour of the surfactant in the long run at an elevated temperature, the test liquid must be agitated or rotated to achieve uniform concentration throughout the bath. Corrosion of the test device by the test liquid can cause contamination of the test solution and precipitations on the test specimens, therefore stainless materials are used for the test device. The time-to-failure is measured and recorded electrically or electronically by a timing clock for each test specimen. Devices are often installed to measure the elongation of the test specimen as a function of time, i.e. the creep curve of the specimen. Parallel bar or dumbbell specimens according to the standard for the tensile test EN ISO 527 (Fig. 3.10) are usually prepared as test specimens with high-speed tungsten carbide saws or high-speed milling machines. The test specimen size has to be chosen in such a way that the specimen contains a representative part of the surface structure. Cut edges must be smooth and free of grooves.

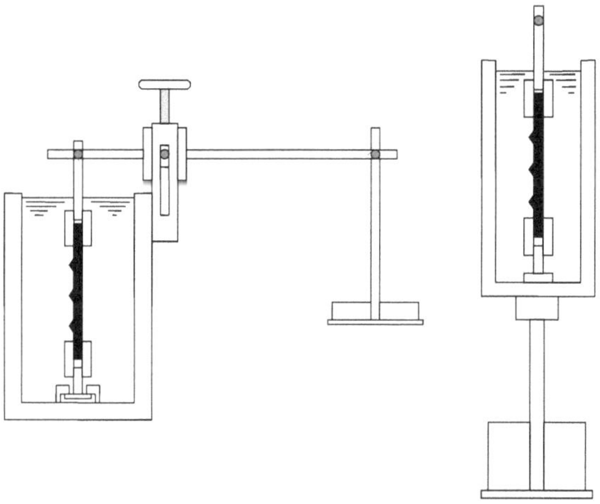

**Fig. 3.19.** Schematic of test devices for the long-term tensile test. The figure shows the test container, the specimen with specimen clamp and two types of force application. Controlled heating maintains a constant temperature in the test device on the left. The test device on the right can be hung in an oven

Textured geomembranes are tested in BAM under the following conditions. The test stress produced by the continuously acting, steady tensile force (range of variation ± 1 %) is 4 N/mm² (4 MPa). The specimen cross-section area is calculated from the specimen width at the narrowest place and the core thickness of the textured geomembrane. The test temperature is adjusted to (80 ± 1) °C. A mixture of deionised water and 2 % by mass of Marlophen 812® surfactant is used as a test liquid which is renewed before each test. After a conditioning the filled device for 24 h at 80 °C the specimens are installed. The test result is the geometrical average of times-to-failure of a minimum of five specimens. Only those times-to-failure are considered where cracking initiation at the clamping or crack initiation form the specimen edges is excluded unambiguously. Figure 5.6 shows schematic fracture patterns observed on textured and smooth geomembranes. According to the requirements of the BAM certification guideline the surface structures must be designed in such a way and the textured geomembranes manufactured so carefully that a minimum of 700 hours is achieved under these test conditions.

### 3.2.17 Friction Properties

Geotechnical structures must be stable. The upper surface of installed geomembranes may not become a sliding surface on which the superstructure above the geomembrane can fail. The geomembrane must not slide on the subgrade or foundation layer either. In this case, stresses would develop in an anchored geomembrane which are usually so great because of the thinness of the geomembrane that it would tear off rapidly. Such a case of damage is discussed for example in (Kayser and Rodatz 1993). Stability problems due to downhill-slope forces not only occur on inclined surfaces on slopes, but also lateral spreading forces or water pressure in superstructures can endanger sliding stability even between horizontal layers. Downhill slope forces and lateral spreading forces must be safely counterbalanced by friction forces at the upper and lower surfaces of the geomembranes. Friction between smooth geomembranes or geomembranes with a textured surface and the materials of the subgrade or foundation layer below and the protective layer above the geomembrane, such as sand, fine gravel, clay, nonwoven geotextiles or other geosynthetics, is therefore an important property which must be tested in each construction project (Blümel 1993; Blümel and Brummermann 1996; Saathoff 1990).

Coulomb's law applies to the friction between two solid surfaces in most cases as a good approximation: friction force $F_F$ is proportional to the normal force $F_N$ which presses the surfaces together:

$$F_F = \mu F_N \, . \tag{3.13}$$

The friction force, which must be overcome between two surfaces at rest, is greater than the friction force, which develops between the surfaces sliding on each another. One speaks of static friction and sliding friction. The respective coefficients of proportionality $\mu_S$ and $\mu_G$ are called coefficient of friction or friction number. This quantity is always a property of two materials, the friction partners. The relationship:

$$\mu = \tan \delta \tag{3.14}$$

defines the so-called angle of friction $\delta$. The angle of static friction is precisely the angle at which a body of one material would start to slide on an inclined plain of another material.

Friction between two geosynthetics or a geosynthetic and a mineral material is analysed by plotting the friction force versus the applied normal force. Usually the data fit to a straight line, which is referred to as friction strength envelope. In some cases one can observe that the strength envelope does not go through the origin but has a positive friction force axis intercept. Apparently still a relatively large force $F_A$ seems to be necessary, even at a very small normal force, to make the friction components slide. Instead of Eq. 3.13, the following relation applies:

$$F_R = F_A + \mu F_N\,. \tag{3.15}$$

Force $F_A$ is called adhesive force. However, this term should not lead to misunderstandings. Adhesion in the proper sense results from electrostatic intermolecular forces, with which the molecules or atoms on the surface of two bodies attract one another when they touch intimately. However, this effect does not play any role for the friction partners of geosynthetics and soil materials (earth, sand, gravel etc.). Geosynthetics or mineral materials do not have a completely smooth friction surfaces. For example, the surfaces of the textured geomembranes show pronounced profiles. The texture can hook into the random array of monofilaments of nonwoven geotextiles and between the slit-films of woven geotextiles. Mineral materials are soft and deformable and the structural components can penetrate into them. The surfaces of the friction partners can thus hook or adapt to each other in a way which could be called "interlocking". Thus a significant force might be necessary to overcome this interlock even if only a small normal force is present. Suction due to capillary forces prevailing even with a small normal force may also have an adhesive effect between smooth geomembranes and soils. The apparent adhesion force $F_A$ thus results from all these interlocking effects.

The interlocking effect changes with the normal force. The relationship Eq. 3.15 with a fixed adhesive force therefore does not apply over the entire range of normal force. The strength envelope can therefore only be linear over a more or less wide range of normal force according to Eq. 3.15.

Dividing the forces by the contact area, the equation of friction strength envelope is obtained in the following form:

$$\tau = a + \sigma_N \tan \delta , \tag{3.16}$$

where $\tau$ is the friction stress due to friction force at the interface, $\sigma_N$ is the normal stress and $a$ is the adhesion stress. Adhesion stress $a$ and angle of friction $\delta$ are the friction parameters which must be determined in a friction property test. As stated, these parameters only apply to certain ranges of normal pressure. In particular, the strength envelope measured at high normal stresses often shows apparently large adhesion components which are then ineffective at low normal stresses (Blümel and Brummermann 1994). Friction parameters at a very low surcharge (e.g. in a landfill cap) cannot be deduced from friction parameters at high surcharges (e.g. in a landfill basal liner).

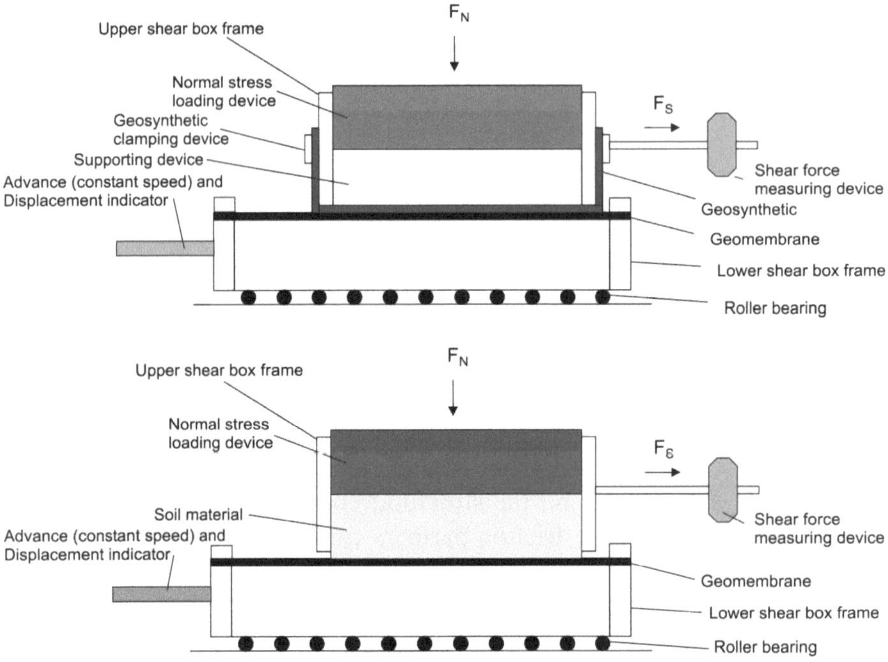


**Fig. 3.20.** Schematic of a shear box apparatus to test the friction properties between a geomembrane and a geosynthetic (top) and between a geomembrane and soil materials (bottom)

Shear tests performed by the shear box apparatus, the so-called direct shear method, have proved particularly suitable for determining friction parameters (Blümel and Brummermann 1994; Saathoff 1990). Figure 3.20 shows a schematic of such devices. The geomembrane specimen is fixed to the stationary bottom shear frame and the geosynthetic specimen is clamped with flat-jaw like clamping device to the top frame and a supporting layer is packed on it into the frame. To test the friction between mineral materials and a geomembrane, the mineral friction partner is placed in the top frame. A specified normal stress is applied by a normal stress loading device. The inside of the upper frame must be lined with a sliding layer so that the normal stress can exert its effect at the interface with only negligible reduction. The height of the gap between the upper and lower frame must be carefully adjusted depending upon the friction partner. The gap must be chosen so small that the sliding surface is not moved into the soil material and so large that the geosynthetic must not be squeezed between the frames. The device must have a constant friction surface of a minimum nominal size of 300·300 $mm^2$. The instructions in the standards have to be followed closely, to achieve test results which are reliable and reproducible under the test conditions.

The frames are moved in relation to each other at a specified normal stress and the shear strength necessary to achieve a constant speed of horizontal displacement is measured as a function of displacement (top diagrams in Figs. 3.21a and 3.21b). First, static friction dominates. The shear stress quickly increases at small displacements due to the deformation of the friction partners to a maximum value which defines static friction force. On reaching a maximum, the friction partners start to slide and the shear stress drops to the value required for continuous sliding at a certain displacement rate which determines the value of the sliding friction force. The shear stress versus horizontal displacement curve does not show a pronounced maximum for all friction partners. When the combination of a geomembrane with protruding surface structures and a clay is tested, the sliding friction is possibly even greater or only slightly smaller than the static friction since the structural components must plough through the clay while sliding (Fig. 3.21a). However, static friction between geomembranes with sprayed-on surface structures (Sect. 6.2) and a mineral layer is usually much greater than the sliding friction (Fig. 3.21b). The shear stress rises to a pronounced maximum and then drops to the value of sliding friction. The opposite can be observed with nonwoven geotextiles as a friction partner. The fibres can hook in the undercuts of the sprayed-on surface structures which result in high sliding friction values. For practical use, however, the maximum value of the shear stress versus displacement curve at small displacement values is relevant for the proof of structural stability.

The values of the static friction stress and the sliding friction stress have to be determined for a minimum of three different normal stresses. Based on a linear regression, the axis intercept yields the adhesion stress and the slope of the line of best fit yields the angle of friction (bottom diagram in Figs. 3.21a and 3.21b).

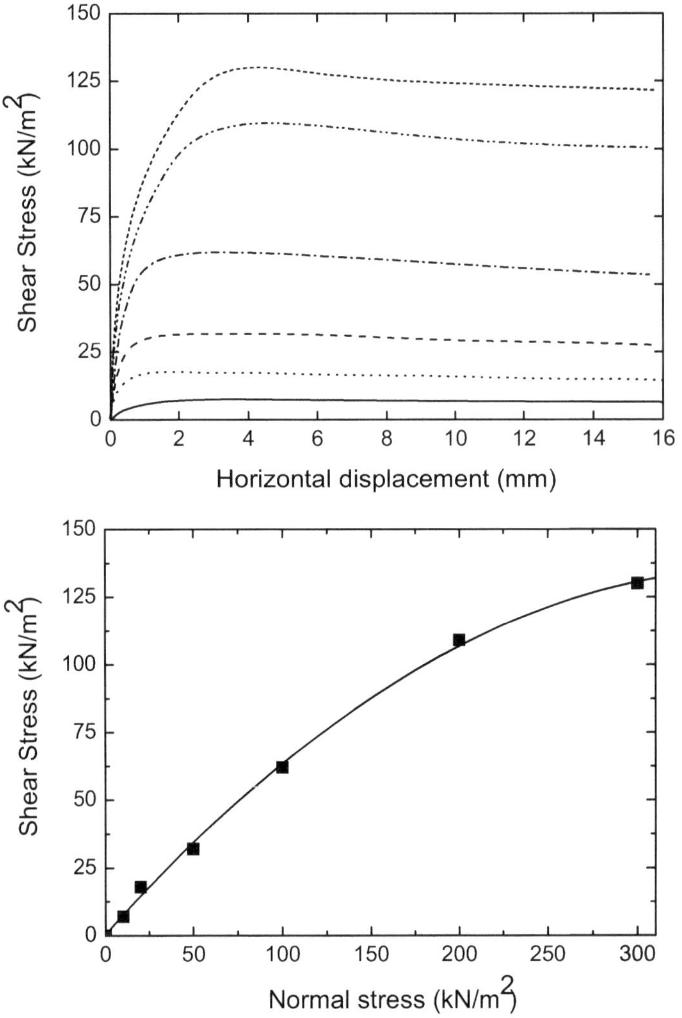


**Fig. 3.21a.** Shear stress as a function of horizontal displacement measured in the shear box test at various normal pressures (10, 20, 50, 100, 200, 300 kN/m²). The bottom diagram shows the static friction stress (maximum of the shear stress-displacement curve) as a function of normal pressure (strength or failure envelope). Shear-stress vs. displacement curves with sliding friction nearly equivalent to static friction are typical for embossed structures in combination with a clay-mineral liner. Source of data is (Blümel and Brummermann 1994)

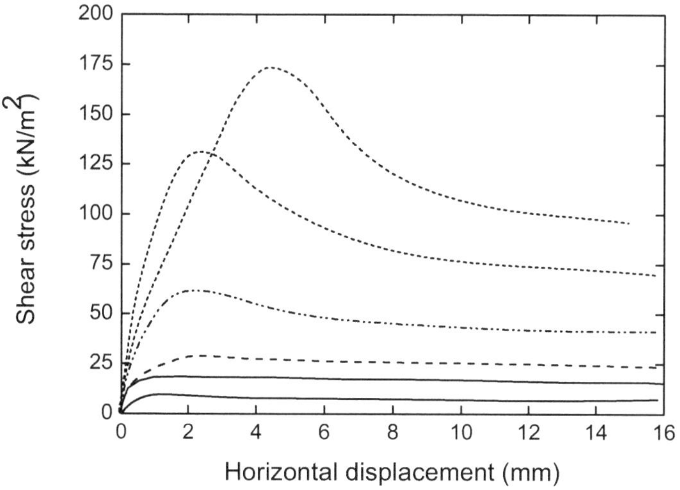


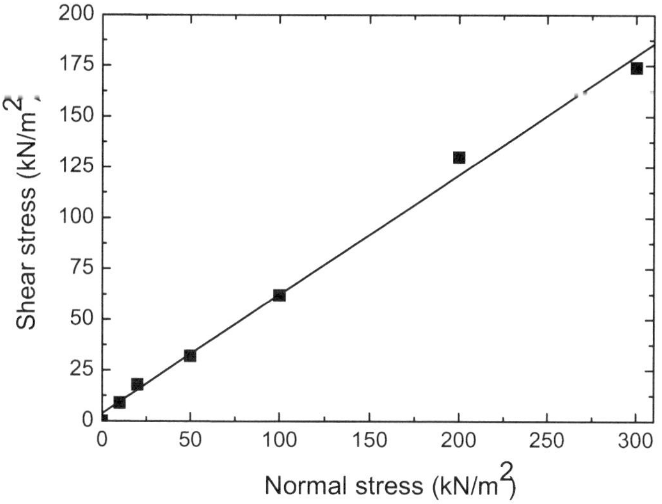


**Fig. 3.21b.** Shear stress as a function of horizontal displacement at 10, 20, 50, 100, 200 and 300 kN/m² normal pressure with a pronounced maximum in the shear stress-displacement curve. The bottom diagram shows the static friction stress (maximum of the shear stress-displacement curve) as a function of normal pressure (strength or failure envelope). Such shear stress-displacement curves are typical for sprayed-on textured geomembranes in combination with a clay-mineral liner. Source of data is (Blümel and Brummermann 1994)

Though the principal scheme of the test method may appear straightforward, execution and evaluation of the test is rather difficult. Many sources of error must be recognised and excluded. The Working Group 5.1 "Plastics in Geotechnics" of the German Society for Geotechnics (DGGt) headed by W. Blümel compiled the Recommendation E 3-8 *Friction Behaviour of Geosynthetics*, which describes the test conditions and procedures in detail. The recommendations were checked and further developed using the results of many comparative tests (Blümel and Brummermann 1996). The recommendation also discusses how design values of friction parameters for proving structural stability in geotechnical structures can be obtained from the experimental values in accordance with geotechnical safety concepts. In addition, shear box testing is described in detail in the standards EN ISO 12957-1:2005 *Geosynthetics - Determination of Friction Characteristics – Part 1: Direct Shear Test* and ASTM D5321-02 *Standard Test Method for Determining the Coefficient of Soil and Geosynthetic or Geosynthetic and Geosynthetic Friction by the Direct Shear Method.*

### 3.2.18 Long-term Shear Strength Test

Geomembrane surfaces are textured to ensure there is sufficient friction for sliding stability in relation to the subgrade or foundation layer below the geomembrane and to the protective layer on top of the geomembrane. The determination of these friction forces was dealt with in the previous section. However, large enough friction forces must usually be guaranteed on a long-term basis from which the requirements for the long-term adhesion of subsequently applied texture particles (Fig. 6.3) on a geomembrane surface can be deduced. The bonding points or the "seams" between texture particles and the smooth geomembrane may be impaired by stress crack formation, ductile failure and oxidative embrittlement. Weak bonds may possibly even emerge during manufacture due to lack of large enough contact time and temperature for melting and welding between texture particles and the smooth geomembrane surface within the contact area. Possible long-term failure mechanisms and defects cannot be tested by the shear box test which is aimed at determining the friction parameters within short test times and at relatively high displacement speed (Sect. 3.2.17). The long-term shear strength of textured geomembranes must therefore be tested in an appropriate long-term test. Based on other long-term stress-rupture test (e.g. long-term tensile test and pipe pressure test), test equipment for long-term shear strength tests was designed and constructed in BAM (Fig. 3.22) to examine the long-term behaviour of textured geomembranes (Müller et al. 2004; Seeger et al. 2000). The test device can also be used for testing the long-term shear strength of other geosynthetics (geosynthetic clay liners or geocomposite drains), in fact it was initially conceived for the investigation of internal shear strength of bentonite mats.

**Fig. 3.22.** Test device for long-term shear test with all components

**Fig. 3.23.** Details of the specimen holder. The top wedge is kept in position just by the friction force between geomembrane on the bottom wedge and nonwoven geotextile fixed on the bottom surface of the top wedge

**Fig. 3.24.** Friction partners, which were mounted onto the wedges ensuring high friction and firm contact to geosynthetic clay liners and geocomposite drain specimens

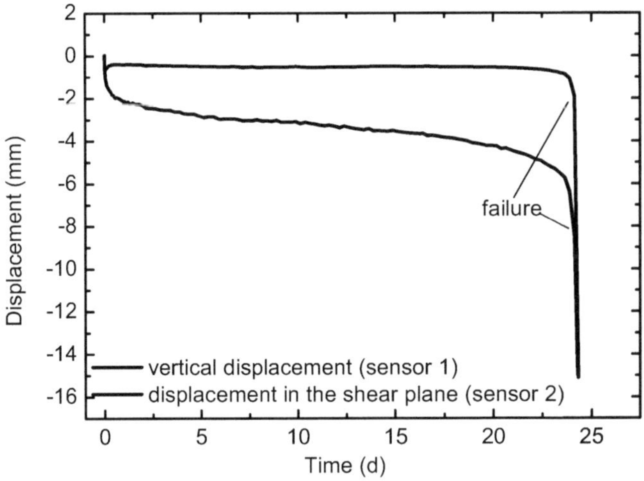


**Fig. 3.25.** Examples of vertical displacement vs. time curves of a geosynthetic clay liner specimen as monitored by displacement sensors (Thies 2002; Müller et al. 2004)

Shear strength testing of geosynthetics is conducted in long-term test stands in which the component is subjected to normal and shear stresses at a high temperature in a liquid or gas test environment until failure occurs (Fig. 3.22). By varying the test conditions, functional relationships can be established with time-to-failure, allowing in principle extrapolation of expected service lifetime under field conditions. A specimen (apprx. 120 mm·130 mm) of the product to be evaluated is fixed between two steel wedges (standard incline of wedges: 21.8° which corresponds to a slope of 1:2.5) as shown in Fig 3.23. A lever mechanism is used to exert a force on the upper wedge, which is representative of loading for example in cap liner applications. The resultant force component down the incline is the shear force between the upper and lower surfaces of the specimen. The testing device is housed in a controlled-temperature water bath ($T_{max}$ = 80 °C). As is the case in long-term pipe pressure testing, the high test temperature accelerates creep deformation, stress crack formation and oxidative degradation.

The test parameters, which can be varied in the test, are vertical load (or the vertical load/shear load ratio) and temperature. Other test environments can be used in place of water if appropriate for specific applications. Two high-precision ($\Delta s \leq 0.1$ mm) displacement sensors automatically monitor vertical displacements of the upper wedge and displacement in the shear plane. The experimental data are used to calculate the compression and shear deformation of the geosynthetic in the incline plane. Failure of the test specimen – e.g. sliding of upper wedge after separation of the texture particles – is documented regardless of whether failure occurs abruptly or over a prolonged period of time. In testing of textured geomembranes, the geomembrane and the contacting geotextile are screwed onto the lower and upper wedges to fix them in place. When the wedges are placed together, sliding is prevented by the friction between the texture particles and the geotextile. In testing of geosynthetic clay liner or geocomposite drains, a firm contact to the wedge is ensured by full-surface anchoring on a textured layer permanently mounted on the surface of the wedges. The textured layers used were textured geomembranes or metal food graters (Fig. 3.24).

Typically, a vertical force per unit of test specimen area of 50 kN/m$^2$ (50 kPa) is applied. The force exerted by the weight of the thrust rod, upper wedge and mounting parts is 28 N and by the weight of the level arm, test weight holder and displacement sensor 1 is 64 N. Since the advantage is 1:3, an additional force of (780 N - 28 N - 64 N)/3 = 229.3 N has to be exerted in this case as test weight in the experiments. Only the vertical displacement is usually monitored by sensor 1. After installation of the specimen holder in the test chamber the load is applied and the displacement measurement is started. The test chamber is then filled with de-ionized water (at room temperature) and the heater is started. Heating time to achieve final test temperature is about 8 h.

Figure 3.25 illustrates typical displacement vs. time plots measured in the test on geosynthetic clay liners (Thies 2002; Müller et al. 2004). As shown by the graphs, an initial phase of rapid deformation (primary creep) in response to the compressive load is followed by a stable phase of gradual, continuous deformation (secondary creep). Shortly before failure, the speed of deformation increases sharply (tertiary creep or materials degradation) and failure occurs. Similarly to the pipe pressure test, these tests on textured geomembranes are carried out at 80 °C over a minimum of 10,000 h.

## 3.3 Other Tests

The most important tests on HDPE geomembranes used in geotechnics were dealt with in the previous sections. However, the number of possible tests on plastic geomembranes used in practice is not yet fully covered. Other mechanical tests are performed within the framework of the BAM certification procedure for plastic geomembranes in landfill liners or caps or within the building-authority certification of plastic geomembranes by DIBt: the static puncture test (CBR test), perforation resistance tests (see also Sect. 8.3.2 on testing of protection layers), a low temperature brittleness test and tear propagation resistance tests may be carried out.

The static puncture test is described in EN ISO 12236 *Geotextiles and Geotextiles-Related Products – Static Puncture Test.* A cylindrical stamp (50 mm in diameter) is pressed through a clamped geomembrane disk of 150 mm free diameter at a speed of 50 mm/min. Thereby the resistance to quasi-static puncture by a small, sharp edged object is determined. The maximum force required (puncture resistance) is the test parameter. The 2.5 mm thick HDPE geomembranes typically reach values > 6000 N.

Dynamic puncture may be tested by dropping an object on the geomembrane. A cone drop test according to DIN 16726, Sect. 5.12, serves to test the puncture resistance to falling loads. A 0.5 kg heavy steel cylindrical drop body, into whose base a ball (diameter 12.7 mm) is pressed, is dropped from a height of 2 m on the geomembrane which lies on an aluminium plate. No leaks may result in the geomembrane. Low temperature brittleness is tested in a bending test according to EN 1876-1:1998 *Rubber or Plastics Coated Fabrics – Low Temperatures Tests – Part 1: Bending Test.* Test specimens and the bending test equipment are cooled to –20 °C in a cold chamber. The cold test specimens are then bent with the test device by approx. 180°. No cracks may develop in the bending edge. Such a test is, however, superfluous for HDPE materials with a glass transition temperature of -120 °C. The force needed to extend the tear in a geomembrane starting from a notch or a cut, is tested in a so-called tear growth test. The test is based on the following two standards: DIN 53515: 1990 *Testing of Rubber and of Plastic Films; Tear Test Using the Graves Angle Test Piece with Incision* or DIN 53356-A:1982 *Testing of Artificial Leather and Similar Sheet Materials; Tear Growth Test.* 2.5 mm thick HDPE geomembranes exhibit tear growth strengths > 300 N in the first indicated test and > 500 N in the second. A detailed description of these tests or other tests can be found in the DIN Pocket Book 150 (DIN 1998) or in the Annual Book of ASTM Standards, Volume 04.13, Geosynthetics (ASTM 2003).

It is beyond the scope of this book to deal with polymer-analytic methods which are used for material identification and characterisation on a molecular level and for the investigation of changes in the polymer molecule and additive properties. An overview about the various methods should be mentioned, which contains further references (Braun 1999; Hoffmann et al. 1977). However, three methods will briefly be discussed here which frequently play a role in the analysis of HDPE materials: viscosimetry, gel permeation chromatography (GC) and Fourier transformed infrared spectroscopy (FTIR).

In viscosimetry, the (dynamic) viscosity of a diluted solution of a special solvent and HDPE resin is determined, e.g. by capillary viscosimetry as per EN ISO 1628-3:2003 *Plastics – Determination of Viscosity of Polymers in Dilute Solution Using Capillary Viscometers – Part 3: Polyethylenes and Polypropylenes*. The flow time of the solution $t$ and of the pure solvent $t_0$ is measured in a capillary immersed in a hot bath at an elevated temperature. The coefficient of viscosity $J$ is determined from this flow times. It is defined as the relative change of the viscosity of the solution $\eta$ (proportional to $t$) with the concentration $c$ related to the viscosity of the solvent $\eta_0$ (proportional to $t_0$):

$$J = \left( \frac{\eta}{\eta_0} - 1 \right) \cdot \frac{1}{c}. \qquad (3.17)$$

The limiting value at an infinite dilution, the so-called intrinsic viscosity $J_0$, also termed Staudinger index, is then determined from the coefficients of viscosity measured at different concentrations:

$$J_0 = \lim_{c \to 0} \left( \frac{\eta}{\eta_0} - 1 \right) \cdot \frac{1}{c} . \tag{3.18}$$

Because of its high chemical resistance, HDPE can only be dissolved at high temperatures. The measurement must be performed on hot liquids and is therefore fairly expensive (135 °C and decahydronaphthaline as a solvent). However, the viscosity measurement does not usually provide any early or better information about a material degradation than the considerably simpler melt-flow rate measurement or tensile tests.

The molecular mass distribution, its number average and mass average and the width of the molecular mass distribution belong to the key parameters of polymeric materials. Let $n_i$ be the number of molecules with the molecule mass $M_i$ according to the distribution of the molecular masses in a certain plastic. The number average molecular mass $M_n$ and the mass average molecular mass $M_w$ can be written as:

$$M_n = \sum_i \frac{n_i M_i}{n_i} \qquad M_w = \sum_i \frac{(n_i M_i) M_i}{n_i M_i} . \tag{3.19}$$

The unit of both quantities is g/mol. Occasionally, the unit Dalton = g/mol is used. Dividing $M_n$ by the molecule mass of the basic module of the polyethylene chain, $CH_2$ (14 g/mol), one obtains the dimensionless average degree of polymerisation. A measure for the range of molecular mass distribution is the relationship of the mass average to the number average, the so-called polydispersity $U$:

$$U = \frac{M_w}{M_n} - 1 . \tag{3.20}$$

The method of choice for the determination of these quantities on HDPE resin is gel permeation chromatography. In this chromatographic method a solvent with the dissolved polyethylene flows through a column filled with a porous material. It is usually a gel of cross-linked polymers. Depending upon the size of the polyethylene coils in the solution, they can fit to the pores of the gel, stick there and block them whereby the transport is retarded. Larger polyethylene chains are transported unhindered with the solvent flow. Thus a size-dependent retardation occurs which justifies the other term size exclusion chromatography (SEC). In the chromatogram, which shows polymer concentration (determined for example via optical measurements (refractive index)) as a function of time, the time coordinate corresponds to a polymer molecular mass coordinate and the concentration is therefore resolved according to the polymer size. The molecule mass distribution curve can then be determined from the chromatogram.

The Fourier transformed infrared spectroscopy (FTIR) serves the identification and quantitative determination of functional groups in the polymer chain, which, due to their specific molecular oscillations and rotation, can absorb energy within the infrared (IR) range of the electromagnetic radiation and show a unique absorption spectrum. Of special importance is FTIR spectroscopy for the investigation of auto-oxidation, since the developing oxidation products, carbonyl groups or hydroperoxide groups, have typical infrared bands in the absorption spectrum of degraded plastics, the carbonyl group e.g. at 1714 $cm^{-1}$ and the free or associated hydroperoxide groups at 3550 and 3410 $cm^{-1}$ (Gugumus 1996).

Finally, an important sub-field of polymer analysis in connection with HDPE geomembranes is the analysis of additives (Freitag 1989; Zweifel 2001). In most cases an extraction procedure (typically an extraction by dissolution followed by precipitation of the polymer and filtration) precedes the chemical, chromatographic, ultraviolet-spectroscopic or infrared-spectroscopic analysis. There have been also attempts to determine additives and their contents directly in the geomembranes, e.g. by pyrolysis using subsequent gas chromatographic and mass-spectrometric determination.

## References

Albertson A-C (1992) Biodegradation of Polymers. In: Hamid SH et al. (eds) Handbook of Polymer Degradation. Marcel Dekker Inc., New York, Basel, Hong Kong, pp 345–364

ASTM (ed) (2003) Annual Book of ASTM Standards, Volume 04.13, Geosynthetics. American Society for Testing and Materials (ASTM), West Conshohocken

Blümel W (1993) Zur Untersuchung der Standsicherheit von oberirdischen Abfalldeponien. Müll und Abfall 25: 739–751

Blümel W and Brummermann K (1994) Reibung zwischen Geokunststoffen und Erdstoffen in Deponiedichtungen. Müll und Abfall 26: 242–259

Blümel W and Brummermann K (1996) Interface Friction Between Geosynthetics and Soils and Between Different Geosynthetics In: De Groot, Den Hoedt and Termaat (eds.) Proceedings of the conference Eurogeo 1 on "Geosynthetics: Applications, Design and Construction", Balkema, Rotterdam, pp 209–216

Braun D (1999) Simple Methods for the Identification of Plastics. Hanser Verlag, Munich

Crank J and Park GS (1968) Diffusion in Polymers. Academic Press, London

DVWK (Deutscher Verband für Wasserwirtschaft und Kulturbau e.V.) (1989) Anwendung und Prüfung von Kunststoffen im Erdbau und Wasserbau, DVWK-Schriften, Heft 76. Verlag Paul Parey, Hamburg und Berlin

DIBt (ed) (2000) Zulassungsgrundsätze für Kunststoffbahnen als Abdichtungsmittel von Auffangwannen, Auffangräumen, Auffangvorrichtungen und Flächen für die Lagerung und das Abfüllen und das Umschlagen wassergefährdender Stoffe (ZG Kunststoffbahnen in LAU-Anlagen). Deutsches Institut für Bautechnik (DIBt), Berlin

DIN (ed) (1998) DIN-Taschenbuch 150, Kunststoff-Dachbahnen, Kunststoff-Dichtungsbahnen, Kunststoff-Folien und kunststoffbeschichtete Flächengebilde (Kunstleder). Beuth Verlag, Berlin

Dolezel B (1978) Die Beständigkeit von Kunststoffen und Gummi. Carl Hanser, München

EPA (ed) (1992) Stress cracking behavior in HDPE geomembranes and its prevention. Environmental Protection Agency (EPA), Cinicinnati, USA

Fleißner M (1987) Langsames Rißwachstum und Zeitstandfestigkeit von Rohren aus Polyethylen. Kunststoffe 77: 45–50

Freitag W (1989) Analyse von Additiven. In: Gächter R and Müller H (eds) Taschenbuch der Kunststoff-Additive. Carl Hanser Verlag, München, pp

Gebler H (1989) Langzeitverhalten und Alterung von PE-HD-Rohren. Kunststoffe 79: 823–826

Gray RL (1990) Accelerated Testing Method for Evaluating Polyolefin Stability. In: Koerner RM (ed) Geosynthetic Testing for Waste Containment Applications ASTM special technical Publication: 1081. American Society for Testing and Materials (ASTM), West Conshohocken, pp 57–74

Greenwood JH et al. (1996) Durability standards for geosynthetics: The tests for weathering and biological resistance. In: De Groot MB et al. (eds) Geosynthetics: Applications, Design and Construction, Proceedings of the first european geosynthetic conference Eurogeo 1. A. A. Balkema, Rotterdam, pp 637–642

GRI (ed) (1998) GRI Standard GM13: Test Properties, Testing Frequency and Recommended Warrant for High Density Polyethylene (HDPE) Smooth and Textured Geomembranes. Geosynthetic Institute (GSI), Folsom, USA

Gugumus F (1990) Lichtschutzmittel. In: Gächter R and Müller H (eds) Taschenbuch der Kunststoff-Additive. Carl Hanser Verlag, München

Gugumus F (1996) Thermooxidative degradation of polyolefins in the solid state: Part 1. Experimental kinetics of functional group formation. Polymer Degradation and Stability 52: 131–144

Hald A (1952) Statistical Theory with Engineering Applications. John Wiley & Sohns, Inc., New York

Hamid HS et al. (1992) Weathering Degradation of Polyethylene. In: Hamid SH et al. (eds) Handbook of Polymer Degradation. Marcel Dekker Inc., New York, Basel, Hong Kong, pp 219–260

Hein S et al. (2003) Stand der Technik und Erfahrungen beim Einbau von Dichtungsbahnen aus PEHD für großflächige Abdichtungen im Bereich des Grundwasserschutzes. In: Knipschild FW (ed) Tagungsband der 19 Fachtagung “Die sichere Deponie, Sicherung von Deponien und Altlasten mit Kunststoffen”. Süddeutsches Kunststoffzentrum (SKZ), Würzburg, pp D1–D14

Hemminger WF and Cammenga HK (1989) Methoden der Thermischen Analyse. Springer Verlag, Berlin

Hoffmann M et al. (1977) Polymeranalytik, Band I und II. Georg Thieme Verlag, Stuttgart

Howard JB (1973) DTA for Control of Stability in Polyolefin Wire and Cable Compounds. Polymer Engineering and Science 13: 429–434

Hsuan YG (2000) Data base of field incidents used to establish HDPE geomembrane stress crack resistance specifications. Geotextiles and Geomembranes 18: 1–22

Hufenus R et al. (1998) Das Geotextil-Handbuch, Kapitel 12, Langzeitverhalten von Geotextilien. Schweizerischer Verband der Geotextilfachleute (SVG), St. Gallen (Schweiz)

Kalbe U et al. (2000) Mineralogische und chemisch-physikalische Auswirkungen der Permeation von Kohlenwasserstoffen in Kombinationsdichtungen und -dichtwänden (Mineralogical, chemical and physical repercussions of migration of VOC in composite liners), Bericht zum Forschungsvorhaben 1461027 des BMBF. Bundesanstalt für Materialforschung und -prüfung (BAM), Labor Kontaminationsbewertung, Berlin

Kamal MR and Huang B (1992) Natural and Artificial Weathering of Polymers. In: Hamid SH et al. (eds) Handbook of Polymer Degradation. Marcel Dekker Inc., New York, Basel, Hong Kong, pp 127–168

Kayser J and Rodatz W (1993) Verhalten einer Kombinationsabdichtung unter extremer Belastung, Analyse eines Schadensfalles. In: Knipschild FW (ed) Tagungsband der 9 Fachtagung “Die sichere Deponie, Wirksamer Grundwasserschutz mit Kunststoffen”. Süddeutsches Kunststoffzentrum (SKZ), Würzburg, pp 291–304

Kempe B (1984) Prüfmethoden zur Ermittlung des Verhaltens von Polyolefinen bei der Einwirkung von Chemikalien. Z Werkstofftech 15: 157–172

Knipschild FW (1985) Werkstoffauswahl und Dimensionierung von Kunststoffdichtungsbahnen für Grundwasserschutzmaßnahmen. In: Knipschild FW (ed) Deponiebasisabdichtungen mit Kunststoffdichtungsbahnen, Müll und Abfall, Beiheft 22. Erich Schmidt Verlag, Berlin, pp 49– 60

Knipschild FW and Tornow K (1979) Großflächen-Dichtungselemente aus Niederdruckpolyethylen. Kunststoffe im Bau 14: 130–134

Koch R et al. (1988) Langzeitfestigkeit von Deponiedichtungsbahnen aus Polyethylen. Müll und Abfall 20: 3–12

Kreiter J and Hutten A (1990) Ergebnisse aus Berstdruckversuchen an Deponiebahnen. Müll und Abfall 22: 497–506

Krishnaswamy RK (2005) Analysis of ductile to brittle failure from creep rupture testing of high density polyethylene (HDPE) pipes. Polymer 46: 11664–11672

Menges G (1990) Werkstoffkunde Kunststoffe. Carl Hanser Verlag, München

Motulsky H (1995) Intuitive Biostatistics. Oxford University Press, Oxford

Mueller WW and Jakob I (2003) Oxidative resistance of high-density polyethylene geomembranes. Polymer Degradation and Stability 79: 161–172

Müller WW (1996) Anforderungen an die Schutzschicht für die Dichtungsbahnen in der Kombinationsdichtung, Teil 2: Zulassungsanforderungen (Requirements on protection layers for geomembranes in landfill liner systems, part 2: certification requirements). Müll und Abfall 28: 90–99

Müller WW (1999a) Die neue BAM-Richtlinie für die Zulassung von Kunststoffdichtungsbahnen für die Abdichtung von Deponien und die Sicherung von Altlasten. In: Egloffstein T et al. (eds) Oberflächenabdichtung von Deponien und Altlasten 1999, Zeitgemäße Oberflächenabdichtungssysteme – Ist die Regelabdichtung nach TA-Si noch zeitgemäß? Erich Schmidt Verlag, Berlin, pp 117

Müller WW (ed) (1992) Richtlinie für die Zulassung von Kunststoffdichtungsbahnen als Bestandteil einer Kombinationsdichtung für Siedlungs- und Sonderabfalldeponien sowie für Abdichtungen von Altlasten. BAM, Labor Deponietechnik, Berlin

Müller WW (ed) (1995) Anforderungen an die Schutzschicht für die Dichtungsbahnen in der Kombinationsdichtung, Zulassungsrichtlinie für Schutzschichten. BAM, Labor Deponietechnik, Berlin

Müller WW (ed) (1999b) Richtlinie für die Zulassung von Kunststoffdichtungsbahnen für die Abdichtung von Deponien und Altlasten. Wirtschaftsverlag NW, Verlag für neue Wissenschaften GmbH, Bremerhaven

Müller WW et al. (1997) Stofftransport in Deponieabdichtungssystemen, Teil 1: Diffusions- und Verteilungskoeffizienten von Schadstoffen bei der Permeation in PEHD-Dichtungsbahnen. Bautechnik 74: 176–190

Müller WW et al. (1998) Solubilities, Diffusion and Partition Coefficients of Organic Pollutants in HDPE Geomembranes: Experimental Results and Calculations. In: Rowe RK (ed) Proceedings of the Sixth International Conference on Geosynthetics. Industrial Fabrics Association International (IFAI), Roseville, MN, USA, pp 239–248

Müller WW et al. (2004) Long-term shear strength of multilayer geosynthetics. In: Floss R et al. (eds) Geotechnical Engineering with Geosynthetics, Proceedings of the Third European Geosynthetics Conference. Deutsche Gesellschaft für Geotechnik (DGGt) und Technische Universität München, Zentrum für Geotechnik (TUM-ZG), München, pp 429–434

Rollin A and Rigo J-M (eds) (1991) Geomembranes, Identification and Performance Testing. Chapman and Hall, London

Rumberg E et al. (1982) Untersuchungen über das Verhalten von Abdichtungsfolien gegen Nagetiere, Abschlußbericht (Nr. 10203401) des BMI Forschungsvorhabens Nr. 102 02 401. Umweltbundesamt, Berlin

Saathoff F (1990) Zum Scherverhalten von Geokunststoffen. Bauingenieur 65: 195–207

Schrijver-Rzymelka P (1999) Lichtschutzmittel. Kunststoffe 89: 87–90

Schröder H et al. (2000) Durability of Polyolefine Geosynthetics under Elevated Oxygen Pressure in Aqueous Liquids. In: Cancelli A et al. (eds) Proceedings

of the Second European Geosynthetics Conference. Pàtron Editore, Bologna, pp 459–463
Schulte U (1997) 100 Jahre Lebensdauer, Langzeitfestigkeit von Druckrohren aus bimodalem PE-HD nach ISO/TR 9080. Kunststoffe 87: 203–206
Seeger S et al. (2000) Long term testing of geomembranes and geotextiles under shear stress. In: Cancelli A et al. (eds) Proceedings of the Second European Geosynthetics Conference. Pàtron Editore, Bologna, pp 607–610
Trubiroha P and Schröder H (1997) Klassifizierung von Geotextilien hinsichtlich der Wetterbeständigkeit. Geotechnik Sonderheft 1997: 181–186
White DF and Verschoor KL (1990) Practical Aspects of evaluating the chemical compatibility of geomembranes for waste containment applications. In: Koerner RM (ed) Geosynthetics Testing for Waste Containment Applications, ASTM special technical Publication: 1081. American Society for Testing and Materials (ASTM), West Conshohocken, pp 25–36
Zweifel H (ed) (2001) Plastic Additives Handbook. Carl Hanser Verlag, Munich

# 4 Deformation Behaviour

## 4.1 Stress Relaxation and Creep

Thermoplastics, such as medium to high density, high-molecular polyethylenes, show a pronounced visco-elastic mechanical behaviour (Strobl 1997). This applies to reasonably small deformations, e.g. deformations below the yield point, for the liquid and the solid state phase of the polymeric material.

The visco-elastic mechanical behaviour can be illustrated with the help of simple mechanical models. Consider a spring and a viscous damping device (dashpot), one in serial arrangement (Maxwell's model) and one in parallel arrangement (Voigt-Kelvin's model), see Fig. 4.1. The elastic deformation behaviour[1] of the spring is described by Hooke's law with the spring constant or Young's modulus $E$, and the viscous behaviour of the damping is to obey Newton's law with the viscosity $\eta$:

$$\sigma_{el} = E\,\varepsilon_{el} \qquad\qquad \sigma_{vis} = \eta\,\dot{\varepsilon}_{vis}\,. \tag{4.1}$$

Due to the serial arrangement:

$$\sigma = \sigma_{el} = \sigma_{vis} \quad \text{and} \quad \varepsilon = \varepsilon_{el} + \varepsilon_{vis} \quad \text{thus} \quad \dot{\varepsilon} = \dot{\varepsilon}_{el} + \dot{\varepsilon}_{vis}\,. \tag{4.2}$$

From Eqs. 4.1 and 4.2 one obtains the following equation of motion for the mechanical behaviour of Maxwell's model:

$$\dot{\sigma} + \frac{E}{\eta}\sigma - E\,\dot{\varepsilon} = 0\,. \tag{4.3}$$

Using this equation the stress can be calculated as a function of time at a specified strain or deformation rate. If the model is suddenly elongated by a specified amount and fixed in the new position, then stress $\sigma_0$ develops in the spring corresponding to the initial strain. However, this stress will gradually dissipate due to the piston motion in the dashpot. Since the de-

---

[1] As usual, $\sigma$ designates stress, $\varepsilon$ denotes strain and a dot on the top indicates a derivative with respect to time. Index $_{el}$ refers to the spring and index $_{vis}$ to the viscous damping device.

formation rate is zero after applying the constant elongation, one obtains from Eq. 4.3:

$$\sigma(t) = \sigma_0 \, e^{-t/\tau_{rel}} \qquad \text{with} \quad \tau_{rel} = \frac{\eta}{E} \ . \tag{4.4}$$

Thus stress relaxes at a relaxation time $\tau_{rel}$ characteristic for the parameters of the model.

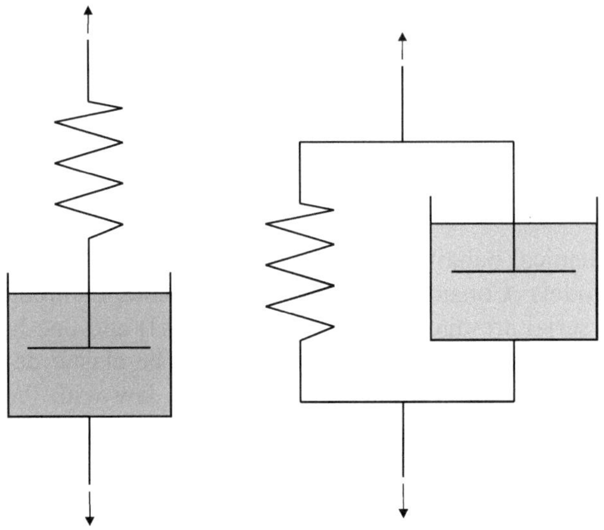

**Fig. 4.1.** Schematic illustration of mechanical models for visco-elastic deformation behaviour composed of a spring and viscous damping device. Left: the so-called Maxwell's model with a spring and a dashpot arranged in series. Right: the so-called Voigt-Kelvin's model with a spring and dashpot arranged in parallel. Maxwell's model shows typical relaxation behaviour: tension in the spring is reduced by the movement of the piston after a strong elongation. Voigt-Kelvin's model shows typical creep behaviour: on weight application the spring gradually expands until the spring force equals the weight force

It holds due to the parallel arrangement:

$$\sigma = \sigma_{el} + \sigma_{vis} \qquad\qquad \varepsilon = \varepsilon_{el} = \varepsilon_{vis} \tag{4.5}$$

and from Eqs. 4.1 and 4.5 one similarly obtains the equation of motion for the mechanical behaviour of Voigt-Kelvin's model:

$$\sigma = E\,\varepsilon + \eta\dot{\varepsilon} \ . \tag{4.6}$$

Creep behaviour can be calculated from Eq. 4.6 for the constant load $\sigma_0$ applied to the model device as follows:

$$\varepsilon = \frac{\sigma_0}{E}\left(1 - e^{-t/\tau_C}\right) \qquad \text{with} \quad \tau_C = \frac{\eta}{E} \, . \tag{4.7}$$

Thus the strain increases gradually. The model creeps with retardation time $\tau_c$ characteristic of the parameters of the model until the load is completely balanced by the restoring force of the spring strained to $\sigma_0/E$.

Stress relaxation at an imposed strain as described by the Maxwell model and creep under a long-term effective load as described by the Voigt-Kelvin model are the characteristic mechanical properties of a visco-elastic material. The actual behaviour of visco-elastic plastics is, however, more complicated. To simulate their stress-strain-relation one has to use a network of various springs and dashpots. A whole spectrum of relaxation and retardation times result. Stress relaxation does not lead to complete stress dissipation. Similarly, creep will gradually succumb. In setting up the spring-dashpot-models, a constant viscosity of the absorber material in the dashpot and a constant Young's modulus of the spring have been assumed. In this case one speaks of linear visco-elastic behaviour. However, thermoplastics often show non-linear visco-elastic behaviour. One would have to use non-linear extensions of the mechanical model for their description. The parameters of the model would become functions of the deformation state. For instance, one would have to describe viscosity as a function of stress. Mechanical models which actually simulate the behaviour of thermoplastics are obviously very complicated. Therefore, in the following section a simple phenomenological dynamical model will be discussed.

It is important to note that all these models of stress-strain-behaviour will not fully describe the true mechanical behaviour in the long-run. Under load acting over a long term, stress might increase in an actual loaded specimen due to the change of cross-sections and become so large that the stress-strain-relation will leave the area of non-linear visco-elastic behaviour: yield processes and finally fracture will occur. In the long-run stress crack formation and aging processes might be relevant which will strongly influence the spectrum of relaxation and retardation times. To predict the long-term behaviour of plastics by extrapolating relaxation or creep curves, for example with the help of the time-temperature superposition principle, might therefore lead to serious mistakes.

To better understand creep and relaxation of polymeric materials these processes can also be considered under another, namely a thermodynamic point of view. This viewpoint will be presented here in a highly simplified, only qualitative manner. A more detailed quantitative description is presented in (Brunt 1966). Thermodynamically, thermoplastics not only exhibit so-called energy elasticity, but also a so-called entropy or rubber elasticity. In the ideal case of a purely energy-elastic body, the work performed by external forces on the deformation is fully utilised to reduce the binding energy in the atomic bonds of the molecules forming the body. When a body, consisting of a network or cluster of polymer chains, is elongated, not only do bond lengths and angles change, but the configuration of polymer chains is also reoriented and re-arranged. The new state which develops in this way will have different configurational entropy. The entropy is understood here to be the measure of the number of microscopic arrangements of the polymer chains, which are compatible with a macroscopic state of the body characterized by a specified deformation or an acting force[2].

If such a body is deformed by an external force very slowly compared to its typical relaxation times of fluctuation on a molecular level, then it goes through a series of equilibrium configurational states with ever decreasing entropy. The reduction of configurational entropy $\Delta S$ is associated with part of the deformation work of the external force being converted into heat. This part is given by $T\Delta S$, where $T$ is the temperature of the body. The heat is transferred to the environment, when the process is performed isothermally, i.e. in contact with a heat bath of constant temperature. If the body is deformed adiabatically, i.e. it is thermally isolated, it will heat up during the deformation. In an ideal pure entropy-elastic body, the deformation work is fully converted into heat. Mechanically considered, the body seems to resist to the restriction of the number of its microscopic configurations: apparently an entropy-elastic stress develops in addition to the energy-elastic stress, against which the external force must perform work[3].

[2] A thermodynamic equilibrium is characterised by the fact that the extensive quantities of the system, such as its actual energy or volume, vary only slightly around an average value, equally, that its intensive quantities such as local temperature, local deformation etc., stay essentially constant within the whole system during these fluctuations. Be $E$ the medium energy and $\Delta E$ the width of the distribution of the fluctuating individual values. The entropy $S$ of the equilibrium state can be defined as the logarithm of the number $P$ of the microscopic configurations of the system, for which the respective energy values are within the range $E \pm \Delta E$: $S = k_B \ln P$, where $k_B$ is Boltzmann's constant.

[3] Be $r = (x^2+y^2+z^2)^{1/2}$ the dimension of a polymeric network. The number of the microscopic configurations $P$, i.e. the number of the various possible configurations of the network within the specified volume of extension $r$, is then proportional to $\exp(-h^2r^2)$ with some parameter $h$. At a deformation from $x$, $y$, $z$ to $x+\delta x$, $y+\delta y$, $z+\delta z$ the deformation work is $E = T\Delta S$, thus $E = k_B T(\delta x^2 + \delta y^2 + \delta z^2)\,(hr)^2$.

The performed work is transformed, as mentioned, not only into a reduction of the bond energies, but into heat.

If this body is very rapidly deformed, as in the relaxation test, it is forced into a state of non-equilibrium whose entropy[4] will even be substantially smaller than the entropy of the equilibrium state belonging to the strain imposed upon it. Correspondingly, the entropy-elastic stress, which develops in the body, is higher. However, by absorption of heat and accompanied thermal motion of the polymer molecules, the state of non-equilibrium gradually relaxes into the equilibrium state. Here the absorbed heat is converted into configuration entropy; therefore the entropy-elastic stress component is also reduced. This stress dissipation is observed in the relaxation test. Finally, an energy-elastic stress component remains (remember the Voigt-Kelvin-model). At which "relaxation speed" the body can relax from the state of non-equilibrium into equilibrium and how long the relaxation times are, is a characteristic of the actual material and depends strongly on the temperature.

In the creep test a hanging weight is attached to the body or a long-term effective tensile stress is applied in some other way. Let us consider the body to be ideally entropy-elastic. Under this new constraint the body is initially in a state of non-equilibrium with low entropy. The arrangement of the polymer molecules only gradually adjust to this constraint. The body relaxes into an equilibrium state by deforming itself according to the external constraint. The increase of entropy due to the relaxation from the state of non-equilibrium is however overcompensated by a further entropy loss which results from the constant shift of the equilibrium state by the work of the external load acting permanently over the long term. Therefore, strain will continuously increase in such a way that the overall entropy reduction and the associated heat production maintain equilibrium with the deformation work of the external force. In an actual thermoplastics body the energy-elastic stress component must be taken into account: this will increase with increasing strain. In principle therefore, creep comes to a standstill when the energy-elastic stress component is large enough to balance the external force.

These thermodynamic considerations clarify why the stress-strain behaviour of a visco-elastic material depends very strongly on the ratio of deformation rate to the "relaxation speed" at which the body can react to ex-

---

[4] The entropy of the state of non-equilibrium can be defined as follows. The body should be divided into cells. The cells are chosen in such a way that each individual cell can be considered to be in a local state of equilibrium at any time, for which the entropy is defined as in footnote 2. The entropy of the body in non-equilibrium state is then the sum of the entropies of the cells.

ternal effects due to its relaxation time spectrum, and on temperature which influences the spectrum of its relaxation times. It also helps to understand why there must be a relation between deformation rate and temperature. Indeed: for a stress-strain curve measured at a low temperature and a low deformation rate, one can generate an identical curve at a higher temperature if a suitably high deformation rate is chosen. Or alternatively: if a particular stress is found after a certain time in the relaxation test carried out at a specified temperature, the same stress will be found after a precisely set earlier test time if the relaxation test is performed at a specified higher temperature. This so-called time-temperature superposition principle is a fundamental tool for the description of visco-elastic material behaviour (Strobl 1997). The law applies not only to elastomers and amorphous plastics, but also to partially crystalline thermoplastics.

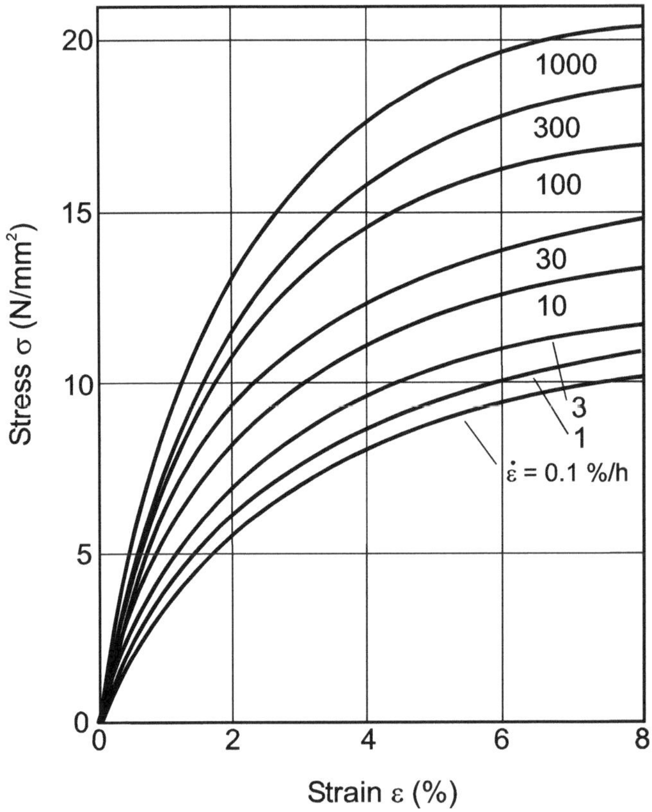


**Fig. 4.2.** Stress-strain curves of a HDPE material in a tensile test at 23 °C determined at various test speeds. Source: (Schmachtenberg 1985)

## 4.2 Phenomenological Dynamical Model

Based on the above considerations, G. Menges and E. Schmachtenberg quantitatively described the deformation behaviour of the HDPE material within the non-linear visco-elastic range below the yield point (Menges and Schmachtenberg 1985). Although the individual HDPE resins differ in the values for Young's modulus, in stresses at yield and elongations at yield over a certain range, these differences are small compared with the typical general dependency of the strains and stresses on temperature and deformation rate, which is the focus of our attention in this section.

Tensile tests on different HDPE resins indicate that the stress-strain behaviour below the yield point can generally be described by the following empirical relationship (Fig. 4.2):

$$\sigma = \frac{E_0 \varepsilon}{1 + D_2 \varepsilon} . \tag{4.8}$$

The moduli $E_0$ and $D_2$ are functions of temperature $T$ and deformation rate $\dot{\varepsilon}$.

Be $E_0(T_{ref}, \dot{\varepsilon}_{ref})$ and $D_2(T_{ref}, \dot{\varepsilon}_{ref})$ the values of these moduli that parameterise the stress-strain curve at a chosen reference temperature $T_{ref}$ and a reference deformation rate or a reference test speed $\dot{\varepsilon}_{ref}$. In accordance with the time-temperature superposition principle there must be a test speed $\dot{\varepsilon}$ at another temperature $T$ which results in an identical stress-strain curve. It must therefore hold:

$$E_0(T_{ref}, \dot{\varepsilon}_{ref}) = E_0(T, \dot{\varepsilon}) \qquad D_2(T_{ref}, \dot{\varepsilon}_{ref}) = D_2(T, \dot{\varepsilon}) . \tag{4.9}$$

By looking at the pair of values $(T, \dot{\varepsilon})$ for a specified reference temperature $T_{ref}$, and reference deformation rate $\dot{\varepsilon}_{ref}$ that satisfy Eq. 4.9, one can quantitatively determine the functional relationship between temperature and deformation rate which indicates what rate must be used to obtain the same stress-strain curve at a different temperature. For a reference temperature $T_{ref}$ = 23 °C, i.e. 296 K, and from the data from tensile tests the following function was derived, which is characteristic of the HDPE material, and can be considered as a quantitative formulation of the time-temperature superposition principle for this material:

$$\log \frac{\dot{\varepsilon}}{\dot{\varepsilon}_{ref}} = 8100K \left(\frac{1}{T} - \frac{1}{296K}\right) . \tag{4.10}$$

Using this relation[5] one can now determine the stress-strain curve at a certain reference temperature even for deformation rates that are too low to be achievable in practice in the tensile test experiment. For this purpose the tensile test is carried out at a deformation rate as low as technically possible and at a corresponding high temperature. Deformation rates and temperatures will be selected in such a way that Eq. 4.10 is fulfilled.

From such experiments one obtains finally the functions $E_0(23°C, \dot{\varepsilon})$ and $D_2(23°C, \dot{\varepsilon})$ at the reference temperature, e.g. $T_{ref} = 23$ °C, over a range of many orders of magnitude of the deformation rate. These functions, the so-called master curves for the moduli, and the time-temperature superposition principle, Eq. 4.10, fully describe the uniaxial deformation behaviour at constant deformation rate and isothermal conditions of the HDPE material in the non-linear visco-elastic range. Figure 4.3 shows the master curves.

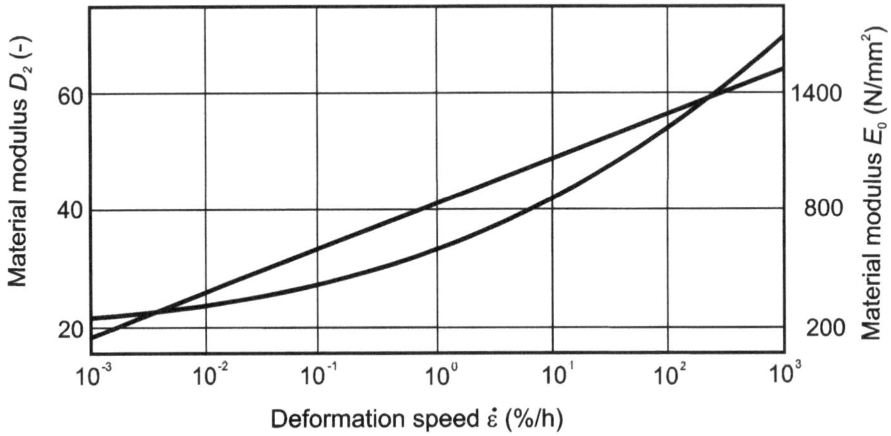


**Fig. 4.3.** The so-called master curves: parameters $E_0$ and $D_2$ as functions of deformation rate at 23 °C for the HDPE material. Source: (Schmachtenberg 1985)

So far the deformation rate served as a parameter to characterize the deformation behaviour. However, time can also be used for it. Let us consider the triple ($\sigma$, $\varepsilon$, $t$) attributed to each point ($\sigma$, $\varepsilon$, $\dot{\varepsilon}$) in the family of curves in Fig. 4.2 over the relationship:

[5] The form of the Eq. 4.10 is obtained because relaxation time and, at a specified deformation, velocity at which a certain stress relaxation is achieved, depends on temperature in accordance with Arrhenius' equation, see Sect. 5.1, Eq. 5.3.

$$t = \frac{\varepsilon}{\dot{\varepsilon}} . \tag{4.11}$$

By combining points with the same value of $t$ into a new curve, one obtains a new family of stress-strain curves, which are now parameterised against time $t$ (Fig 4.4). This so-called isochronous stress-strain diagram initially appears quite artificial. However, one can directly see the behaviour in the relaxation test at the test temperature from such a diagram. For a specified deformation value, one obtains data points $(\sigma_1, t_1)$, $(\sigma_2, t_2)$, $(\sigma_3, t_3)$ … etc. from this family of curves, which form the relaxation curve. Furthermore, by determining the stress which belongs to a specified strain in the relaxation test after a long test time, one can experimentally verify the only calculated tensile test stress-strain-curves for very small deformation rates. The isochronous stress-strain diagram also directly indicates the non-linearity of the visco-elastic behaviour; otherwise the curves would be straight lines.

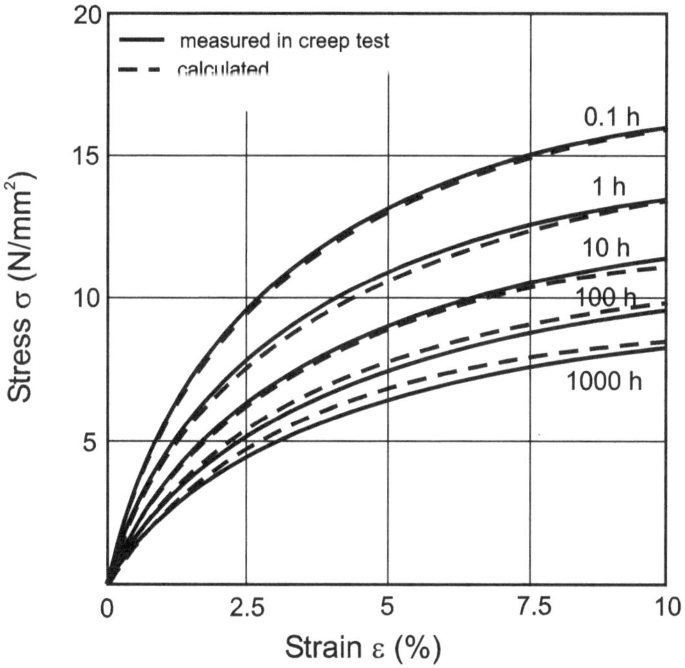


**Fig. 4.4.** Isochronous stress-strain diagram of HDPE material for uniaxial deformation. Continuous line: calculated from tensile test, dashed line: data measured in creep tests. In an uniaxial deformation force is applied along one direction and only stress and strain components along this direction are considered, e.g. as in the tensile test (Sect. 3.2.8). Source: (Schmachtenberg 1985)

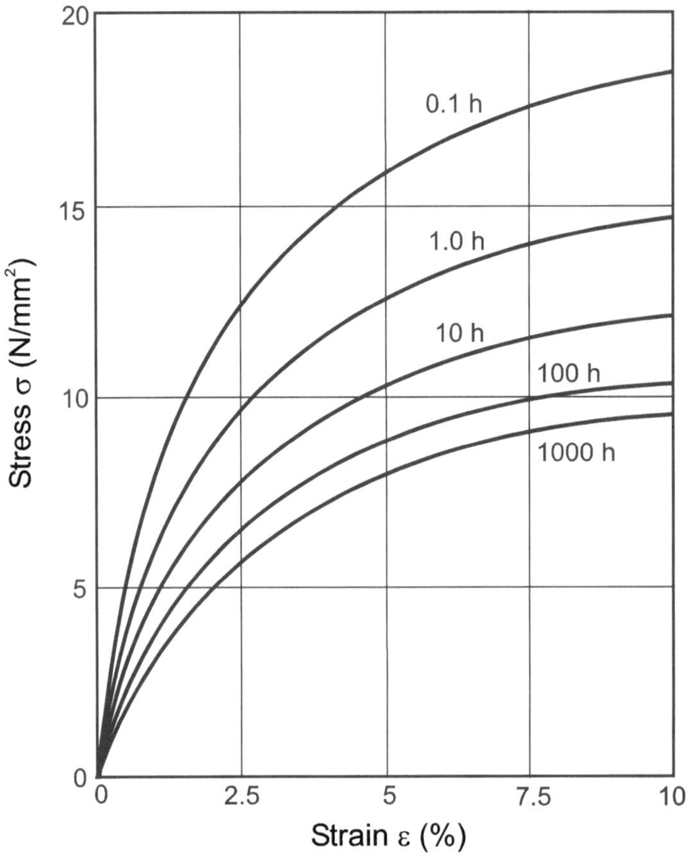


**Fig. 4.5.** Calculated isochronous stress-strain diagram of HDPE material for plane 1:1 state of stress, as it develops e.g. in the burst test (Sect. 3.2.9). Source: (Schmachtenberg 1985)

So far only a uniaxial deformation was discussed, i.e. the external force acts along a well defined direction and only the stress and strain components along this axis are considered. However, the deformation of a geomembrane during and after the installation is practically never uniaxial. Under field conditions the geomembrane is forced to fit to the shape of the subgrade, which may change for example due to settlements, or of building structures. Typically a plane stress state will be caused in the geomembrane (Sect. 5.3.3). Strictly speaking it is a three-dimensional planar state of stress. However, one can usually neglect the effect of the compressive stress by the overburden. In order to obtain a practically applicable description of the deformation behaviour, one has to include the effect of such a stress state. The moduli for uniaxial deformation as determined from the tensile test cannot be used directly for the calculation of the deformation behaviour at a plane state of stress because they depend on the state of stress. It is clear that, in a planar state of stress, a stress component, which corresponds to a certain strain along an axis, is substantially higher than in the case of uniaxial deformation, since contraction in the perpendicular direction is now prevented by another force component.

However, stress increases are not as serious as one would expect at first sight. The actual deformation behaviour in a multiaxial load was parameterized by G. Menges and E. Schmachtenberg using an additional conversion factor for the tensile test moduli dependent on the state of stress. Further details are described in (Menges and Schmachtenberg 1985). Figure 4.5 shows the isochronous stress-strain diagram determined for a plane 1:1 state of stress, i.e. $\sigma_1 = \sigma_2$ (= $\sigma$), $\sigma_3 = 0$ and $\varepsilon_1 = \varepsilon_2$ (= $\varepsilon$). Comparing Figs. 4.4 and 4.5, one can see, for example, that at a strain of 2.5 %, the stress is about 6 N/mm$^2$ for $t$ = 1000 hours in a plane 1:1 state of stress, while under uniaxial deformation, the stress would only be about 4.5 N/mm$^2$. The stress is thus higher by about a factor of 1.3. R. Koch and colleagues have measured stress increase factors between 1.1 (at a strain of 6 %) and 1.35 (at a strain of 1 %) (Koch et al. 1988).

Even with this correction the entirety of the mechanical behaviour of plastic geomembranes is not yet fully described. Isothermal conditions and an approximately constant deformation rate were assumed in the previous considerations. The model can be extended in such a way that also non-isothermal conditions and dynamic effects due to time-dependent deformation rates can be considered. For this, however, one should refer to the special literature (Bonten et al. 2000; Schöche 1997; Wanders 1999).

## 4.3 Deformation Behaviour in Tensile and Burst Testing

A HDPE geomembrane, as a thermoplastic product, consists of crystalline and amorphous regions below the melting temperature (see Sect. 2.2). At small deformations up to 0.25 % (limit of linear visco-elasticity), a linear and completely reversible visco-elastic deformation behaviour can be observed which is characterized by energy-elastic and entropy-elastic properties of the amorphous range. At larger strains, up to approximately 10 %, the non-linear visco-elastic deformation behaviour can be described by phenomenological models such as those of G. Menges and E. Schmachtenberg. However, in this strain range, deformations are already irreversible: the morphology is permanently changed. In particular, at strains above approximately 3 %, microcracks are formed along spherulite borders, which may trigger stress crack formation (see Sect. 5.3.4). At a strain of about 10 % a necking of the test specimen starts and a continuous "plastic flow" or "yielding" with rapidly increasing extension occurs (Strobl 1997). In the stress-strain curve of a tensile test (see Sect. 3.2.8, Fig. 3.9) a maximum is formed. The necking and the associated cross-section contraction are so large that the local stress further increases even when the force needed to maintain the constant elongation rate decreases. The maximum is called upper yield point; its coordinates are denoted as stress at yield and elongation at yield. With further increase of the strain the curve goes through a broad minimum which is called lower yield point. Above the lower yield point deformation occurs at essentially constant tensile force. The ratio of stress at yield to Young's modulus has about the same value for all polymers with high ductility: 0.025. One therefore assumes that the van der Waals bonds between the polymer chains break above the yield point and the chain segments can slip by one another. The morphology is thereby completely changed. The amorphous region becomes oriented and the fibrils within the spherulites are pulled apart and stretched. This process is often denoted as "cold drawing". Finally the force necessary to maintain the deformation rate increases again and the test specimen breaks after having reached several times its original length. Figure 3.9 shows typical stress-strain diagrams from a tensile test.

In the burst test, where a plane state of stress is forced upon the geomembrane test specimen, the maximum in the stress-strain curve, i.e. the yield point, is less pronounced and a tear occurs rapidly within the yielding region of the test specimen (see Sect. 3.2.9, Fig. 3.12).

From what has already been discussed in the sections above, it is sufficiently clear that all characteristic parameters derived from a stress-strain diagram (Young's modulus, (upper and lower) stress at yield and elongation at yield, stress at break and elongation at break) are functions of the deformation rate, test temperature and imposed state of stress.

## 4.4 Determination of Local Strain from the Contour Line

The extent of settlement in the subgrade of a geomembrane is described in various ways. Depth of settlement, radii of curvature along the contour line of settlement or local deformations is used. Ambiguity often exists, e.g. how a permissible local strain of HDPE geomembranes can be converted into a radius of curvature or in a depth of settlement, which are still acceptable.

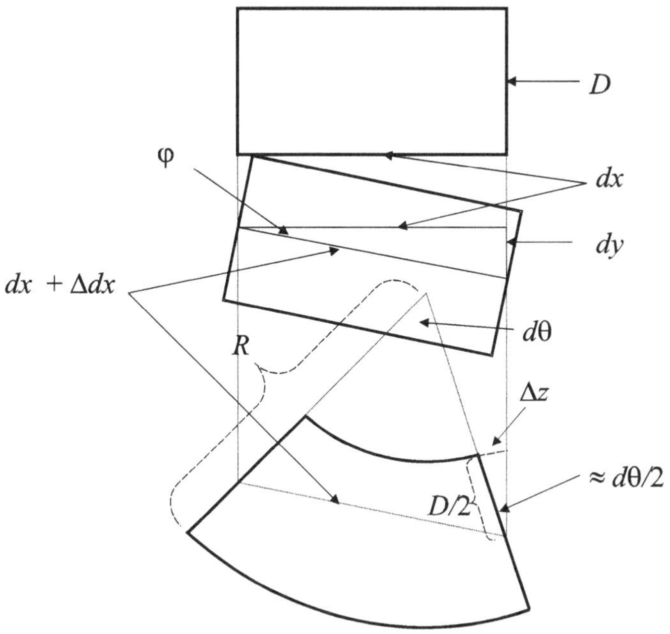


**Fig. 4.6.** Deformation of a volume element of the length $dx$ from a liner component of thickness $D$ when a settlement with contour line $y(x)$ is forced upon it. Note that the symbol $d$ is used here for the differential and not for thickness. The volume element experiences an elongation by some increment $\Delta dx$ and a bending by a small angle of $d\theta/2$ (associated with a radius of curvature $R$ of the contour line), which increases the outer fibre (and contracts the inner fibre) by two times a small increment $\Delta z$. The outer boundary fibre thus experiences an overall elongation from $dx$ to $dx + \Delta dx + 2\Delta z$

In the following the maximum local strain $\varepsilon_{max}$ in a liner component of thickness $D$ will be determined which arises when a new contour line $y(x)$ is forced upon the originally even liner by a settlement. Depths of settlement and radii of curvature can be read from the contour line. Therefore a relation between the extent of settlement and the local strain in the liner material is established. For this purpose a volume element of the liner material with a small edge length of $dx$ is considered, see Fig. 4.6. The figure illustrates that the volume element experiences an elongation $\varepsilon_L$ and a bending $\varepsilon_B$, which is characterised by the local radius of curvature $R$, when a new contour line is forced upon the liner. In addition, the following relations hold as can be easily seen in Fig. 4.6:

$$dx + \Delta dx = \frac{dy}{\sin \varphi}, \tag{4.12}$$

$$\varphi = \tan^{-1}\left(\frac{dy}{dx}\right) \tag{4.13}$$

and

$$\Delta z = \frac{D}{2}\frac{d\theta}{2}. \tag{4.14}$$

Finally one has, using the usual expression for the local radius of curvature of a contour line:

$$d\theta = \frac{dx + \Delta dx}{R} \approx \frac{dx}{R} \approx \frac{1}{\dfrac{\left(1+\left(\dfrac{dy}{dx}\right)^2\right)^{3/2}}{\dfrac{d^2y}{dx^2}}} dx \approx \frac{d^2y}{dx^2} dx. \tag{4.15}$$

Here $y'(x)$ as well as $dy/dx$ denote the first derivative of curve $y(x)$ and $y''(x)$ or $d^2y/dx^2$ stand for the second derivative.

The maximum local strain $\varepsilon_{max}$ in the volume element is given by:

$$\varepsilon_{max} = \frac{(dx + \Delta dx + 2\,\Delta z) - dx}{dx}. \tag{4.16}$$

Inserting Eqs. 4.12–4.15, a simple calculation yields:

$$\varepsilon_{max} = \left(\frac{y'(x)}{\sin(\tan^{-1}(y'(x)))} - 1\right) + \frac{D}{2} y''(x). \tag{4.17}$$

Where the elongation strain and the bending strain are given by:

$$\varepsilon_L = \frac{y'(x)}{\sin(\tan^{-1}(y'(x)))} - 1 \qquad \varepsilon_B = \frac{D}{2R} = \frac{D}{2} y''(x). \tag{4.18}$$

These are well-known formulae from the technical theory of bending of beams and bars (Housner and Vreeland 1966). They may be used for the calculation of maximum local strains in liners, for instance in the case of settlement in compacted clay liners (Scherbeck and Jessberger 1992; Scherbeck 1993). In applying these equations it is assumed that the neutral fibre lies in the middle of the liner element. This, however, is not necessarily always the case. Thickness *D* in Eqs. 4.17 and 4.18 should therefore be corrected by some factor *a* if necessary, which is, however, of the order of magnitude $10^0$. For example, R. Scherbeck and H. L. Jessberger give a value $a = 4/3$ for compacted clay liners.

**Table 4.1.** Limits of local tensile strain $\varepsilon_{max}$ for various liner materials (Simon and Müller 2004). Tensile strain may be caused by bending or elongation due to subsidence or indentations. The data were taken from the references indicated

| Sealing material | $\varepsilon_{max}$(%) | Remarks |
|---|---|---|
| Compacted clay liner | 0.1–3 | Strongly varying with soil properties (Scherbeck and Jessberger 1992) |
| Asphaltic concrete (base course) | 1.75 | Limit estimated from tensile tests, aging effects were not taken into account (Deutsches Asphaltinstitut 1996) |
| Asphaltic concrete (sealing layer) | 0.85 | From tensile tests, Aging effects were not taken into account (Deutsches Asphaltinstitut 1996) |
| HDPE-GM | 6 | Strain limit to exclude stress cracking (Müller, 2001; Seeger and Müller, 2003) |
| HDPE-GM | 10–15 | Strain limit with respect to yielding (Müller, 2001) |
| Geosynthetic clay liner | 10–15 | Limit estimated from short term tensile tests, aging effects or repercussions on shear strength were not taken into account (Koerner and Daniel 1994) |

Regardless of such details, however, the following conclusions result from this general formula 4.17 for the strain in a liner component. When settlement in the subgrade of a liner occurs, the radii of curvature will always be very large compared to the few millimetres thickness of the geomembranes. The bending stress can then be discarded for the geomembrane and the developing local strain equals $\varepsilon_L$. This generally does not apply to compacted clay liners, to asphaltic concrete layer or other liner components with thicknesses of a few centimetres up to some decimetres. Depending on the settlement process and the radii of curvature, the bending stress $\varepsilon_B$ can substantially contribute to the total local strain in these liners. The very conservatively specified value for the limiting strain in the HDPE geomembrane (GM) of 3 % (basal liner) and 6 % (capping systems), permissible over the long term, is usually much larger than acceptable strains for compacted clay liners (0.1 to 3 %, highly dependant on the earth material) or for the asphaltic concrete liners (0.85 %), see Table 4.1. An HDPE geomembrane is therefore capable of coping without any damage with substantially larger settlements depths and settlement troughs with large curvatures than all other liner components except perhaps geosynthetic clay liners.

With a thickness of the geomembrane of $D = 2.5$ mm and a limiting strain value of 3 % the permissible radius of curvature is about 4 cm for pure bending stress. This value is surely not exceeded by normal processing and installation impacts. However, it is important to note that it may be exceeded, when indentations in the geomembrane are caused by pieces of gravel or other parts of the subgrade material below or the backfilling material above the geomembrane. Small indentations caused by the grains of a gravel layer result namely in substantial local bending stresses in thin geomembranes. The analysis of typical contour lines of indentations in the geomembrane show that, due to the emerging radii of curvature, the strain $\varepsilon_B$ provides even the substantial contribution to the total local strain (Fig. 8.3).

The aforementioned can be summed up as follows: far larger settlements are permissible over the long term for the HDPE geomembranes than for nearly all other liners components, in particular for compacted clay liners or asphaltic concrete liners. Geomembranes must, however, be reliably protected from penetrative deformations with small radii of curvature, such as those due to pieces of gravel. Types, appropriate design and tests of protective layers will be dealt with in Chap. 8.

## References


Bonten et al. (2000) Spritzgußteile mit FEM auslegen. Kunststoffe 90: 99–104

Brunt NA (1966) Statistical Theory of the Glas - Rubber Transition of High Polymers. Kolloid-Z u Z Polymere 209: 5–19

Deutsches Asphaltinstitut (ed) (1996) Asphaltbeton für Deponieabdichtungen, Deutsches Institut für Bautechnik: Allgemeine bauaufsichtliche Zulassung "Deponieasphalt für Deponieabdichtungen der Deponieklasse II". Deutsches Asphaltinstitut e.V., Bonn

Housner GW and Vreeland T (1966) The Analysis of Stress and Deformation. The Macmillan Company, New York

Koch R et al. (1988) Langzeitfestigkeit von Deponiedichtungsbahnen aus Polyethylen. Müll und Abfall 20: 3–12

Koerner RM and Daniel ED (1994) A suggested methodology for assessing the technical equivalence of GCLs to CCLs. In: Gartung E (ed) Geokunststoff-Ton-Dichtungen (GTD). Grundbauinstitut der Landesgewerbeanstalt Bayern (LGA), Nürnberg, pp 59–85

Menges G and Schmachtenberg E (1985) Das Verformungsverhalten von Kunststoffdichtungsbahnen bei mehrachsiger Beanspruchung. In: Knipschild FW (ed) Deponieabdichtungen mit Kunststoffdichtungsbahnen, Müll und Abfall, Beiheft 22. Erich Schmidt Verlag, Berlin, pp 68–74

Scherbeck R (1993) Verformungsnachweis für mineralische Abdichtungsschichten – Erläuterung der Grundlagen, Hinweise zur Anwendung. In: Schmidt W (ed) Oberflächenabdichtung für Deponien, LWA-Materialien Nr 4/93. Landesamt für Wasser und Abfall, Nordrhein-Westfalen (LAW), Düsseldorf, pp 15–42

Scherbeck R and Jessberger HL (1992) Zur Bewertung der Verformbarkeit mineralischer Abdichtungsschichten. Bautechnik 69: 497–506

Schmachtenberg E (1985) Die mechanischen Eigenschaften nichtlinear viskoelastischer Werkstoffe. Fakultät für Maschinenwesen der Rheinisch Westfälischen Technischen Hochschule Aachen, Aachen

Schöche N (1997) Wärmespannungen in Bauteilen aus Thermoplasten. Shaker Verlag, Aachen

Simon FG and Müller WW (2004) Standard and alternative landfill capping design in Germany. Environmental Science & Policy 7: 277–290

Strobl G (1997) The Physics of Polymers. Springer Verlag, Berlin, Heidelberg

Wanders M (1999) Beitrag eines Modells zur Beschreibung des mechanischen Verhaltens nichtlinearer viskoelastischer Werkstoffe mit mehrachsiger Beanspruchung. Shaker Verlag, Aachen

# 5 Long-term Behaviour

## 5.1 Ageing

The long-term behaviour of a construction product, which must fulfil a certain function in a building structure, is determined by the external forces acting upon the product and the specific ageing processes in the selected material. Interaction of these forces (deformations, chemical attack, weathering etc.) and changes in the material properties due to ageing processes finally lead to failure. Failure here generally means that the attributed function is no longer guaranteed. The time taken until failure is called the service lifetime (or working life) of the product.

The service life of a product is determined by functional reserves on the one hand. Functional reserve is defined here as the difference between the capacity of the material to resist an action and the actual action imposed on the product i.e. by the extent to which it can resist without suffering any damage. The properties of the material and the functional design of the product determine its reserves. These reserves are often investigated using short-term destructive tests in which resistance of the material in the "ultimate limit state" of failure is determined. On the other hand, service lifetime depends on the ageing processes in the material and their influence on relevant product properties, so called degradation. Under ageing we understand the totality of all irreversible chemical and physical processes which occur in a material in the course of its use (in accordance with DIN 50035-1:1989 *Terms and Definitions used on Ageing of Materials; Basic Terms and Definitions* and DIN 50035-2:1989 *Terms and Definitions used on Ageing of Materials; Examples Concerning Polymeric Materials*). The ageing phenomena may have internal causes (e.g. additional crystallisation in partially crystalline thermoplastics or distillative and structural ageing of bitumen in asphaltic concrete): materials age when composition and structure is not in thermodynamic equilibrium. Ageing processes are however caused primarily by external agents such as chemical effects, mechanical loadings, heat, moisture change or frost. For example, they encourage polymer oxidation in geomembranes or in geotextile components of geosynthetic clay liners or bitumen oxidation in asphaltic concrete, stress crack formation in partially crystalline thermoplastics, Na-Ca ion exchange in geosynthetic clay liners, corrosion of steel sheet piles etc. Interaction between various causes of ageing may substantially accelerate the ageing processes.

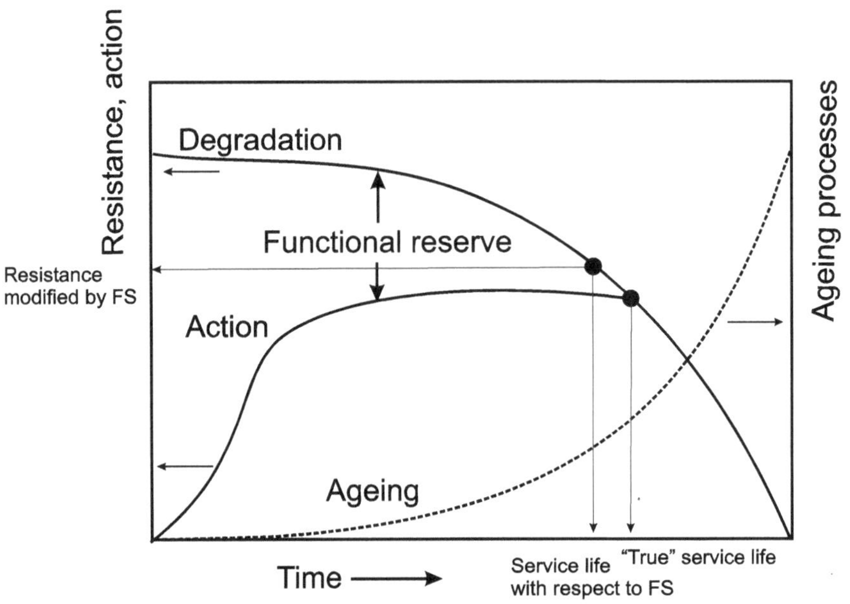


**Fig. 5.1.** Ageing impairs functional reserves. Depending on the level of action, failure occurs sooner or later. Slow ageing or large functional reserves result in long service lives. Applying a factor of safety (FS) to the resistance means to implicitly define a certain service lifetime of the construction product. The questions whether the "true" service lifetime of a certain geosynthetic product and its service lifetime with respect to the choice of FS are actually larger than the required service life are seldom considered explicitly in geotechnical design

The lower the functional reserves and the faster the degradation process by ageing, the shorter the service lifetime. The larger the functional reserves and the more "resistant" the material is to the ageing processes, the longer the product can withstand stresses and fulfil its functioning. Figure 5.1 illustrates the decisive relationships for service life. When the actions are complex and cannot be reliably specified in detail, as is the case with landfill liners, proper material selection and design must provide as much functional reserves for the component as possible (high chemical resistance, generous limiting strains and robustness under mechanical loadings) and the ageing processes may have an effect on the materials properties only after very long periods of time.

The term durability frequently emerges in connection with the characterisation of the long-term behaviour of construction products. Durability is a somewhat ambiguous term. Generally, we understand it as resistibility in the long term of a product of a certain material to changes by physical, chemical and biological influences in performance properties. The attribute "durable" is used to express expectations of long service life under normal actions. However, it is often used specifically as meaning the extent of functional reserves under extreme action, i.e. synonymous to material resistance. The attribute "resistant" is used to express expectations of very high functional reserve against certain actions. One should distinguish between durability and resistance. However, they are interrelated as can be seen by Fig. 5.1 and it is often not possible to classify test procedures unambiguously as testing durability or testing resistance. Based on the general definition, one attributes durability or resistance to atmospheric effects (weathering), to ionizing radiation, to thermal and thermal-oxidative degradation, to chemical effects, to biological effects and to mechanical actions. Biological effects generally mean effects caused by organisms (in civil engineering, they are primarily plant roots, burrowing animals and micro-organisms). Time-to-failure determined in so-called long-term tests is often used to quantitatively characterise durability of a material. In a long-term test, test specimens prepared from the material using a specified technique are subjected to certain test conditions (representative for the actions under field conditions) and the aging processes are accelerated by elevated test temperatures. The time is measured until a specified damage criterion is fulfilled, the time-to-failure. Service lives can sometimes be extrapolated from times-to-failure, and, conversely, requirements for the durability of the material in behavioural terms over the long-term test can be derived from required service lives.

This chapter deals with ageing processes of HDPE geomembranes, with their durability and the resulting service lives. Implicitly, statements on material resistance and functional reserves are always implied. Explicit statements about resistances of HDPE geomembranes to individual influences (chemical resistance, resistance to thermo-oxidative degradation, resistance to stress crack, resistance to weathering, biological resistance) have been made in Chap. 3 in connection with the respective resistance test methods. A concise summary of durability and resistance of plastics and rubber is published by (Dolezel 1978).

The evaluation of long-term behaviour of polymeric material products must be preceded by ideas about type and duration of stresses. In geotechnical engineering, very long service lives are often required. A wide range of stresses may occur and change over time. The use of geomembranes in landfill engineering may serve as an example. In landfills, various biological, chemical and physical degradation and conversion processes can take place over many decades. The waste body contains a large amount of water: rainfall infiltrates the still open waste body or is supplied in a planned manner in order to enhance the processes. Heat, landfill gas and leachate contaminated with organic and inorganic pollutants are generated in this phase. The increasing weight of the waste body leads to subsidence in the landfill base. The surface profile of the waste body also changes due to the conversion processes. Chemicals, deformations and heat exert influences on the geomembrane. These processes gradually fade away. In the waste body, however, a considerable amount of pollutant remains (especially heavy metals). Therefore, one can differentiate between two phases: an "active" phase (which more or less corresponds to the operational and aftercare phases) which is followed by a temporally unlimited final disposal phase. Even the first phase lasts for a very long time (many decades).

Entirely different approaches about the requirements on the service life of construction products can be deduced from such stress scenarios. Table 5.1 shows a categorisation of service lives for buildings and construction products used in accordance with the Guidelines of the European Organisation for Technical Approvals (EOTA), www.eota.be, (EOTA 1999). A building can achieve a very long service life when it is regularly maintained and its components are repaired or replaced, even if repair is difficult to perform in certain cases. Under these conditions, the construction products used only need to exhibit medium (25 years) to normal (50 years) service lifetimes.

In Germany the following approach is mandatory for landfill lining. If a so called temporary landfill liner is installed, which is considered reparable and replaceable (after a few decades), control, repair and replacement of the liner belong to the operational phase of the landfill and the landfill operator is free to chose appropriate plastic geomembranes. However, final landfill closure and transition into the aftercare phase is then not possible according to the German regulations.

"Final" capping systems and, in particular, basal liners of landfills must be classified in this sense as buildings with very long service lives (≥ 100 years) which are, with the transition into the aftercare phase, irreparable for a very long time from the operator's and the permitting authority's point of view, and by no means is their replacement economically feasible. Therefore building components and products used, including geomembranes, must exhibit a long service lifetime and certified HDPE geomembranes have to be used.

For lining tunnels, canals, reservoirs or dikes, long service lives must be envisaged and an economically justifiable repair is only possible in exceptional cases. Therefore, plastic geomembrane with approved long service life should be used in these circumstances. In many other geotechnical applications medium to normal service lives are sufficient, since the structures can easily be controlled and repaired. An economic calculation may, however, lead to the conclusion that construction products with long service lives should be used even in these cases.

**Table 5.1.** Service life (or working life) of buildings (or works) and required service lifetime of the construction products used (EOTA 1999)

| Assumed working life of works (years) | | Working life (service lifetime) of construction products (years) | | |
|---|---|---|---|---|
| Category | Life time | Category | | |
| | | Repairable or easily replaceable | Repairable or replaceable with greater effort | Lifelong[2] |
| Short | 10 | 10 | 10 | 10 |
| Medium | 25 | 10 | 25 | 25 |
| Normal | 50 | 10 | 25 | 50 |
| Long | 100 | 10 | 25 | 100 |

[2]) When not repairable or replaceable "easily" or "with some effort"

The identification of high resistance with high durability leads to the "factor of safety" design approach. Data from resistance tests involving short-term simple experimental procedures is frequently used as dimensioning data in geotechnical structural design[1]. Degradation effects are then allowed for by modifying the resistance by a factor of safety for each relevant ageing process. The definition of a safety factor implies that during the service life of a construction a certain property of the construction material – e.g. shear strength – will not fall short of a well defined fraction of its initial resistance value. However, the essence of the ageing processes is that material properties will steadily or suddenly change with time and sooner or later will fall below such well defined limits. Therefore, with the definition of a certain factor of safety necessarily a certain lifetime of the construction product under consideration is implicitly defined (Fig. 5.1). The use of safety factors with geosynthetics relies on the (often unspoken) assumption that in normal geotechnical applications the lifetime of all plastic material used in geosynthetics is much larger than the required service lifetime of the geotechnical construction (typically some decades).

[1] Ageing and degradation are phenomenons still quite unknown to the geotechnical design. Among other things this has historical reasons, because the gravedigger was the first geotechnician and his constructions are, as everybody knows, of highest durability: „The houses he makes last till doomsday" (Hamlet, V, 1).

However, there are long-term geotechnical applications, like landfill lining, where the above assumption is certainly not valid and, hence, the ageing behaviour of the geosynthetic products has to be investigated explicitly.

Long-term behaviour, i.e. the relationships between action and degradation due to ageing for a certain construction product can be assessed using three methods:

1. *Case histories:* systematic quantitative and qualitative evaluation of previous experience; reference (Hutten 1992) provides an example for geomembranes and the book "Geosynthetic Case Histories" (Raymond and Giroud 1993) for geosynthetics in general,
2. *Durability tests*: performance based long-term tests under accelerated test conditions; pipe pressure tests (Gebler 1989) and long-term shear strength test (Seeger et al. 2000) discussed in Sects. 3.2.13 and 3.2.18 can serve as illustrations
3. *Modelling*: theoretical qualitative and quantitative model considerations based on the precise knowledge of material properties and the physics and chemistry of potential ageing mechanisms; theoretical investigations into the desiccation of compacted clay liners within the BMBF (German Federal Ministry for Education and Research) integrated research programme "Advanced Landfill Liner Systems" can be mentioned here as an example (August et al. 1997).

Theoretical model considerations are also involved when the results of long-term tests obtained under accelerated test conditions are analysed and extrapolated to application conditions. Increase in concentration of influencing substances or increase in mechanical loads may be used to accelerate ageing processes. However, the most important accelerating test condition is increase in temperature. The following consideration can illustrate this (Landau and Lifshitz 2000).

Ageing can manifest itself gradually and continuously over time – when the chemical and physical processes associated with ageing directly affect the function of the material. Ageing can also occur very abruptly, when degradation is initially hidden and affects function only after a certain degree of accumulation has occurred. An example of the latter case is gradual depletion of antioxidants as is frequently found in polyolefin materials. In both cases the change in a property $P$ observed over time can often be expressed as an exponential function:

$$P(t) = P_0 e^{-kt} . \qquad (5.1)$$

When a certain threshold value $P_S$ is attained under the stress encountered, failure occurs. The time-to-failure $t_S$ can therefore be expressed as:

$$t_S = \frac{1}{k} \ln \frac{P_0}{P_S}. \tag{5.2}$$

If the ageing process mechanism does not change over the considered temperature interval, the time-to-failure versus temperature function is governed essentially by the speed of ageing $k$. The function is normally a simple Arrhenius law dependency:

$$k = k_0 e^{-E/RT}. \tag{5.3}$$

Such simple equations can satisfactorily characterise the macroscopic property changes even if the molecular processes causing them on a microscopic scale are extremely complicated and by no means obey simple kinetic laws. This can be illustrated (more so than proved) by a simple chain of reasoning. A process causing a given ageing manifestation can be considered basically to consist of many localized degradation processes occurring in discrete small, but still macroscopic subsystems. In each local process a transient ordered non-equilibrium state in the subsystem considered undergoes transition with a certain degradation rate $r$ to a stable equilibrium state. Each local process begins in its subsystem only after the subsystem has a certain internal energy $E$, the apparent activation energy. $N(E)$ is the number of subsystems activated in this manner. In any given time interval $\Delta t$, $\Delta N$ subsystems will have undergone transition:

$$\Delta N = r \cdot N(E) \cdot \Delta t. \tag{5.4}$$

When the subsystem is at thermal equilibrium with its surroundings at temperature $T$, the probability $w$ of a given subsystem to have the internal energy $E$ can be expressed as:

$$w = \frac{e^{-E/RT}}{Z(T)}. \tag{5.5}$$

Where $R$ is the universal gas constant (8.314 J/mol K) and $Z(T)$ is a (temperature-dependent) normalisation factor. This equation is a universally valid thermostatistic relationship. $N(E)$ can then be expressed as:

$$N(E) = \frac{e^{-E/RT}}{Z(T)} N. \tag{5.6}$$

Where $N$ is the total number of subsystems still in the initial state. Substituting Eq. 5.6 into Eq. 5.4, the rate of change of $N$ is:

$$\frac{dN}{dt} = \frac{r \cdot e^{-E/RT}}{Z(T)} N . \tag{5.7}$$

$P$ is proportional to $N$. The exponential relationship for $P$ as a function of time given in Eq. 5.1 with the relationship for the speed $k$ as a function of temperature given in Eq. 5.3 therefore follows from Eq. 5.7.

Time-to-failure as a function of temperature is obtained by substituting the Arrhenius equation 5.3 into Eq. 5.2:

$$t_S(T) = \left( \frac{1}{k_0} \ln \frac{P_0}{P_S} \right) e^{E/RT} . \tag{5.8}$$

As can be seen from Eq. 5.8, a plot of logarithm of time-to-failure versus reciprocal absolute temperature (such a plot is referred to as an Arrhenius diagram) is a linear relationship, which can be extrapolated to application temperatures. Knowing the activation energy, service lifetime $t_1$ for application temperature $T_1$ can be calculated from service lifetime $t_2$ measured at high temperature $T_2$:

$$t_1 = t_2 \, e^{(E/R)[1/T1 - 1/T2]} . \tag{5.9}$$

In accordance with Eq. 5.7, $k_0$ in Eq. 5.8 should also be temperature dependent. However, when $RT < E$ – which is normally the case – this temperature dependence is very slight and is almost always negligible in comparison to the factor exp(-$E/RT$). In the event of slight divergence from the Arrhenius linear relationship, a better data fit is obtained by the equation $k = T^n \exp(-E/RT)$, where $n$ is a small negative or positive coefficient.

As noted previously, an essential prerequisite for any extrapolation is that the mechanism of the ageing process does not change over the temperature interval considered. For many plastics this is not the case at temperatures above 100 °C. In polyethylene for example, the degree of crystallinity – a characteristic which substantially affects many processes – decreases considerably above 100 °C; in some polypropylene resins the effectiveness of certain stabilizers changes significantly at around 100 °C. Other abrupt changes at or near this temperature are known as well and could be given as further examples. As a result, Arrhenius extrapolation of data determined at test temperatures above 100 °C is not normally meaningful for prediction of performance at normal application temperatures. Testing must be conducted at lower temperatures as well – with the attendant disadvantage of increased test duration.

The apparent activation energy of many chemical and physical processes is somewhere in the vicinity of 50 kJ/mol. At normal application temperatures Eq. 5.9 gives rise to the Van't Hoff rule for chemical reactions – known since the end of the 19th century – which states that increasing the temperature by 10 °C increases reaction rate by a factor of 2–3. This rule is astonishingly accurate for many ageing processes in plastics. For thermoplastics a factor is often empirically found to be around 2.5. Therefore to reliably demonstrate service lifetimes of at least 100 years at typical geotechnical application temperatures of 15–20 °C, tests conducted at 80 °C must be carried out for at least 1 year. An overview of the application of the Arrhenius extrapolation for geosynthetics can be found in (Koerner et al. 1992).

However, in order to reliably ensure very long service lives of one hundred or even some hundred of years, extensive experience with the plastic material and results of various special long-term tests are needed as a rule. Limited field experience and pure model considerations alone are not sufficient in view of the complexity of the processes in order to obtain more than mere speculative considerations on long-term behaviour.

## 5.2 Oxidative Degradation

Polyolefins, such as polyethylene or polypropylene, can be oxidized by oxygen. Oxidation of hydrocarbons by molecular oxygen is also generally called auto-oxidation, because, once initiated, there is an intrinsic mechanism which "fuels" the oxidation process. This includes very complex radical reaction chains which lead to decomposition and cross-linking of polymer chains and to the formation of low-molecular reaction products. These changes at the molecular level will induce changes of the macroscopic material properties. Finally, embrittlement and a drastic loss of strength and flexibility ensue. One speaks of oxidative degradation. A comprehensive and detailed overview on this topic is given in (Gugumus 1990; Zweifel 2001) and (Emanuel and Buchachenko 1987).

The oxidation process depends on the type of material, its manufacturing process and morphology. Oxygen cannot penetrate into the crystalline ranges of polyethylene; therefore oxidation takes place only in the amorphous phase. At the beginning of oxidation, branching and cross-linking of long polymer chains dominate. In polypropylene, polymer chain decomposition occurs more frequently from the beginning, crystalline ranges are also affected but to a lesser extent then the amorphous phase. Generally, morphology has a very great influence on the oxidation process. Orientation of the polymers, caused for example by the cold drawing of fibres and slit films, obstructs the kinetics of the oxidation process. One speaks of structural or physical stabilisation against oxidative degradation.

Since the oxidation process is initiated by free chemical radicals, which are only very rarely formed under normal conditions, it starts only extremely slowly. However, the initiation of oxidation process is greatly accelerated by the influence of UV radiation or high temperatures, e.g. those used at polymer processing. One speaks of photo-oxidative degradation or thermal-oxidative degradation. Addition of antioxidants, which are effective even at low concentrations, stabilizes plastics against oxidative degradation as discussed in Chap. 2.

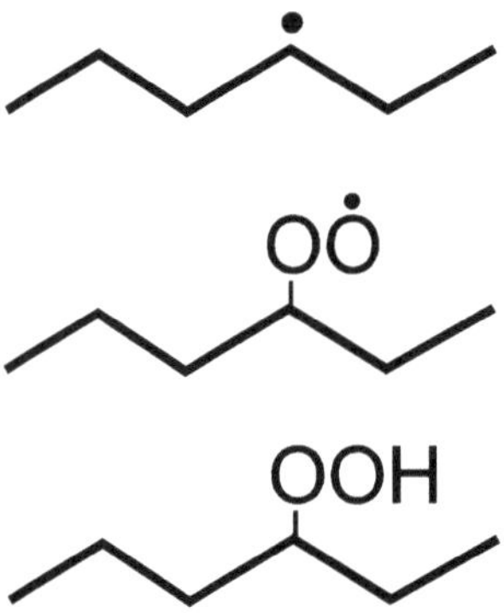


**Fig. 5.2.** Functional groups in the polyolefin polymer chain relevant for auto-oxidation: Alkyl-radical (above), which reacts very fast with molecular oxygen, forming the peroxy radical (middle). Further abstraction of hydrogen leads to the hydroperoxide (below). Decomposition of hydroperoxides initiates new reaction chains

## 5.2.1 Auto-Oxidation of Non-Stabilised Polyolefins

Kinetics of the complex reaction process of auto-oxidation can be described by the mechanism of the so-called kinetic chain reaction[2]. In a kinetic chain reaction, reactive reaction products (radicals) are first formed in one or more initiation reactions, in the so-called chain start or chain initiation, and these products can then enter so-called propagation reactions. The propagation reactions are characterised by the fact that the radicals are

[2] The term chain here sometimes refers to a sequence or chain of chemical reactions and sometime to the polymer chain. This difference must be kept in mind in the following discussion.

first consumed and then reproduced. In this way a reaction cycle results which can be iterated many times (Fig. 5.3). The kinetic chain reaction comes to a standstill only when the reactive products are finally consumed in a so-called chain termination reaction (Frost and Pearson 1961). A propagation reaction can branch into new propagation reactions where new reaction cycles emerge. Kinetic chain reactions can be classified depending on the number of reaction carriers, number of branches and order of the termination reaction. Special kinetic laws describe concentration of the reactant and reaction products as a function of time and the apparent activation energy of the overall reaction.

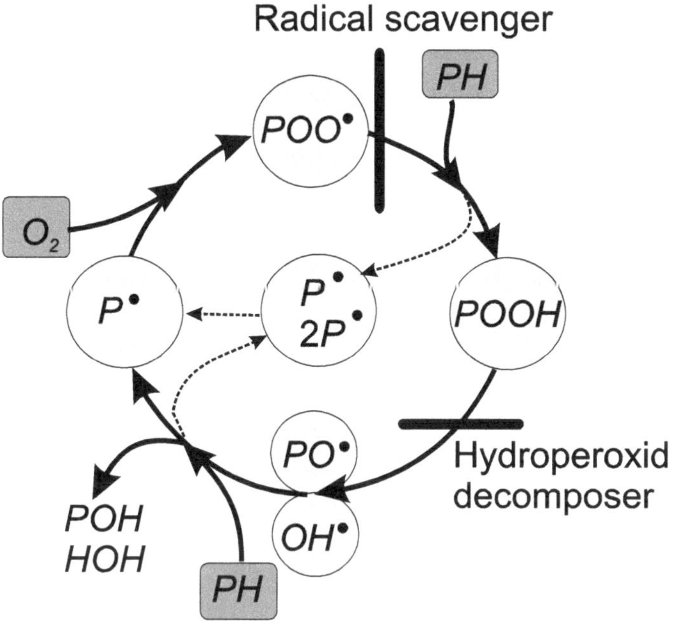


**Fig. 5.3.** Schematic illustration of the auto-oxidation reaction cycle. The cycle can be interrupted by radical scavengers reacting with peroxy radicals and by hydroperoxide decomposers forming inert reaction products with hydroperoxides

In the auto-oxidation of polyethylene or polypropylene, both chain initiation and chain propagation, chain branching and chain termination consist of several potential chemical reactions. In the following, only the most significant reaction steps will be indicated. *P* denotes here a polymer chain. $^{\bullet}$ is used as a symbol for an unsaturated chemical bond with an unpaired electron. Polymer molecules, where such unpaired electrons occur, are called macro radicals. The remaining letters denote the chemical elements in the common way.

The start or initiation reaction consists of the formation of alkyl radicals $P^{\bullet}$, e.g. by decomposition of a polymer chain or hydrogen abstraction due to thermal decay, radiation, or the reactivity of oxygen or catalyst residues. The alkyl radicals react with molecular oxygen to a peroxy radical $PO^{\bullet}_2$, which again forms an alkyl radical and a hydroperoxide $POOH$ with a polymer chain. This propagation reaction can repeat itself. The reaction chain branches off over reactions which lead to the decay of hydroperoxides, since an alkyl radical is formed again which becomes the starting point of a new chain propagation reaction. Thus new reaction cycles emerge permanently and the decay reactions of hydroperoxides become the actual trigger for the auto-oxidation process. Metal ions can very effectively catalyze the decay of hydroperoxide (Müller 1990). Contact with metals, e.g. during processing or wetting by water with dissolved metallic ions (e.g. copper ions), can therefore accelerate oxidation. Chain termination is finally chiefly actuated by reactions which consume peroxy radicals. All in all, the following, very simplified reaction pattern results:

Chain start:

$$(\text{Trigger (therm., mech. stress, } O_2, \text{catalyst residues)} \xrightarrow{w0} P^{\bullet} + \ldots \quad (5.10)$$

Chain initiation:

$$\begin{array}{lllll} P^{\bullet} & + & O_2 & \xrightarrow{k1} & PO_2^{\bullet} \\ PO_2^{\bullet} & + & PH & \xrightarrow{k2} & POOH + P^{\bullet} \\ & \ldots & & & \end{array} \quad (5.11)$$

Chain branching:

$$\begin{array}{l} POOH \; + \; PH \xrightarrow{\delta k3} P^{\bullet} \; + \; PO^{\bullet} \; + \; H_2O \\ \ldots \end{array} \quad (5.12)$$

Chain termination:

$$\begin{array}{l} PO_2^{\bullet} \; + \; PO_2^{\bullet} \xrightarrow{kt} POOP \; + \; O_2 \\ \ldots, \end{array} \quad (5.13)$$

where $w_0$ is the rate at which radicals, which initiate a kinetic chain reaction, emerge. $k_1$, $k_2$, $k_3$, $k_t$ are the reaction constants of the propagation, branching and termination reactions. $\delta$ denotes the fraction of alkyl radicals, which result in actual new reaction cycles in the branching reaction Eq. 5.12.

The radicals formed in various start, propagation and branching reactions are not only capable of binding with molecular oxygen or abstracting hydrogen from the polymer chains, as in reaction Eq. 5.11 and Eq. 5.12, but also splitting the polymer chain in decomposition reactions. The chain termination reactions can lead to cross-linking of polymer chains. The decay of the peroxy radical $PO^{\bullet}$ is primarily responsible for decomposition of the polymer chains. Again an alkyl radical $P^{\bullet}$ and a polymer chain with a carbonyl group are formed as fragments.

$$PO^{\bullet} \xrightarrow{k} \overset{|}{\underset{|}{C}}=O + P^{\bullet} . \tag{5.14}$$

Thus the continuous formation of radicals and reaction products in the oxidative kinetic chain reaction induces changes in the molecular mass distribution, and this affects the mechanical properties in the long run. Figure 5.4 shows an example for polyethylene and polypropylene samples, which were aged in an oxygen atmosphere at 130 °C. In addition to the measuring of the change in molecular mass distribution and in mechanical properties, the oxidation process can also be directly observed by the measurement of the change in the concentration of the basic reaction products. Development of functional groups, such as carbonyl groups and the hydroxyl group of hydroperoxides is characteristic of the oxidation process. The hydroxyl group and the carbonyl groups absorb infrared radiation at characteristic wavelengths. In HDPE resins, the maximum of the infrared bands of free and associated hydroperoxides is at 3410 $cm^{-1}$ and 3550 $cm^{-1}$ and that of the carbonyl group 1710 $cm^{-1}$. The change of concentration of these reaction products can be measured with the help of infrared spectroscopy.

The auto-oxidation of non-stabilised polyethylene and polypropylene shows a typical function of time (Gugumus 1996b, 1997; Zweifel 2001). During an initial phase the concentration of the functional groups increases only gradually. There are no changes in the properties of the plastic. The period of this phase is called induction time $t_{i,1}$. This induction time must be distinguished from an induction time $t_{i,2}$ – associated with stabiliser depletion – to be discussed below, which additionally applies to stabilised materials. After the induction phase the concentration of the reaction products increases exponentially, and finally, carbonyl groups and hydroperoxides develop at a constant rate. The properties of the plastic then change drastically. In particular, elongation at break reacts very sensitively to the accelerating oxidation process. Elongation at break drops drastically after the induction phase. The material becomes more and more brittle and finally, completely loses its strength (Fig. 5.29).

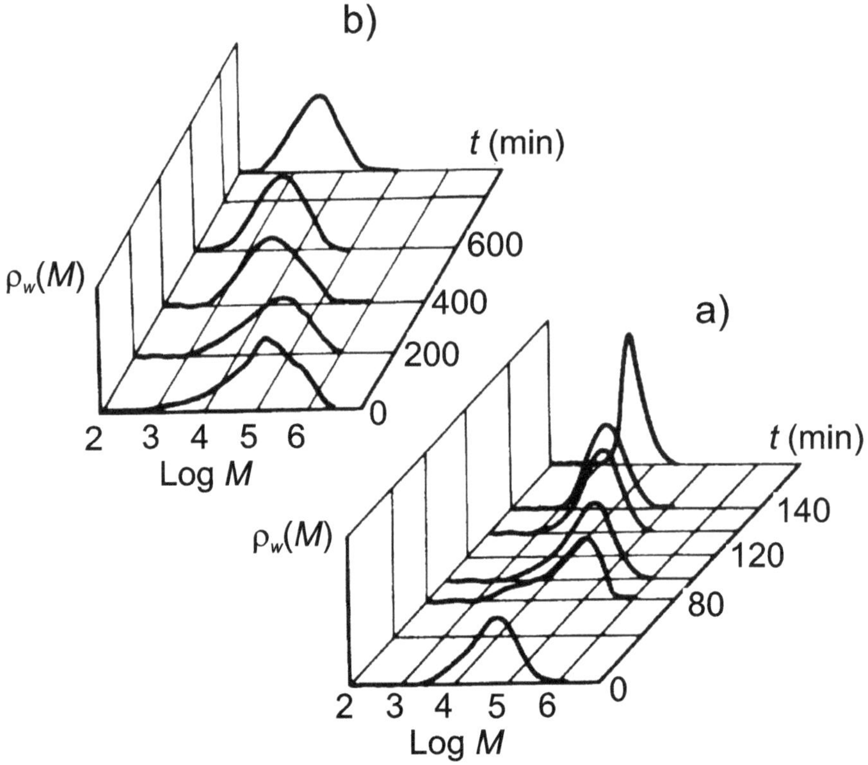


**Fig. 5.4.** Change of the molecular mass distribution after oven ageing (*t*: ageing time) at 130 °C in an oxygen atmosphere, $p_{O_2}$= 760 mm Hg. *a*: polypropylene, the distribution is not only shifted towards low molecular masses, but it also becomes narrower. *b*: polyethylene, the distribution is shifted towards low molecular masses, the width changes only slightly. Source: (Emanuel and Buchachenko 1987)

The details of the temporal progress, e.g. duration of the induction period, and the oxidation rates depend on a number of boundary conditions, such as manufacturing process and, in connection with it, catalyst residues, oxygen partial pressure, temperature, sample configuration, degree of orientation, morphology etc. The partially crystalline polymer in the solid state consists of amorphous ranges and crystalline ranges. The oxidation process starts in the individual amorphous ranges at different rates. Oxidation spreads by the diffusion of reaction products from ranges of advanced state of oxidation into other amorphous ranges. This spatial heterogeneity in the oxidation process is reflected by the integral temporal progress and the influence of sample thickness and temperature (Gugumus 1997). Activation energy of the reactions participating in the oxidation process is in the range of 100 kJ/mol and that of oxygen diffusion around 40 kJ/mol (Gugumus 1996a). Therefore at application temperatures, usually sufficient oxygen is available for the oxidation process, even in thick samples.

### 5.2.2 Chemical Stabilisation

Non-stabilised polyethylenes and polypropylenes age by oxidative degradation under normal conditions within a few years to a few decades depending on the properties of the individual materials. However, due to the effect of UV light or during the high processing temperatures (200 °C) necessary to produce end or intermediate products, such as geomembranes, fibres, slit films etc., oxidative degradation may develop very rapidly. The materials must therefore be protected against oxidation for processing, long term storage and extended use. Addition of certain chemical substances, so-called antioxidants or stabilisers, can impede the kinetic chain reaction prevent the start reaction from beginning or activate termination reactions, which leads to a substantial reduction in the reaction rate and increase in induction time. For this purpose a package of small quantities of special chemicals, typically between 0.1 and 0.5 % by weight for each chemical, are added to the base polymer. Based on their effect, two types of antioxidants can be distinguished (Fig. 5.3). The so-called primary antioxidants (or radical scavengers or chain breaking donors) intervene in the chain propagation and force their termination by the consumption of peroxy radicals. Typically, an $H$ ion is abstracted from the primary antioxidant (denoted here by $IH$). The general reaction pattern is as follows:

$$PO_2^{\bullet} + IH \xrightarrow{ki} POOH + I^{\bullet} \tag{5.15}$$

$$PO_2^{\bullet} + IH \xrightarrow{ki} POOH + I^{\bullet} \tag{5.16}$$

$$I^{\bullet} + I^{\bullet} \xrightarrow{k6} \text{stable reaction products.} \tag{5.17}$$

The so-called secondary antioxidants $X$ convert the hydroperoxides into innocuous non-radical compounds. Thus the chain branching by the start of new kinetic chain reactions is prevented. Therefore these substances are also called hydroperoxide decomposers. The general reaction pattern reads as:

$$POOH + X \xrightarrow{k4} \text{stable reaction products.} \tag{5.18}$$

Typical representatives of primary antioxidants are sterically inhibited phenols and aromatic amines. The reduction of hydroperoxide to alcohol by a phosphite, which changes into a phosphate, is a typical example of secondary stabilisation. Various chemical compounds used as antioxidants and their designation were discussed in Sect. 2.1.

The highly simplified reaction equations 5.9 to 5.18 enable a model-like quantitative explanation of the functioning of stabilisers (Emanuel and Buchachenko 1987). It is assumed that a steady state is achieved during the kinetic chain reaction, where the concentrations of the intermediate products $P^{\bullet}$ and $I^{\bullet}$ do not change. As soon as they emerge and are consumed: the rate of concentration change is thus approximately zero. The termination reaction Eq. 5.13 can also be neglected.

Based on these assumptions, it follows from Eqs. 5.9, 5.10, 5.11 and 5.12:

$$\frac{d[P^{\bullet}]}{dt} = w_0 - k_1[P^{\bullet}][O_2] + k_2[PO_2^{\bullet}][PH] + \delta k_3[POOH] = 0\,. \tag{5.19}$$

Similarly, it follows from Eqs. 5.13, 5.16 and 5.17:

$$\frac{d[I^{\bullet}]}{dt} = k_i[PO_2^{\bullet}][IH] - k_5[PO_2^{\bullet}][I^{\bullet}] - k_6[I^{\bullet}]^2 = 0\,. \tag{5.20}$$

Solving this equation for [ $I$ ]:

$$[I^{\bullet}] = \frac{k_i[PO_2^{\bullet}][IH]}{k_5[PO_2^{\bullet}] + k_6[I^{\bullet}]} = \frac{\alpha k_i[IH]}{k_5} \quad \text{with} \quad \alpha = \frac{k_5[PO_2^{\bullet}]}{k_5[PO_2^{\bullet}] + k_6[I^{\bullet}]}\,, \tag{5.21}$$

$\alpha$ can be interpreted as a relative portion or probability of a chain termination reaction by the inhibitor radical.

From Eqs. 5.10, 5.11, 5.15 and 5.16, it is obtained for the change of concentration of peroxy radicals and hydroperoxides:

$$\frac{d[PO_2^{\bullet}]}{dt} = k_1[P^{\bullet}][O_2] - k_2[PO_2^{\bullet}][PH] + \\ + k_i[PO_2^{\bullet}][IH] - k_5[PO_2^{\bullet}][I^{\bullet}], \tag{5.22}$$

or with Eqs. 5.11, 5.12, 5.15 and 5.18:

$$\frac{d[POOH]}{dt} = k_2[PO_2^{\bullet}][PH] - \delta k_3[POOH][PH] - \\ - k_i[PO_2^{\bullet}][IH] - k_4[POOH][X]. \tag{5.23}$$

By replacing the terms with $[P^{\bullet}]$ and $[I^{\bullet}]$ in Eq. 5.22 with Eqs. 5.19 and 5.21, one obtains:

$$\frac{d[PO_2^\bullet]}{dt} = w_0 + \delta k_3[POOH][PH] - mk_i[PO_2^\bullet][IH], \tag{5.24}$$

with $m = 1 + \alpha$.

Eqs. 5.23 and 5.24 form a set of linear differential equations describing the concentrations of the peroxy radicals and hydroperoxides as a function of time. The general solution of such a system has the following form:

$$[PO_2^\bullet] = \sum_{i=1,2} A_i e^{\lambda_i t} + C \quad \text{and} \quad [POOH] = \sum_{i=1,2} B_i e^{\lambda_i t} + D, \tag{5.25}$$

where constants $\lambda_1$ and $\lambda_2$ are the roots of the characteristic equation of the linear set of equations, which is obtained when Eq. 5.25 is substituted into the set of differential Eqs. 5.23 and 5.24. The characteristic equation reads:

$$\begin{aligned}&\lambda^2 + (k_3[PH] + k_4[X] + mk_i[IH])\lambda + \\ &+ mk_i k_3[PH][IH] + mk_i k_4[IH][X] - \\ &- \delta k_2 k_3[PH]^2 - \delta k_i k_3[PH][IH] = 0.\end{aligned} \tag{5.26}$$

An analysis of the quadratic equation shows that either both roots are negative or both are positive. The sign depends on the constant term in Eq. 5.26:

$$\begin{aligned}&mk_i k_3[PH][IH] + mk_i k_4[IH][X] - \\ &- \delta k_2 k_3[PH]^2 - \delta k_i k_3[PH][IH] = \omega.\end{aligned} \tag{5.27}$$

If $\omega > 0$, then both $\lambda_1$ and $\lambda_2$ are negative. Concentrations then rapidly decrease to a time-independent constant value. The system is in a steady state being stabilised by the antioxidants where oxidation proceeds at a very low constant rate. This solution is consistent with the assumptions made. However, if $\omega < 0$, then both $\lambda_1$ and $\lambda_2$ are positive. Concentrations increase exponentially. The system is no longer in a stable state. It shows an oxidation which rapidly accelerates. Therefore the stabiliser exerts its effect only up to a critical concentration for which $\omega = 0$. The critical concentration is given by:

$$[IH]_{cr} = \frac{\delta k_2 k_3[PH]^2}{(m-\delta)k_i k_3[PH] + mk_i k_4[X]}. \tag{5.28}$$

As long as concentration of stabiliser is sufficient, oxidation is strongly suppressed and oxidation products are formed only to a small extent but the stabiliser is gradually consumed. If after a certain time $t_{i,2}$ the stabiliser content drops below a critical threshold, accelerating auto-oxidation starts which finally leads to the degradation of the plastic. The overall induction time for the degradation of a stabilised polyolefin or its service lifetime $t_i$, here interpreted as the period up to the loss of the performance properties, is composed of two induction times: induction time that characterises stabiliser consumption $t_{i,2}$ and, as discussed in Sect. 5.2.1, induction time that characterises the first stage of auto-oxidation:

$$t_i = t_{i,1} + t_{i,2} \,. \tag{5.29}$$

The exponential increase of oxidation products, when stabiliser concentration falls below a well defined "critical" level, is, however, only an artefact in this model consideration due to the incomplete treatment of the reaction equations. The termination reaction Eq. 5.13 can no longer be neglected at low stabiliser concentrations which would slow down an exponential increase. In reality, the rise of oxidation rate is not sudden but gradual, and no sharp transition can be observed at a certain "critical" stabiliser concentration. On the one hand it is fairly reasonable for the interpretation and description of experimental data to separate the phase of stabiliser consumption from the phase of incipient auto-oxidation. On the other hand, such a separation should, in a strict sense, not be considered as one specified by a well-defined stabiliser concentration.

In this simple model description of antioxidant function, it can be seen, that a combination of primary and secondary antioxidants ensures a particularly effective stabilisation. Synergies resulting from mixing different antioxidants are still the subject of scientific investigation. Specific packages of stabilisers have been used as accepted standards in practice for a considerable time. A typical stabiliser package for polyethylene and polypropylene contains a phosphite and a phenolic antioxidant. (Zweifel 2001) reports on developments in antioxidants offered.

Under normal conditions, oxidative reactions progress extremely slowly. Correspondingly, the associated oxidative stabiliser consumption is also slow. However, not only oxidation, but other chemical degradation processes can also destroy antioxidants. For example, phosphates and other antioxidants are hydrolysis-sensitive (Gugumus 1990). In addition to such chemical "antioxidant depletion", there are physical depletion processes: The concentration of the added antioxidants is reduced by extraction and migration processes while the plastic is stored and used (Pfahler and Lötzsch 1988). Such processes are usually the main cause of gradual loss of stabilisers at application temperatures. All these depletion processes determine induction time $t_{i,2}$ and thus service lifetime $t_i$ of the plastic. They indirectly accelerate oxidative degradation and may be considered as an external cause of ageing.

The velocity of extraction or migration processes and consumption by oxidation or other chemical degradation processes or the antioxidant depletion rate can be approximately assumed to be proportional to the available amount of stabiliser. The antioxidant content $[A]$ as a function of time can then be described by an exponential function, already introduced in Eq. 5.1:

$$\frac{d[A]}{dt} = -k_{eff}\,[A] \qquad [A](t) = [A]_0 e^{-k_{eff}}. \tag{5.30}$$

When stabiliser concentration $[A]_{cr}$ drops below a certain level, the end of the induction phase $t_{i,2}$, characterised by stabiliser depletion, is reached and it holds:

$$t_{i,2} = \frac{1}{k_{eff}} \ln \frac{[A]_0}{[A]_{cr}}. \tag{5.31}$$

We call $k_{eff}$ the antioxidant depletion rate and its reciprocal $\tau_{eff}$ the depletion time constant. Both are characteristic parameters of the depletion process. Constant $k_{eff}$ as a function of temperature $T$ is given by the Arrhenius law Eq. 5.3.

Induction time $t_{i,2}$ as a function of temperature can be written according to Eq. 5.8:

$$t_{i,2} = t_0\, e^{\frac{E}{RT}} \qquad \text{with} \quad t_0 = \frac{1}{B} \ln \frac{[A]_0}{[A]_{cr}}. \tag{5.32}$$

In this respect $t_{i,2}$ is called antioxidant depletion time and $E$ the apparent activation energy of antioxidant depletion. Again values for $E$ in the vicinity of 50 kJ/mol are experimentally found. As discussed in Sect. 5.1, the Arrhenius equation leads to the old rule of thumb that an increase in temperature by about 10 °C doubles or trebles the depletion rate. This rule can also be used as an initial guide to estimate service life. If, after ageing of geomembrane samples in water or air at 80 °C test temperature over a period of at least one year, "effective" stabiliser substance can still be found, for example by OIT-measurements, a service lifetime of many decades can be expected at normal ambient temperatures.

More than a rough guide, however, cannot be achieved in this way. Reliable estimation of service life must take into account the characteristics of the ageing process. Even in the case of the simple model in Eq. 5.30, strictly speaking, the reaction constant $k_{eff}$ comprises different constants

$k_{eff} = k' + k'' + \ldots$ for the different processes which affect the stabiliser content, where the individual constants are characterised as functions of temperature by different activation energies $E$, $E'$, $E''$... and amplitudes $B$, $B'$, $B''$... Therefore, when Arrhenius lines are measured at high temperatures so that the polymer is in a molten state or at a high oxygen partial pressure, one may not simply extrapolate to temperature ranges below the melting temperature or estimate the service lifetime in air. Different ambient conditions, for instance ageing of samples in an oven, in an oven with air exchange or in a water bath, and different sample geometries, such as planar structure (geomembrane sample), linear structure (fibre sample) or intermediate structure (needle-punched nonwoven geotextiles) lead to different stabiliser depletion as functions of time, with different Arrhenius lines for their temperature dependence.

In some cases, induction time as a function of temperature of the oxidation process under given test conditions deviates very markedly from the Arrhenius equation. Such a deviation can be observed, for example, for the time up to embrittlement as a function of the temperature of stabilised and non-stabilised polypropylene in an air oven ageing test with air exchange (Gugumus 1999). Times-to-failure shorten towards lower temperatures far more markedly than proportionality to the inverse absolute temperature would suggest. Polypropylene stabilised with phenolic antioxidants exhibits an even more complicated behaviour: times-to-failure strongly increase again below about 100 °C. An S-shaped curve is obtained for the logarithm of times-to-failure versus the inverse temperature (Fig. 5.5). An Arrhenius extrapolation of times-to-failure measured at high temperatures (> 100 °C) would overestimate the actual times-to-failure at low temperatures by about one order of magnitude. The repercussion of Ti catalyst residues in non-stabilised polypropylenes on the oxidation process and an additional interaction of catalyst residues and phenolic antioxidants in stabilised polypropylenes is assumed to be the cause of the deviations from the Arrhenius line. LDPE polyethylene manufactured using the high-pressure method, i.e. without any catalyst residues and HALS-stabilised polypropylene namely show no such deviations.

Reliable statements of service life at normal application temperatures usually require tests on representative samples under close-to-application test conditions (e.g. oven ageing in air or in a water bath) at reasonable test temperatures (i.e. test temperature below 100 °C). Therefore, test times of some years might become necessary for a reliable estimation of service lives of hundreds of years. A more detailed estimation of the service lifetime of HDPE geomembranes will be dealt with in Sect. 5.4.

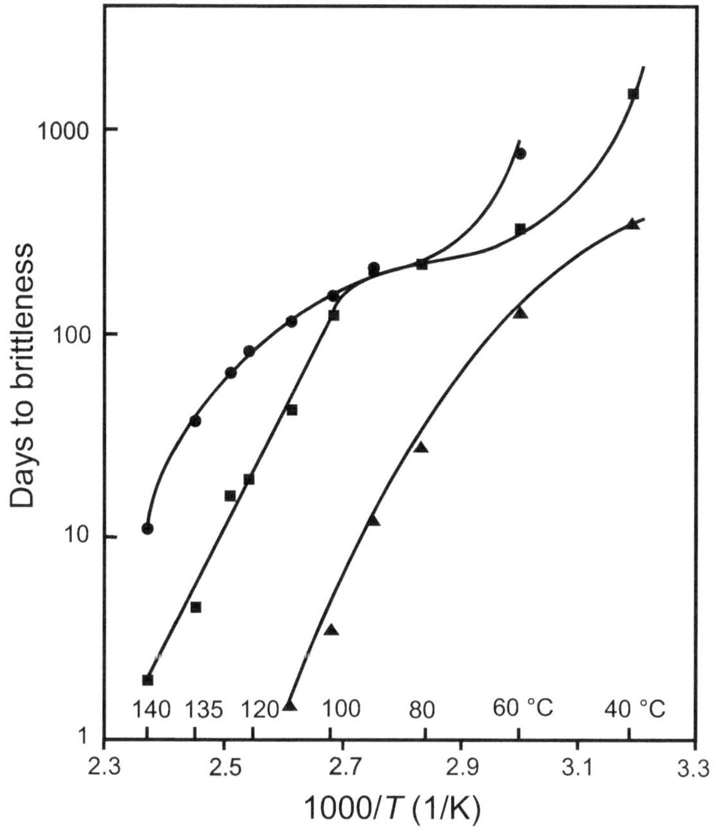


**Fig. 5.5.** Arrhenius diagram for ageing of polypropylene samples stabilised using different stabilisers in an oven with air exchange. Days to brittleness have been plotted against the reciprocal value of ageing temperature. ▲: un-stabilised polypropylene, ●: stabilised using 0.05 % AO-1, ■: stabilised using 0.05 % AO-2. Source: (Gugumus 1999)

It is usually assumed that antioxidant depletion time $t_{i,2}$ will be substantially larger than auto-oxidation induction time $t_{i,1}$ thus service lifetime $t_i$ is basically determined by $t_{i,2}$. For instance, times of up to some hundred hours can be measured for $t_{i,1}$ on 0.5 mm thick films from non-stabilised HDPE and LLDPE in oven aging at 80 °C. Well stabilised 2.5 mm thick geomembranes made from these materials still exhibit sufficient stabilisation even after 10,000 hours of ageing in water or air at 80 °C, therefore $t_{i,2}$ is larger by two orders of magnitude. With polyethylene or polypropylene fibres, however, the relation seems to reverse. The large ratio of surface to volume can lead to a rapid stabiliser depletion, while the large orientation imposed on fibres by cold drawing in the manufacturing process seems to lead to a so-called structural stabilisation, which can increase $t_{i,1}$ many times (Müller and Jakob 2000). This structural stabilisation will be dealt with in the next section.

### 5.2.3 Structural Stabilisation

Mobility of the reactants and reaction products of the oxidative kinetic chain reaction, of stabilisers and of the polymer molecules themselves affects the kinetics of the radical reactions. Morphology of a polymer material and its physical state, e.g. stress, strain and orientation, has an effect on the mobility and therefore on the process of oxidative degradation. Fibres or slit films of polyethylene or polypropylene are cold-drawn in the production. The orientation of the cold-drawn polymer material produced here has a particularly strong repercussion on oxidation stability.

The extent of orientation of a fibre or a slit film is defined by the degree of cold drawing or of orientation $\lambda$, which is the ratio of the length after cold drawing $L$ to the original length $L_0$:

$$\lambda = \frac{L}{L_0}. \tag{5.33}$$

Experiments on cold drawn materials indicate that induction time $t_{i,1}$ increases exponentially with the degree of orientation:

$$t_{i,1}(\lambda) = t_{i,1}(0)\exp(a\lambda), \tag{5.34}$$

$a$ is an experimental parameter which depends on the experimental conditions (e.g. oxygen partial pressure) and the materials. The rate of the oxidation reaction, starting after the induction period, is also slowed down. Reference can be found in (Emanuel and Buchachenko 1987; Mueller et al. 2003; Popov et al. 1991).

Orientation seems to hinder formation and propagation of hydroperoxides (Emanuel and Buchachenko 1987; Popov et al. 1991). Chain propagation and branching are thereby suppressed as it is achieved by the addition of chemicals as stabilisers. One therefore speaks of a structural stabilisation, which has a particularly marked effect in partially crystalline, strongly oriented polymer materials.

# 5.3 Stress Crack Formation

## 5.3.1 Description of Crack Phenomena and Terms

Stresses caused by external continuously acting forces or imposed strains may lead to crack formation in a partially crystalline thermoplastic and, finally, to the fracture of the plastic, even if the stresses are clearly below the short-time stress at break point or stress at yield of the material. This so-called stress crack formation (Orthmann 1983) differs from failure by creep which also occurs in thermoplastics. Their strength with respect to ductile failure depends on the deformation rate. When loaded by an external force the material starts to creeps as described in Chap. 4. In weak points, a reduction in cross-section with increasing local stress occurs. The material expands further at these points and finally yields and tears. The result is a so-called ductile fracture with strong plastic deformations of the material at the crack. During stress crack formation, however, an apparently smooth fracture surface gradually grows into the cross-section of the material until eventually the local stress increased by the cross-section reduction exceeds the strength and the material yields and tears at the edge of the fracture surface. The failure is characterised by a very smooth, even fracture surface. Therefore the term brittle fracture is used. Brittle fracture is characteristic of stress crack formation (Figs. 5.6 and 5.8).

Initially, the formation of microscopically small cracks primarily between the spherulite interfaces precedes the macroscopic stress crack formation (Lustiger and Markham 1983; Menges 1973a). The microcracks run across the direction of maximum strain. These microscopic defects then overlap and open to a disk-shaped area, the so-called craze or pseudo crack which is filled with strongly yielded, fibre-like amorphous ranges of the polymer (fibrils). The pseudo crack surfaces are still connected by the fibrils (Figs. 5.7 and 5.19). The pseudo crack surfaces or craze edges also run across the stress direction. Finally, the fibrils tear and a macroscopic crack develops. At the tip of the stress crack, where it moves into the material, a craze pattern develops. Therefore, the apparently smooth fracture surface shows the characteristic pattern of torn off fibrils when viewed in the enlarged image of a scanning electron microscope (Fig 5.8).

The resistance to stress crack formation of different polyethylene resins can be extremely different. The larger the molecular mass, the number of branches in the polymer chain, the co-monomer content and the molecular mass of the co-monomer, the larger the resistance. Roughly speaking it is therefore true that the resistance decreases with increasing density and increasing degree of crystallisation. The proper high-density HDPE materials of linear polymer chains, which crystallise with high density and high degree of crystallisation, can therefore be extraordinarily susceptible to stress crack formation. LLDPE materials, however, can be very resistant and their resistance mainly differs according to the content and molecular mass of co-monomers. The manufacturing condition of a thermoplastic product and the resulting morphology also has an influence on its stress crack resistance. When a geomembrane is loaded across the direction of extrusion, stress crack formation is more pronounced than when loading occurs in the direction of extrusion. Orientation of the thermoplastic materials, e.g. in fibres or slit films, considerably reduces sensitivity to stress cracks. Points of discontinuity and disturbance in morphology due to local melting and recrystallisation, as they arise in the case of welding seams or subsequent application of hot texture particles onto the surface of the smooth geomembranes, create points of attack for stress cracks. Also, geometrical surface structures with sharp edges and abruptly changing cross-sections, which lead to local stress concentrations, encourage stress crack formation (Fig. 5.6).

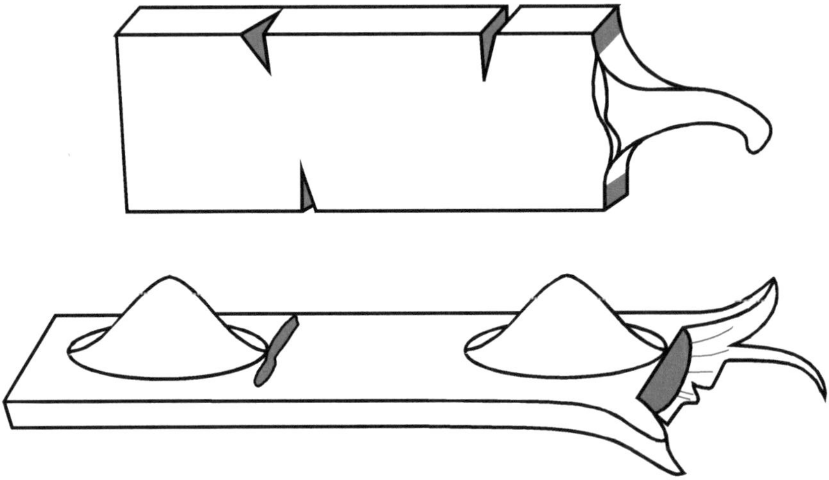

**Fig. 5.6.** Schematic illustration of the damage of tensile bars from a smooth geomembrane (top) and a textured geomembrane (bottom), which have been tested in a long-term tensile test at 80 °C in a solution of surfactant (2 % by weight of Marlophen 812®) under a load of 4 N/mm². Stress cracks grow in the smooth geomembrane starting from the edge until the material cross-section is so strongly reduced that yielding and fracture occur. In textured geomembranes stress cracks emerge at typical weak points (e.g. points of stress concentration) in the surface structure

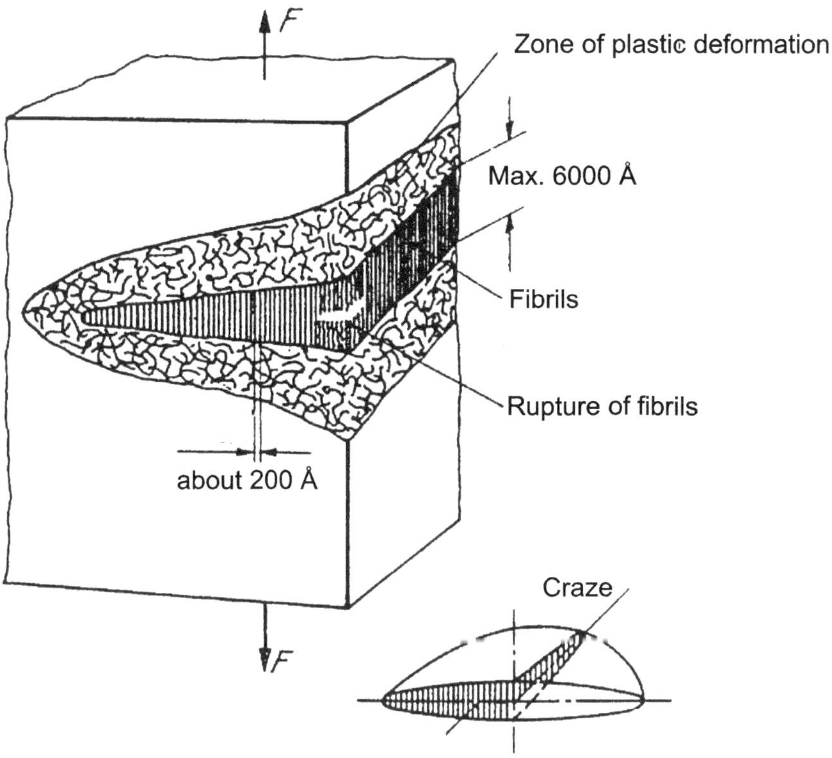


**Fig. 5.7.** Schematic representation of a craze which emerges at the crack tip in the stress crack initiation process and which precedes the actual crack growth. The representation is based on R. P. Kambour's work. Source: (Menges 1973a)

The effect of liquid surfactants can powerfully accelerate stress crack formation. Nevertheless, stress crack formation in plastics must be distinguished from stress crack corrosion as known in particular in metallic materials. Corrosion is understood as the erosion of atoms from the material by chemical processes and in metals particularly by electro-chemical reactions. Additional influence by stresses leads to crack formation and brittle fracture which often resembles of the failure of stress cracks in plastics. Stress crack formation in thermoplastics is, however, a purely physical process. No chemical changes take place in the material even under the influence of surfactants. The terminology is nevertheless not completely uniform. The accelerating effect of liquids on stress crack formation in plastics is occasionally described as stress crack corrosion; although no real corrosion process is connected with it.

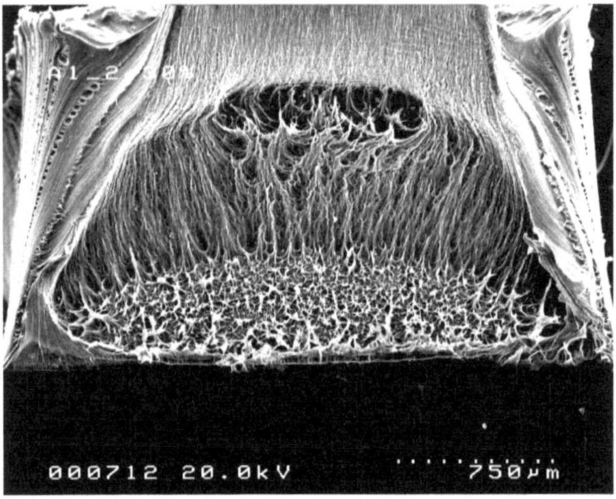


**Fig. 5.8a.** Scanning electron microscopic image of the fracture surface of a brittle crack from a NCTL test specimen. The view is directed from the front top on the top of the lower part of the 3.2 mm wide ruptured test specimen (see also Fig. 3.17). The edge of the notch is fully recognisable in the foreground. A semicircular fracture surface emerges from the edge. Only when the remaining cross section reduces to such an extent that the local stress exceeds stress at yield, the material yields and breaks after large plastic deformation

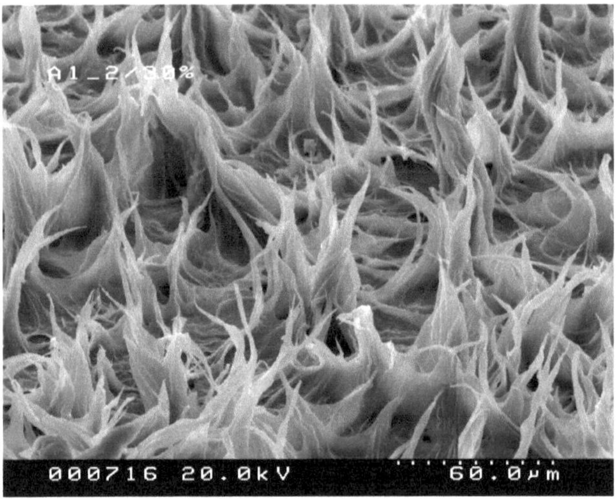


**Fig. 5.8b.** Scanning electron microscopic image of the fracture surface of a brittle crack shown in Fig. 5.8a, which resulted from a NCTL test. The fracture surface, apparently smooth to the naked eye and brittle, shows a number of torn fibrils in this enlargement

### 5.3.2 Test Method for Stress Crack Resistance

Damage by stress crack formation arose occasionally in HDPE geomembranes in the past, in particular in weld seam areas (EPA 1992; Hsuan et al. 1993). Stress crack resistance is therefore a substantial criterion in the selection of polyethylene resins for geomembranes. For this reason, in the following, test methods used will be dealt with in detail. At the same time the phenomenon should be specified more precisely. Table 5.2 gives an overview of the test methods.

In principle, the methods can be distinguished where a certain strain is imposed (relaxation tests) or a certain constant load is applied (creep test). Furthermore the methods may differ depending on whether a predetermined crack initiation point is created by notching the specimen using a precisely specified method or un-notched specimens are used. The technique used to produce the notch – e.g. machining methods (milling, sawing or planing) or cutting methods (e.g. cutting or stamping by a razor blade) – has a substantial influence on the test results. Stress crack formation is strongly accelerated by the provision of a notch. In addition, high test temperatures and the application of surfactants which both encourage stress crack formation resulting in the shortening of times-to-failure. Apart from pipe pressure test, all test methods are index tests, on whose basis the stress crack resistance of materials can only be compared. However, no service lives can be derived under application conditions.

The pipe pressure test and the NCTL test (notched constant tensile load test) have been dealt with in Sect. 3.2.13 in detail. The long-term tensile test was also described, as far as it is used for testing textured geomembranes (Sect. 3.2.16). If the long-term tensile test is performed on test specimens (typically dumbbell shaped or parallel bar specimens) taken from smooth geomembranes, failure usually occurs by stress crack formation starting from the edges of the specimen, see Fig. 5.6. Random scoring, notching or other treatment traces, which are typical for the respective manufacturing process of the specimen, trigger stress cracking as preferred points of crack initiation. Therefore times-to-failure depend very strongly on specimen preparation. Punched specimens fail rapidly. Sawn specimens using fast running hard metal saws or milled by high-speed milling exhibit very long times-to-failure. The range of "true" times-to-failure of smooth geomembranes, where edge influences do not determine the failure, can only be achieved when specimens are used whose edges have been planed off by a microtome knife. Here a surface, perfectly smooth by visual inspection, must be produced. This manufacturing process is however very expensive. Times-to-failure of PE resins of medium density used for geomembranes are then very long (>> 1000 h at 80 °C).

**Table 5.2.** Test methods for geomembrane stress crack resistance

| Test method | Standard | Type | Test conditions | Test duration | Test parameter |
|---|---|---|---|---|---|
| Pipe pressure test (resistance to hydrostatic pressure) | DIN 8075:1999, DIN 16887:1990 | Constant load method | Planar state of stress elevated temperature, water other liquid media | ≥ 10.000 h | Hoop stress versus time-to-failure curves |
| Notched Constant Tensile Load Test | ASTM D5397-95 | | Notched specimen, tensile force, 50 °C, surfactant | ≥ 100 h | Tensile force versus time-to failure curve or mean time-to-failure at 30% of yield stress |
| Long-term tensile test | DIN EN ISO 6252:1998 | | Tensile force, elevated temperature, Surfactant or water | Depending on specimen preparation | Mean time-to-failure at specified tensile force |
| Bell-Test | ASTM D1693-98 | Constant strain method | Notched specimen, strain across the notch, 50 °C, surfactant | >>1000 h | Median time-to-failure (time at which 50% of the specimens have failed) |
| Ball or pin impression method | BAM-certification guideline, edition 1992, DIN EN ISO 4600:1998 | | Notched specimen, strain across the notch, 40 °C, liquid test media | >>10.000 h | Median time-to-failure (time at which 50 % of the specimens have failed) |
| Bent strip method | DIN EN ISO 4599:1997 | | Bending | Depending on specimen preparation | Median time-to-failure (time at which 50% of the specimens have failed) |

Thus the long-term tensile test on plain specimens cannot be used to characterise stress crack resistance of smooth geomembranes. In particular, long-term welding factors cannot be determined for geomembranes either as suggested in (Hessel and John 1987), see Sect. 10.2.

Since failure of smooth geomembrane specimens in the long-term tensile test practically always start from weak points predetermined by specimen preparation, a well defined notch must be applied right at the beginning using a precisely specified method to achieve reproducible times-to-failure. The long-term tensile test on a notched smooth geomembrane specimen, i.e. the NCTL test, was introduced by the Geosynthetic Research Institute, Philadelphia as a test method. It was soon standardised as ASTM D5397-95 and became an established standard test for stress crack resistance of geomembranes (see Sect. 3.2.13).

In the pipe pressure test, NCTL test and long-term tensile test, a well-defined load condition (internal pressure and tensile force) is specified under which the stress crack develops. In reality, however, geomembranes are predominantly stressed by imposed deformations. Therefore, methods which characterise stress crack resistances under constant strain have also been used for geomembrane tests.

The so-called Bell test or bent strip test was developed at the end of the 1950s in the laboratories of Bell Telephone Company to test stress crack resistance of polyethylenes under strong bending deformation. It was standardised as ASTM D1693-98 *Test Method for Environmental Stress-Cracking of Ethylene Plastics*. Ten rectangular specimens (38 mm long by 13 mm wide) are longitudinally slit in the middle, the notch depth being about 20 % of the specimen thickness, bent into a u-shape and fixed within the flanges of a small metal channel. The specimen holder is submerged in a solution of surfactant in a glass tube and immersed at 50 °C. Figure 5.9 shows the test device. The cracks usually develop from the edge of the notch within the range of the strongest bend and move perpendicularly to the notch outwards and through the specimen. The test result may be described in three ways: the test duration is determined when a certain percentage of the specimens are torn, i.e. test duration when 5 of 10 specimens have failed. These durations may be plotted in a probability graph and the time-to-failure is then graphically determined as the time at the 50 % line. The fraction of failed specimen is determined for specified test duration.

This test has been routinely used by geomembrane manufacturers (on geomembrane specimens) and polyethylene manufacturers (on specimens from compression moulded sheets from resin granules) over many years to test their products. However, geomembranes made from polyethylene resins used today do not exhibit any failure at a test temperature of 50 °C even after many thousands of hours. These test methods are therefore no longer suitable for providing an assessment of the differences of stress crack resistance of resins and geomembranes manufactured from them. Thus, to a large extent, the Bell test has been replaced by the NCTL test.

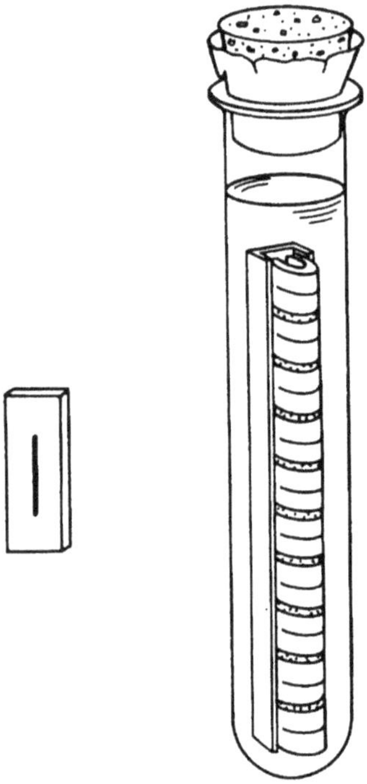

**Fig. 5.9.** Test assembly and specimen for the stress crack resistance test based on the Bell test. The specimen that is slit in the middle has the dimensions 38 mm · 13 mm · thickness of the geomembrane. The test tube is stored at 50 °C (temperature of the test fluids) in a constant-temperature bath

Another test method, the so-called pin impression method, has been developed and standardised to test stress crack resistances of containers and packaging. The method is intended first of all to characterise the resistance under the influence of chemicals transported and stored in these containers. It has also been used for some time in the BAM certification procedure to test geomembranes for landfill liners (Müller 1992).

A hole is drilled into a rectangular geomembrane specimen, which is then notched along the long side parallel to the direction of extrusion using a razor blade. An oversize conical steel pin is pressed into the hole. The specimens are immersed at 40 °C in a solution of surfactant or, depending on application, in another liquid chemical. Figure 5.10 shows the specimen and the steel pin. The crack develops in the notch base and moves from there through the specimen. Specimens are taken at regular intervals from the test bath, the pin is removed and they are cut in half by a sawn cut parallel to the notched side through the centre of the hole. A tensile test is carried out on the notched half. Then the time-to-failure can be calculated, at which the tensile strength drops to a certain percentage. Alternatively, the time interval can be determined, when for example 50 % of the specimens are broken, the appearance of which are regularly controlled visually.

Finally, the so called "bent strip method,, EN ISO 4599:1996 *Plastics – Determination of resistance to environmental stress cracking (ESC) – Bent strip method*, is worth mentioning. Using this method, test samples, e.g. dumbbell shaped bars or parallel bars, are fixed onto a bending template. Usually a circular arc template is used. The imposed strain is then given by: $\varepsilon = d/(d+2R)$, where $d$ denotes the specimen thickness and $R$ the outside radius of curvature of the template. Figure 5.11 shows the test device with a fixed dumbbell shaped bar. The fixed specimens are immersed at a specified test temperature and, depending on test purpose, in paste like, liquid or gaseous media. The change of a typical property, such as external appearance, tensile strength, elongation at break etc., can then be used as an indicator of failure under specified test conditions.

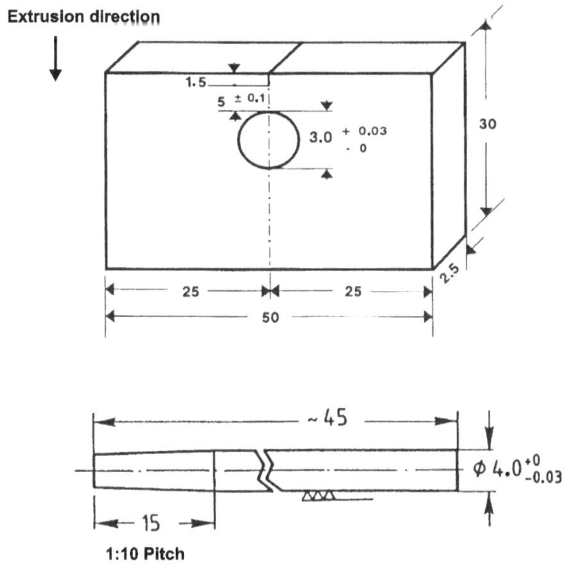


**Fig. 5.10.** A schematic not-to-scale drawing of the specimen from the geomembrane and the steel pin to widen the specimen using the pin impression method. The dimensions are indicated in mm. The arrow shows the direction of extrusion of the geomembrane. In each case a new razor blade is used to punch a 1.5 mm deep notch. A conical steel pin with an end diameter of 4 mm is pushed into a hole with a diameter of 3 mm

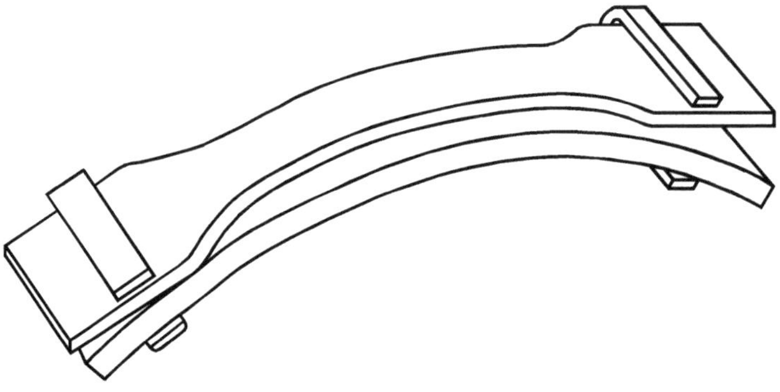

**Fig. 5.11.** Schematic illustration of the bending template with a clamped specimen from the geomembrane for the determination of stress crack resistances using the bent strip method

The same critical remarks must be made in connection with this test method with un-notched specimens as concerning long-term tensile tests on smooth geomembrane specimens: stress crack formation proceeds from random weak points in the boundary region of the specimen. The type of specimen preparation will have a significant influence on the test results.

The Bell test, the pin impression method, which can also be performed in a ball impression test version, and the bent strip method were initially developed to quantitatively describe the influence of chemicals on stress crack formation for a given strain imposed on a standard test specimen manufactured from a certain resin material or a finished product (e.g. a plastic transport container). Based on the behaviour of a certain test parameter (e.g. external appearance, tensile strength, bending strength, elongation at break, tensile impact resistance) a failure criterion is defined and times-to-failure measured accordingly. Variation of times-to-failure can be investigated for various chemicals under specified test conditions (deformation, temperature and medium). A resin is rated as resistant against stress crack formation due to a certain chemical if the time-to-failure exceeds a given threshold value.

However, methods applying specified strain at moderate temperatures do not usually show any damage in the PE resins of medium density to high density used today for geomembranes within manageable test periods (100–1000 h). This applies even if very high strains are imposed in the range of yield point. Rapid damage by stress crack formation can only be observed if very high test temperatures (80–95 °C) are chosen.

In a test under an imposed strain (relaxation test) crack initiation and crack growths, i.e. craze formation, is slowed down by simultaneous stress relaxation. Crack initiation and growth and stress relaxation are competitive processes both of which depend very strongly on temperature. Crack initiation can only proceed sufficiently rapidly at a certain high temperature, so that, despite stress relaxation, damage develops within reasonable test times. This temperature, however, seems to depend strongly on the material. When a high test temperature is chosen, either the process of stress crack formation wins and the material fails rapidly, or stress relaxation wins and no failure will be observed at all. In the conflict of stress relaxation and stress crack formation there is a narrow "test window" for each material within the range of specified strain and test temperature values (Gaube 1959). Only if one reaches this test window can a failure criterion be fulfilled within a practically measurable time-to-failure. For HDPE geomembranes the tests under imposed strain, therefore, is not suitable to obtain a simple and properly differentiating criterion for the evaluation and comparison of the stress crack resistances.

Although constant force acting over the long term do not correspond to any typical loading of geomembranes under application conditions – they must even be consistently avoided in geotechnical constructions to prevent long-term failure – the test of the stress crack resistances under constant load on notched samples (NCTL test) has proven its worth as a realistic method for the evaluation of stress crack resistance of different HDPE resins and for quality control of geomembrane manufacturing.

Long-term behaviour of geomembranes can indirectly be deduced from the results of pipe pressure tests, see Sections 3.2.13 and 5.4. For this purpose, the plane stress state due to typical deformations of geomembranes (e.g. due to settlements) and its temporal change by relaxation will be compared with the hoop stress versus time-to-failure curves of pipe sections made of the geomembrane material. Long-term behaviour however can also be tested directly. The multi-axial tension test (or burst test) (Sect. 3.2.9), in which a deformation is imposed which is typical for a settlement, can be performed as a long-term test by analogy with the pipe pressure test. D. E. Duvall and D. B. Edwards report on such tests (Duvall and Edwards 1992). So far, however, there are no data of "genuine" long-term burst tests with a specified constant arc height and strain available, i.e. data, which would enable establishment of arc height versus time-to-failure curves for different test temperatures.

### 5.3.3 Excursion into Fracture Mechanics

Before the theoretical concepts for the description and understanding of stress crack formation will be discussed in detail, some fracture-mechanics terms and relationships should be reviewed and explained to help understand these theories (Knott 1973).

If a crack develops in a solid body due to an external force acting at its surface, the distribution and amount of strain and stress (the stress and strain fields) in the body are changed. Large strains and stresses primarily develop near the crack tip. Part of the work done by the external cracking force is stored in the form of elastic energy associated with the strain and stress field. Cracking, on the other hand, produces two new opposing surfaces in the body. Formation of a new surface requires energy: atoms or molecules in the surface lose their bonding partners. Another part of the work of the external force is therefore used to break up these bonds. Specific energy connected with the formation of a surface (in a vacuum) is called surface tension and is a characteristic quantity of the material.

In the theory of linear elasticity, the general statement that work done by constant external forces in deforming an elastic body is exactly twice the elastic energy associated with the strain and stress field, can be proved[3]. With this in mind, let's considered the case, where a crack develops in a test specimen under the effect of constant external forces (constant load condition). The amount of work available for the formation of crack surfaces then corresponds at most to the amount of elastic energy which must be applied during crack formation to build up the associated stress and strain fields. Obviously, a criterion for cracking can be obtained from these trains of thought (Griffith 1921): a crack can develop when the required surface energy is smaller than the elastic energy of the strain and stress field to be applied in connection with the crack. If the surface energy is greater, the work of the external constant forces cannot produce any crack, since the reduction in potential energy of the external forces is necessarily a certain function of the stress and strain in the body during cracking and

---

[3] This statement can be explained by the simple example of a spring with the spring constant $k_S$ which is stressed by the force $F_W$ of an attached weight $W$. In equilibrium, let the spring be elongated by length $d$. As well known, elastic energy in the spring is $\frac{1}{2}k_W d^2$. Work performed by the force $F_W$ is $F_W d$. The weight force $F_W$ is equal to the spring force $k_W d$ in equilibrium (action = reaction), therefore $F_W = k_S d$. Thus work performed by the force, i.e. the reduction of the potential energy of the weight, is given by $k_S d^2$ and therefore exactly twice the deformation energy stored in the spring. The constant weight force cannot deform the spring without setting it in motion. Consequently, a part of the work is necessarily transformed into kinetic energy. The spring then swings into equilibrium and kinetic energy is eventually converted into heat. This means that when a spring is deformed by a constant weight force, a part of the energy, which equals the elastic energy, is lost as heat. When a crack is formed, a part of the energy, which equals the elastic energy, is converted into surface energy of the newly formed crack surface (provided that crack formation proceeds very slowly).

can only be exactly twice the elastic energy stored in the developed strain and stress field[4].

Prerequisite of a quantitative description is that the elastic energy stored in the stress and strain fields in the vicinity of a crack can be calculated. Even if it is assumed that the theory of linear elasticity can be applied, nevertheless, the calculation in a general case of any load condition, any geometry of the test specimen and resulting complicated crack geometry remains a very difficult task (nowadays solved by fast computers programmed according to the numerical Finite Element Method). One can, however, gradually simplify the problem and finally establish simple formulae for the criterion.

First, one can distinguish three simple modes of loading and associated modes of fracture. The possible modes of crack propagation in elastic solids can be classified by these three types as well. Figure 5.12 shows the three modes (usually denoted by using Roman numerals) schematically. In mode I the direction of the external tensile forces runs perpendicular to the crack surface, one speaks of an opening mode and fracture under tension. In modes II and III the direction of the force is in the plane of the crack surface. The forces work across the crack front in mode II, this is called in-plane shear mode with respect to the loading condition and sliding mode with respect to crack propagation, and they are parallel to the crack front in mode III, which is the out-of-plane shear mode and tearing mode respectively. Both loading modes lead to cracking under shear. If different simultaneously acting forces each lead to the same mode of load, the stress and strain fields can be added according to the law of superposition. However, it is difficult to treat the mixed-mode cracking when two modes of crack propagation occur simultaneously.

The calculations of stress and strain fields can be simplified by taking into account special symmetries of the test specimen and configuration of the forces. Let us consider e.g. a very thin plate of large dimension. The stress and strain field only depends on the two coordinates within the plane of the plate. Let $x$- and $y$-axes be in the plane of the plate and the $z$-axis

---

[4] Instead of this practice-relevant but somewhat complex consideration, a less practice-relevant, but theoretically more understandable analysis can also be used. It is assumed that the ends of a test sample, at which tensile forces are applied, are fixed in their spatial situation (constant grip condition). The external forces cannot then perform any work. If the crack in the test sample now increases, elastic energy is released, since energy stored in the strain and stress field is reduced under these boundary conditions as a function of crack growth. Only this energy is available for the formation of surfaces. This consideration also leads to the crack formation criterion.

perpendicular to it (see Fig. 5.13). Forces should be applied only at the narrow sides. The stress components on the two large surfaces of the plate ($\sigma_{zz}$, $\sigma_{zx}$, $\sigma_{zy}$ and, consequently, $\sigma_{xz}$ and $\sigma_{yz}$) are zero, because no forces are applied there. Since the plate is very narrow, these components cannot take large values within the plate either, therefore, one can generally set these components to zero. To characterise the stresses at point ($x$, $y$) an infinitesimal quadratic element whose sites are perpendicular to the coordinate axes is considered (Fig. 5.14). Only three stress components must then be calculated which only depend on $x$ and $y$: a tensile or a compressive stress perpendicular to $x$ and perpendicular to $y$ ($\sigma_{xx}$ and $\sigma_{yy}$) and a shear stress at the two-dimensional element perpendicular to $x$ in the direction of $y$ ($\sigma_{xy}$). The shear stress $\sigma_{yx}$ perpendicular to $y$ in the direction of $x$ must then necessarily be equal to $\sigma_{xy}$ at equilibrium.

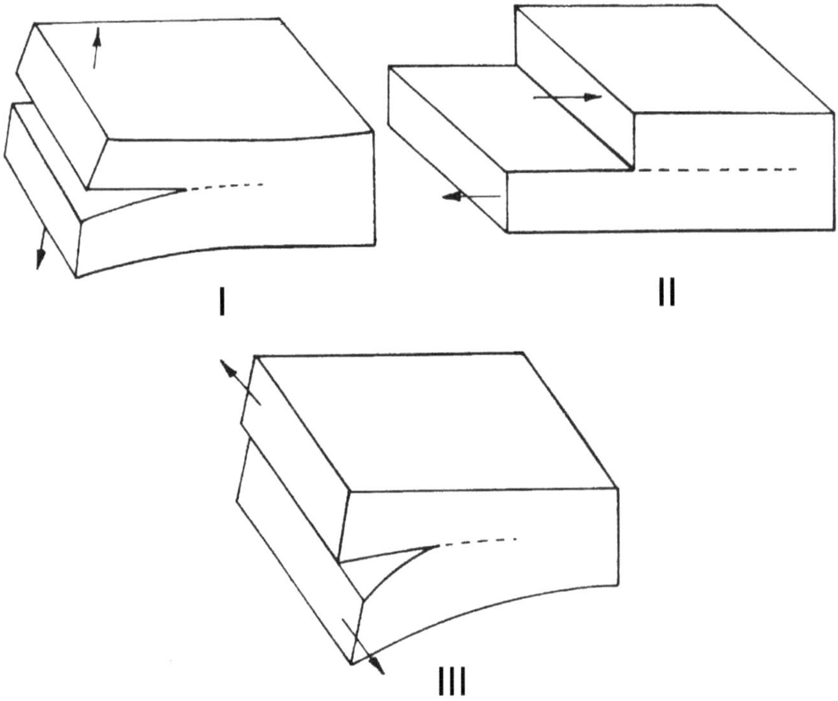


**Fig. 5.12.** Schematic illustration of pure fracture modes. Top left: opening mode with fracture under tension (mode I), the forces are perpendicular to the crack surface. Top right: in-plane shear mode and fracture under shear (mode II), the forces are in the crack surface, but act perpendicularly to the crack front. Bottom: out-of-plane shear mode and fracture under torsion (mode III), the forces are in the crack surface, but act tangentially to the crack front

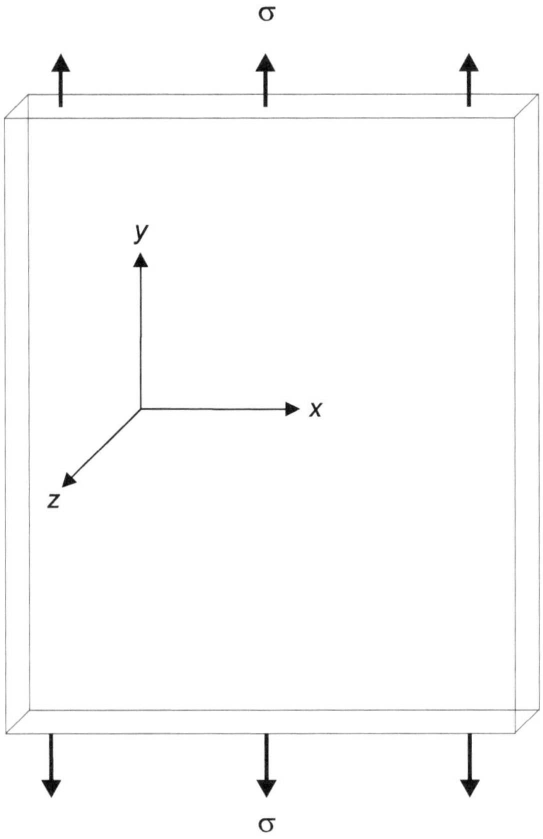


**Fig. 5.13.** In a thin plate of large dimension, a plane state of stress exists where the forces are applied on the upper and lower narrow sides. Stress components only depend on coordinates $x$ and $y$. All stress components in the $z$ direction are negligible. Under this condition the stress field in the area of a sharp notch can be calculated analytically

The state of a thin plate loaded on the narrow sides is an example of a so-called plane state of stress: one of the components of the stress is all over identical to zero. Similar simplifications are possible in the case of the so-called plane state of strain. In such a state, one of the components of the displacement vector, that describes the spatial displacement of the points of the body at elastic deformation, is all over identical to zero and other components again depend only on $x$ and $y$. A large but very thick plate can be taken as an example. In the plane state of strain, in addition to $\sigma_{xx}$, $\sigma_{yy}$ and $\sigma_{yx}$, component $\sigma_{zz}$ does also exist. This component stems from the condition that no displacement should take place in $z$ direction.

For plane states of stress and strain the methods of function theory can be used that enable the analytical calculation of the stresses and deformation fields for various fracture modes and various forms of notch and crack tips (Westergaard 1939).

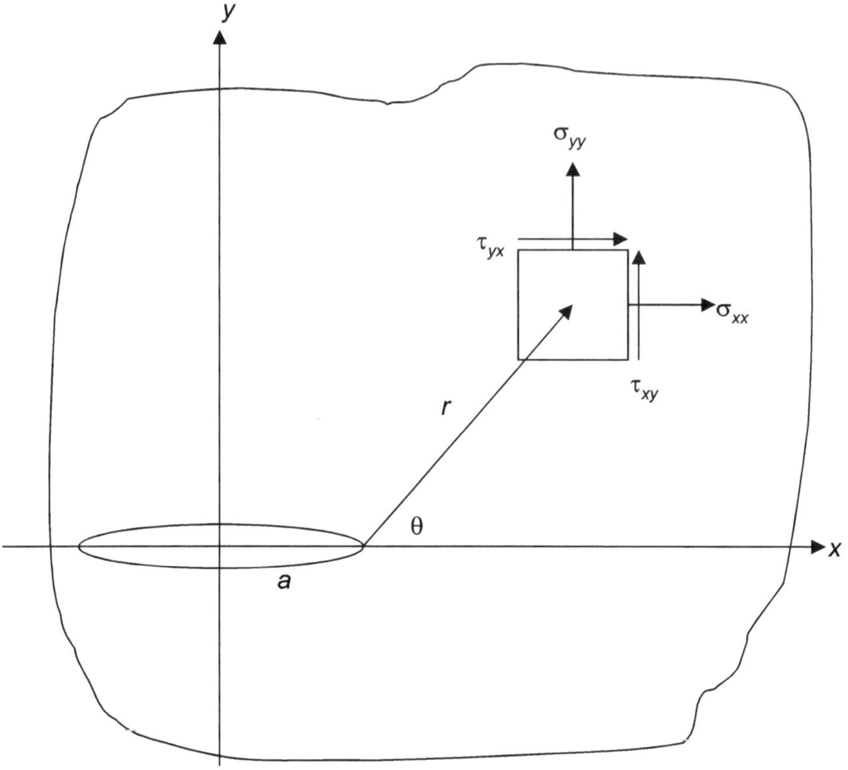


**Fig. 5.14.** A coordinate system ($r$, $\theta$) with the origin in the crack tip is used to describe the stress field in the area of a crack of the length $2a$

Stresses are large near the crack tip but drop rapidly at distances which are large in comparison to the dimension of the crack. It is therefore suitable to select a coordinate system with the origin in the notch or crack tip to describe the stress field. A two-dimensional coordinate system suffices for plane states of stress or strain. A point in the crack area can be described by distance $r$ from the crack base and angle $\theta$ from the plane of the crack ($x$-axis) (Fig. 5.14). Terms with the powers of $r/a$ can be neglected in the expression of the stress field, where $a$ is the depth of the notch or (half) length of the crack. The stress field or the strain field can be represented as the product of two factors: one of them characterising the functional dependence on the spatial coordinates, another factor expressing the intensity of the stress field which stems from the extent of loading, the fracture mode, geometry of the test sample and the form of the notch tip[5]. This factor is called stress intensity factor $K$ (Irwin 1957).

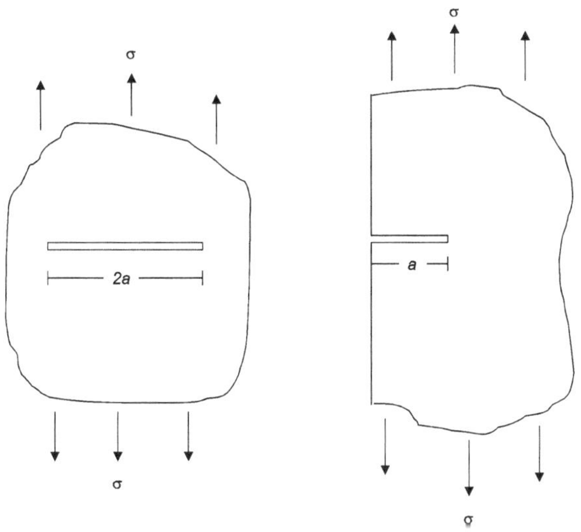


**Fig. 5.15.** Simple expressions for stress intensity factor $K_I$ can be obtained for a crack of width $2a$ in a plate of infinite dimension and for a notch of depth $a$ in a plate of semi-infinite dimension

For an opening mode fracture, stresses $\sigma_{xx}$ and $\sigma_{yy}$ in the plane of the crack ($\theta = 0$) can be given by:

$$\sigma_{xx} = \sigma_{yy} = \frac{K_I}{\sqrt{2\pi r}} \tag{5.35}$$

[5] After all these preparatory remarks, the stress field should here be completely rewritten, at least for mode I. It holds for $(r/a) < 1$:

$$\begin{pmatrix} \sigma_x \\ \sigma_y \\ \tau_{xy} \end{pmatrix} = \frac{K_I}{\sqrt{2\pi r}} \cos\frac{\theta}{2} \begin{pmatrix} 1 - \sin\frac{\theta}{2}\sin\frac{3\theta}{2} \\ 1 + \sin\frac{\theta}{2}\sin\frac{3\theta}{2} \\ \sin\frac{\theta}{2}\cos\frac{3\theta}{2} \end{pmatrix} + \ldots$$

$\sigma_z = \nu(\sigma_x + \sigma_y)$ plane state of stress

$\sigma_z = 0$ planar state of stress

They decrease inversely proportionally to the root of the distance from the crack tip. For a crack of length $2a$ in a plate of infinite dimension (Fig 5.15) in a plane state of stress or deformation, one obtains for the stress intensity factor:

$$K_I = \sigma\sqrt{\pi a} \tag{5.36}$$

and for a notch of depth $a$ in a semi-infinite plate (Fig. 5.15):

$$K_I = 1.12\sigma\sqrt{\pi a}\,. \tag{5.37}$$

The importance of the stress intensity factor can be illustrated in a simplified way using an analogy to an electric or magnetic field. The formula $\boldsymbol{E} = (q/r)\,\boldsymbol{n}$ for an electrical field ($\boldsymbol{n}$ is the radial unit vector pointing outwards) is composed of term $\boldsymbol{n}/r$ which describes the general geometrical field pattern and charge $q$ on which the electrical field strength depends. In a similar way as the electrical or magnetic charges characterise the field strength, the stress intensity factor characterises the strength of the stress field around a crack tip, which, as already mentioned, belongs to a certain fracture mode whose approximate geometrical pattern far outside the direct vicinity of the crack tip is specified independently to the details of the fracture state.

The elastic energy $U$ in connection with the crack (per unit length of crack width) can be calculated from the stress and strain field. Let the cases shown in Fig. 5.15 be considered: a crack of length $2a$ and a notch of depth $a$ in a plate with a comparatively large dimension. One obtains:

$$\begin{aligned} U_{el} &= \frac{1}{2}\frac{K_I^2}{E}a && \text{(plane state of stress)} \\ U_{el} &= \frac{1}{2}\frac{K_I^2}{E}(1-\nu^2)a && \text{(plane state of strain).} \end{aligned} \tag{5.38}$$

$E$ is Young's modulus and $\nu$ Poisson's ratio of the considered material. A quantitative form can now be derived for the criterion of crack formation which was described only qualitatively above. Let us consider that the crack grows by a small length $\Delta a$. The energy needed for surface enlargement per unit length of crack width can be expressed as:

$$\Delta U_S = 2\gamma\Delta a \tag{5.39}$$

$\gamma$ is the surface tension of the material. The change in elastic energy (i.e. the fraction of reduction in potential energy of external forces available for surface formation) can be obtained from Eq. 5.38 taking into account Eq. 5.36 and Eq. 5.37:

$$\Delta U_{el} = \frac{dU_{el}}{da}\Delta a = \frac{K_I^2}{E}\Delta a \qquad \text{(plane state of stress)}$$
$$\Delta U_{el} = \frac{dU_{el}}{da}\Delta a = \frac{K_I^2}{E}(1-\nu^2)\Delta a \qquad \text{(plane state of strain).} \tag{5.40}$$

A crack can only develop or grow if:

$$\Delta U_{el} \geq \Delta U_S. \tag{5.41}$$

The stress at break $\sigma_B$, i.e. the stress at which crack formation can occur, can now be calculated for the crack of width $2a$ in the plate of large dimension. Substituting Eqs. 5.36, 5.37, 5.39 and 5.40 into 5.41 one obtains:

$$\sigma_B = \sqrt{\frac{2E\gamma}{\pi a}} \qquad \text{(plane state of stress)}$$
$$\sigma_B = \sqrt{\frac{2E\gamma}{\pi(1-\nu^2)a}} \qquad \text{(plane state of strain).} \tag{5.42}$$

This criterion of crack formation, first published by A. A. Griffith, is a thermodynamic criterion eventually reflecting the first law of thermodynamics. It considers the change in the total energy of an elastic body when a crack grows, without taking into account the details of the crack formation process in the crack tip. The criterion can therefore only be used for very brittle materials where the plastic deformation in the vicinity of the crack tip may be neglected. However, in visco-elastic thermoplastics a craze develops with large plastic deformations in the strain areas around notch or crack tip.

However, the effect on crack formation by liquid or gaseous surfactants can often be parameterised based by Eq. 5.42, even in thermoplastic materials. The craze is "softened up" and its toughness is decreased by the effect of surface-active substances. The interface energy, i.e. the energy needed to form an interface between the material and this substance, is then the basic factor of crack formation, and the stress at break is described correctly by Eq. 5.42, as a function of interface energy.

The crack formation in non-brittle materials, i.e. fracture with relevant plastic deformation in the notch or crack tip, can similarly be parameterised using an appropriate parameter for plastic deformation. A considerable opening of the notch or crack due to craze formation and plastic deformation in the notch or crack tip is usually observable for these materials. A crack opening displacement (COD) as a measurable quantity, denoted here by $\delta$, is connected with this opening (see Fig. 5.19) and is used as a parameter to describe the crack formation process for plastic deformations in the crack tip.

Let $\delta_0$ be the value of the crack opening displacement, which is due to the formation of a plastic deformed zone in the notch and crack tip when applying a load. In the plane state of stress, the effective stress in the boundary region of the plastically deformed craze is identical to the stress at yield $\sigma_Y$ of the material[6]. When the craze increases by the value $\Delta a$ per unit length of crack width, the work of plastic deformation becomes $\Delta U_p = \delta_0\, \sigma_Y\, \Delta a$. Based on the above energetic considerations, the energy available for this work of plastic deformation is exactly equal to the increase in elastic energy: $\Delta U_{el} = \Delta U_p$, or:

$$\sigma_Y \delta_0 = \frac{K^2}{E} . \tag{5.43}$$

If the crack opening displacement exceeds a certain "critical" value $\delta_c$, being characteristic of the material, the crack keeps on growing. It follows for the criterion of stress at break at a plane state of stress:

$$\sigma_B = \sqrt{\frac{E \sigma_Y \delta_c}{\pi a}} . \tag{5.44}$$

This quick glance into elastic-plastic fracture mechanics completes the short "primer" of some terms and basic relationship of fracture mechanics.

### 5.3.4 Models for the Description of Stress Crack Formations

Stress crack formation in thermoplastics is an extraordinarily complex process. It is determined by various quantities: molecular structure of the polymer (molecular mass distribution and average molecular mass, type and size of co-monomers) and additives (e.g. carbon black), morphology and related potential internal stresses stemming from this structure and the additives, and from the processing history of the plastic material, plus geometry and surface properties, loading conditions and influence of liquid or gaseous media and temperature. Consequently, a uniform theory, which would include all aspects and would enable a detailed forecast about damage processes, is not available. There are, however, models which describe specific aspects of the phenomenon of stress crack formation. Simple theo-

---

6 This applies only to the plane state of stress. In a state of plane strain, all three main stress components occur and the stress profiles within the plastic deformation zone may be more complicated.

retical considerations based on them can then help establish quantitative relationships. Three kinds of considerations or three concepts which emphasise basic aspects of stress crack formation can be distinguished.

1. Models on a microscopic level describe the behaviour of the so-called tie molecules which are attributed a special role in stress cracking. These models take into account the effect of molecular structure.
2. Phenomenological models describe the phenomena of craze and crack formation and crack growth in notched tensile bars in the long-term tensile test. The failure behaviour in the tensile tests is quantitatively parameterised and analysed using fracture-mechanics theories.
3. The so-called structure particle model explains craze formation and crack formation by microcracks along the interface of structural units ("particles") of the morphology. The formation of microcracks is determined by the interface energy between these particles. Microcracks develop if a critical deformation limit, which depends on the interface energy, is exceeded. This model primarily provides a quantitative description of the effect of liquid or gaseous media on stress crack formation.

The model theories will be briefly discussed in the following on the basis of a few important references. Research work that had chosen the long-term tensile test on notched samples (versions of the NCTL tests) as their test method have mainly been selected since this test plays an important role for geomembranes. A complete and exhaustive overview of the technical literature would go beyond the scope of this book.

### *To 1:*

A. Lustiger and R. L. Markham emphasised the role of tie molecules in stress crack resistance (Lustiger and Markham 1983). Tie molecules are molecules anchored in different crystal lamellae. They bridge the amorphous range between the lamellae. Molecule loops protruding from the lamellae, which are entangled among them can also act as tie molecules. Figure 5.16 shows the authors' simple graphical description for stress cracking. If an increasing tensile force is applied the tie-molecules are stretched until they can be pulled out no further. The lamellae break up and the smaller units or crystal blocks are re-arranged into a fibril-shaped structure. Such a ductile deformation leads to a new fibre-like morphology. However, if the tensile force remains constant on a low level the already stretched tie molecules gradually start to untangle and relax. After some time most of the tie-molecules are slowly pulled out from one of the lamellae, the remaining tie-molecules are no longer able to support the tensile stress and a brittle crack emerges. The extraction force at a molecule due to the tensile forces is counteracted by a friction force which stems from both the attractive forces of molecules among themselves and steric hindering.

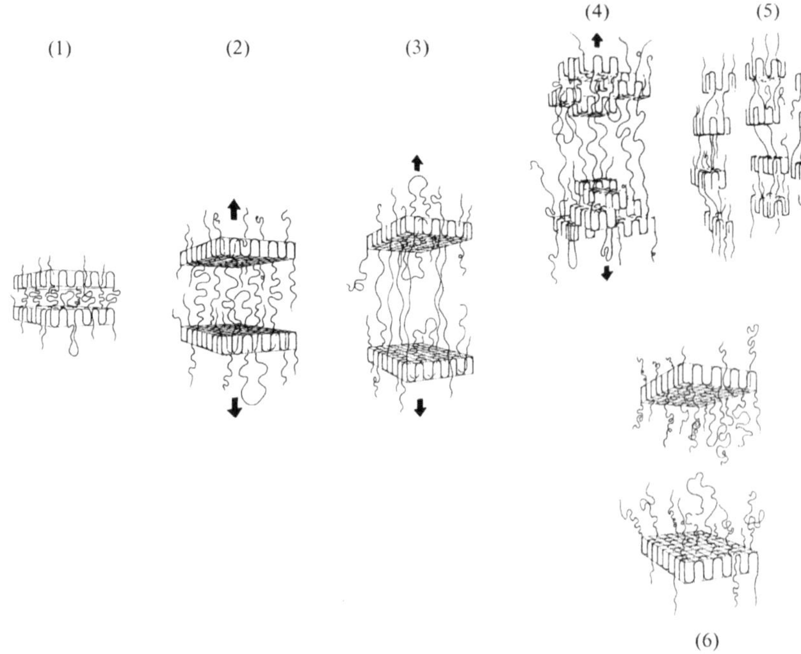


**Fig. 5.16.** Graphic illustration of role of tie-molecules in ductile fracture and brittle fracture characteristic of stress crack formation. (1) shows the starting situation: crystalline lamellae, amorphous intermediate range and tie molecules. (2) The amorphous range opens under a tensile stress. (3) Tie molecules are stretched by the tensile stress. Ductile fracture: (4) crystalline lamellae break up. (5) the fragments are re-arranged into a morphology of long-drawn fibrils. Brittle fracture: (6) under the continuous action of a low level tensile force the tie molecules gradually untangle from the anchorage in the lamellae. The main difference between ductile fracture and brittle fracture from a macroscopic point of view is that the zone of large plastic deformation on one hand extends over the whole material cross section and on the other hand is limited to a small zone near the crack tip where strongly drawn fibrils develop. The drawing is taken from (Lustiger and Markham 1983)

This model concept can be quantified in a simple way (Ramsteiner 1990). The friction force is proportional to the overall length $s$, with which the molecule is anchored in the crystal lamellae, and the pull-out speed $ds/dt$. The proportionality factor $\eta$, a kind of intrinsic viscosity, depends on type and structure of the molecule. Long molecules with many side chains exhibit a higher viscosity than short linear chains. If $\sigma/N$ denotes the pull-out strength per single tie molecule and $N$ is the number of tie molecules, it holds:

$$\frac{\sigma}{N} = -\eta \cdot s \cdot \frac{ds}{dt}. \tag{5.45}$$

The average time $t_0$, until a complete extraction of the molecules with an average anchoring distance $s_0$ has occurred, is then proportional to the fracture time $t_B$. An integration of Eq. 5.45 yields:

$$\frac{\sigma}{\eta \cdot N} \int_0^{t0} dt = -\int_{s0}^{0} s ds, \tag{5.46}$$

from which

$$s_0^{\,2} \propto \frac{\sigma}{\eta \cdot N} \cdot t_B \tag{5.47}$$

follows, or since the medium anchoring distance can be assumed to be proportional to the molecular mass $M$:

$$t_B \propto \eta \cdot N \cdot M^2. \tag{5.48}$$

Already this simple consideration, which leads to Eq. 5.48, makes the experimental observations plausible that stress crack resistance strongly depends on molecular mass, number of tie molecules and type and number of co-monomers which determine the "intrinsic viscosity".

N. Brown's research group at the University of Pennsylvania has investigated the relationships in detail. X. Lu, N. Ishikawa and N. Brown (Lu et al. 1996) divided a LLDPE material (polyethylene hexene co-polymer) with a typical molecule mass distribution ($M_w = 206{,}000$, $M_n = 26{,}900$, $U = 7$) into different fractions with low, medium and high molecular mass. Stress crack investigations based on the NCTL test were carried out on test samples produced from the different fractions. Times-to-failure differed by several orders of magnitude. The authors conclude from the test results that the mass average of molecule mass $M_w$ has a substantial effect on stress crack resistance and only the part with molecular masses greater than $1.5 \cdot 10^5$ Dalton contributes to stress crack resistance of the tested material. Table 5.3 shows the results of the investigation.

The influence of the density of butyl side chains in polyethylene hexene co-polymers on stress crack resistance has been investigated by Y. Huang and N. Brown (Huang and Brown 1991). Figure 5.17 shows the results. The rate of crack opening displacement in a NCTL test (see the discussion under point 2) served here as a measure of stress crack resistance. This rate is proportional to the time-to-failure at specified specimen geometry. The test results of Fig. 5.17 show the strong influence of the density of side chains on time-to-failure.

**Table 5.3.** Stress crack resistance as a function of molecular mass

| Fraction | Short-chain branching ($CH_3$/1000C) | Range of molecular mass ($10^3$ Dalton) | Notch depth (mm) | Time-to-failure at 80 °C, 1.8 MPa (min) |
|---|---|---|---|---|
| F1 | 7.4 | 1–30 | 1 | 0 |
| F2 | 5.1 | 30–80 | 1 | 6.1 |
| F3 | 3.3 | 80–150 | 2 | 7.0 |
| F4 | 1.9 | 150–510 | 2 | 360,000 |
| F5 | 1.4 | 410–4000 | 2 | >520,000 |
| Whole polymer | 4.5 | 1–4000 | 2 | 250,000 |

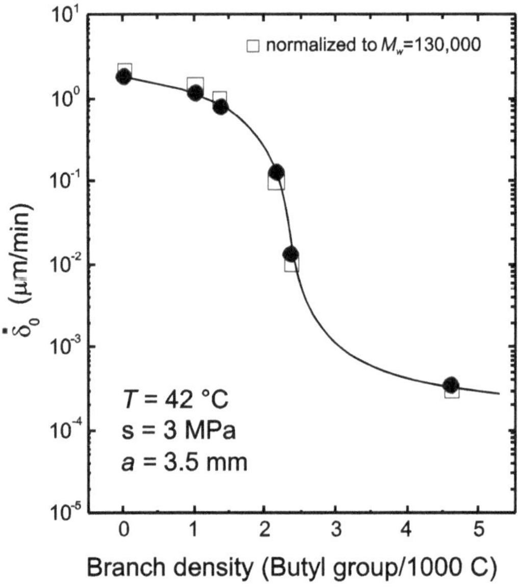


**Fig. 5.17.** The rate, at which the COD value changes in the phase of fracture initiation (see Eq. 5.49), was measured for different densities of butyl side groups of a polyethylene hexene copolymer in a NTCL test. Test parameters (tensile stress and temperature) and notch depth are indicated. The rate decreases drastically with the number of side chains per 1000 carbon atoms. Since the rate is proportional to the time-to-failure, the results indicate that times-to-failure will change by four orders of magnitude as well. Source: (Huang and Brown 1991)

These results suggest that stress crack resistance is eventually determined by the number of tie molecules and the strength of anchorage of these molecules in the crystalline lamellae. The following train of thought was used by the authors to show how molecule mass and the density of co-monomer side chains can affect this number. Within the molten phase a polymer molecule has an average distance between its molecule ends. The average distance between the molecule ends increases with the molecule length. In crystallisation, lamellae are formed with a characteristic thickness and a related thickness of the intermediate amorphous ranges. The thickness of the lamellae and the amorphous intermediate range becomes ever smaller with increasing density of co-monomer side chains. It is highly probable that the molecule will become a tie molecule if the average distance between the molecule ends is greater than the thickness of a package of two lamellae and the intermediate amorphous range. Therefore, a large fraction of very long molecules and a high density of side chains results in a large number of tie molecules.

Another important observation should also be mentioned in the context of our discussion: stress crack resistance in the long-term tensile test on specimens taken from geomembranes in the direction of extrusion was usually substantially greater than that obtained on specimens taken across, which indicates that the orientation has a substantial influence on stress crack resistance. Z. Zhou and N. Brown report for example results of NCTL tests on cold drawn dumbbell-shaped test specimen from compression moulded plaques of a linear PE homo-polymer (Zhou and Brown 1994). The specimen was stretched at 80 °C at a strain rate of $10^{-3}$ $s^{-1}$ with degrees of orientation of up to $\lambda = 3$. Figure 5.18 shows the times-to-failure of specimens from the plaque with notches across and parallel to the direction of orientation. While times-to-failure of cold drawn specimens with the notches across the direction of stretching increase exponentially with the degree of orientation, time-to-failure of a specimen with a notch parallel to the draw direction drops dramatically even when poorly oriented. The interpretation can rely on the model of tie molecules in two aspects. On the one hand, one can argue that just those crystalline lamella fragments that are extensively connected by tie molecules are oriented by cold drawing in such a way that they lie perpendicularly to the direction of orientation (see Fig. 5.16). There are thus a large number of tie molecules acting preferentially in this direction. On the other hand, orientation of amorphous ranges, into which the tie molecules are embedded, can reduce the pull-out force on the individual tie molecule.

Stress crack resistance of PE materials is characterised by times-to-failure in long-term tensile tests on notched specimens not only in the above mentioned but also in many other technical papers. The phenomenological description of crack formation in this test with the help of fracture-mechanics terms and relationships was therefore the object of a number of investigations and will be considered in the following section.

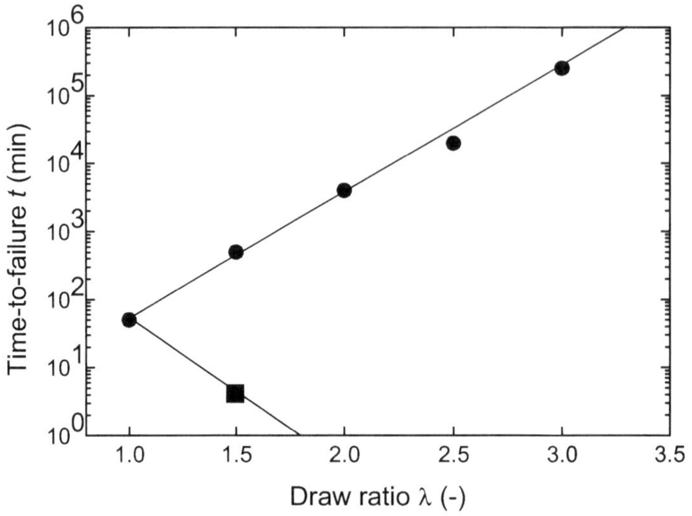


**Fig. 5.18.** Time-to-failure $t$ in the long-term tensile test (NCTL test) on cold drawn test specimen as a function of the degree of orientation $\lambda$. Circles: tensile stress parallel to the direction of stretching and the notch across it, rectangle: applied tensile stress across the direction of stretching and the notch parallel to it. In the first case the time-to-failure increases exponentially with the degree of orientation, in the second case it decreases exponentially. Orientation has a drastic effect on stress crack resistance. Source: (Zhou and Brown 1994)

### *To 2:*

In the NCTL test (see Sect. 3.2.13) ductile failure occurs at tensile stresses above 50 % of the stress at yield. Under about 30 % of the stress at yield the specimen fails through a brittle fracture. There is a range between 30 % and 50 %, where both mechanisms – creep and plastic deformation of the entire material cross-section as well as stress cracking – determine the fracture behaviour. This intermediate range and the range of ductile failure will be briefly dealt with below. The attention is first directed to stress crack formation that results in an unambiguous brittle failure.

This crack formation on a specimen notched by a razor blade – a method used in the NCTL test – is described in detail by X. Lu, R. Qian and N. Brown (Lu et al. 1991). Figure 5.19 shows sketches based on photographs of thin sections of the cracked specimen in the transmission microscope which illustrate the different phases of crack formation (Lu and Brown 1990).

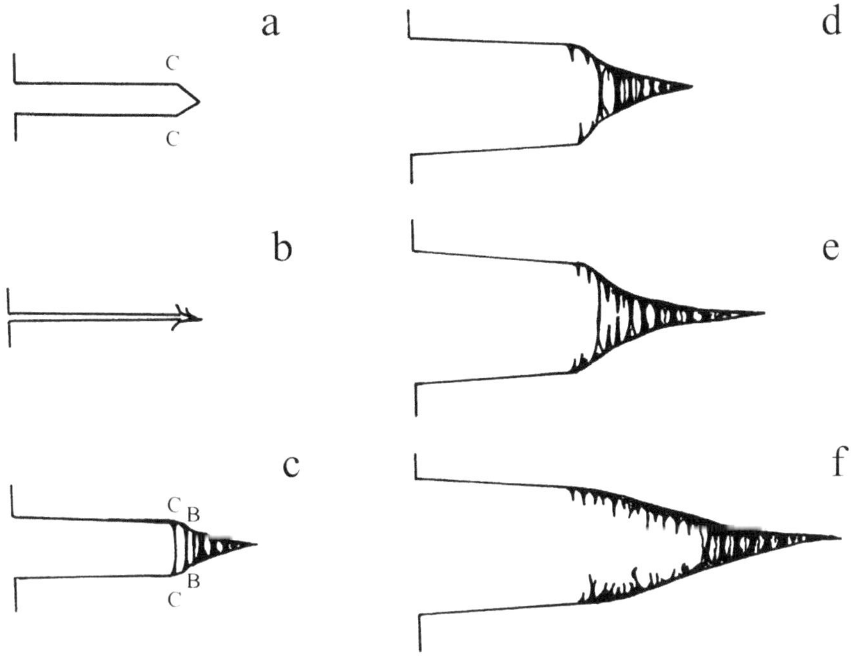


**Fig. 5.19.** Schematic representation of the phases of crack formation in the NCTL test. The morphology of the different phases can be represented by transmitted light images on thin microtome slices in the microscope or by Scanning Electron Microscope micrographs (Lu and Brown 1990). These drawings, based on figures in (Lu et al. 1991), imitate such images. Figure a shows the notch which results from the penetration of a razor blade. The notch tip suffers a strong plastic deformation. Figure b shows the notch after pulling the razor blade out: the notch closes, severe damage is indicated in the notch tip. Figure c shows the notch immediately after tensile stress application: a craze is formed in the notch tip. The crack opening displacement can be measured in two different spots: as distance CC, on the one hand, which gives the opening of the notch tip, and as distance BB, on the other hand, which describes the opening of the actual craze. CC and BB are closely related and are usually used for the determination of the COD value, which was measured with a microscope by viewing the interior of the notch. Figures d, e and f show the consecutive growth of the craze. Figure e shows the first fibrils already torn and represents the condition prevailing at time $t_B$ of fracture initiation (see next figure). The DOC value now increases more rapidly and a brittle crack with a craze around the crack tip moves through the material (Figure f)

Due to the cut of the razor blade penetrating into the material, a sharp notch results with strong deformations in a narrow boundary region around the blade (Fig. 5.19a). On removing the blade, the damage zone collapses (Fig. 5.19b). When a tensile force is applied, the notch opens again, the notch base expands and a zone of plastic deformation is formed, whose outer front (CC in Fig. 5.19) is formed by the fibrils of the stretched, strongly deformed razor blade damage zone and whose inner part is formed by fibrils of the craze which develops during loading (Fig. 5.19c). When the load is fully applied, the fibrils in the outer front occasionally tear. Of importance, however, is the extent of the plastic deformation in the craze zone that is characterised by a typical crack opening displacement (COD value) $\delta_0$ in the notch base (distance CC or BB in Fig. 5.19c). Based on the fracture-mechanics theory described in Sect. 5.3.3, $\delta_0$ can be brought in connection with stress intensity factor in accordance with Eq. 5.43 and determined from the specific load condition.

After the load has been fully applied, the phase of the actual fracture initiation follows. The craze continues to move further inside the material and slowly opens up (Fig. 5.19d). The COD value $\delta(t)$ gradually increases. After a certain "crack initiation" time $t_B$, after which the craze has fully developed, the fibrils tear in this strongly stretched zone (Fig. 5.19e). The crack starts to move now ever more rapidly through the material and the COD value increases over-proportionally (Fig. 5.19f) until the remaining thickness finally becomes so small that the entire material in the remaining cross-section is stretched and tears. Figure 5.20 shows the COD value as a function of time. Time needed until failure is the actual time-to-failure $t_F$.

Crack initiation time and time-to-failure $t_B$ and $t_F$ are proportional to each other for a specified material and specified test conditions; the proportionality factor depends strongly on the material tested. A rate of crack growth can be attributed to the crack formation process, or more precisely to fracture initiation:

$$\dot{\delta} = \frac{\delta(t_B) - \delta_0}{t_B}. \tag{5.49}$$

$\dot{\delta}$ is inversely proportional to crack initiation time $t_B$ or time-to-failure $t_F$. All three quantities are used to characterise stress crack resistance of a material.

Generally, the following relationship can be observed between quantities $\dot{\delta}$ or $t_F$ and test conditions (notch depth: $a$, tensile stress applied: $\sigma$ and test temperature: $T$) in the NCTL test in the parameter range of stress crack formation (Brown and Lu 1995):

$$\dot{\delta} = C\,\sigma^n a^m e^{-Q/RT} \quad \text{and} \quad t_F = C'\,\sigma^{-n} a^{-m} e^{Q/RT}\,. \tag{5.50}$$

$n = 5$ and $m = 2$ have been found for PE homopolymers and $n = 3.2–2.6$ and $m = 1.3$ for PE copolymers. Activation energy $Q$ is similar for all linear polyethylenes and is in the range of 100 kJ/mol. The constants $C$ and $C'$ are material parameters which can differ by some orders of magnitude depending on the molecular structure and morphology of a PE material.

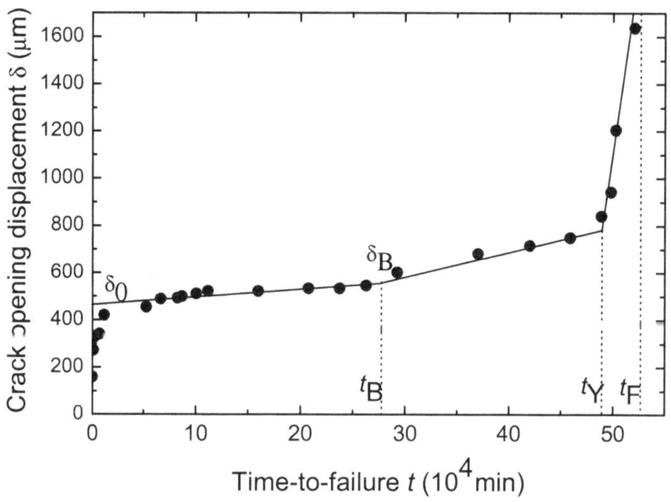


**Fig. 5.20.** Crack opening displacement (COD) $\delta$ as a function of time measured at the notch tip in a NCTL test. On fully applying a tensile stress, an initial value $\delta_0$ develops. This value gradually increases due to further formation of the craze. The fibrils start tearing at time $t_B$ and a macroscopic crack moves ever faster through the cross-section. Starting from $\delta_B$ i.e. the COD value at $t_B$ crack opening displacement grows more steeply until the remaining cross-section of the specimen is finally as small at $t_Y$ that the material starts to yield over the whole cross section under the tensile stress and tears at $t_F$. Source: (Brown and Lu 1995)

N. Brown and X. Lu have suggested a simple consideration which may explain the relationship Eq. 5.50 (Brown and Lu 1995). In the transient area $A$ from the material to the craze zone, where the formation to fibrils begins, just the stress at yield $\sigma_Y$ of the material prevails. Local stress $\sigma_f$ which arises in the fibrils themselves will however be substantially greater because of their small cross-section $A_0$. Constricted fibrils with a small cross-section now carry the force acting in the transition zone. Balance of forces requires $\sigma_Y A = \sigma_f A_0$. Since the volume of fibrils in the craze $A_0{\cdot}\delta_0$

is formed from some primordial material volume $A \cdot d_0$ one may assume that $A_0 \cdot \delta_0 = A \cdot d_0$. Therefore, the reduction in cross-section is proportional to $\delta_0$ and the local stress is proportional to $\sigma_Y \cdot (\delta_0/d_0)$. Newton's theory is assumed for the relationship between deformation rate $\dot{\varepsilon}$ and stress $\sigma_f$ in the further development of the craze. With an intrinsic viscosity $\eta$, $\dot{\varepsilon}$ can be given by:

$$\dot{\varepsilon} = \frac{\dot{\delta}}{\delta_0} = \frac{\sigma_f}{\eta} . \tag{5.51}$$

From this

$$\dot{\delta} \propto \frac{\sigma_f}{\eta} \delta_0 \propto \frac{\sigma_Y}{\eta} \delta_0^{\ 2} \tag{5.52}$$

follows. Equation 5.43 gives the relationship between crack opening displacement $\delta_0$ and stress intensity factor $K$. Using expression 5.37 for stress intensity factor of a sharp notch, one obtains the relationship:

$$\dot{\delta} \propto \frac{1}{\eta E^2 \sigma_Y} K^4 \propto \frac{1}{\eta E^2 \sigma_Y} \sigma^4 a^2 . \tag{5.53}$$

Based on this consideration, one expects $n = 4$ and $m = 2$, which corresponds to the observed exponents. Constant $C$ contains the reciprocal value of an intrinsic viscosity in accordance with Eq. 5.53 and $C'$ contains the viscosity itself. The measured numerical values of these constants enable the order of magnitude for the viscosity to be estimated. The value is substantially greater than one would expect for amorphous ranges (melt viscosity of a linear polyethylene). The high viscosity values can only be the result of the strong anchorage of the tie molecules in the lamellae. Based on this, in addition to the number of tie molecules, the strength of the crystalline anchorage would be the decisive quantity that determines stress crack resistance. It is, however, still unknown how the pull-out process from the crystalline bond occurs both structurally and dynamically.

The processes in the range of ductile failure and in the intermediate range will now be briefly discussed to complete the description of the materials behaviour in the long-term tensile test. The branch of ductile failure in the stress versus time-to-failure diagram (or more precisely log $\sigma$ versus log $t_F$) runs considerably flatter than the branch of brittle failure. Figure 3.18 shows typical stress versus time-to-failure curves. Within the range of high stresses:

$$t_F \propto \sigma^{-n}. \tag{5.54}$$

can be observed where $n = 20–40$. Failure behaviour is characterised by the following processes: increasing deformation due to creep, increasing local stress due to reduction in the material cross-section and, finally, yielding and shear flow in the entire material on exceeding a critical deformation limit.

Many materials exhibit a quite unusual behaviour in the intermediate range. Entering the intermediate range from the brittle range by increasing tensile stress in the test, the time-to-failure becomes actually greater with increasing stress until the branch of ductile failure is finally reached. A "nose" like anomaly can be seen in the failure curve (Fig. 3.18). The width of this intermediate range and the extent in which the time-to-failure increases again with tensile can be very different for different materials. In the intermediate range the plastic deformation in the notch tip area are already so large extended due to creep that the notch base becomes blunted (Brown et al. 1987). The actual effective stress intensity factor decreases due to the blunting effect. The time until fracture initiation and thus time-to-failure increases, but still remains shorter than the time needed to perform plastic deformation by creep in the entire material cross-section. The brittle branch slope therefore dips towards longer times with an increasing tensile stress, until ductile failure finally dominates and times-to-failure become shorter again if the test stress increases further. To ensure that a sharp notch is not blunted by creep, a certain "critical" value may not be exceeded by the stress at a specified notch depth or by the notch depth at a specified stress or, briefly, by the stress concentration factor of the sharp notch (Lu and Brown 1991). The intermediate range typically starts above about 35 % of the stress at yield (measured at standard climate) in the NCTL test (at 50 °C) on 2.5 mm HDPE geomembranes with a notch depth of 0.5 mm (20 % of the nominal thickness). At a stress at yield of 18 N/mm² (measured at standard climate) the critical value of the stress intensity factor would thus be 0.25 MPa·m$^{1/2}$ at 50 °C. Ductile failure occurs above this limiting value and brittle fracture below it.

The question arises whether there is a lower threshold for the stress intensity factor too, below which no craze can develop at all. No crack could then be initiated and stress crack formation would no longer be possible. There are indications in the literature that such a threshold value may be around a tenth of the discussed critical stress concentration factor providing an upper limit for the range of pure stress crack formation. This issue about critical local stresses and critical deformations, which must be exceeded, so that stress crack formation occurs, leads to the third group of publications, i.e. those which deal with the so-called structure particle model.

### *To 3:*

As early as at the beginning of the 1970s G. Menges' research group at the Institute for Plastics Processing (IKV) at RWTH Aachen university (www.ikv.rwth-aachen.de) had dealt with the causes and the process of crack formation in plastics (Menges 1973a, 1973b). A general model was developed for the mechanical behaviour of plastic materials which was based on the existence of material-specific critical strains (structure particle model).

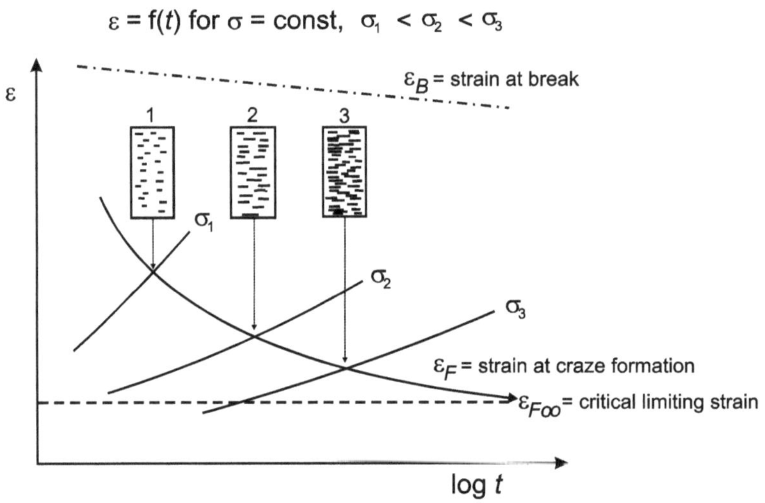


**Fig. 5.21.** Schematic illustration of craze formation in the creep test on transparent amorphous thermoplastics. Visible crazes occur at a certain time and strain during the creep test. These times are indicated in the creep curves measured at different stresses. The connecting line of these points provides a curve which describes the strain limit at which craze formation occurs as a function of time or deformation rate respectively. An extrapolation of the curve towards great times yields the critical limiting strain for craze formation

One of the starting points for the development of this model was the behaviour of amorphous transparent plastics in the long-term tensile test. The specimen exhibit different elongation rates depending on the extent of the load. Fracture occurs after long times and at high elongation values. Long before this elongation at break is reached, however, local crazes develop at a certain strain, which can be observed in the transparent samples by the naked eye. In the strain-time diagram (Fig. 5.21) each creep curve corresponds to a certain load. Crazes arise at a certain point of the creep curve. Connecting these points on the different creep curves, a curve is obtained for strain $\varepsilon_F$, at which crazes occur as a function of time. This curve does not go towards zero but approaches a certain limiting value $\varepsilon_{F\infty}$ at large times. This limiting value must obviously be exceeded so that, independently of type and extent of load, a craze can form at all.

**Fig. 5.22.** Schematic illustration of microcrack formation along spherulite interfaces, which are perpendicular to the direction of stress. The illustration is based on light-microscopic images on cold drawn thin sections of a PP material by G. Menges and E. Alf (Menges and Alf 1972)

Similar observations were made on partially crystalline plastics which were only translucent or opaque. Here a strong milky clouding occurs inside the material which indicates the formation of crazes. A limiting strain can be extrapolated here as well. Microscopic investigations on thin sections from tensile specimens showed that microcracks developed along the interfaces between spherulites when this limiting strain was exceeded. The cracks open preferentially along interface sections that are approximately perpendicular to the direction of processing and are in the range where several spherulites meet. Figure 5.22 shows a schematic illustration of the described situation.

Depending on the constitution of the polymer, more or less large and pronounced spherulites and their interfaces form obviously basic structural components in partially crystalline thermoplastics. Even amorphous thermoplastics are not homogeneous but exhibit, under detailed investigation, a pattern of morphological units with associated interfaces. It was therefore assumed that microcracks can only develop along the boundaries of these units and the concept of limiting strain can also applied to amorphous thermoplastics. In plastics filled with a mineral powder or other powdered fillers, a separation of the matrix from the filler particle can be observed above a certain deformation. Therefore, the approach discussed here, can also explain such crazing phenomena in filled plastics. In all these examples the plastic material can be considered as composed of structural units or structure particles, similar to the brick-work of a wall. The interfaces between the particles are the weak points, like the joints of the brick-work in the analogy, where cracks open at certain critical deformations. The formation and accumulation of such microcracks under the mechanical loads determines the entire mechanical behaviour.

The structure particle model describes the craze formation, being the next stage of crack development, as follows. If static or dynamic loads act over the long term, microcracks open preferentially at the longest particle interfaces which lie approximately across the direction of deformation and stress. The adjacent structural particles can relax to some extend and the stress is concentrated in the microcrack tips. Therefore, the microcracks grow, until they run into a structural particle that lies across the crack front which will then take over the stress while being stretched at the same time. A craze is thereby developed of accumulated microcracks forming a cracked zone, which is bridged by many stretched structural particles lying crosswise. The actual macroscopic crack only emerges when the load conditions force these stretched structural particles to break. Such load conditions has to be considered as the risky ones. Apart from continuously acting stresses, loadings with high deformation rate and energy and moderate loadings that, however, result in pronounced multiaxial states of stress belong to this group.

When the critical limiting strain for microcrack formation is exceeded under such "dangerous loads", the brittle failure of the plastic material will occur. Only where the loads are temporary and sufficiently small and slow, e.g. in common machining processes or application conditions with limited strains, can microcracks be stopped by structural particles lying crosswise which take over the stress and ensure a sufficient internal cohesion of the material. When only a limited number of microcracks and associated crazes are formed under such "harmless loads", the serviceability of the plastic material is not impaired. Therefore, depending on the type of loading, there may still be some reserves available in the material even when deformation exceeds the limiting strain.

The structural particle model, so far only discussed at a qualitative level, can be quantified to some extent with the help of the fracture criterion of A. A. Griffith discussed above in Sect. 5.3.3 (Menges 1973b). The principle of energy conservation in crack formation can also be directly used here. Let $L$ be the greatest length of critical interfaces between structural particles, which runs across the direction of deformation, $E^0$ Young's modulus of structural particles and $\gamma$ the interface energy, i.e. the energy needed to break the interface between the structural particles. One can then set for the stress at break in accordance with Eq. 5.42:

$$\sigma_B = C\sqrt{\frac{E^0\gamma}{L}}, \tag{5.55}$$

where $C$ is some constant. Using Hooke's law for the relationship between local stress and strain, one obtains for the critical deformation or, more precisely, for the limiting strain of craze formation:

$$\varepsilon_{F\infty} = C\sqrt{\frac{\gamma}{E^0 L}}. \tag{5.56}$$

With the help of this equation one may explain the effects of surfactants or other liquid substances on stress crack formation.

Even an initially harmless loads which only lead to a limited number of microcracks and initial crazes, can have a devastating effect under the influence of liquid or gaseous media. Certain media can penetrate into the microcracks and the particles as well. The stressed, crack-stopping particles may become "softer" and their mechanical strength may change. Wetting of the microcrack surfaces can reduce the interface energy and thus the critical limiting strain as follows from Eq.5.56. Both effects encourage further crack growth. These effects are especially pronounced when the media have a low viscosity and a high solubility in the plastic material, i.e. when the solubility parameters of the plastic and the medium are close to each other. The effect of such media considerably impairs stress crack resistance especially when the critical deformation limit is exceeded.

Using Eq. 5.56 the already discussed effect of the orientation in the morphology of plastics on stress crack resistance can be interpreted in the context of the particle model (Menges and Rieß 1974). The size of structural particles (spherulites) in the partially crystalline thermoplastic geomembrane depends not only on the constitution of the polymer, but also on the conditions of the manufacturing process. Orientation makes the structural particles align and elongate and thus length $L$ – included in Eq. 5.56 – of the interfaces, which are responsible for microcrack formation, will depend on the direction of orientation. The length of fracture-susceptible interfaces is smaller across than along the direction. Accordingly, the critical limiting strain should also be different. The limiting strain increases with increasing degree of orientation and it is greater when the strain is in the direction of orientation than across it.

The investigations and theoretical considerations of G. Menges and his colleagues presented here suggest that the critical limiting strain is an important material parameter, which must be considered when designing with plastics. The exact determination of critical limiting strains should be therefore of great interest. However, in spite of the large number of thermoplastic materials and a wide range of applications, there are only very few data available on limiting strain values in the literature. The methods of determination are difficult and, in addition, not always unambiguous and comparable with each other. Material changes of transparent amorphous plastics in the long-term tensile test, which are visible using a lens or even the naked eye, can serve as a starting point for the extrapolation of the strain limit. The procedure is schematically illustrated in Fig. 5.21. Its application is however limited or not feasible at all for translucent or opaque plastics. G. Menges and his colleagues have found that the critical limiting strain is identical with the limiting value up to which the material shows an exact linear visco-elastic behaviour. The behaviour is linear visco-elastic as long as proportionality can be assumed between stress and strain for a given deformation rate and all visco-elastic effects can be expressed by the proportionality constant alone as a function of deformation rate. Chapter 4 has dealt with the visco-elastic deformation behaviour in detail. This identification defines the critical limiting strain only with some uncertainty, since there is no sharp experimentally defined transition between linear and non-linear visco-elastic behaviour, but within this uncertainty it serves as a starting point for an experimental method to determine the critical limiting strain (Menges et al. 1975). For this the so-called specific "damage work" is measured. This is the work which must be applied to induce irreversible microscopic crack formation when the critical limiting strain is exceeded. This work differs from the dissipated energy due to configurational changes and internal friction of the visco-elastic material. The measurement is carried out as follows: The overall deformation work is determined for each cycle of loading in a uniaxial cyclic tensile tests on isotropic specimens of the material: a cycle includes a loading of the specimen up to a certain stress and strain value (loading level) and a consecutive relief. Deformation work applied during loading and released during relief and the dissipated energy being their difference can be calculated from the measured stress-strain curve. If several cycles are run in sequence each with the same loading level below the critical limiting strain, the deformation work of the cycles should be equal. However, if the critical limiting strain is exceeded, a specific damage work is applied in the first cycle in addition to the dissipated energy. The deformation work of the first cycle is therefore greater than the deformation work of the following cycles where the damage has already occurred. The difference between the deformation work of the first cycle and the average of the consecutive cycles is plotted as a function of the strain value of the loading level. The point of deviation of the curve from the zero-line indicates the limiting strain of the material.

Using this method, critical deformations have been found for various polyethylene materials in the range of 4–5 %. Due to the different size and shape of the structural particles (spherulites) of different PE materials, the range of potential values will scatter somewhat more broadly. LLDPE materials should be in the upper range, since the spherulites are relatively small in comparison to HDPE materials. Critical deformations of 2 % have been measured in polypropylenes with large spherulites, easily visible by the optical microscope. The critical limiting strain values are independent of the temperature over a wide range from 0 to 100 °C.

Having briefly presented the three concepts to analyse and interpret stress crack formation in thermoplastics, the question arises whether an internal connection exists between the models and a uniform description of the phenomenon is feasible. Such a connection can be established if one considers the models arranged in a hierarchy of their spatial resolution. At the highest resolution, i.e. on the molecular level, the constitution of the polymer determines the formation of crystalline lamellae and their linkage over amorphous ranges by tie molecules. At a lower resolution this structure is in a simplified way described by the structure particle model. It is in a figurative sense seen as similar to the structure of brick work and joints of the mortar. The structure particle model describes how microcracks can form and grow in this structure to macroscopic crazes. Finally, crack growth and rupture as a function of time on a macroscopic level due to external forces is quantitatively described using fracture-mechanics considerations.

It has to be emphasized, however, that the models are not fully compatible among themselves. The tie molecule model often emphasises that the constitution of the polymer determines the number and strength of the anchorage of the tie molecules. Resistance to cracking directly results from the effect of these tie molecules. In the particle model, however, the constitution of the polymer defines its morphology and type and extent of critical interfaces between crystalline superstructures. Resistance to cracking stems from the type and extent of the interfaces, tie molecules play an indirect role only since their number is linked with a certain morphology.

Stress crack formation of thermoplastics is still an active and wide field of research, therefore, a comprehensive review cannot be given here. However, following all these preparatory considerations on ageing processes, before the chapter about ageing is completed, the question should be answered which actual service lifetimes HDPE geomembranes exhibit.

## 5.4 Service Lifetime of HDPE Geomembranes

Polyethylene plastics may have different ageing and degradation properties. Classification based on density (see Sect. 2.1) and the restriction to so called high-density materials is insufficient as a criterion to characterise those polyethylene resins which have passed the long-term tests in the field of plastic pipes and geomembranes. Therefore it must be emphasised that the following comments and conclusion on the service lifetime of HDPE geomembranes applies only to those products which have been certified on the basis of long-term tests and which have been properly manufactured and installed according to the state-of-the-art installation technique (see Chap. 9). In the following, we will concentrate on stress cracking and oxidative degradation as the main causes of possible long-term failure. The discussion to be presented is based on the investigations of S. Seeger (Seeger and Müller 2003) and W. Müller and coworkers (Mueller and Jakob 2003).

### 5.4.1 Stress Cracking

Under normal conditions in a geotechnical application stress in the geomembrane arises due to imposed deformations and there are two main sources of long-term deformation in the geomembrane: firstly, strain due to subgrade settlement, and secondly, indentations by grains or crushed aggregates or other particles, i.e. from the "roughness" of adjacent layers. Subsoil settlement occurs on a relatively large area of at least several metres over a period of some decades, depending on the compliance of the subsoil. Indentations are punctiform and associated with growing loads on the geomembrane, e.g. with the accumulation of waste in the landfill, which may usually be completed after a few decades. Any of these deformations will result in tensile stresses and potential damage to the geomembrane due to stress cracking. There are two solutions for the problem to avoid long-term stress cracking in a geomembrane: firstly, application of geomembrane resins with approved very high resistance to stress cracking and, secondly, limitation of strains over the entire service life of the geomembrane by suitable geotechnical design. The latter is essentially a task of identifying a limiting strain criterion and designing a subgrade with sufficient bearing capacity and a protective system. What is the upper long term limit for acceptable stresses and strains in a geomembrane? This question has already been addressed in Sect. 5.3.4. Within the structure particle model of stress crack formation an upper principal limiting strain of 3–5 % was determined for HDPE resins, based on an analysis of the strain-dependent irreversible formation of microcracks in the polymer matrix (Menges 1973a; 1973b; Menges et al. 1975). In this section we will consider the approach of R. Koch and coworkers since estimates of service life may be derived from their chain of thought. R. Koch analysed the stress relaxation behaviour of HDPE resins and compared residual stress levels with those leading to stress cracking in internal pressure tests on HDPE pipes in order to derive a strain limit (Koch et al. 1988).

Figures 3.14a and 3.14b shows data from long term hydrostatic pressure creep tests on pipes. This test under constant stress was part of the BAM certification procedure for geomembranes and was frequently executed on pipe specimens made of several HDPE resins for geomembranes (Sect. 3.2.13). As a result of such tests, the hoop stress is plotted against the geometric average value of the failure times at different temperatures (stress-rupture curves). These curves characterise the stress cracking behaviour of an HDPE geomembrane resin. At high stress levels pipe specimens typically show a ductile failure mode, i.e. yielding starts at a local weak point of stress concentration in the direction of hoop stress. The yield zone grows in the direction of the pipe axis. The yielded material bulges outwards and finally tears. The times-to-failure increase drastically with decreasing stress, as indicated by the nearly horizontal portions of the curves (ductile failure section). On the other hand, a brittle failure mode, i.e. stress cracking without yielding, is observed at lower stress levels where the specimen fails after considerably longer times as indicated by the steep portion of the curves (brittle failure section). Such pipe pressure tests can be used for an Arrhenius extrapolation of data, measured over a few years at elevated temperatures, in order to predict the acceptable stress level at ambient temperatures for an HDPE geomembrane resin over several decades.

Figure 5.23 shows the extrapolated hoop stress versus time-to-failure curve at 40 °C for pipes of a HDPE resin for geomembranes which were frequently used in the past in Germany and which had a time period to the ductile to brittle transition of about 100 h in the NCLT-test on geomembranes. The curve section in the brittle region at 40 °C has been extrapolated from measured data at 80 °C, according to the extrapolation procedure described in the German standard DIN 16887:1990 *Testing of Thermoplastics Pipes; Determination of the Behaviour to Long Term Internal Pressure* and in ISO 9080:2003 *Plastics Piping and Ducting Systems – Determination of the Long-term Hydrostatic Strength of Thermoplastics Materials in Pipe Form by Extrapolation.* An estimate can be read from this stress-rupture curve of the service lifetime of geomembranes exposed to a permanent stress level at a given temperature. In the example shown, the resin complies with the original BAM certification requirement for stress crack resistance (extrapolated service life > 50 years at 4 $N/mm^2$ and at 40 °C in the hydrostatic pressure creep test) and there is a service life expectancy of more than 100 years at 40 °C, provided the stress level does not exceed approximately 3 $N/m^2$.

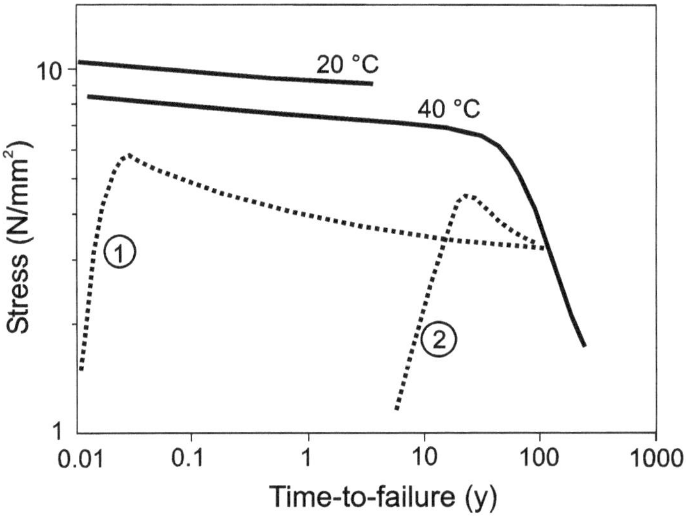


**Fig. 5.23.** Time-to-failure data from the long-term hydrostatic pressure creep test on HDPE pipes at 40 °C (continuous line). Calculated relaxation curves for planar state of stress (dotted line). (1): Relaxation after rapid straining (3 % strain in 3 hours). (2): Stress relaxation after slow straining (3 % strain in 30 years)

A temperature of 40 °C, quite unusual for geotechnical applications, was chosen within the BAM certification because geomembranes in landfill liners may be exposed to elevated temperatures when the waste body of the landfill contains biodegradable materials and especially when it is run as a bioreactor landfill.

The next step towards developing a strain criterion is the quantification of those deformations leading to a long lasting stress state of maximum stress of at most 3 $N/mm^2$ in the long run. For this purpose stress relaxation curves are also shown in Fig. 5.23. These curves result from a biaxial deformation (i.e. bending and in plane elongation as it is imposed by a settlement or simulated by the burst test (Sect. 3.2.9)) and different deformation histories: the deformation was assumed to increase linearly as a function of time (constant deformation speed) until an in plane elongation of a given value is reached and then kept constant. The time dependent stresses assigned to each of the deformation histories have been calculated with the help of a fully visco-elastic numerical materials model for uniaxial deformation (Menges and Schmachtenberg 1987; Michaeli et al. 1989) and correction factors for biaxial deformation (Sect. 4.2). The resulting curves reveal an initial stress increase followed by a stress relaxation typical for HDPE resins. Both high and low strain-speed curves converge sooner or later to a master curve. From this relaxation behaviour one may learn the following: the in plane stress component of a 1:1 planar state of stress in a HDPE geomembrane at 40 °C, generated by an imposed in plane strain of 3 %, converges to a residual value of ~3 $N/mm^2$ for a time period of at least 100 years. Thus, in a worst-case analysis 3 % may be regarded as a good estimate for the long-term strain limit, in order to rule out stress cracking for a period of at least 100 years.

This procedure for the estimation of the limiting strain belonging to a certain service lifetime at a specified temperature is rather conservative for the following reasons. The procedure relies on the assumption that relaxing tensile stresses in a geomembrane section subjected to a constant strain may cause stress crack initiation and propagation in the same way as a constant stress under conditions of creep. There is no rigorous proof of this assumption for HDPE materials. However, there is experimental evidence available from pipe testing and research, which supports this assumption (Akiyama et al. 1991). S. Seeger reports on the following investigation. An extrapolation of uniaxial relaxation data of a widely used HDPE geomembrane resin revealed that the stress relaxation curve after 20 % strain at 80 °C follows the curve for 3 % strain at 40 °C almost exactly with respect to stress height and relaxation speed. This coincidence was used in long-term "pipe deformation" experiment: pipe specimens made of this resin were immersed in hot water and strained for 15 minutes over a conical mandrel until the circumferential strain reached 20 %. The specimens were stored underwater at 80 °C at this constant strain while the hoop stress was allowed to relax. The calculated stress relaxation characteristics and the pipe pressure test data indicated that stress cracking was likely to occur after 2.7 years at the earliest. No failure was observed in these specimens over more than four years although the stress was not yet fully relaxed. This experimental result may support the fact that the above assumption and therefore the derived strain criterion is conservative.

How does the deviation of the individual relaxation curves from the master curve influence the time-to-failure? The expected times-to-failure for a geomembrane in a constant strain situation were calculated using the rule of linear damage accumulation – Miner's Rule (Gaube et al. 1976). To simplify matters the relaxation curves of Fig. 5.23 have been approximated by discrete time-steps. $\sigma_n$ ($n = 1, 2, 3...$) is defined as the average stress level in each step and $t_n$ is the step width. The specific failure times $t_S(\sigma_n)$ attributed to the above discrete stress levels $\sigma_n$ may be obtained from the hoop stress versus time-to-failure curve. The time-to-failure $t_F$ (or service lifetime) of the geomembrane under the strain controlled conditions in a geotechnical application may then be estimated by the following equation:

$$\sum_n \left\{ \frac{t_n}{t_S(\sigma_n)} \right\} \leq 1, \tag{5.57}$$

which simply states that failure is imminent when the sum equals one. Let $n_{max}$ be the maximum value for which Eq. 5.57 can just be fulfilled. $t_F$ is then given by:

$$t_F = n_{max} \cdot t. \tag{5.58}$$

Such a calculation shows that the service life of a geomembrane is predominantly determined by the average stress level as indicated by the master curve. A high initial stress does not substantially shorten the service life as there is only a 10 % difference in the calculated service lives between the relaxation histories (1) and (2) in Fig. 5.23. In both cases the service life clearly exceeds 100 years. On the other hand, the service life drastically decreases if the average stress level is increased.

Applying such a concept for the derivation of limiting strain as a design criterion, one has to keep a few other aspects in mind. Since there are no accurate data available for the relaxation after imposing a deformation which results in a planar state of stress in the HDPE geomembrane, experimentally determined corrections factors from stress relaxation have been used (Koch et al. 1988; Menges and Schmachtenberg 1987). One should also take into account that the 2:1 planar state of stress of pipes in hydrostatic pressure tests does not precisely match the 1:1 plane state of stress from deformations of geomembranes. Hence, a detailed calculation of limiting strain for given service lifetimes under various conditions is far more complex.

The application temperature of the geomembrane was assumed to be 40 °C in order to simplify matters, although it may vary in both directions in landfill liners or other geotechnical application. At lower temperatures the time-to-failure at a given stress level is drastically increased, but the relaxation of stress decelerates, leading to a higher long-term stress level. Vice versa, the times-to-failure are much shorter at higher temperatures but the stress relaxation is accelerated. None of these cases have been considered in detail here but as a general rule lower ambient temperatures will considerably reduce the problem of stress cracking in deformed geomembranes. An increase of the limiting strain to 6 % is expected. Variation in service lifetimes according to temperature variation may be estimated using Eq. 5.57 and the temperature dependence of the specific time-to-failure $t_S(\sigma_n,T_n)$.

In addition to pipe pressure test at 80 °C from pipes made of geomembrane resins, the NCTL-test is currently used in the BAM certification of plastic geomembranes (Müller 2001) for a direct assessment of the stress crack resistance in geomembranes. However, the NCTL test is a pure index test. In the NCTL-test a ductile to brittle transition time of at least 100 h is required which seems to correspond to the BAM requirement of a ductile to brittle transition in the pipe pressure test above 10,000 h at 80 °C and 4 N/mm$^2$ hoop stress. On the other hand, to perform an Arrhenius extrapolation for each HDPE resin of the full hoop-stress-versus-time-to-failure curve at a low temperature on the basis of pipe pressure tests at elevated temperatures would be extremely time consuming and expansive. It is therefore reasonable to apply the above strain criterion to the geotechnical design with HDPE geomembranes, which have passed one of the above tests. Such a design approach guaranties that their service lifetime expectancy with respect to stress cracking is well above 100 years.

The approach seems to be conservative, as already mentioned, but one has to keep in mind that seams – hot wedge weld seams and extrusion seams – are weak points in a geomembrane liner. Indentations and elongation due to settlement might cause high local stresses in seams, because of the seam geometry. In fact, in all cases where geomembranes highly prone to crack formation had been installed, the observed stress crack failures were always connected to seams. It is therefore highly recommended to use such a conservative approach in the derivation of a design criterion. Long-term behaviour of seams will be considered in detail in Chap. 10.

## 5.4.2 Oxidative Degradation of HDPE Geomembranes

Along with stress cracking, oxidative degradation can also impair the integrity and/or performance of polyolefin geomembranes. It is generally accepted that oxidative degradation of HDPE geomembranes is not a relevant factor for normal geotechnical applications, in which service lifetimes of around 30 to 50 years are expected and repair is possible. However, it may become relevant when the geomembranes are integral parts of building structures that have to perform over extremely long service lives (> 100 y), as is the case for geomembranes in landfill liner systems. Long-term slope stability may be impaired by oxidative degradation of textured geomembranes especially in capping systems (Seeger et al. 2000).

German building authorities have defined two durability categories for materials in landfill liner systems. The durability of a material is classified as long-term if it can be assumed that no relevant change in the functional engineering properties of the material will occur over “a period of 50–100 years as predicted in accordance with accepted scientific and engineering knowledge and experience”. Durability is classified as permanent if it can be predicted that no relevant change will occur over “several hundred years”. Landfill liner systems must contain at least one component that is permanent according to this definition. G. Hsuan and R. Koerner estimated the typical service lives of landfills as follows: regulatory minima (e.g. post closure) typically 30 years, non hazardous (municipal) waste at least 100 years, hazardous/low level radioactive waste ≥ 1,000 years (Hsuan and Koerner 1998). Accordingly, the service lifetime of HDPE geomembranes in landfill liner systems is more likely to be in the range of centuries than of decades. For such a long time scale oxidative degradation might substantially contribute to the ageing process and failure modes. Often HDPE geomembranes are classified only as long-term, i.e. not permanent – because of the problem of possible oxidative degradation.

The service lifetime with respect to oxidative degradation of geomembranes made of certain HDPE-resins was first estimated in (Koch et al. 1988). They used hydrostatic pressure tests on pipes of a resin, from which geomembrane were manufactured as well, and measured the time of the transition from brittle failure due to stress crack to brittle failure due to oxidation at 80 °C and 60 °C (see Fig. 3.14a). From an Arrhenius diagram (Fig. 3.15) they extrapolated a service lifetime of about 700 years at ambient temperature with respect to oxidative degradation. Hsuan and Koerner studied the ageing of a commercially available HDPE geomembrane in an incubation device that simulated the service environment of a geomembrane beneath 30 m of waste (Hsuan and Koerner 1998). The ageing process was accelerated by temperature and mechanical properties and OIT were measured as a function of ageing time. No oxidative degradation was achieved. However, it was argued that the time of the depletion of antioxidants should be the main component of service lifetime. This oxidative depletion time was monitored by OIT measurements which led to a lower estimate for the service lifetime at ambient temperature of 200 years. Similar results were obtained by Rowe and Sangam who emphasized the influence of the immersion liquid on the antioxidant depletion rate (Sangam and Rowe 2002). An overview of experimental results is presented in (Sangam and Rowe 2002).

However, none of these experiments were pursued to the end of the depletion process and oxidative degradation was not actually achieved, since this requires very long test times even at high temperatures. Therefore, the depletion rates have to be considered with caution because they have been estimated from short-term experiments, and the assumption that the antioxidant depletion time is the most relevant part of the service lifetime, needs experimental justification.

As German certification authority for plastic geomembranes in landfill liner systems (Müller 2001), the Landfill Engineering Working Group of the German Federal Institute for Materials Research and Testing (BAM) has performed oven ageing and immersion tests on various HDPE geomembranes since the late eighties. These results will be discussed in the following and an attempt will be made to answer the following questions: How long is this service life of HDPE geomembranes with respect to oxidative degradation under normal operational conditions, and how does the ageing process evolve? What are the best test methods to select and compare HDPE geomembrane resins with high oxidation stability under normal operational conditions?

Polyolefin components used in geosynthetics are often stabilised to a lesser degree than HDPE geomembranes. They are made of polymers which are more sensitive to oxidative degradation (e.g. PP or LDPE as opposed to HDPE) or have a much higher surface to volume ratio (e.g. fibres or slit films). The oxidative long-term behaviour of HDPE geomembranes should therefore set a time frame for the oxidation stability of geosynthetics and, accordingly, and the following results are of significance for the general topic of oxidative resistance of these products as well.

Oxidative degradation and the importance of antioxidant additives have already been discussed in detail in Sect. 5.2.2. Typically, phosphite and a hindered phenol are used as the main components of the antioxidant package for most of the HDPE geomembranes available on the market. Normally, carbon black is added as UV stabiliser. The concentration of the antioxidant components is typically in the range of a few thousand ppm (mg/kg). Depletion of these antioxidants was identified as the relevant process determining service life with respect to oxidative degradation. These antioxidant depletion processes depend highly on the environmental conditions. The induction time $t_2$ of antioxidant depletion was defined in Eq. 5.31 and the induction time $t_1$ of the auto-oxidation was discussed in Sect. 5.2.1. Therefore, we expect that the service lifetime $t_L$ of a geomembrane can be written as Eq. 5.29, namely:

$$t_L = t_1 + t_2 \tag{5.59}$$

The oxidation stability of a geomembrane material at high temperatures (above the melting point) can be directly measured using thermoanalytical methods. The thermoanalytical OIT measurement determines the time interval (oxidative induction time (OIT)) to the onset of exothermic oxidation of a polymer at a specified OIT test temperature $T_M$ in a specified oxygen atmosphere (see Sect. 3.2.7). For some stabilizer packages the change in OIT values of a given HDPE geomembrane sample during ageing can be used to monitor the change in the level of stabilisation and to estimate the induction time due to antioxidant depletion $t_2$ of the sample. Especially for the above mentioned stabiliser package the OIT increases uniformly with the antioxidant concentration (Chen and Rånby 1990; Gray 1990; Howard 1973). Therefore, it is possible to approximate the oxidative induction time roughly by a linear function $OIT$ of antioxidant concentration $[AO]$ within a wide range of medium concentration excluding the sectors of low and very high concentration values. We may assume:

$$OIT = \alpha [AO] + \beta, \tag{5.60}$$

where $\alpha > 0$. $\alpha$ and $\beta$ highly depend on the specific properties of the resin, the nature of the stabiliser package and the used OIT measurement conditions. Normally negative values have to be attributed to $\beta$ to interpolate measured relations between OIT and antioxidant concentration (Chen and Rånby 1990; Gray 1990) because zero OIT value might be achieved at high OIT test temperatures even with a certain amount of stabilizer present. However, a high oxidation stability of the unstabilised material might come into play at a very low OIT test temperature which would lead to an offset in the OIT-value even at zero stabiliser content. In this case, $\beta$ might also become positive. Inserting Eq. 5.60 into Eq. 5.30 we expect that OIT as a function of ageing, $OIT(t)$, might be fitted with a declining exponential function with an amplitude $A$ and an offset $B$, and a depletion rate $k_{eff}$ which is characteristic for the depletion process:

$$OIT(t) = A\, e^{-k_{eff}} + B\,. \tag{5.61}$$

Parameters $A$ and $B$ differ significantly for various materials and measurement conditions.

The oxidative stability of 9 commercially available HDPE geomembranes (samples 1 and 48 (same resin, same manufacturer but several years difference in dates of manufacture), 12 and 136 (ditto), 82, 123, 139, 146, 257) made of 7 different resins by 5 manufacturers was tested against this theoretical background. These geomembranes are widely used in geotechnical engineering. The tested samples were 2.5 mm thick. UV stabilisation was ensured in all cases by carbon black with a typical content of 2–2.5 % by mass. Table 5.4 shows the specifications for melt mass-flow rate (MFR) and density of the geomembranes. The ratio between MFR (190/5) and MFR (190/21.6), which reflects the width of the molecular weight distribution and the type of co-polymer, is also indicated. The co-polymer content is normally relatively low, i.e. < 10 % by mass[7].

**Table 5.4.** Resin specifications of the HDPE geomembrane (GM) materials studied. All values measured at the samples were within the specifications. Values stated in parentheses are measured values, i.e. no specification was available

| Sample | Specification | | | |
|---|---|---|---|---|
| | MFR (190/5) g/10min | MFR Ratio (190/21.6)/(190/5) (-) | Density (GM) (g/cm³) | Co-polymer |
| 1<br>48 | 0.85 ± 0.15 | 19 | 0.946 ± 0.002 | ? |
| 12<br>136 | 1.6 ± 0.2 | 9 | 0.942 ± 0.003 | butene |
| 82 | 23 ± 3* | 18 | 0.947 ± 0.003 | hexene |
| 123 | (0.5) | (26) | (0.950) | hexene |
| 139 | 2.5 ± 0.4 | 9 | 0.943 ± 0.002 | octene |
| 146 | 0.80 ± 0.15 | 23 | 0.950 ± 0.005 | hexene |
| 257 | (1.7) | (9) | (0.940) | blend |

*)MFR 190/21.6

All resins were stabilised at least with a few thousand ppm of phenolic and phosphite antioxidants. However, there are considerable differences in the initial OIT-values (Table 5.5) due to the details of the make-up of the stabiliser package and the intrinsic polymer properties. The exact make-up

[7] Most of the resins used for HDPE geomembranes are ethylene-α-olefin-copolymers and should therefore rather be classified as linear low density or linear medium density HDPE. Although only the carbon black additive places them formally into the high-density range, the geosynthetic community customarily refers to them as HDPE geomembranes.

of the stabiliser package is normally a well-kept secret of the resin manufacturer and the information is either not available or must be treated confidentially.

The following experimental methods were used to characterise the oxidation stability of the geomembrane:

1. Oven ageing (in all but one case: gravity convection), see Sect. 3.2.12
2. Immersion in hot water, see Sect. 3.2.11
3. Tensile test, measurement of melt mass-flow rate (MFR) and density, see Sects. 3.2.4 and 3.2.8
4. DSC measurement for high temperature OIT values and melting curves, see Sect. 3.2.7

Description of the details of experimental procedures were presented elsewhere (Mueller and Jakob 2003).

**Table 5.5.** Initial OIT value of HDPE geomembrane samples at $T_M$= 160 °C and 180 °C measured in Cu pans and at $T_M$= 200 °C and 210 °C measured in Al pans

| Sample | OIT (min) at $T_M$ | | | |
|---|---|---|---|---|
| | Cu pan | Al pan | | |
| | 160 °C | 180 °C | 200 °C | 210 °C |
| 1 | 61 | 7 | - | |
| 48 | 72 | 13 | 11 | |
| 12 | - | 48 | 79 | |
| 82 | - | 113 | 81 | |
| 123 | - | 56 | 102 | |
| 136 | - | 123 | 138 | |
| 139 | - | 203 | 81 | |
| 146 | - | 46 | 70 | |
| 257 | - | - | | 41 |

The tensile properties of the GM were tested according to EN ISO 527-3. Figure 5.24 shows typical stress-strain diagrams from the tensile test. Using these data, the relative value of the elongation at break, $\delta\varepsilon_B(t) = \frac{\varepsilon_B(t)}{\varepsilon_B(0)}$, was determined as a function of ageing time or immersion time $t$. Figure 5.25 shows some examples of measured OIT curves (difference of heat flow $\Delta Q$ versus measuring time $t_m$). The OIT value was determined as the difference between the time of the change to oxygen atmosphere and the time when $\Delta Q$ exceeded the threshold value of 0.01 W/g.

The OIT value of a specimen was set to zero at a specified $T_M$ when it was below the detection limit. From these measurements the relative OIT value, $\delta OIT(t) = \dfrac{OIT(t)}{OIT(0)}$, was obtained as a function of ageing time or immersion time $t$.

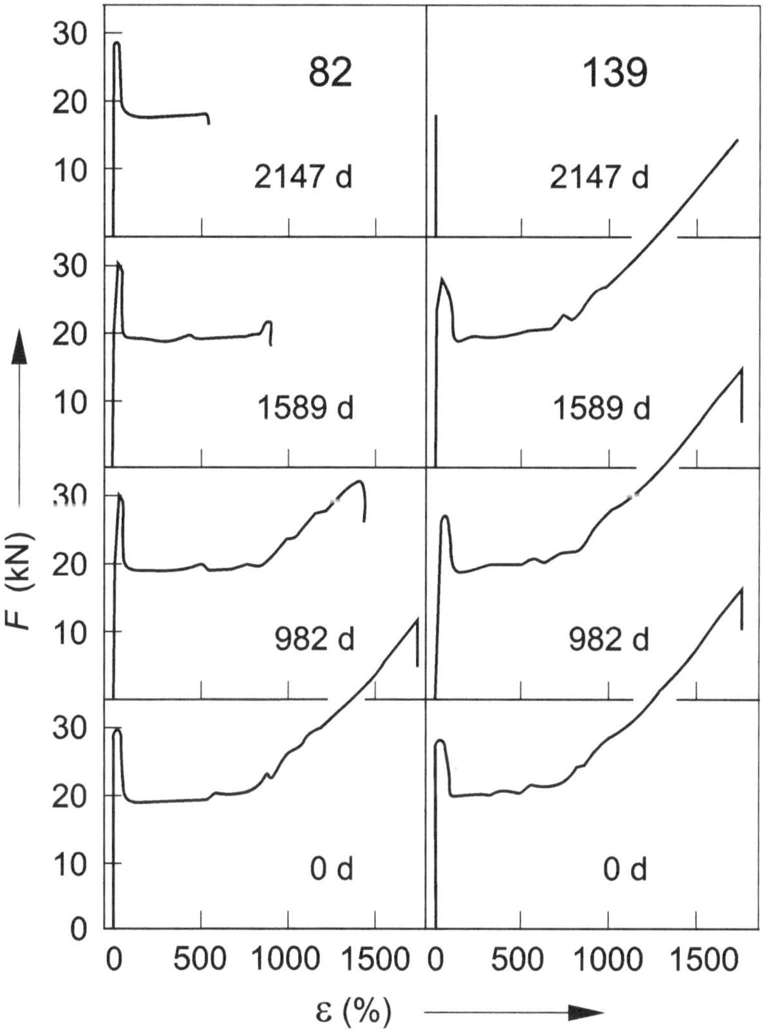


**Fig. 5.24.** Examples of tensile force ($F$) versus strain ($\varepsilon$) diagrams from tensile tests on sample 82 and 139 after immersion in hot water. The immersion times are indicated

Before extrapolating service lifetimes the test results should be discussed in detail. First, let us consider the oven ageing data. Fig. 5.26 shows the change in relative OIT values, δOIT, of geomembrane samples 1 and

48, sample 12 and sample 139 during air oven ageing at 80 °C with gravity convection, and of geomembrane sample 257 during oven ageing in forced air circulation. The corresponding measured OIT values can be calculated using the initial values shown in Table 5.5. The relative OIT values of all samples slowly decrease. After 13.6 years of ageing in air the OIT values were very low: 6 min at $T_m$ = 140 °C with Cu pans for sample 1 and about 2 min at $T_m$ = 200°C with Al pans for sample 12. None of the samples showed any significant change in the tensile properties during air oven ageing.

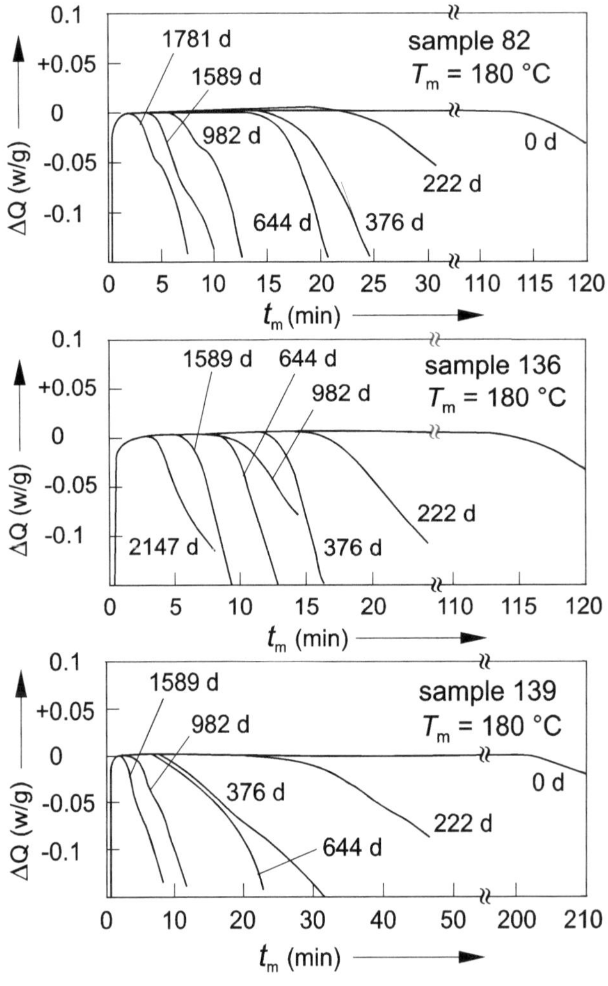


**Fig. 5.25.** Examples of OIT curves measured at $T_M$= 180 °C with Cu pans on specimens from samples 82, 136, 139 after immersion in hot water. The immersion times are indicated

However, for the specimens of samples 1 and 12, which were aged for the whole period of 13.6 years, a reduction in the melt mass-flow rate was observed (Table 5.6). The MFR(190/5) and MFR(190/21.6) after an ageing time of 4978 days were 0.5 g/10min and 14 g/10min for sample 1 and 1.2 g/10min and 12 g/10min for sample 12. Though overall crystallinity changed only slightly (Table 5.7), recrystallisation led to a small but distinct second peak in the melting curve, known as memory peak attributable to the heat treatment at 80 °C.

**Table 5.6.** Change of melt mass-flow rate, melt flow ratio and density of the HDPE geomembrane samples during air oven ageing (samples 1 and 12) and water immersion (samples 82, 123, 136 and 146) at 80 °C

| Sample | Parameter | Ageing or immersion time (d) | | | | |
|---|---|---|---|---|---|---|
| | | 0 | 982 | 2147 | 2419 | 4978 |
| 1 | 190/5 (g/10min) | 1.0 | - | - | 0.8 | 0.5 |
| | 190/21.6 (g/10min) | 18 | | | 16 | 13 |
| | ratio (-) | 18 | | | 20 | 27 |
| | density (g/cm³) | 0.950 | | | 0.949 | 0.951 |
| 12 | 190/5 (g/10min) | 1.6 | | - | 1.5 | 1.2 |
| | 190/21.6 (g/10min) | 15 | | | 14 | 12 |
| | ratio (-) | 9 | | | 9 | 10 |
| | density (g/cm³) | 0.943 | | | 0.943 | 0.946 |
| 82 | 190/5 (g/10min) | 1.4 | 0.9 | 1.5 | - | - |
| | 190/21.6 (g/10min) | 27 | 25 | 37 | | |
| | ratio (-) | 20 | 28 | 25 | | |
| | density (g/cm³) | 0.947 | - | 0.954 | | |
| 123 | 190/5 (g/10min) | 0.5 | 0.5 | - | - | - |
| | 190/21.6 (g/10min) | 13 | 13 | | | |
| | ratio (-) | 26 | 26 | | | |
| | density (g/cm³) | 0.950 | 0.950 | | | |
| 136 | 190/5 (g/10min) | 1.6 | - | 1.6 | - | - |
| | 190/21.6 (g/10min) | 14 | | 15 | | |
| | ratio (-) | 9 | | 9 | | |
| | density (g/cm³) | 0.943 | | 0.950 | | |
| 146 | 190/5 (g/10min) | 0.83 | - | 0.7 | - | - |
| | 190/21.6 (g/10min) | 19 | | 18 | | |
| | ratio (-) | 23 | | 26 | | |
| | density (g/cm³) | 0.951 | | 0.952 | | |

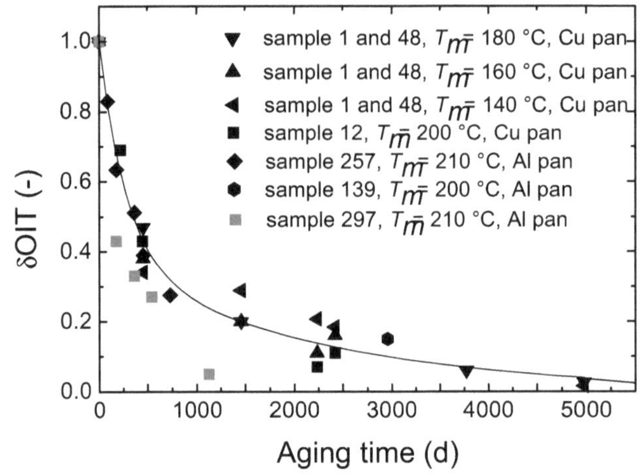


**Fig. 5.26.** Change in the relative OIT value δOIT during ageing in air at 80 °C of various samples. Sample 257 was tested in forced air circulation, all other samples in air circulating by gravity convection

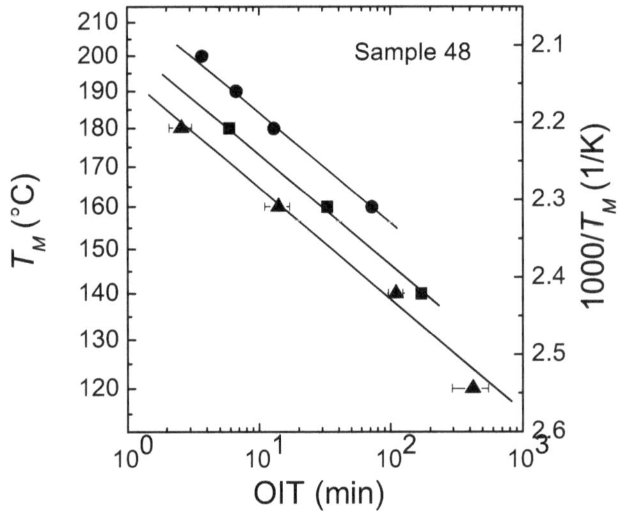


**Fig. 5.27.** Arrhenius plot of the reciprocal absolute OIT test temperatures ($1/T_M$) versus the logarithm of the corresponding initial OIT values of sample 48

At low OIT test temperatures the OIT values become very large. Still, for the oven aged specimen of sample 48, it was possible to determine the OIT value even at $T_M$ = 140 °C. The initial OIT value of sample 48 measured at 200 °C, 180 °C and 160 °C was used to extrapolate to its initial OIT value at 140 °C (Fig. 5.27). The data (closed circles) follow clearly an Arrhenius law. Using the result of 381 min at $T = 140$ °C the relative OIT value was calculated and added to Fig. 5.26. With all data included, Fig. 5.26 illustrates the experimental finding that the relative OIT values show a fairly common decline as a function of ageing time, regardless of the resin or the OIT test temperature. The decline vs. time graph is actually a superposition of two exponential declines, as can easily be seen in a logarithmic plot of the data. The ageing period to date of sample 257 is only 730 days. Within this period the change in the relative OIT value due to ageing in forced air circulation follows closely the course of the data of ageing samples in air circulated by gravity convection.

**Table 5.7.** Crystallinity (%) of HDPE geomembranes studied at various immersion times

| Immersion | Sample | | | | |
|---|---|---|---|---|---|
| Time (d) | 82 | 123 | 136 | 139 | 146 |
| 0 | 52 | 52 | 51 | 48 | 54 |
| 222 | 54 | 55 | 52 | 49 | 57 |
| 376 | 56 | 54 | 52 | 50 | 57 |
| 644 | 58 | 55 | 51 | 51 | 57 |
| 982 | 55 | 55 | 53 | 51 | 58 |
| 1589 | 58 | 56 | 51 | 51 | 57 |
| 2147 | 56 | 57 | 53 | 57 | 59 |

The behaviour of the geomembrane samples immersed in water is different to those exposed to air oven ageing. The effect of a two-step exponential decline is drastically enhanced in the immersion data. At the beginning of the immersion test the relative OIT values of all samples decrease sharply (Fig. 5.28). The corresponding reduction in the absolute OIT values can be obtained by comparing the OIT values plotted in Figs. 5.29 to 5.34 with the initial values from Table 5.5. Depending on the initial OIT value, the decrease is enormous (e.g. sample 139). However, the curve of the OIT versus immersion time levels off at the latest after 200 days for all but one sample (namely 48) and the further decline in oxidation stability is only small.

Figs. 5.29 to 5.34 show the change of the OIT value itself and the corresponding change in the relative elongation at break $\delta\varepsilon_B$ as a function of immersion time in 80 °C de-ionized water for samples 48, 82, 123, 136, 139 and 146. Sample 136 was manufactured from the identical resin as sample 12 which was tested in air oven ageing. Elongation at yield is the tensile test parameter most sensitive to oxidative degradation.

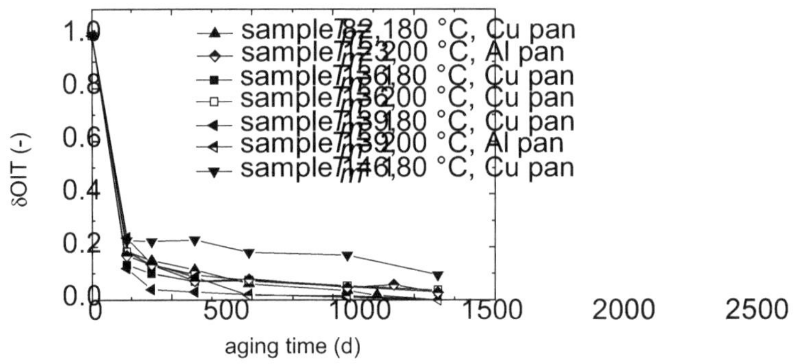


**Fig. 5.28.** Change in the relative OIT value δOIT of various samples during immersion in water at 80 °C. The OIT test conditions are indicated

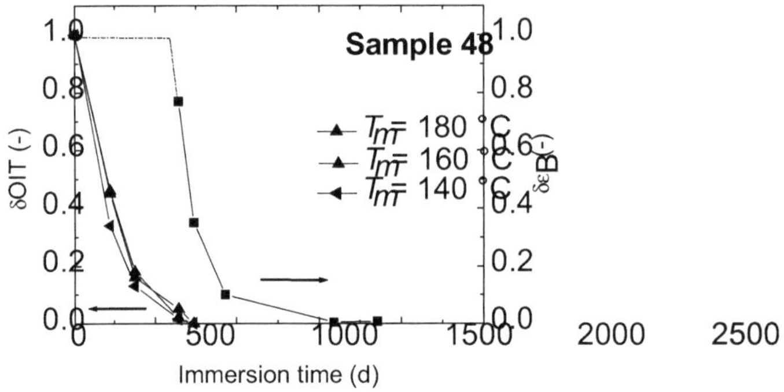


**Fig. 5.29.** Change in the relative OIT value δOIT and relative elongation at yield $\delta\varepsilon_B$ during immersion in water at 80 °C of sample 48. The OIT test temperatures are indicated. Cu pans were used

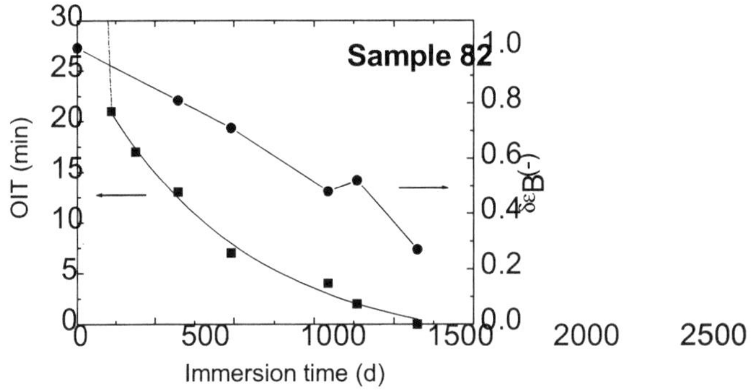


**Fig. 5.30.** Change in the OIT value ($T_M$= 180 °C, Cu pans) and relative elongation at break $\delta\varepsilon_B$ during immersion in water at 80 °C of sample 82

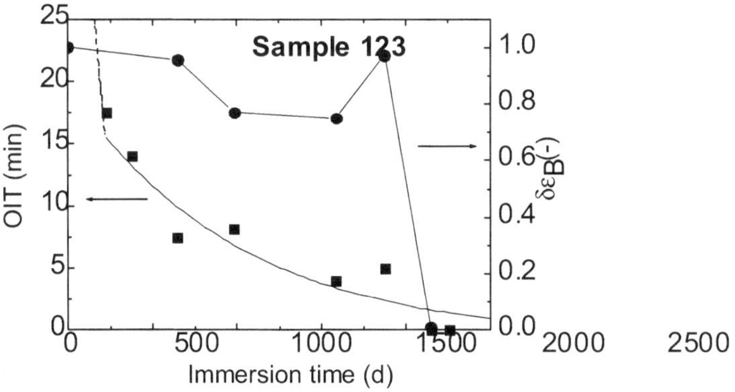


**Fig. 5.31.** Results of the immersion test on sample 123, see caption Fig. 5.30

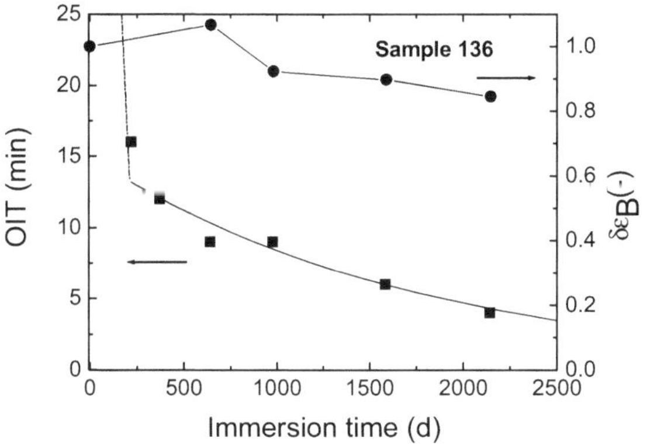


**Fig. 5.32.** Results of the immersion test on sample 136, see caption Fig. 5.30

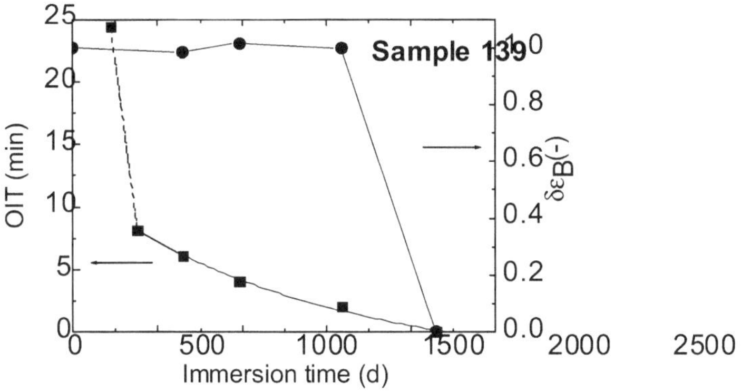


**Fig. 5.33.** Results of the immersion test on sample 139, see caption Fig. 5.30

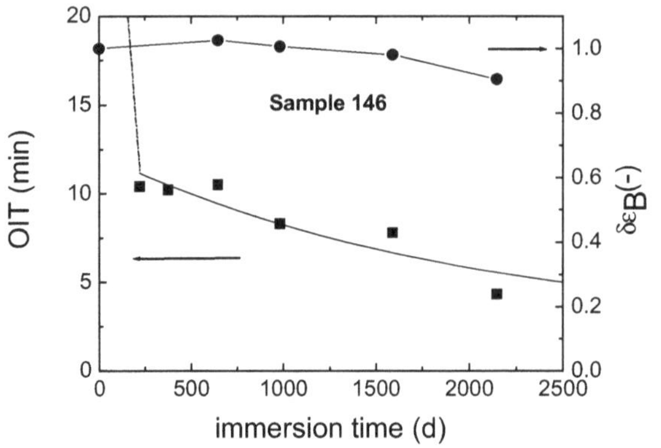


**Fig. 5.34.** Results of the immersion test on sample 146, see caption Fig. 5.30

Sample 48 (Fig. 5.29) is obviously made of a poorly stabilised resin when compared with other resins tested. The initial OIT value is extremely low (Table 5.5). For geomembranes made of this resin the rapid decrease in OIT value continues. After about 2 years of immersion in water it was not possible to measure an OIT value. Even at $T_M$ = 140 °C the OIT value was practically zero. The intrinsic oxidation stability of this resin appears to be very low. At the time when the OIT value reached very low values (580 d), the oxidative degradation process started. A drastic reduction in the elongation and stress at break was observed within months. Yet, at that time there was no change in yield stress and strain and therefore the field performance of the GM would not have changed. However, the oxidation process has continued and after 6 years of immersion the elongation at break is well below the yield point and as low as about 6 %: the samples are very brittle. The decrease in $\varepsilon_B$ was accompanied by a reduction in the melt mass-flow rate indicating cross linking as the main early degradation effect. The identical oxidative behaviour as shown in Fig. 5.29 (see Sect. 3.2.13) was observed in pipe pressure tests at 80 C with pipes made of the same resin (Koch et al. 1988). According to the authors, stress rupture at 2 MPa was due to oxidation and not due to stress cracking. Indeed, the failure time of the pipes at 80 °C agrees perfectly with the immersion time at which the drastic change in elongation at break of the immersed geomembrane starts (see Fig. 5.29).

For sample 139 (Fig. 5.33) the decline in the OIT value evolves much more slowly after the initial strong reduction. The remaining OIT value after 982 days was 5 min at $T_M$ = 180 °C with a Cu pan. However, after 2147 days (i.e. about 6 years) the OIT value has dropped below the detection limit both at $T_M$ = 180 °C and $T_M$ = 200 °C both with Cu pans and with

Al pans. Similar to the behaviour of sample 48, the elongation at break of sample 139, which is fairly constant as long as there is a measurable OIT value, was reduced to values below the elongation at yield by oxidative degradation within several months (see Fig. 5.24 for the stress-strain curves). The behaviour of the mechanical properties and the course of the OIT curve of sample 123 (Fig. 5.31) are similar to sample 139. After 2147 days of immersion and a considerable reduction in the OIT value, the material suddenly became brittle.

The behaviour of the mechanical properties of sample 82 is quite unique (Fig. 5.30). The decline in OIT of this sample is accompanied by a continuous decline of its mechanical strength and of the melt mass-flow rate over several years. However, the elongation at break is still far above the yield point (see Fig. 5.24 for the stress-strain curves). The MFR (190/5) and MFR (190/21.6) decreased from 1.4 g/10min and 27 g/10min (initial values) to 0.9 g/10min and 25 g/10min (at 982 d) and then increased to 1.5 g/10min and 37 g/10min (at 2147 d) (Table 5.6). After about 6 years test time the OIT value has dropped below the detection limit both at $T_m = 180$ °C and $T_m = 200$ °C when measured either with Cu pans or with Al pans. The behaviour of sample 82 gives rise to the question of morphological changes during ageing. Table 5.7 shows the crystallinity of the samples at different immersion times, which was determined by the heat of melting assuming 292 J/g for fully crystallised HDPE. A small recrystallisation occurred in some of the samples in the first year of immersion which creates a so-called memory peak in the melting curve. The position of this small peak is characteristic for the immersion temperature. Afterwards crystallinity remained constant within the measurement errors.

The OIT value at $T_M = 180$ °C of samples 136 (Fig. 5.32) and 146 (Fig. 5.34) at 982 d was 9 min (136) and 8 min (146). After 2147 days the OIT value is still 3 min ($T_M = 180$°°C, Cu pans) and 8 min ($T_M = 180$ °C, Al pans) for sample 136 and 4.5 min ($T_M = 180$ °C, Cu pans) and 15 min ($T_M = 180$ °C, Al pans) for sample 146. No change was observed in the tensile properties of these samples (Fig. 5.32 and 5.34), melt mass-flow rate or density (Table 5.6) during the entire 6 years of immersion. Obviously we have to assume that a considerable amount of stabiliser is still available even after this long period of time.

Table 5.8 assembles the data of the elongation at break, tensile strength and stress at yield, if the yield point was reached, and of the OIT values at $T_M = 180$ °C for all geomembrane samples after 2147 days of immersion in hot water. These data show that the mechanical properties of HDPE geomembranes are not substantially deteriorated as long as a significant OIT value is measurable. When the OIT value reaches zero or very low values as compared with the initial OIT, the oxidative deterioration evolves rapidly.

What conclusion about service lifetimes can be drawn from these long-term test results? For our discussion it is crucial that the change in OIT value provides at least a rough estimation for the depletion time $t_2$. This assumption was established for the antioxidant package commonly used in HDPE geomembranes and it explains the oxidation behaviour in the OIT measurement itself as described in Sect. 3.2.7. In the following we will show that a consistent explanation of all our experimental results is possible within this interpretation scheme.

**Table 5.8.** OIT values and mechanical properties of HDPE geomembranes determined after 5.9 years of immersion in 80 °C water for samples 82, 123, 136, 139, 146 and after 5.1 years for sample 48

| Sample | Properties after 2147 d (5.9 y) of immersion in hot water | | | |
|---|---|---|---|---|
| | $\varepsilon_B$(%) | $\sigma_B$/ $\sigma_Y$ (N/mm²) | OIT value $T_M$= 180 °C Cu pan | OIT value $T_M$= 180 °C Al pan |
| 48 | 6 | 22 | 0 | 0 |
| 82 | 500 | 13 / 21 | 0 | 0 |
| 123 | 5 | 12 / - | 0 | 0 |
| 136 | 1650 | 29 / 19 | 3 | 8 |
| 139 | 3 | 16 / - | 0 | 0 |
| 146 | 1400 | 25 / 22 | 4.5 | 15 |

Common to the ageing behaviour of all geomembrane samples investigated is that as long as there is a significant OIT value measurable at temperatures above the melting point, the mechanical properties remain intact. This includes even the mechanical behaviour of sample 82 since elongation at break in tensile testing still remains well above the yield point. When the OIT value has dropped to zero or to very low values, the oxidative degradation starts and the materials become brittle within several months. Brittleness means that bending the geomembrane specimen by hand, one can easily break it without yielding in the fracture surface. However, the HDPE material remains essentially intact and can sustain high stress. There is no indication of complete deterioration as is sometimes observed in the ageing of PP materials.

Significant is the difference in the behaviour between ageing in air and immersion in hot water. During ageing in air the OIT-value slowly decreases in an exponential-like steady fashion. After 13.6 years the melt mass-flow rate indicates that finally oxidative degradation with cross linking effects might have started but there is still enough oxidative resistance to prevent any change in the mechanical properties. During ageing in water a steep decrease in the OIT value occurs before it levels off. For most samples it approaches then very low values within a few years and oxidation of the polymer itself starts. However, closer inspection of the decline of the OIT value during oven ageing clearly shows that these data also follow a two-step exponential decline with ageing time. Fig. 5.26 shows that the data are nicely fitted by a superposition of two exponential declines, numbered 1 and 2 and each described by Eq. 5.61, with a short-term antioxidant depletion time constant of $\tau_{eff,1} = 280$ d and a long-term one of about $\tau_{eff,2} = 2200$ d. The corresponding depletion rates are $k_{eff,1} = 0.11$ month$^{-1}$ and $k_{eff,2} = 0.014$ month$^{-1}$. The offset $B = B_1 + B_2$ is zero and the amplitudes are $A_1 = 0.6$ and $A_2 = 0.4$. The superposition of two different depletion rates is obvious for the immersion in water. Here the first antioxidant depletion time is so short (< 100 d) that the long-term behaviour is essentially described by the second component alone. The curves obtained by fitting the long-term antioxidant depletion data with Eq. 5.61 are plotted in Figs. 5.30 to 5.34. Reasonably accurate fits were obtained with antioxidant depletion times in the range of 1000 d to 2000 d. As expected, the constant $B$ is either zero or negative having a value of minus a few minutes.

Two mechanisms might contribute to depletion. The first one is stabiliser migration from the bulk HDPE material which is a combined process of stabiliser diffusion to the surface driven by a concentration gradient followed by evaporation or leaching and possibly accompanied by chemical degradation of the stabiliser (i.e. hydrolysis of phosphites). The second one is the consumption of stabiliser by inhibiting the oxidative reaction chain. We may therefore write:

$$k_{eff} = k_{migration} + k_{consumption}. \tag{5.62}$$

If stabiliser depletion is dominated by antioxidant consumption, it should be controlled by oxygen supply and therefore highly depend on the ageing conditions, i.e. nearly stationary air or forced air circulation or immersion in water. Even if the oxygen supply is high enough to establish equilibrium, antioxidant consumption for ageing in air could not be slower than for immersion in water, as we expect the equilibrium concentration of oxygen in the geomembrane to be at least similar for both ageing conditions. Therefore the consumption should proceed more rapidly in air than in water. On the contrary, both short-term and long-term depletion rates are higher for immersion in water than ageing in air. Furthermore, there is no difference in the change in OIT value between ageing in nearly stationary air and ageing in forced air circulation. All these findings support the view that the stabiliser depletion observed in our cxperiments where the partial pressure of oxygen is no higher than in normal air and fresh air supply is limited to some extent, is essentially determined by migration processes and not by consumption.

However, the oxidative induction time of the high-temperature oxidation process in a pure oxygen atmosphere, i.e. in the DSC-apparatus, should be due to consumption of antioxidants in the oxidative reaction chain, because of a high rate of free radical initiation and a good oxygen supply. This assumption is confirmed by the much shorter OIT values obtained in DSC measurements with Cu pans than with Al pans: transition metals such as copper enhance radical generation by catalytic effects and therefore increase the antioxidant consumption rate (Hawkins et al. 1971). Nevertheless, the activation energy of the antioxidant migration process is smaller (see below) than the activation energy of the high-temperature oxidation process (see Sect. 3.2.7). By decreasing the temperature, the rate of oxidative consumption will decrease much faster than the depletion rate due to migration processes. Therefore at low temperatures and reduced oxygen supply the condition should be fulfilled:

$$k_{consumption} < k_{migration} \text{ and therefore } k_{eff} \approx k_{migration}. \tag{5.63}$$

It is well known from pipe pressure tests that immersion in water enhances antioxidant migration in comparison to air, as was observed in our experiments. Smith et al. attributed this to the influence of carbon black (Smith et al. 1992). Antioxidants are adsorbed at the carbon black surface, whereby the migration process is retarded considerably. During immersion in water some moisture is also present in the HDPE bulk material and this is preferentially adsorbed by the carbon black aggregates. The adsorption of water thereby supersedes that of the antioxidants and decreases the retardation coefficient and increases the apparent diffusion coefficient of the latter. In addition, it was suggested that the antioxidant forms loosely bonded clusters within the dry PE bulk material which may disperse slowly in contact with diffusing water thus enhancing the antioxidant diffusion coefficient (Le Poidevin 1977).

Since the phosphite stabiliser substantially determines the extent of the initial OIT value of an HDPE GM product and since this stabiliser forms the largest part of the antioxidant package, it is assumed that depletion especially of the phosphite component is the major factor in the first short-term depletion process. Most of this stabiliser seems to be lost in the first year of immersion in hot water. Within this context it is reasonable to assume that not only leaching but also hydrolytic deterioration of the phosphite stabiliser, possibly even within the bulk material, is reflected in the steep decrease of the OIT value during immersion in hot water. The long-term antioxidant depletion time would then be determined by the migration of the remaining phenolic stabiliser. Therefore, a high initial OIT value does not necessarily correlate with good long-term oxidation stability.

Since geomembranes are parts of geotechnical constructions, we should assume as a worst case that they are under conditions of high moisture or even under a permanent water head for most of the time. Therefore, we consider the failure time for immersion in hot water or the corresponding long-term antioxidant depletion rate as a starting point to extrapolate a lower limit of the relevant service life for most geotechnical applications at normal temperature. For all our samples, except the purely stabilised material 48, oxidative degradation starts only after 5 years of immersion in water at 80 °C and much later for ageing in air at the same temperature. We can take this time interval of 5 years as a lower limit of the antioxidant depletion time $t_2$ at $T = 80$ °C or, alternatively, we can take the corresponding long-term antioxidant depletion rates 0.015 to 0.03 month$^{-1}$ at 80 °C as lower limit for an estimate of the depletion rate at ambient conditions.

Knowing the time-to-failure at 80 °C, we need the activation energy of the depletion process to be able to estimate the service life. For Sample 48, the poorest stabilised HDPE resin in our study, Koch et. al. extrapolated an activation energy of 100 kJ/mol from extensive pipe burst pressure test data generated at 60 °C, 80 °C and 95 °C and at low pressures (see Sect. 3.2.7) (Koch et al. 1988). G. Hsuang presented oven ageing data for antioxidant depletion of a commercially available HDPE geomembrane, extrapolating 200 years and activation energies of about 60 kJ/mol (Hsuan and Koerner 1998). From high-temperature OIT measurements they obtained an antioxidant depletion rate at 85 °C of about 0.14 month$^{-1}$ for mixed ageing conditions with air below the geomembrane, and immersion in water above the geomembrane, which lasted for about 2 years. This depletion rate coincides nicely with the depletion rates observed in our experiments for the short-term depletion process within the first two years. However, our data clearly show that long-term oxidation stability is determined by a long-term depletion process with a rate about half an order of magnitude smaller: 0.015 to 0.03 month$^{-1}$. Therefore we expect that service lifetimes and activation energies are also much higher than estimated by G. Hsuan and R. Koerner from 2 years of testing (Hsuan and Koerner 1998). The activation energy of the diffusion process of organic molecules with high molecular weight in HDPE bulk material is typically in the range of 100 kJ/mol, e.g. the activation energy of Irganox 1010, a phenolic stabiliser frequently used in geosynthetics, was determined to be 115 kJ/mol in LDPE (Moisan 1980), 100 kJ/mol in PP homopolymer and 113 kJ/mol in a PP-PE copolymer (Ferrera et al. 2001). Smith et al. found an activation energy of the overall migration process of a combined phosphite and phenolic antioxidant package of 80 kJ/mol for ageing in air and 100 kJ/mol for immersion in water (Smith et al. 1992).

The Van't Hoff rule for temperature dependence of the antioxidant depletion time $t_2(T)$ is given by Eq. 5.9. Taking $t_2(T') = 5$ y at $T' = 80$ °C and a very small value of the activation energy of 60 kJ/mol (or $E/R = 7200$ K), the lower limit of the service life with respect to oxidative degradation under ambient conditions ($T = 20$ °C) is expected to be at least 300 years. However, we obtain more than 1000 years with 100 kJ/mol (or $E/R = 12{,}000$ K). The time until the functional engineering properties of an HDPE geomembrane are significantly affected by oxidation is therefore so long that oxidative degradation is not relevant for design considerations even in landfill liner systems. The durability of HDPE geomembrane can be classified as permanent in this respect.

In the light of these results test procedures and specifications should be established to assess the oxidation stability of HDPE geomembranes. Obviously, high initial OIT values need not correlate with long-term oxidation stability. It was therefore suggested to combine OIT measurements with oven ageing procedures which accelerate stabiliser migration. The GR-13 Standard requires an initial OIT value at $T_M = 200$ °C of 100 min or a high pressure OIT value of 400 min at $T_M = 150$ °C and $p_{O2} = 38$ bar. The reduction in OIT value must be less than 45 % following 90 d oven ageing at 85 °C. The BAM certification requirements (Müller 2001) for geomembrane specify initial OIT values at $T_M = 200$ °C of > 20 min and > 10 min after 0.5 year of air oven ageing at 80 °C. Continuing the ageing for 1 year, the relative change in OIT value between 0.5 year and 1 year must be less than 30 %. A combination of requirements for the initial value and the change during ageing seems to be a reasonable approach even if the test conditions for accelerating antioxidant depletion are still rather poor.

Oven ageing not too close to the melting temperature requires a long test time to monitor changes in the materials properties. Testing with high oxygen pressure was therefore suggested (Salman et al. 1998; Schröder et al. 2000; Vink and Fontijn 2000). However, the extrapolation of high temperature OIT values, see Sect. 3.2.7, coincides with findings (Vink and Fontijn 2000) that the change from air to flowing pure oxygen atmosphere at ambient pressure by itself causes a dramatic decrease in time-to-failure. Therefore, the results of high pressure tests have to be interpreted with caution. An increase in oxygen partial pressure and supply can easily change the regime from antioxidant migration to antioxidant consumption. Oven ageing at high oxygen pressure obviously measures antioxidant consumption rate (under the specific test conditions) and completely neglects antioxidant depletion effects, since test periods are usually only a few weeks:

$$k_{consumption} >> k_{migration} \text{ and therefore } k_{eff} \approx k_{consumption}. \quad (5.64)$$

However, the consumption regime is not relevant for typical geotechnical field conditions where the migration regime always prevails. Although the change in OIT value clearly reflects antioxidant depletion, it is not reasonable to use the absolute OIT values themselves, even if measured at an elevated oxygen pressure below the melting point, to extrapolate service lives or to assess the oxidation stability of HDPE geomembranes.

## References

Akiyama S et al. (1991) A method for life prediction of PE under the constant strain condition. In: Proceedings of the Twelfth Plastic Fuel Gas Pipe Symposium. American Gas Association, Boston, USA

August H et al. (eds) (1997) Advanced landfill liner systems. Thomas Telford, London

Brown N et al. (1987) The transition between ductile and slow-crack-growth in polyethylene. Polymer 28: 1326–1330

Brown N and Lu X (1995) A fundamental theory for slow crack growth in polyethylene. Polymer 36: 543–548

Chen YL and Rånby B (1990) Photocrosslinking of Polyethylene. III. Thermal Oxidative Stability of Photocrosslinked Polyethylene. Journal of Polymer Science: Part A: Polymer Chemistry 28: 1847–1859

Dolezel B (1978) Die Beständigkeit von Kunststoffen und Gummi. Carl Hanser, München

Duvall DE and Edwards DB (1992) Creep and Stress Rupture Testing of Polyethylene Sheets under Equal Biaxial Tensile Stresses. In: Proceedings of the Annual Technical Conference 1992 (ANTEC 92) of the Society of Plastic Engineers. Society of Plastic Engineers (SPE), Brookfield, USA, pp 128–131

Emanuel NM and Buchachenko AL (1987) Chemical Physics of Polymer Degradation and Stabilization. VNU Science Press, Utrecht, The Netherlands

EOTA (ed) (1999) Assumption of working life of construction products in Guidelines for European Technical Approval, European Technical Approvals and Harmonized Standards, Guidance Document 002. European Organisation for Technical Approvals (EOTA), Brussel, Belgium

EPA (ed) (1992) Stress cracking behavior in HDPE geomembranes and its prevention. Environmental Protection Agency (EPA), Cinicinnati, USA

Ferrera G et al. (2001) Diffusion coefficient and activation energy of Irganox 1010 in poly(propylene-co-ethylene) copolymers. Polymer Degradation and Stability 73: 411–416

Frost AA and Pearson RG (1961) Kinetics and Mechanism. John Wiley & Sons, Inc., New York

Gaube E (1959) Zeitstandfestigkeit und Spannungsrißbildung von Niederdruck-polyäthylen. Kunststoffe 49: 446–454
Gaube E et al. (1976) Rohre aus thermoplastischen Kunststoffen – Erfahrungen aus 20 Jahren Zeitstandprüfung. Kunststoffe 66: 1–8
Gebler H (1989) Langzeitverhalten und Alterung von PE-HD-Rohren. Kunststoffe 79: 823–826
Gray RL (1990) Accelerated Testing Method for Evaluating Polyolefin Stability. In: Koerner RM (ed) Geosynthetic Testing for Waste Containment Applications ASTM special technical Publication: 1081. American Society for Testing and Materials (ASTM), West Conshohocken, pp 57–74
Griffith AA (1921) The Phenomena of Rupture and Flow in Solids. Phil Tran Roy Soc A221: 163–198
Gugumus F (1990) Antioxidantien. In: Gächter R and Müller H (eds) Taschenbuch der Kunststoff-Additive. Carl Hanser Verlag, München, pp 1–103
Gugumus F (1996a) Thermooxidative degradation of polyolefins in the solid state. Part 2: Homogeneous and heterogeneous aspects of thermal oxidation. Polymer Degradation and Stability 52: 145–157
Gugumus F (1996b) Thermooxidative degradation of polyolefins in the solid state: Part 1. Experimental kinetics of functional group formation. Polymer Degradation and Stability 52: 131–144
Gugumus F (1997) Thermooxidative degradation of polyolefins in the solid state: Part 5. Kinetics of functional group formation in PE-HD and PE-LLD. Polymer Degradation and Stability 55: 21–43
Gugumus F (1999) Effect of temperature on the lifetime of stabilized and unstabilized PP films. Polymer Degradation and Stability 63: 41–52
Hawkins WL et al. (1971) Factors Influencing the Thermal Oxidation of Polyethylene. Polymer Engineering and Science 11: 377–380
Hessel J and John P (1987) Langzeitfestigkeit von Schweißverbindungen an Dichtungsbahnen aus Polyethylen. Werkstofftechnik 18: 228–231
Howard JB (1973) DTA for Control of Stability in Polyolefin Wire and Cable Compounds. Polymer Engineering and Science 13: 429–434
Hsuan YG and Koerner RM (1998) Antioxidant Depletion Lifetime in High Density Polyethylene Geomembranes. Journal of Geotechnical and Geoenvironmental Engineering 124: 532–541
Hsuan YG et al. (1993) Stress Cracking Resistance of High Density Polyethylene Geomembranes. J of Geotechnical Engineering 119: 11
Huang Y-L and Brown N (1991) Dependence of Slow Crack Growth in Polyethylene on Butyl Branch Density: Morphology and Theory. J Poym Sci, Polym Phys 29: 129–137
Hutten A (1992) Langzeiterfahrungen mit Deponiebasisabdichtungen aus PE-HD am Beispiel der Deponie Glurns (I). Müll und Abfall 24: 314–323
Irwin GR (1957) Analysis of Stresses and Strains Near the End of a Crack Traversing a Plate. Journal of Applied Mechanics 24: 361–364
Knott JF (1973) Fundamentals of Fracture Mechanics. Butterworth & Co. Ltd., London
Koch R et al. (1988) Langzeitfestigkeit von Deponiedichtungsbahnen aus Poly-

ethylen. Müll und Abfall 20: 3–12

Koerner MR et al. (1992) Arrhenius Modeling to Predict Geosynthetic Degradation. Goetextiles and Geomembrane 11: 151–183

Landau LD and Lifshitz EM (2000) Statistical Physics, 3rd Edition, Part 1. Reed Educational and Professional Publishing Ltd., Oxford

Le Poidevin GJ (1977) The oxidation of polyethylene in aqueous solution, Report ECRC/N1063. The Electricity Council Research Centre, Capenhust Chester

Lu X and Brown N (1990) The ductile-brittle transition in a polyethylene copolymer. J Mater Sci 25: 29–34

Lu X and Brown N (1991) Unification of ductile failure and slow crack growth in an ethylene-octene copolymer. J Mater Sci 26:612-620

Lu X et al. (1996) The Critical Molecular Weight for Resisting Slow Crack Growth in a Polyethylene. J Polym Sci Part B: Polym Phys 34: 1809–1813

Lu X et al. (1991) Notchology – the effect of the notching method on the slow crack growth failure in a tough polyethylene. J Mater Sci 26: 881–888

Lustiger A and Markham RL (1983) Importance of tie molecules in preventing polyethylene fracture under long-term loading conditions. Polymer 24: 1647–1654

Menges G (1973a) Das Verhalten von Kunststoffen unter Dehnung, Teil 1. Phänomenologie der Rißerscheinungen. Kunststoffe 63: 95–100

Menges G (1973b) Das Verhalten von Kunststoffen unter Dehnung, Teil 2. Deutung der kritischen Dehnung und Verhalten der Kunststoffe bei überkritischer Dehnung. Kunststoffe 63: 173–177

Menges G and Alf E (1972) Beziehung zwischen der verformungsbedingten Spannungsrißbildung und dem Versagen von Polypropylen. Kunststoffe 62: 259–267

Menges G and Rieß R (1974) Verarbeitungs- und Umgebungseinflüsse auf die kritische Dehnung von Kunststoffen. Kunststoffe 64: 87–92

Menges G and Schmachtenberg E (1987) Das Deformationsmodell. Kunststoffe 77: 289

Menges G et al. (1975) Ermittlung der kritischen Dehnung teilkristalliner Thermoplaste. Kunststoffe 65: 368–371

Michaeli W et al. (1989) Beschreibung des nichtlinearen viskoelastischen Verhaltens mit dem Deformationsmodell. Kunststoffe 79: 1356

Moisan JY (1980) Diffusion des additifs du polyethylene – I Influence de la nature de diffusant. European Polymer Journal 16: 979–987

Mueller WW et al. (2003) Comparison of the oxidative resistance of various polyolefin geotextiles. Geotextiles and Geomembranes 21: 289–315

Mueller WW and Jakob I (2003) Oxidative resistance of high-density polyethylene geomembranes. Polymer Degradation and Stability 79: 161–172

Müller H (1990) Metalldesaktivatoren. In: Gächter R and Müller H (eds) Taschenbuch der Kunststoff-Additive. Hanser Verlag, München.

Müller WW (ed) (1992) Richtlinie für die Zulassung von Kunststoffdichtungsbahnen als Bestandteil einer Kombinationsdichtung für Siedlungs- und Sonderabfalldeponien sowie für Abdichtungen von Altlasten. BAM, Labor Deponietechnik, Berlin

Müller WW (ed) (2001) Certification Guidelines for Plastic Geomembranes Used to Line Landfills and Contaminated Sites. Laboratory of Landfill Engineering, BAM, Berlin

Müller WW and Jakob I (2000) Comparison of Oxidation Stability of various Geosynthetics. In: Cancelli A et al. (eds) Proceedings of the Second European Geosynthetics Conference. Pàtron Editore, Bologna, pp 449–454

Orthmann HJ (1983) Das Spannungsrißverhalten von thermoplastischen Kunststoffen. Kunststoffe 73: 96–101

Pfahler G and Lötzsch K (1988) Stabilisierung von Polyolefinen für Anwendungen in extrahierenden Medien. Kunststoffe 78: 142–148

Popov A et al. (1991) Oxidation of stressed polymers. Gordon and Breach Science Publishers, New York

Ramsteiner F (1990) Zur Spannungsrißbildung in Thermoplasten durch flüssige Umgebungsmedien. Kunststoffe 80: 695–700

Raymond GP and Giroud JP (eds) (1993) Geosynthetic case histories, Thirty five years of experience. BiTech Publishers Ltd, Richmond, British Columbia, Canada

Salman A et al. (1998) The Effect of Oxygen Pressure, Temperature and Manufacturing Processes on Laboratory Degradation of Polypropylene Geosynthetics. In: Rowe RK (ed) Conference Proceedings of the Sixth International Conference on Geosynthetics. Industrial Fabrics Association International (IFAI), Atlanta, USA, pp 683–690

Sangam HP and Rowe KR (2002) Durability of HDPE geomembranes. Geotextiles and Geomembranes 20: 77–95

Schröder H et al. (2000) Durability of Polyolefine Geosynthetics under Elevated Oxygen Pressure in Aqueous Liquids. In: Cancelli A et al. (eds) Proceedings of the Second European Geosynthetics Conference. Pàtron Editore, Bologna, pp 459–463

Seeger S et al. (2000) Long term testing of geomembranes and geotextiles under shear stress. In: Cancelli A et al. (eds) Proceedings of the Second European Geosynthetics Conference. Pàtron Editore, Bologna, pp 607–610

Seeger S and Müller WW (2003) Theoretical approach to designing protection: selecting a geomembrane strain criterion. In: Dixon N et al. (eds) Geosynthetics: Protecting the Environment. Thomas Telford, London, pp 137–152

Smith GD et al. (1992) Modeling of Antioxidant Loss From Polyolefins in Hot Water Applications I: Model and Application to Medium Density Polyethylene Pipes. Polymer Engineering and Science 32: 658–667

Vink P and Fontijn HFN (2000) Testing the resistance of oxidation of polypropylene geotextiles at enhanced oxygen pressure. Geotextiles and Geomembranes 18: 333–343

Westergaard HM (1939) Bearing Pressures and Cracks. Journal of Applied Mechnics 61: A49–A53

Zhou Z and Brown N (1994) The effect of orientation on slow crack growth in high density polyethylene. Polymer 35: 1948–1951

Zweifel H (ed) (2001) Plastic Additives Handbook. Carl Hanser Verlag, Munich

# 6 HDPE Geomembranes with Textured Surface

## 6.1 Type and Manufacture of Surface Textures

The friction force between a plastic geomembrane and its subgrade, e.g. the mineral liner or a fine sand layer, as well as between the geomembrane and an overlying protective layer, e.g. a nonwoven geotextile, can be increased by texturing the surface on one or both sides of the geomembrane[1] (Donaldson 1994; Foik and Günther 1989; Krause 1990). Varying types of texture can be found depending on the manufacturing method. The following methods are applied (Koerner 1994; Müller 1996):

1. Embossing rollers can imprint a structure on the surfaces of the polyethylene sheet as it emerges from the flat die still in a viscous state. For this an appropriate pattern is engraved into the surface of the cooling rollers, which therefore also act in the capacity of calender rollers. The bearings and mountings of the embossing rollers are exposed to heavy loads during this calendering process. This method is often called structuring or embossing.
2. In a subsequent processing step, texture particles (like thin threads or foam scraps) made from the same material as the geomembrane or another thermoplastic material are thermally bonded to the surface of a smooth geomembrane. There can be a lot of variation in the technical realisation of this method. In the so called impingement method hot thermoplastic particles are impinged or sprayed onto the surface of the smooth geomembrane by a high pressure hot air spray gun. In the lamination method thermoplastic foam is laminated onto the surface of the smooth geomembrane.

---

[1] The "visual or tactile characteristics of a surface" is called texture. Geomembranes, whose surfaces are treated in some way to increase friction, are called textured geomembranes. Since the surface characteristics is often very marked and of considerable depth the term structure and structured geomembrane are often used synonymously.

3. An internal and a gas-infused external layer of geomembrane material or two gas-infused external layers on both sides of the internal layer are co-extruded through a multi-slot co-extrusion die. On emerging from the die, the gas bubbles in the hot material burst on and just beneath the surface of the gas-infused external layers and a coarse surface, covered with small craters, develops. This is the so called co-extrusion method.

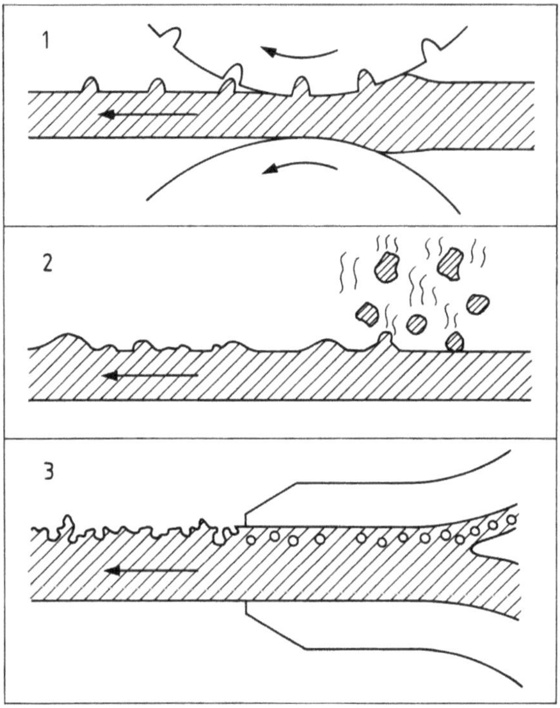


**Fig. 6.1.** Schematic illustration of the manufacturing processes for textured geomembrane surfaces: embossing and embossed structure (1), impingement and impinged texture (2) and co-extrusion and co-extruded texture (3). All three processes are capable of manufacturing double-sided textured geomembranes

Figure 6.1 schematically illustrates the three manufacturing processes. Processes 1 and 2 (embossed and impinged or laminated texture) can easily be implemented in terms of process engineering in such a way that an edge strip on each side of the geomembrane remains smooth (which is then covered with a protective tape). The geomembrane will later be welded on this protected smooth edge strip. In Process 3 (co-extruded texture) a smooth-edge strip is very difficult to produce. The geomembranes are usually textured throughout, so textured areas must be welded.

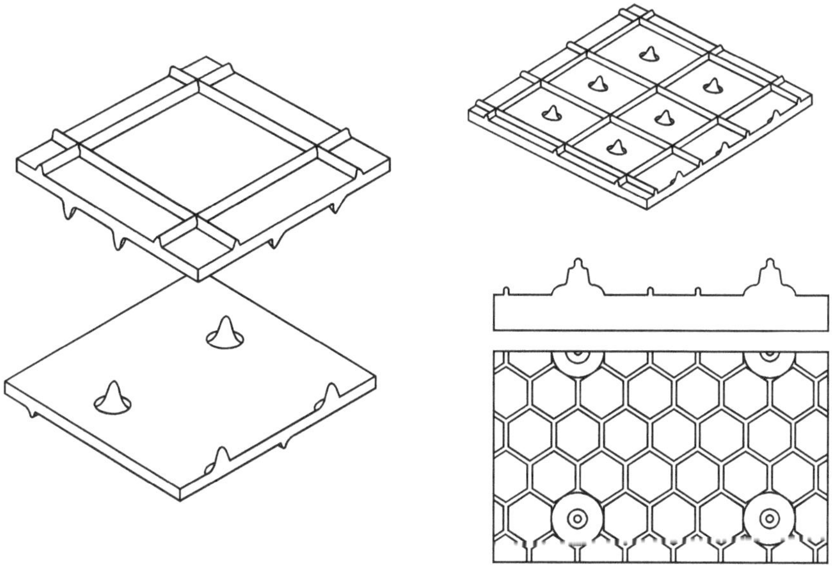

**Fig. 6.2.** Drawings of HDPE geomembranes with embossed surface structures. The manufacturers often give special names to their structure. Left and right top: structures called "spike", "grid" and "diamond-nub". Right bottom: view and cross-section of a structure called "megakron"

Each manufacturing process has specific pros and cons. The embossing procedure for wide width geomembranes requires particular calender roller equipment with bearings and mountings of high mechanical stability and an appropriate effectiveness of the cooling of the embossing rollers. Textured geomembranes are stronger oriented by this procedure compared to smooth geomembranes. Adherence to the permissible dimensional stability (Sect. 3.2.5) therefore requires a very slow manufacture speed. Abrupt changes and sharp edges in the cross-section of the structure may lead to points where stress concentration occurs and crack initiation will preferably start, as can be observed in stress crack tests. Considerably reduced times-to-failure in the long-term tensile test are sometimes observed. The stress crack resistance of these structured geomembranes must therefore be evaluated by long-term tensile test (Sect. 3.2.16). On the other hand, such structures can provide large friction forces, especially in combination with compacted clay liners of high water content. Large spikes (Fig. 6.2) are used, for example, to insure satisfactory friction in cases where a slippery thin clay layer between compacted clay liner and impervious geomembrane develops due to discharge of consolidation water. The embossed textures cannot be over-welded using the hot wedge welding method, which is necessary at T-connections despite the smooth edge strip. Therefore, the structure must be removed in the weld area by hand held rotary grinders. So a careful, professional seam preparation may require a large amount of work. However, these geomembranes have the great advantage that the questions of long-term adhesion of the structural elements and of the aging behaviour of the texture particles do not arise. Double-sided textured geomembranes, with various geometrical patterns with different friction resistance on either surface, can be manufactured without problem. Figure 6.2 shows examples of embossed textures.

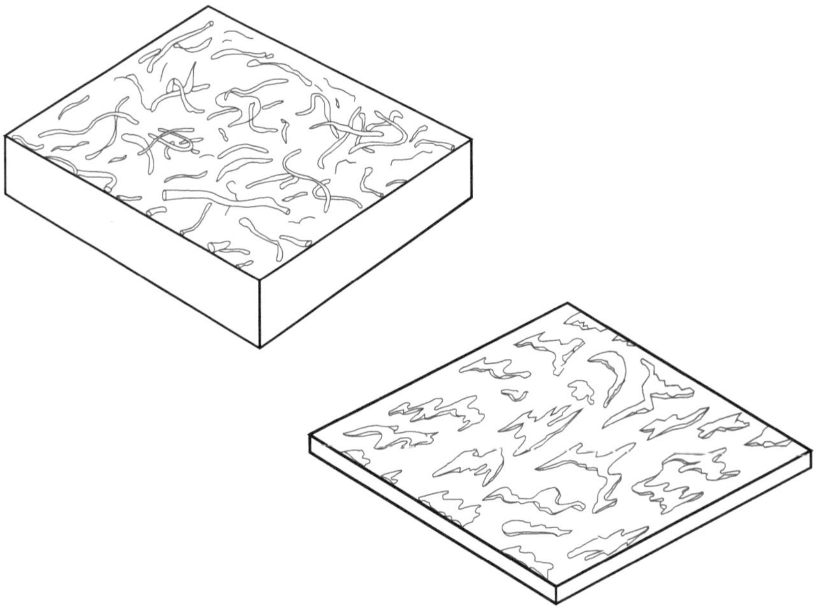

**Fig. 6.3.** Drawings of 2.5 mm thick HDPE geomembranes with impinged and laminated surface texture (different scales). Top left: texture particles sprayed with a hot air spray gun. Bottom right: HDPE foam laminated onto the smooth geomembrane. Such textures are often called “rough” or “sand-rough”

With the textures applied in a second process step, the base geomembrane of the textured product has identical properties to a smooth geomembrane. However, the above mentioned questions arise of how effective the adhesion of the texture particles is and of their specific long-term behaviour, i.e. the question of the long-term shear strength of the textured surfaces (Sect. 3.2.18). The problem of stress crack resistance must also be considered in connection with these textures. The joint between texture particles and the base geomembrane can be a "weak point" which encourages stress crack formation. Top sketch in Fig. 6.3 gives an impression how texture particle look like, which were sprayed onto the surface at a very high temperature and pressure by a hot air piston. The bottom sketch shows a texture where gas-infused HDPE foam was laminated after preheating the smooth geomembrane. When double-sided structures are applied, different friction resistances can be achieved on the upper and lower surfaces by changing the weight per unit area of the material applied. The uniformity of the texture distribution and the applied mass has to be continuously monitored within manufacture quality control. These textures have the special advantage that the texture particles may develop undercuts where non-woven fabrics may grip. Therefore, this combination of textured geomembrane and geotextile can provide very high friction forces. When textured parts have to be welded for T-connections, the low height and relatively small mass per area of the textures normally enables direct welding if suitable hot wedge welding machines are used.

Texture manufactured by the co-extrusion process (Process 3) is quite uniform and it offers good friction values in combination with fine grained mineral materials. The texture is evenly distributed over the surface, but their friction resistance cannot be varied or at least only to a small extent. The problem of stress crack resistance must be considered as with the other textures. Since the geomembranes are usually textured throughout the whole surface, two textured areas must be welded for each hot wedge weld seam. For the individual co-extruded textures as well as for impinged or laminated textures proof must be provided showing that this is possible without any loss of quality in the properties of the weld seams (Sect. 10.3).

Each manufacturing process and each type of texture, whether embossed, impinged, laminated or co-extruded, has special features and problems in evaluating their performance which have to be (and can be) solved (see Sect. 6.2). All processes enable the manufacture of geomembranes with textured surfaces which meet the BAM certification requirements, in particular concerning long-term behaviour. Therefore, a suitably textured HDPE geomembrane can usually be found in each single project for a wide range of friction values over a wide range of friction partners. An overview (in German) of HDPE geomembranes with a surface texture certified by BAM for landfill liners can be viewed on the internet at the website of BAM's working group "Plastics in Geotechnical and Geoenvironmental Engineering": www.bam.de\deponietechnik.htm.

## 6.2 Tests on Textured Geomembranes

The characteristics of a textured surface affect the physical properties and the construction method (installation and welding) of the geomembrane. An optimum between conflicting requirements must be sought. On the one hand as much friction as possible should be achieved with different friction partners, on the other hand all other properties, especially the long-term behaviour, should change as little as possible in comparison to the smooth geomembrane. In principle, the desired friction resistance should be achieved primarily by a large number of smaller texture particles or structure elements and to a lesser extent by the height of individual structure elements. A few, although large and high structural components, lead to an intimate but very localized connection with the subgrade below and/or the protective layer above the geomembrane. Thus strong local stress concentration may develop generated by the shear forces. The material of the texture particles should correspond to that of the base geomembrane as much as possible, processing should be gentle and the joints should not create especially "weak points", so that the statements of long-term performance of smooth geomembranes should equally apply to textured geomembranes. However, these principles can only be more or less adhered to if large friction parameters are to be simultaneously achieved. Therefore special test methods and standards must be specified for the textured geomembranes which enable the evaluation of the properties of textured geomembranes. The requirements largely depend on whether the friction forces act only temporarily during installation and transient operating conditions (which may occur in base landfill liners) or whether shear stresses will act over the entire service lifetime of the geomembrane (like in landfill capping systems).

Under the procedure of BAM certification of plastic geomembranes for landfill liners, special test methods and requirements have been specified for geomembranes with textured surface. Table A1.4 in Appendix 1 shows the test procedures and requirements from the certification guidelines (Müller 2001), which aim to secure service lifetimes comparable to that of a smooth geomembrane. To what extent the requirements may be reduced if a shear stress transfer is required for only small or normal service lifetimes (see Table 5.1), is difficult to assess.

An obvious challenge in testing of textured geomembranes manufactured by processes 1 and 3 is the determination of geomembrane thickness. The measurement of thickness cannot usually be performed using the standardised procedures described in Sect. 3.2.2. Sometimes individual test methods fitting to the special geometrical properties of a texture must be implemented as internal test standards of the manufacturer following already existing standards or guidelines. For embossed textures the thinnest place in the contour of the geomembrane cross-section, the "low spots" or "valleys", can be determined as reference points and the local thickness is measured by a mechanical feeler gauge. Such a procedure using a mechanical thickness gauge is described in the standard ASTM D5994-96 *Standard Test Method for Measuring Core Thickness of Textured Geomembrane.* According to the standard the mean value of thickness should be determined over such a large number $n$ of randomly taken specimens that the interval of ± 5 % around the mean value is its 95 % confidence interval[2]. This mean value is referred to as core thickness of the geomembrane.

In some embossed textures and, in particular, in co-extruded textures this is not so easy to perform since there is no unambiguously definable position where to apply the thickness gauge. Here the thickness of the geomembrane can only be optically determined. For this purpose the cross-section of geomembrane specimens is assessed by a profile projector in a laboratory test (10 to 20-fold enlargement, accuracy of the mechanical positioning: 1 μm). The thinnest spot in the profile is located and the thickness measured. On the installation site the cross-section of the geomembrane sample can be investigated using a simple hand microscope of constant enlargement (30-fold) and

[2] The standard gives a formula for the calculation of $n$ assuming that a reliable estimate of the true coefficient of variation is known, for example, from quality control of the manufacturer. Let $c$ be the coefficient of variation of a very large number $N$ of laboratory quality control measurements of thickness on randomly taken specimens. Let $t_{97.5}$ be the 97.5-fractile of the $t$-distribution for $N$ degrees of freedom. The number $n$ can then be calculated according to $n = (t_{97.5}c/0.05)^2$. A reliable estimate is obtain for $n$, if $N >> n$.
Usually no value for $c$ is available and one only wants to control whether there is, for example, a 95 % chance that the "true" mean value is above a certain specification. Assume that values of thickness measurements are normally distributed. Let $\mu$ be the "true mean thickness" of the geomembrane, $m$ the measured mean value and $s$ the standard deviation of a sample of an agreed upon number $n$ of randomly taken specimen. (For example 10 specimens is a reasonable number.) The $t$-ratio is given by $t = \frac{m-\mu}{s/\sqrt{n}}$. This ratio is distributed according to the $t$-distribution for $(n\text{-}1)$ degrees of freedom and there is a 95 % probability that this $t$-ratio is smaller then the 95 %-fractile $t_{95}$ (For $n = 10$, $t_{95} = 1.833$). Therefore, there is a 95 % chance that the true mean value is $\mu > m - t_{95}\frac{s}{\sqrt{n}}$. I.e. check whether $m - t_{95}\frac{s}{\sqrt{n}}$ obtained from your measurement is larger than the specification value.

a clamping device and the thickness (of the cross-section image) can be measured using a hair cross and a scale (3 mm) in the eyepiece with an accuracy of at least ± 0.05 mm. With a somewhat higher accuracy it can be determined whether the minimum local thickness of the geomembrane specimen falls below a specified minimum value, for example 2.50 mm in landfill liners. The thickness of the base geomembrane is decisive for impinged or laminated textures (Process 2).

When textures are embossed, the geomembrane material is usually considerably oriented. The extent depends on the manufacturing process parameters. In particular the manufacturing speed has a substantial influence. In Sect. 3.2.5 the measuring of the dimensional stability was discussed. A narrow range for the permissible values of the dimensional stability test parameter, i.e. that for the absolute value of shrinkage or extension, should also be set in geomembranes with embossed textures, although more "latitude" can be allowed than in smooth geomembranes, so that optimum characteristics of the textured geomembranes is reached at justifiable cost of manufacture.

Textured HDPE geomembranes must agree in all substantial mechanical properties, such as yield stress, elongation at yield and arc elongation at break in burst testing within the typical variation of the measured values with those of the smooth geomembrane from the same resin. However, the achievable elongation at break is closely linked to the kind of surface texture. Depending upon the texture it can decrease very strongly. However, since high elongations at break are never relevant for practical applications, this does not represent any reduction in performance quality, as long as the actual value is not below a certain minimum (e.g. 100 %) which in turn is substantially greater than the elongation at yield. The mean value and width of the distribution of elongation at break can, however, be considered as a parameter for the evaluation of the quality of the manufacturing process. For this purpose the properties of the tensile test specimen – e.g. the texture details included in the cut-out of the test specimen must be the same for all specimens – and therefore the sampling procedure must be specified exactly for each individual texture. A minimum value can then be specified either for the average value or for all individual values of a sample of the elongation at break, which serves as a quality assurance criterion for a uniform and careful production. In accordance with the regulations of the certification procedure, the minimum quality assurance value of elongation at break attached to the respective texture should not be less than 100 %.

In impinged or laminated textures, attention must be paid to the uniformity of the texture. The distribution in the mass per area and in the form of the texture particle may not be so different along the geomembrane that an additional significant variation develops in the friction parameters determined in a shear box test (Sect. 3.2.17) on 30 cm by 30 cm samples taken from different parts of the geomembrane. The uniformity can initially be visually assessed. The foreman regularly inspects the uniformity of the texture during production. In cases of doubt reference samples can be used as a comparison. Uniformity can also be characterised quantitatively to certain extent by determining the distribution of the mass per area of the impinged or laminated material. Specimens with an area of 100 $cm^2$ are randomly taken over the geomembrane and the mass of the texture material applied to their surface is determined by weighing. Uniformity can be checked by comparing the mean value and standard deviation of the mass per area against a pre-set specification. The manufacturers have well defined in house standards for the test method and relevant specifications for the individual manufacturing process parameters.

Similarly to the testing of the required uniform distribution of texture particles, the quality of their adhesion to the smooth base geomembrane should also be checked carefully. However, a test standard to regularly check the adhesion of the texture particles on geomembranes in manufacture quality assurance does not yet exist. Only a few mechanical short-term tests have been applied so far. Therefore no quality control tests or only very simple qualitative "organoleptic tests", e.g. scraping off with the thumb nail and similar processes, are performed during the manufacturing process. One relies on the assumption that the exact specification of the process steps and parameters leads repeatedly to the same quality of adhesion when applying texture particles.

Obviously adhesion must be so strong that no problems occur in the friction parameter determination using the shear box test. Structural components may not be torn or pulled off in the transition from static friction to sliding friction. If any such effects occur, then the whole geomembrane delivery should be carefully inspected and possibly be rejected. G. Heimer and K. Müller of the Technical University (State Materials Testing Office (MPA)) of Darmstadt have developed a planing procedure in which a sharpened steel wedge planes off the texture particles from a geomembrane specimen at a specified pressure and speed, the planing force is measured in the sliding state. L. Glück of the SKZ (Süddeutsches Kunststoffzentrum), Würzburg has used abrasion tests, in the form as used for the characterisation of abrasion behaviour of wall-to-wall carpets, to provide a quantitative indication of the adhesion.

If a texture geomembrane is installed on a steeply inclined external slope with some backfilling on top of it, where down-hill slope forces act continuously and permanent shear stresses and friction forces emerge in the geomembrane and in its interface to neighbouring components, it is in principle possible that a sliding surface develops on the intact smooth base geomembrane due to failure of the weld joints between impinged or laminated texture particles and base geomembrane. This can not only be caused by a transition to a ductile failure in the joints because of creep, but also by a premature ageing of the load-transferring, fine texture particles (threads, foam scraps) and accompanied brittle failure, which may then cause the entire liner structure to slide. The long-term shear strength of impinged or laminated textured geomembranes must therefore be tested. Section 3.2.18 deals with the test procedure developed by BAM (Seeger et al. 2000). In (Seeger and Müller 2001) the test results for certain impinged textures are reported in detail. Such tests in principle enable an estimation of the service lifetime of textured geomembrane, on the basis of the temperature dependence of friction force versus time-to-failure curves.

BAM's certification procedure includes an immersion test of textured geomembranes in concentrated, liquid chemicals to investigate the chemical resistance of the adhesion of the impinged or laminated texture. For this purpose samples of textured geomembranes are immersed for some time in liquid chemicals and the change in force required for a specified planing of the texture is determined in accordance with the above mentioned MPA Darmstadt's laboratory method. In this test, particularly corrosive liquid chemicals are used, which are the same as used for testing the chemical resistance of smooth geomembranes (see Sect. 3.2.11).

Every manufacturing process of textured geomembranes will have some repercussions on the sensitivity to stress crack formation. For this reason only textured geomembranes, which exceed a specified time-to-failure in the long-term tensile test under specified test conditions, will be approved for the use in landfills in accordance with BAM's certification guideline. This has to guarantee that the formation of surface textures or structures in the production process does not cause substantial weak points in the material where stress crack formation might occur. The long-term tensile test on textured geomembranes is dealt with in detail in Sect. 3.2.16.

## 6.3 Properties of Textured Geomembranes, Slope Stability

Multi-layered liner systems (e.g. subgrade, geomembrane, protective layer, drainage layer, restoration layer) must be structurally stable especially on steep slopes. The issue of structural stability needs careful consideration, as is clearly illustrated by occasional, but sometime severe cases of failure and considerable damage due to sliding. The internal shear strength of the layers is usually quite large and the problems occur on the interfaces between the different layers. Therefore, for the geotechnical design of structurally stable liner systems, not only knowledge of the loading conditions and internal shear parameters of the materials[3] is necessary but also the knowledge of friction parameters in the interfaces between the materials lying on top of each other. On each interface the static friction force achievable within the relevant range of the forces acting normally to the interface must be greater than the lateral force (down-hill slope force or lateral spreading force of cover materials), then the liner structure is stable against sliding in the interface, irrespective of the relation of the actual size of these friction forces in the different interfaces[4].

In Sect. 3.2.17 the testing procedure for determining the friction parameters is discussed in detail. The calculation procedures to prove structural safety against sliding in the liner interfaces are quite simple and are described in the literature, see e.g. (Blümel and Brummermann 1995; Koerner 1994; Terzaghi et al. 1996) or in the GLR-recommendation E2-7 *Resistance to Sliding of the Liner Systems* (DGGt 1997). Only a few special aspects shall be discussed here which are important for geomembrane liner systems (Müller 1997).

---

[3] In "sandwich-like" geosynthetics, such as geosynthetic clay liners or geocomposite drains, the internal shear strength is due to intrinsic plastic structures and it is necessary to demonstrate the stability of these plastic structures over the long term. Therefore the repercussions of creep and aging on internal shear strength must be considered by long-term tests. This issue is considered in detail in (Müller 2004).

[4] In more general terms, the sum of the "driving" forces or the "actions" on a mass element on a sliding surface – which is not only the down-hill slope force of a dead weight, but also earth pressures, water flow pressures and possibly dynamic loads due to installation procedures – must be greater than the sum of the achievable maximum values of reacting forces in the materials just before the "ultimate-limit-state" of failure occurs, i.e. greater than the so called material "resistances" (in addition to friction forces, earth pressures can also contribute to the resistance). There are different concepts as to which structural safety is to be considered. Based on probabilistic concepts, individual fractiles are calculated or determined as design values for the actions and resistances and it is shown that the sum of these design values for the action is greater than the sum of the design values for the resistance, i.e. it is explicitly shown that the probability of failure is below a specified limit (probabilistic safety concept). However, distributions are normally not available and expert opinion based all-inclusive factors of safety are used to estimate characteristic values for the actions and resistances "on the save side", whereby the sum of the actions must be greater by a global safety factor than the sum of the resistances (global safety concept), see (Blümel and Brummermann 1995; Driscoll and Simpson 2001; DGGt 1997).

**Table 6.1.** Collection of friction parameters from shear box tests (adhesion and friction angle)

| Texture | Friction partner | Adhesion | Friction angle | Friction force at $\sigma_N$=100 kN/m² | Slope inclination (1 : $n$)* |
|---|---|---|---|---|---|
| | | $a$ (kN/m²) | $\varphi$ (°) | $\tau_R$(kN/m²) | $n$ (-) |
| a | CCL[1] | 19 | 23.6 | 63 | 2.4 (3.0) |
| a | CCL[1] | 8 | 25.3 | 55 | 2.7 (2.7) |
| a | CCL/sand[2] | 12 | 25.3 | 59 | 2.5 (2.7) |
| a | CCL[1] | 9 | 19.9 | 45 | 3.3 (3.6) |
| a | CCL[5] | 45 | 12.5 | 67 | 2.2 (5.8) |
| b | CCL[3] | 12 | 17.1 | 42 | 3.5 (4.2) |
| b | CCL[1] | 15 | 20 | 51 | 2.9 (3.5) |
| b | CCL[4] | 13 | 12.2 | 34 | 4.3 (6.0) |
| b | CCL[5] | 30 | 23 | 72 | 2.0 (3.0) |
| smooth | CCL[5] | 35 | 11 | 54 | 2.7 (6.6) |
| smooth | CCL[4] | 0 | 8 | 14 | (9) |
| smooth | CCL[3] | 0 | 8 | 14 | (9) |
| smooth | Silt[6] | 11 | 26 | 60 | 2.5 (2.6) |
| smooth | CCL[7] | 8 | 17 | 38 | 3.8 (4.3) |
| c | GTnw | 0 | 11 | 19 | (6.6) |
| c | GTnw/sand | 0 | 12 | 21 | (6.1) |
| b | GTnw | 15 | 31 | 75 | 2.0 (2.2) |
| b | GTnw | 15 | 22 | 55 | 2.7 (3.2) |
| smooth | GTnw | 0 | 12 | 21 | (6.1) |

The table has been compiled from test reports provided to BAM, a, b and c denote various surface textures, CCL abbreviates compacted clay liner and GTnw non-woven protection geotextile.
*) To illustrate the range of attainable slope inclinations, the value of $n$ was calculated as $\tau_R/(\eta\sigma_N)$ from the friction stress $\tau_R$ at a testing pressure of $\sigma_N$= 100 kN/m² and a safety factor of $\eta$ = 1.5. On the other hand (values in parentheses) $n$ was derived from the slope angle $\psi = \tan^{-1}(\tan\varphi/\eta)$ as calculated from the friction angle $\varphi$ with $\eta$ = 1.3 without taking into account the adhesion.
[1]) No detailed data. [2]) 70 % clay/30 % sand. [3]) Clay of medium plasticity; shear test at 34 % water content gave a cohesion $c'$ = 15 kN/m² and internal shear angle $\varphi'$ = 22.2°. [4]) High plasticity clay; shear test at 34 % water content gave $c'$ = 13 kN/m², $\varphi'$ = 14.0°. [5]) Clay, shear test: $c'$ = 45 kN/m², $\varphi'$ = 24.5°. [6]) Silt, shear test at 16 % water content: $c'$ = 27 kN/m², $\varphi'$ = 38°. [7]) Clay, shear test at 27 % water content, $c'$ = 55 kN/m², $\varphi'$ = 17.5°; $c'$: effective-stress cohesion, $\varphi'$: effective-stress internal friction angle.

The various manufacturing methods of texturing or structuring the geomembrane surfaces offer numerous possibilities to achieve high friction parameters necessary for a stable structure. Table 6.1 shows a collection of friction parameters and friction forces of textured geomembranes against a set of mineral liners and nonwoven protective geotextiles. This table is compiled from incidentally available test reports and therefore the data are not statistically representative. Nevertheless, it does show that the friction angle of a smooth geomembrane is small where an adhesion component either apparently does not exist or is present only under very small loads. The friction angle is typically about 10° for sand, nonwoven geotextiles or mineral liners. The friction resistance of a smooth geomembrane in the interface to neighbouring components therefore remains far below the internal shear resistance of mineral layers and also usually of other geosynthetic liner components. The friction resistance of geomembranes with a textured surface, however, may reach values near the internal shear resistance of the typical neighbouring liner materials; the friction resistance to some mineral subgrades can even exceed their internal shear resistance.

The attainable friction parameter depends on both the kind and coarseness of the texture and on the moisture content and grain size distribution of the mineral layer. Friction resistances are strongly material dependant and structural stability must be proven in each case. Therefore, figures applicable universally on slope angles which are still permissible for various liner combinations cannot be specified. However, the data in Table 6.1 show, in close agreement with examples from many construction sites, that slope inclinations up to about 1 : 3 can be implemented without significant constraints, up to 1 : 2.5 with special selection and adjustment of the liner components. In special cases stable liners with geomembranes have been installed at inclinations considerably greater than 1 : 2.5.

If the lateral force acting on a layer in a liner system exceeds the friction force in the interface to the layer below and the static equilibrium condition is violated, the whole structure above this interface will start to slide. Sliding will typically occur in the interface with the lowest value of friction force. If this sliding surface lies beneath the geomembrane, which is usually fixed in an anchor trench, it will become stressed by the lateral force acting on the overlying layers. Due to the small cross-section of the geomembrane and the relatively large value of the forces involved, the imposed tensile stress is almost always greater than the stress breakpoint of the geomembrane, so that it will rapidly yield and tear.

Sometimes knowledge of the lateral forces acting on a liner system is so uncertain that a well defined equilibrium condition for the interfaces could not be formulated. Examples are the lateral forces on inside slopes of deep landfills or waste pits and the lateral spreading force in a large waste pile. Sliding within a waste body can not be completely excluded or may even be tolerable. However for safety reasons, it should always be guaranteed that the liner system remains intact, i.e. that the sliding surface is always above the geomembrane. In these cases one must ensure that the static friction force at the interface between geomembrane (GM) and subgrade (*S*) $\tau_{GM,S}$ in the relevant load range must be greater than the static friction force between geomembrane and protective layer (*P*) $\tau_{GM,P}$:

$$\tau_{GM,S} = \eta \cdot \tau_{GM,P} \qquad \eta > 1 \tag{6.1}$$

The uncertainty in determining friction parameters is taken into account by using a factor of safety $\eta$ in the design equation. Clearly friction forces in interfaces and the internal shear strength of the layers beneath the geomembrane must be correspondingly larger than $\tau_{GM,P}$. Such a requirement does not only apply specially to geomembrane liners. An analogous requirement may be imposed on other liner systems to effect that sliding always takes place above and not in or under the liner. Beside the general equilibrium condition for action and resistance, requirement Eq. 6.1 imposes an additional constraint for prove of structural stability.

When are the loading conditions so unclear, that failure must be taken into account, and under what conditions must damage to the liner systems be prevented under any circumstances? These questions must be answered with respect of the conditions of the specific building project. It is difficult to generalise the conditions under which Eq. 6.1 has to be applied. Recommendations for landfill liners are given in (DGGt 1997). In the case of landfill caps, the friction parameters of the materials and the lateral forces can be reliably determined with sufficient accuracy for any building project. This may be accompanied by considerable expenditure and difficulties, which are however, surmountable based on the state of the art technology and testing methods. If errors were made and the structure actually slides, in most cases the entire capping system must be repaired. It does not help much if the sliding surface is above the geomembrane. Therefore, requirement Eq. 6.1 does not provide any additional safety for capping systems.

Such arguments do not apply to basal liners. Degradation and accompanied settlement processes and movements in the waste body can induce substantial forces into the liner which cannot reliably be quantified and included in proof of structural stability. That such force emerges during settling and subsidence is indicated both by practical experience and through numerical studies on the mechanical behaviour of waste bodies. In most case basal liners cannot be repaired, therefore, it must be guaranteed that sliding always occurs above the liner, i.e. above the geomembrane. For basal liners on inside slopes of landfills and waste pits, the adherence to requirement Eq. 6.1 is an important additional condition for the long-term structural safety of a liner system. To what extent this rule must be applied in gently inclined areas of the basal liner, where down-hill slope forces are weak but lateral spreading forces exert their effect, or in special cases, e.g. in intermediate liners between different waste bodies as for piggybacking landfills, must be judged on a case-by-case basis. The decision criterion is the reliability and precision with which the acting forces and the equilibrium condition can be specified.

## References


Blümel W and Brummermann K (1995) Standsicherheit von Dichtungssystemen – Grundlagen der Berechnung. In: Knipschild FW (ed) Tagungsband der 11 Fachtagung "Die sichere Deponie". Süddeutsches Kunststoffzentrum (SKZ), Würzburg, pp 307–334

DGGt (ed) (1997) GDA-Empfehlungen. Verlag Ernst & Sohn, Berlin

Donaldson JJ (1994) Texturing Techniques. In: Hsuan G and Koerner RM (eds) Proceedings of the 8th GRI Conference, Geosynthetics Resins, Formulations and Manufacturing. Industrial Fabrics Association International (IFAI), St. Paul, USA, pp 113–122

Driscoll R and Simpson B (2001) EN 1997 Eurocode 7. Geotechnical design. Civil Engineering 144: 49–54

Foik G and Günther K (1989) Bemessungsgrundlage für PE-HD Bahnen auf Böschungen. In: Knipschild FW (ed) Tagungsband der 5 Fachtagung "Die sichere Deponie". Süddeutsches Kunststoffzentrum (SKZ), Würzburg, pp 149–183

Koerner RM (1994) Designing with Geosynthetics. Prentice Hall International, London

Krause R (1990) Profilierte Dichtungsbahnen. In: Knipschild FW (ed) Tagungsband der 6 Fachtagung "Die sichere Deponie". Süddeutsches Kunststoffzentrum (SKZ), Würzburg, pp 103–125

Müller WW (1996) Scherverhalten zwischen Kunststoffdichtungsbahnen und Boden. Geotechnik 19: 35–42

Müller WW (1997) Zum Standsicherheitsnachweis bei Abdichtungssystemen. Geotechnik 20: 297–299

Müller WW (ed) (2001) Certification Guidelines for Plastic Geomembranes Used to Line Landfills and Contaminated Sites. Laboratory of Landfill Engineering, BAM, Berlin

Müller WW et al. (2004) Long-term shear strength of multilayer geosynthetics. In: Floss R et al. (eds) Geotechnical Engineering with Geosynthetics, Proceedings of the Third European Geosynthetics Conference. Deutsche Gesellschaft für Geotechnik (DGGt) und Technische Universität München, Zentrum für Geotechnik (TUM-ZG), München, pp 429–434

Seeger S et al. (2000) Long term testing of geomembranes and geotextiles under shear stress. In: Cancelli A et al. (eds) Proceedings of the Second European Geosynthetics Conference. Pàtron Editore, Bologna, pp 607–610

Seeger S and Müller WW (2001) Langzeitbeständigkeit strukturierter Dichtungsbahnen, BAM-Forschungsbericht 256. Wirtschaftsverlag NW, Verlag für neue Wissenschaft GmbH, Bremerhaven
Terzaghi K et al. (1996) Soil Mechanics in Engineering Practice. John Wiley & Sons, New York

# 7 Mass Transport

## 7.1 Introduction

The mass transport through geomembrane liners is described with the help of mathematical models. However, modelling will only supply reasonable results if all the relevant aspects of the basic physical processes are covered by the model, if mass transport parameters relevant for the model calculation and properties of the liner materials relevant for the mass transport modelling are sufficiently well known and if the boundary conditions above and beneath the liner, under which the pollutants exert their effect, can be specified. Difficulties arise specifically with the last point, since it is usual for a mixture of several pollutants in a complex and temporally changing contamination situation to be present. On the other hand, the importance of the effect of the interaction of different substances on the mass transport is sometimes overestimated, and as a rule it is sufficient to look at the behaviour of a few "guide substances".

In addition, often too little attention is paid to the correct representation of the relevant physical processes in the mathematical model. Two examples should be mentioned here:

1. Available model calculations for the mass transport through landfill liners mainly refer to the mineral layer and consist of numerical solutions of a diffusion-dispersion-equation (see Eq. 7.24) for such substances, typically the chloride anion and metal cations, for which the necessary parameters of mass transport quantities in soils or mineral materials are well known. The results are often applied to composite liners, although these ions cannot diffuse through the geomembrane at all. In addition, an incorrect assumption is made that the geomembrane "disappears" in few decades and does not realistically contribute to emission reduction through the landfill liner system. For the class of polyethylene materials which are used in BAM-certified geomembranes for example, application experience over decades (from pipes and chemical apparatus) and results of extensive scientific investigations into their long-term behaviour is available, see Sect. 5.4. They show that the extrapolated service lifetime under landfill condi-

tions is not just a few decades, but lies somewhere within the range of the logarithmic scale between 100 and 1000 years or even beyond 1000 years, i.e. in a range where a reliable prediction of any liner properties is no longer reasonable. Geomembranes contribute significantly to the prevention of pollutant emissions since the pollutants are mainly mobilized in landfills during the operational and aftercare phases when rainwater enters and leachate leaves the landfill and gas production takes place. Therefore model calculations aiming to properly describe the mode of operation of the composite liner system must first of all take into account the mass transport in the geomembrane and the composite effect of the geomembrane and mineral layer.

2. Holes in the geomembrane can have a considerable influence on mass transport depending on the properties of the subgrade. However, different views exist as to how water flow through a hole should be correctly modelled.

This chapter will deal with the basic physical aspects of mass transport in geomembranes and porous mineral materials which serve as a subgrade for geomembranes. Data will be compiled about mass transport parameters for HDPE geomembranes and mineral liners. Mass transport will be described in an intact composite liner. Finally, the problem of modelling the flow through holes in the geomembranes will be dealt with. The chapter is based on a series of three papers about mass transport published in the German journal Bautechnik (Civil Engineering). A summary of some results was given on the Sixth International Conference on Geosynthetics (Müller 1999; Müller et al. 1997a; Müller et al. 1997b; Müller et al. 1998).

## 7.2 Mass Transport in Geomembrane

Diffusion coefficient and partition coefficient or solubility are the parameters which characterise pollutant transport in the plastic geomembrane. The partition coefficient has a special importance in composite systems. First of all, the definition of these quantities and the description of the relevant physical processes will be dealt with in greater detail.

The non-equilibrium state of a physical system, in which mass transport will take place, may be described by the varying spatial distributions of a set of intensive thermodynamic variables. Any gradient in the distribution of these variables will initiate transport processes which finally lead to equilibrium. In this respect the so-called chemical potential is of central importance for our consideration: similarly as temperature gradients determine the heat conduction processes, the chemical potential $\mu$ is the very physical quantity whose gradient initiates diffusive mass transport and solution processes. However, chemical potential is a pretty "academic" quantity in comparison to temperature or pressure. In the thermodynamic theory of solution and diffusion processes, however, its use can hardly be avoided. One can understand $\mu$ of a certain substance in a physical system as the derivative of the system's internal energy with respect to mole number of the substance, i.e. as the specific energy, which must be applied or is released when a molecule of the substance is added to or removed from a system under otherwise unchanged conditions (entropy, volume, mole number of other substances). Obviously, the chemical potential defined in this way plays a central role in the thermodynamic description of chemical reactions and this is where its name comes from: the affinity of a chemical reaction is the sum of the chemical potentials of all reaction participants weighted by the stoichiometric coefficients of the reaction equation. The sign of the affinity decides in which direction the chemical reaction will take place and the state of reaction equilibrium is defined by an affinity equal to zero, from which condition the law of mass action is derived. In the following discussions, however, chemical reactions during mass transport will be ignored from our considerations.

Each thermodynamic system strives to equalise gradients in temperature, pressure and chemical potentials which induces heat flow, liquid or gas flow or diffusive mass flows. The entropy production associated with a diffusive mass flow is proportional to the product of mass flow density and gradient in chemical potential. Gradients in the chemical potentials are therefore considered within the scope of thermodynamics of irreversible processes as the generalised driving force for diffusive mass flow (Prigogine 1961). In a thermodynamic equilibrium state not only temperature and pressure are equal, but the chemical potentials are also.

The thermodynamic state of a system of molecules, which diffuse in a medium, e.g. pollutants in a geomembrane, is usually (fully) described by the spatial temperature distribution $T(x)$, pressure distribution $p(x)$ and concentration distribution $c(x)$, since these quantities can be directly measured, at least in principle. The chemical potential is then a function of these three quantities:

$$\mu = \mu\big(T(x), p(x), c(x)\big). \tag{7.1}$$

We assume that the system is not in equilibrium and a gradient in the chemical potential generates a flow of diffusing particles. The flow density $\boldsymbol{J}$ is the mass of particles which diffuse in a certain direction per unit time and through unit surface. In the linear approximation for such irreversible processes, it is assumed that the mass flow density is proportional to the driving force of mass transport, namely the gradient[1] in the chemical potential (Landau and Lifshitz 2000):

$$\boldsymbol{J} = -\alpha \cdot \nabla\mu(T, p, c). \tag{7.2}$$

Using the gradients of the independent variables concentration, pressure and temperature one obtains:

$$\boldsymbol{J} = -\alpha \cdot \left( \left( \frac{\partial \mu}{\partial c} \right)_{T,p} \nabla c + \left( \frac{\partial \mu}{\partial T} \right)_{c,p} \nabla T + \left( \frac{\partial \mu}{\partial p} \right)_{T,c} \nabla p \right). \tag{7.3}$$

Diffusive mass transport can be classified according to the terms on the right-hand side of Eq. 7.3. The particle flow due to concentration gradient (first summand in Eq. 7.3) is usually called diffusion, the proportionality coefficient *D*, defined by:

$$D = \alpha \cdot \left( \frac{\partial \mu}{\partial c} \right)_{T,p}, \tag{7.4}$$

is then the so-called diffusion coefficient. Our consideration indicates that the diffusion coefficient is not a constant, but depends (very strongly) on temperature, pressure and, in addition, on concentration. The latter aspect will be looked into more closely below. The unit of $c$ is g/cm$^3$ and that of $\boldsymbol{J}$, by definition, is g/(s·cm$^2$). $D$ has the unit cm$^2$/s or m$^2$/s.

Mass transport released by a temperature gradient (second summand in Eq. 7.3) is called thermodiffusion[2]. However, it is remarkable that even a pure pressure gradient (without any temperature or concentration gradient)

---

[1] For recapitulation: vectors are denoted by bald face letters. In an $x$, $y$, $z$ Cartesian co-ordinate system $\boldsymbol{J}$ has the components: $\boldsymbol{J} = (J_x, J_y, J_z)$. Applying the gradient operator to a function $c$, the vector $\nabla c = (\frac{\partial c}{\partial x}, \frac{\partial c}{\partial y}, \frac{\partial c}{\partial z})$ is obtained, applying the scalar Laplace operator to a function $c$ yields: $\Delta c = (\frac{\partial^2 c}{\partial x^2} + \frac{\partial^2 c}{\partial y^2} + \frac{\partial^2 c}{\partial z^2})$ and, finally, it holds in Cartesian coordinates: $\mathrm{div}\boldsymbol{J} = \frac{\partial J_x}{\partial x} + \frac{\partial J_y}{\partial y} + \frac{\partial J_z}{\partial z}$.

[2] A temperature gradient always induces a heat flow. Heat flow and material flow are coupled with one another. This aspect will not be dealt with here since it will be assumed that the system considered is in equilibrium with respect to temperature.

can initiate a diffusive mass transport in a nonporous liner material, the so-called barodiffusion (third summand in Eq. 7.3). This effect can be observed when water accumulates under high pressure on a geomembrane and the pore water pressure is small in comparison in the subgrade beneath the geomembrane: diffusion of water is then induced although there is no concentration gradient (Faure et al. 1989; Giroud and Bonaparte 1989a). However, the barodiffusion coefficient is usually very small. It can cause confusion when a hydraulic conductivity is formally attributed to the geomembrane because of the barodiffusion: the physical process of barodiffusion has nothing to do with the flow of liquids in porous media as described by Darcy's law. In our further considerations, however, we will assume that no temperature or pressure gradient exist.

For the mass transport processes considered here the law of mass conservation must be applied: the mass of molecules diffusing from a volume element (or the mass diffusing into the element) must be equal to the reduction (or increase) of molecule mass in the volume element. Mathematically, a continuity equation results from such a conservation law:

$$\frac{\partial c}{\partial t} + \mathrm{div}\boldsymbol{J} = 0\,, \tag{7.5}$$

Using definition 7.4, a substitution of Eq. 7.3 into Eq. 7.5 yields the well-known diffusion equation:

$$\frac{\partial c}{\partial t} = D\,\Delta c\,. \tag{7.6}$$

Before a pollutant molecule can diffuse in the geomembrane, it must first penetrate from the surrounding medium into the geomembrane. If one submerges a geomembrane in an aqueous solution of a chemical (immersion test), the molecules of the chemical can potentially become dissolved in the plastic. Depending on the extent of their enrichment, molecules may also leave the geomembrane. At a specified temperature $T$ and specified pressure $p$ an equilibrium state is established where a certain mole fraction[3] $x_1$ of the chemical in the plastic is in equilibrium with a certain mole fraction $x_0$ in the aqueous solution. The partition coefficient $\sigma'$ is defined as the ratio of the two mole fractions:

[3] The mole fraction of the amount of material of the chemical dissolved in the plastic is understood in such a way that the mole number of the chemical is related to the mole number of the structural unit of the polymer chain, e.g. $CH_2$ in polyethylenes, and not to the mole number of polymer chains themselves.

$$\sigma' = \frac{x_1}{x_0}. \tag{7.7}$$

However, for the characterisation of a solution the volumetric concentration $c$ (mass per volume) or the mass fraction $w$, i.e. the ratio of the mass of the dissolved chemical to the mass of the geomembrane or the solvent, is more often used as a concentration measure. The relevant partition coefficients are denoted as $\sigma = c_1/c_0$ and $\sigma'' = w_1/w_0$. The connection between the partition coefficients based on different concentration measures can be expressed as:

$$\sigma' = \frac{x_1}{x_0} \approx \frac{V_P}{V_L}\sigma = \frac{M_P}{M_L}\sigma''. \tag{7.8}$$

$V_L$ and $V_P$ and $M_L$ and $M_P$ are the mole volumes and mole masses of solvent and structural unit of the polymer respectively.

If a geomembrane is immersed in a liquid chemical and if the plastic is resistant to this liquid, then the plastic will take up the chemical to a certain limit, called the limit of saturation. The mass fraction in the state of saturation is called solubility $s$ of the chemical in the geomembrane.

The partition coefficient $\sigma$ or the solubility $s$ together with the diffusion coefficient $D$ of the chemical molecules in the geomembrane determine the permeation rate $J$, i.e. the pollutant mass diffusing in steady state per unit time and unit area, and the induction time $t_{ind}$, i.e. the time the pollutants need to diffuse through the geomembrane, and thus the imperviousness of a geomembrane.

Let us consider the following typical situation of a permeation test (see Sect. 3.2.6): an aqueous solution of a chemical of concentration $c_0$ is continuously applied above the geomembrane with thickness $d$. The concentration of the chemical under the geomembrane is always kept at zero by constant removal of the chemical. The geomembrane is thus exposed to a constant external concentration difference[4] $\Delta c = c_0$ or a concentration gradient $\Delta c/d$. Thus a concentration gradient is also formed within the geomembrane which drives the diffusive mass transport. This internal concentration gradient, however, is not equal to the concentration gradient acting outside the geomembrane, rather it is affected by the partition coefficient. If the adsorption equilibrium develops fast enough at the interface

[4] In the following, $\Delta$ will no longer stand for Laplace's operator as in Eq. 7.6, but simply for the difference between the concentrations above and beneath the geomembrane. The symbol is in italics to indicate the distinction. The context makes clear the respective meaning of the symbol anyway.

of geomembrane and liquid, the concentration gradient in the geomembrane can be given by $\sigma(\Delta c/d)$. Permeation rate and induction time can then easily be calculated from the solution of the diffusion equation 7.6 for these simple boundary conditions:

$$J = \frac{\sigma\, D\, \Delta c}{d} \qquad t_{ind} = \frac{d^2}{6\,D}. \tag{7.9}$$

The imperviousness of the plastic geomembrane material with respect to a certain chemical can be characterised by the permeability coefficient $P = Jd/\Delta c = \sigma D$ with the dimension m²/s which indicates the ratio of the product of permeation rate and depth to the concentration difference and therefore the product of the material parameters diffusion coefficient and partition coefficient. The value $J/\Delta c = P/d$ with the dimension m/s, may be called diffusive "conductivity"[5], and is used sometimes as analogous to the hydraulic conductivity of Darcy's law. Its reciprocal value, $d/P$, may be called the diffusion resistance.

The formulae and terms also apply when the permeation experiment is performed with the pure liquid chemical. The partition coefficient $\sigma$ must be replaced with solubility $s$. Since organic substances can be fairly extensively enriched in a thermoplastic geomembrane, it must be considered in the analysis of mass transport through geomembranes exposed to pure substances that the diffusion coefficient can depend greatly on the concentration as will be discussed later.

Similar relationships apply for the description of the permeation of gases through geomembranes. However, some differences in quantities and units used must be pointed out. A detailed analysis is presented in (Piringer 1993). The gas is not characterised by a concentration, but by its partial pressure $p$. Eq. 7.7 then is replaced by Henry's law, according to which concentration $c_1$ of gas molecules dissolved in the geomembrane is proportional to the partial pressure of the gas outside the geomembrane. The proportionality factor is mainly called Henry's constant or gas solubility coefficient $S$:

$$S = \frac{c_1}{p}. \tag{7.10}$$

Equation 7.9 is then replaced by:

[5] Confusion with the hydraulic conductivity and the associated intrinsic permeability or fluidity of a porous medium should also be avoided here.

$$J = \frac{S\,D\,\Delta p}{d} = \frac{P\,\Delta p}{d}, \tag{7.11}$$

where $\Delta p$ is the difference of the partial pressure of the gas above and beneath the geomembrane. $P$ is the gas permeability coefficient. This material-specific quantity characterises the permeability of a geomembrane material against a certain gas.

The pressure of a gas is measured in the units Pa or bar. The amount of the diffusing substance in the permeation of gases is usually measured in volume units related to standard conditions[6](STP). Therefore special units result for the permeability coefficient and Henry's constant. Based on Eqs. 7.10 and 7.11:

$$\begin{aligned}[P] &= \left[\frac{\text{Amount of substance}\cdot\text{Tickness}}{\text{Area}\cdot\text{Time}\cdot\text{Pressure difference}}\right] = \\ &= \frac{\text{cm}^3(\text{STP})\cdot\text{cm}}{\text{cm}^2\cdot\text{s}\cdot\text{Pa}} = \frac{10^5\cdot\text{cm}^3(\text{STP})\cdot\text{cm}}{\text{cm}^2\cdot\text{s}\cdot\text{bar}} \quad \text{etc.}\end{aligned} \tag{7.12}$$

$$[S] = \frac{\text{cm}^3(\text{STP})}{\text{cm}^3\cdot\text{Pa}} \quad \text{etc.} \tag{7.13}$$

Permeability (or permeability coefficient for gases) of a geomembrane is thus given by the product of diffusion coefficient and partition coefficient and/or solubility (Henry's constant for gases) where the partition coefficient, Henry's constant and diffusion coefficient depend on the respective impact (gaseous or liquid pure chemical, chemical in aqueous solution of small or high concentration). These parameters $\sigma$, $s$, $S$ and $D$ can be determined experimentally by immersion tests or permeation tests. The tests are described in Sect. 3.2.6. Table 7.1 shows the solubility measured in such tests, Table 7.3 displays the measured diffusion coefficients for various hydrophilic and hydrophobic organic liquids. Table 7.2 illustrates the partition coefficients for aqueous solutions of organic substances and Tables 7.3 and 7.4 contains the relevant diffusion coefficients. Table 7.5a gives examples of measured permeability coefficients of gases in LDPE and HDPE geomembranes and Table 7.5b of water vapour permeability coefficients in HDPE pipes.

[6] Standard conditions comprise a temperature of 273.15 K and a pressure of $10^5$ Pa. Before introducing the SI units, the standard pressure was 1 atm. The correction factor is, however, negligible within the accuracy of measured permeability coefficients.

At a very low concentration the diffusion coefficient characterises the mobility of an individual foreign molecule in the amorphous polymer surroundings. One can imagine the diffusion process in such a way that the foreign molecule jumps randomly from one position in the molecular surroundings to the next possible position. The diffusion coefficient can then be estimated by (six times) the product of the average jump rate and the average of the square of the jump width of the molecule and is determined by the size of the molecule and its interaction with the polymer matrix (Einstein 1905). The more "bulky" the molecule and the smaller its solubility, i.e. the less it integrates into its molecular environment, the smaller the diffusion coefficient is. This quantity can only be accessed by measurement methods that are capable of measuring the molecular motion at extremely low concentrations, e.g. by measurements using radioactive tracers. At greater concentrations the dissolved molecules interact with each other and change the surroundings and conditions for the migration within the polymer which may even appear as a macroscopic swelling of the polymeric material. The diffusion coefficient therefore depends on the concentration: it explicitly follows from both immersion tests and the permeation rates and induction times in permeation tests that the diffusion coefficient is substantially smaller in tests with aqueous solutions than when measured under an impact by the pure test liquids. Induction times on the other hand are considerably longer. The diffusion coefficient, which parameterises the results of immersion and permeation tests with concentrated media on the assumption of Fick's law, is thus a quantity which may depend greatly on the test conditions selected.

The repercussion of the concentration dependence of the diffusion coefficient on the induction time can mathematically be described in some cases. For a permeation process where the diffusion coefficient $D$ increases as a function of concentration following a simple exponential law, $D(c) = D(0)\, e^{bc}$, the induction time was, for example, calculated[7]. A large reduction, dependent on the concentration in the surface of the geomembrane, results compared to the induction time that would develop at a very small concentration and diffusion coefficient $D(0)$.

[7] Frisch, HL (1957) J Phys Chem, Was. 61:93. Induction time in this case is given by:

$$t_{ind} = \frac{6(4\,e^{bc_1} - 1 + e^{2bc_1}(2\beta\, c_1 - 3))}{4(e^{bc_1} - 1)^3}\,\frac{d^2}{6D(0)},$$

where $c_1 = \sigma c_0$ is the concentration which develops in equilibrium on the surface of the geomembrane. The ratio of the induction times $t_{ind}/(d^2/6D(0))$ is approximately $(1 - bc_1/4)$ for $0 < bc_1 < 3$ and approaches zero as $3bc_1 \cdot e^{-bc_1}$ for $bc_1 > 6$.

Considering Tables 7.1 and 7.2 as well as 7.3 and 7.4, it is noticeable that the variation of the partition coefficient and the solubility for different substances extends over many orders of magnitude. The diffusion coefficient changes only little in comparison. This indicates that the solubility of the chemicals in the polyethylene ultimately determines their permeation rates. Roughly speaking, only non-polar, low-molecular substances can diffuse to any relevant extent. Geomembranes are practically impermeable against ions in an aqueous solution.

At BAM, from 1987 to 1997, investigations were carried out into permeation of metal salts, metal cations and nitrate anion, through HDPE geomembranes (Müller et al. 1997a). At this time no indication was found by the measurements for any permeation and a comparison of test and control experiment with pure water. From the test times and test conditions an upper limit can be estimated for the permeation rate. For example the permeability for cadmium is estimated to be less than $1.3 \cdot 10^{-18}$ m²/s. R. K. Rowe and collaborators (Rowe et al. 1995b) report on similar experiments about chloride permeation through an HDPE geomembrane. An upper limit could be specified for the permeability as $3 \cdot 10^{-17}$ $m^2/s$ from their test conditions.

The extraordinarily low permeability can be explained by the fact that polyethylene as a non-polar medium can only be very weakly polarized and diffusion cannot lead to a separation of charge carrier. The ions are surrounded in the aqueous solution by a cloud of water molecules shielding the ion's charge. Cations and anions would therefore have to recombine from this hydrate shell to the molecule and become dissolved in the polyethylene or both become dissolved with their hydrate shell and diffuse. Such processes are thermodynamically rather unfavourable. The importance of dissociation of inorganic molecules for the migration becomes clear by permeation tests performed with concentrated hydrochloric acid. Undissociated HCl molecules are found to some extent in concentrated hydrochloric acid while the molecules are fully dissociated in aqueous NaCl or metallic salt solution. The available undissociated HCl molecules can become dissolved in the polyethylene and only then diffuse similarly to water molecules or undissociated acetic acid molecules. While no permeation of chlorine can be observed in permeation experiments with metal salts, diffused chlorine can be proven when using concentrated hydrochloric acid.

Since a very large variety of chemicals exist for which knowledge of permeation rates might become relevant, it would be helpful to have a theoretical method for a systematic estimation of the relevant parameters. Partition coefficients and diffusion coefficients of organic compounds in

HDPE geomembranes can indeed be calculated by the so-called group contribution methods.

**Table 7.1.** Calculated and measured solubilities (mass fractions) of test liquids in four different HDPE geomembranes (sample 48 (crystallinity 54 %), 139 (crystallinity 48 %), a (crystallinity 56 %) and b (crystallinity 51 %)) (Müller et al. 1997b)

| Test liquid | Measured values of mass fraction of absorbed molecule with respect to the amorphous part of the HDPE-geomembrane | | Calculated mass fraction according to the method of Takeru Oishi and John. M. Prausnitz |
|---|---|---|---|
| | Sample 48<br>[a]) Sample a | Sample 139<br>[b]) Sample b | |
| Water | 0.002 | 0.002 | |
| Methanol | 0.002[a] | 0.002[b] | 0.004 |
| Acetone | 0.022 | 0.022 | 0.021 |
| Methylethylketone | 0.039 | 0.037 | 0.032 |
| Acetic acid | 0.019 | 0.017 | 0.022 |
| Propanoic acid | 0.046 | 0.040 | 0.030 |
| Acetic acid ethyl ester | 0.048[a] | 0.047[b] | 0.065 |
| Chloroform | 0.305[a] | 0.312[b] | 0.325 |
| Carbon tetrachlorid | 0.409[a] | 0.408[b] | 0.222 |
| Trichloroethylene | 0.424 | 0.385 | 0.450 |
| Tetrachloroethylene | 0.432[a] | 0.443[b] | 0.442 |
| Chlorobenzene | 0.220[a] | 0.220[b] | 0.285 |
| Xylene | 0.189[a] | 0.194[b] | 0.133 |
| Toluene | 0.182[a] | 0.184[b] | 0.439 |
| Pentane | 0.139 | 0.121 | 0.090 |
| Hexane | 0.144 | 0.124 | 0.105 |
| Heptane | 0.144 | 0.124 | 0.113 |
| Octane | 0.142 | 0.122 | 0.122 |
| Isooctane | 0.107[a] | 0.104[b] | 0.117 |
| Decahydronaphtaline | 0.238 | 0.207 | 0.227 |

For the calculation of $\sigma'$ one proceeds from the thermodynamic condition for the sorption equilibrium. The adsorption equilibrium is characterised, as mentioned earlier, by the condition that the chemical potential $\mu(T,p,x)$ for the substance dissolved in water and for the substance dissolved in the plastic geomembrane are equal. The dependence of the chemical potential of the dissolved substance on the concentration can formally be written as:

$$\mu(T,p,x) = \psi(T,p) + RT\ln(\gamma(T,p,x)x)\,, \tag{7.14}$$

where $R$ is the general gas constant. $\psi$ is the chemical potential of the pure liquid substance, a certain function of only the variables $T$ and $p$. $\gamma(T, p, x)$ is the so-called activity coefficient which depends on the mole fraction, pressure and temperature. The function $a = \gamma(x,T,p)\, x$ is called activity.

In the sorption equilibrium it thus applies

$$\psi(T,p) + RT\ln(\gamma_1(x_1)x_1) = \psi(T,p) + RT\ln(\gamma_0(x_0)x_0)\,. \tag{7.15}$$

The equality of the activities of the solutions at equilibrium follows from this:

$$\gamma_1(x_1)\,x_1 = \gamma_0(x_0)\,x_0\,. \tag{7.16}$$

Therefore, knowing the activities or coefficients of activity as functions of concentration for the solution of an organic compound in water and in the polymer matrix, the two unknown quantities, partition coefficient $\sigma'$ and mole fraction $x_1$ in the plastic geomembrane, can be calculated from Eq. 7.7 and 7.16 for a specified $x_0$ in the aqueous solution.

There are numerous references to the calculation of activities and activity coefficients. They are usually determined using versions of the so-called group contribution method. The basic idea of this method is that a large number of organic compounds exist that are formed by a small number of functional groups and that the thermodynamic properties of a compound can be obtained by summing up the contributions of the functional groups involved (Gmehling and Kolbe 1992; Piringer 1993; Reid et al. 1977). A wide-spread and widely applicable procedure for the calculation of the activity or the activity coefficient of non-electrolytic liquid mixtures is the so-called UNIFAC (Universal Functional Group Activity Coefficient) method (Reid et al. 1977). T. Oihsi and J. M. Prausnitz reported about an extension of the procedure which can be used to determine the activity of substances dissolved in amorphous polymers (Oishi and Prausnitz 1978). R. Goydan, R. C. Reid and H.-S. Tseng (Goydan et al. 1989) have shown that this procedure provides sufficiently precise results applicable over a wide range for the determination of the solubility of organic substances in polymers in comparison with various similar methods. The details of the method can be found in the literature listed. The calculations for HDPE geomembranes were carried out in (Müller et al. 1997b) and the results are displayed in Tables 7.1 and 7.2.

O. G. Piringer suggested a simple contribution method for the direct calculation of the partition coefficient σ. A detailed explanation can be found in (Piringer 1993). The method is based on the comparison of the retention times in gas chromatography and is denoted as the Retention-index-method. The polyolefin will be simulated here by the non-polar material Silicon OV 101. Increments for important structural units in the non-polar material Silicon OV 101OV 101 and in the polar water are given in (Piringer 1993). Increments for further structural units were obtained from a private communication with O. G. Piringer. The calculation results are displayed in Table 7.2.

**Table 7.2.** Calculated and measured partition coefficients between diluted aqueous solutions and HDPE geomembrane (Müller et al. 1997b)

| Test liquid | Partition coefficient, measured and estimated from the ratio of measured solubilities (square brackets) | Partition coefficient, calculated according to the UNIFAC-method | Partition coefficient, calculated according to the Retention-index-method |
|---|---|---|---|
| Methanol-water | - | 0,004 | 0,001 |
| Acetone-water | 0.032 [a] | 0.016 | 0.106 |
| Methylethylketone-water | [0.06] | 0.173 | 0.403 |
| Acetic acid, diluted | 0.015 [b] | 0.017 | 0.020 |
| Propanoic acid, diluted | - | 0.043 | 0.076 |
| Acetic acid ethyl ester-water | [0.28] | 0.42 | 0.930 |
| Formaldehyde-water | 0.015 [c] | - | 0.009 |
| Chloroform-water | [18] | 19 | 22 |
| Carbon tetrachloride-water | [238] | 252 | 141 |
| Trichloroethylene-water | 189 [d] | - | 114 |
| Trichloroethylene-water | [135 [f]] | | |
| 1,2-dichloroethane-water | [7 [f]] | 10 | 20 |
| Tetrachloroethylen-Wasser | [1357] | - | 9972 |
| Chlorobenzene-water | [209] | 350 | 487 |
| Benzene-water | [57 [f]] | 55 | 37 |
| Xylene-water | [499 [f]], [556] | 1157 | 517 |
| Toluene-water | [192 [f]], [160] | 54 | 206 |
| Pentane-water | [1,512] | 3,600 | 2,800 |
| Hexane-water | [5,800] | 22,200 | 10,700 |
| Heptane-water | [24,348] | 131,000 | 40,600 |

[a]) 10 % by volume acetone-water [b]) 0.5 g/l acetic acid
[c]) 37 % by weight formaldehyde [d]) 0.5 g/l trichloroethylene-water
[f]) Data from reference (Prasad et al. 1994) for the partition coefficient estimated from the ratio of measured solubilities (Piringer 1993; Reid et al. 1977)

**Table 7.3.** Diffusion coefficients of substances in three different HDPE geomembranes (see caption of Table 7.1) obtained from immersion and permeation tests with pure test liquids (Müller et al. 1997b)

| Test liquid | Immersion test Sample 48 $D$ ($10^{-12}$ m²/s) | Immersion test Sample 139 $D$ ($10^{-12}$ m²/s) | Permeation test Sample a§ $D$ ($10^{-12}$ m²/s) |
|---|---|---|---|
| Water | 0.82 | 0.90 | - |
| Acetone | 0.87 | 0.91 | - |
| Methylethylketone | 0.75 | 0.86 | - |
| Acetic acid | 0.58 | 0.52 | - |
| Propanoic acid | 0.30 | 0.32 | - |
| Acetic acid ethyl ester | - | - | 1.1 |
| Chloroform | - | - | 5.9 |
| Carbon tetrachloride | - | - | 2.4 |
| Trichloroethylene | 7.70 | 8.40 | 10.8 |
| Tetrachloroethylene | - | - | 3.8 |
| Chlorobenzene | - | - | 3.6 |
| Xylene | - | - | 4.7 |
| Toluene | - | - | 6.1 |
| Pentane | 2.44 | 2.90 | - |
| Hexane | 2.08 | 2.47 | - |
| Heptane | 1.52 | 1.74 | - |
| Octane | 1.08 | 1.31 | - |
| Isooctane | - | - | 0.44 |
| Decahydronaphthaline | 0.35 | 0.36 | - |

§) Calculated from the permeation rate, see reference (August et al. 1984; August et al. 1992), and from measured solubilities according to Eq. 7.9

The partition coefficients calculated according to the UNIFAC method are in fairly good agreement with those calculated by the Retention-Index-Method in every case.

O. G. Piringer published an empirical approximation formula for the determination of the diffusion coefficient as well (Piringer 1993). The formula reads:

$$D = \exp(A_p - 0.008\ M_r - 10450\,\frac{1}{T}) \qquad (\mathrm{m^2/s})\,. \tag{7.17}$$

$A_p$ is a parameter characteristic of the respective plastics, $M_r$ is the mole mass of the diffusing substance. $A_p = 9$ provides a good agreement with measured values for LDPE and $A_p = 5$ does for HDPE. For medium density PE-co-polymers the value should be in between and with $A_p = 7.5$ it is in good agreement with the measured values indicated in Table 7.4.

**Table 7.4.** Diffusion coefficients of substances in three different HDPE geomembranes (see Table caption 7.1) obtained from immersion and permeation tests with aqueous solutions (Müller et al. 1997b). Values for aqueous solutions from reference (Prasad et al. 1994) are indicated in round parentheses

| Test liquid | Immersion test Sample 48 and data from literature $D$ ($10^{-12}$ m²/s) | Immersion test Sample 139 $D$ ($10^{-12}$ m²/s) | Permeation test Sample a§ and data from literature $D$ ($10^{-12}$ m²/s) |
|---|---|---|---|
| Acetone-water 10 % by volume | 0.66 | 0.84 | - |
| Acetone-water 50 % by volume | 0.88 | 0.87 | - |
| Acetic acid, 0.50 kg/l | 0.11 | 0.18 | |
| Acetic acid, 0.70 kg/l | 0.15 | 0.22 | |
| Acetic acid, 0.90 kg/l | 0.25 | 0.29 | - |
| 1,2-dichloroethane-water | (6.8) | - | - |
| Trichloroethylene-water | 0.20 | 0.30 | 0.6 |
| Trichloroethylene-water | (0.52) | | (0.50) |
| Benzene | (0.037) | - | - |
| Xylene | (0.10) | - | 0.2 (0.10) |
| Toluene | (0.51) | - | 0.2 (0.23) |

**Table 7.5a.** Permeability coefficient $P$ for the permeation of gases in PE

| Polymer | Gas | Measurement temperature (°C) | Permeability $P$ $\left(\frac{cm^3(STP)\cdot mm}{m^2\cdot d\cdot bar}\right)$ |
|---|---|---|---|
| LDPE | $O_2$ | 25 | 448 |
| | $CO_2$ | | 1810 |
| | $N_2$ | | 137 |
| HDPE | $O_2$ | 25 | 71 |
| | $CO_2$ | | 279 |
| | $N_2$ | | 21 |
| HDPE | $O_2$ | 25 | 76 |
| | $CO_2$ | | 290 |
| | $N_2$ | | 21 |
| | Air | | 30 |
| | Methane | 20 | 56 |
| | Ethane | | 89 |
| | Propane | | 35 |
| | Ethylene | | 110 |
| | Propylene | | 76 |
| | Sulphur dioxide | | 430 |

**Table 7.5b.** Permeability coefficient $P$ for the permeation of water vapour (quantity of diffused water vapour measured in mass units) in HDPE pipes

| Polymer | Gas | Measurement temperature (°C) | Permeability $P$ $\left(\frac{g \cdot mm}{m^2 \cdot d}\right)$ |
|---|---|---|---|
| | | 20 | 0.034 |
| | | 25 | 0.043 |
| HDPE | Water vapour | 30 | 0.068 |
| | | 40 | 0.140 |
| | | 50 | 0.324 |

## 7.3 Mass Transport in Soil Materials (Geomembrane Subgrade)

Soil or more generally mineral materials used for liners and subgrades are porous media containing pores of various size and geometry (voids, capillaries, dead end pores and so on) which form an interconnected network of pore channels. Water flows through this pore space driven by hydrostatic pressure, the force of gravity and by surface tension in pores only partially saturated with water. Pollutants contained in the water are thereby also transported. In view of this passive pollutant transport by the flowing water, one talks of advection or advective mass transport. However, the pollutants also diffuse in soil water and even if the flow velocity is very small, the pollutants spread by diffusion. At a high flow velocity a comparable "spreading out" or so-called dispersion of a pollutant front occurs due to large variation in local flow velocity and the associated different individual "flow histories" of pollutant particles. Sorption at the solid phase of the soil material and pollutant decay, e.g. by microbial degradation processes, must also be considered in the description of pollutant transport in soil materials.

In a pore space only partly saturated with soil water, mass transport in the gaseous phase must additionally be considered, i.e. through the pore space filled with soil air. Temperature differences result in the movement of water and water vapour and thus also in pollutant transport. These effects are not considered here. In the following, only the simplest mathematical model is discussed for the description of the mass transport (van Genuchten and Wagenet 1989): a homogeneous, water-saturated soil material of constant temperature is assumed. All initial values, concentration, hydraulic gradient etc. should only depend on one coordinate $x$ and the coordinate axis is assumed to be perpendicular to the plane of the liner, see

Fig. 7.1. Such a model naturally becomes useless when preferred migration pathways through gaps, cracks or interconnected large voids etc. predominantly determine mass transport.

Let $\Theta$ be the volumetric water content or, as saturation is assumed, the porosity, i.e. the ratio of pore space volume to the total volume of the soil material and $c$ the concentration of the substance in soil water. $\Theta c$ is then the concentration of a pollutant with respect to the volume of the soil material. The temporal change of the concentration of a pollutant dissolved in water in a volume element of the porous material is given by the net flow of the pollutant in or out of a volume element with flow density $J_{adv}$, by diffusion in or out of the volume element with flow density $J_{diff}$, by the sorption processes taking place at rate $J_s$ within the volume element and by the chemical degradation of the pollutant in the pore water of the volume element at rate $J_{d,L}$. The concentration of pollutant adsorbed at the solid surface in the volume element may be parameterised as $\rho s$, where $\rho$ is the dry bulk density of the soil material and $s$ the concentration of the adsorbed pollutant related to the mass of the solid phase of the soil material. The temporal change of the concentration of adsorbed pollutants is given by the rate of the sorption process $J_s$ already mentioned as well as by the rate of chemical degradation process $J_{d,S}$ at the solid phase of the soil material. Because of the law of conservation of mass, the following continuity equations are valid:

$$\frac{\partial(\Theta c)}{\partial t} = -\frac{\partial}{\partial x}\left(J_{adv} + J_{diff}\right) - J_s - J_{d,L} \tag{7.18}$$

and

$$\rho \frac{\partial s}{\partial t} = J_s - J_{d,S} \,. \tag{7.19}$$

For further calculation, confining assumptions and simplifications must be made about the relevant aspects of physics of the mass transport processes for the respective pollutant in the respective soil material and a mathematical description, derived from it, can then only be applied as an approximation. In the simplest case the following assumptions are made: Darcy's law is assumed for the advective mass transport, i.e.:

$$J_{adv} = k_f \cdot i \cdot c = q \cdot c \,, \tag{7.20}$$

with $k_f$ as hydraulic conductivity, $i$ as local hydraulic gradient or $q$ as discharge velocity. Diffusion in the pore water and dispersion are described by Fick's law, see Eq. 7.3, with diffusion coefficient $D_e$ and dispersion coefficient $D_{hd}$, i.e. :

$$J_{diff} = -\Theta(D_e + D_{hd})\frac{\partial c}{\partial x}. \tag{7.21}$$

For the sorption process linearity and reversibility are assumed, i.e. the concentration of the pollutant adsorbed in the soil material is at every instant proportional to its concentration in the soil water:

$$s = k \cdot c, \tag{7.22}$$

with $k$ as distribution or partition coefficient. Finally, a reaction equation of the first order is used for the degradation:

$$J_{d,S} = \lambda_S \cdot \rho \cdot s \tag{7.32a}$$

$$J_{d,L} = \lambda_L \cdot \Theta \cdot s, \tag{7.23b}$$

where $\lambda_S$ and $\lambda_L$ are the reaction constants of the degradation reaction in the soil matrix and in soil water. For a homogeneous porous mineral layer with a steady-state Darcy-like flow of water Eqs. 7.18 and 7.19 are simplified on the assumptions 7.20 to 7.23 to the transport equation for a pollutant in a water-saturated porous medium, the so-called diffusion-dispersion equation:

$$R\frac{\partial c}{\partial t} = (D_e + D_{hd})\frac{\partial^2 c}{\partial x^2} - v\frac{\partial c}{\partial x} - \lambda c, \tag{7.24}$$

where the parameters characteristic of mass transport (apart from diffusion coefficient $D_e$ and dispersion coefficient $D_{\mathrm{hd}}$) are given by:

$$R = 1 + \frac{\rho \cdot k}{\Theta} \quad \text{(retardation coefficient)}, \tag{7.25}$$

$$\lambda = \lambda_L + \frac{\rho \cdot k}{\Theta}\lambda_F \quad \text{(effective reaction constant)}, \tag{7.26}$$

$$v = \frac{k_f \cdot i}{\Theta} \quad \text{(seepage velocity)}. \tag{7.27}$$

Finally, the diffusion coefficient $D_e$ of the pollutant in soil water of the soil material, or the so-called effective diffusion coefficient, may be traced back to the diffusion coefficient $D_0$ of the pollutant in free water, which is usually known, with the help of the so-called tortuosity factor $\Gamma$ (Bear 1988):

$$D_e = \Gamma D_0 \,. \tag{7.28}$$

The tortuosity of a porous medium and the associated tortuosity factor is one of the important geometrical quantities, which, in addition to its porosity and its specific internal surface, characterises the physical properties of a porous medium (Scheidegger 1963). They are defined as the ratio of the actual length of either the "true" flow path or of the "true" diffusion path of a fluid or pollutant particle to the length of the apparent pathway along which the groundwater flow or the particle diffusion respectively seems to proceed. It therefore expresses the "complexity" of the pore structure. The tortuosity factor can easily be calculated for the case of a linear concentration gradient $\Delta c$ along the $x$-axis in an isotropic and homogenous porous medium (Bear 1988). The diffusion coefficient is defined in accordance with Eq. 7.21 as a proportionality factor that indicates the ratio of the $x$ component of particle flow density $J_x$ to the concentration gradient $(\Delta c/\Delta x)$. However, the particles in the soil matrix actually do not diffuse along the $x$-axis, but along winding pore channels which can take all possible directions relative to the $x$-axis. Let us consider a path element $\Delta l$ of a water-filled pore channel at an angle $\vartheta$ to the $x$-axis. The local particle flow density $J = D_0(\Delta c/\Delta l)$ is determined by the concentration gradient $\Delta c/\Delta l$ with respect to the pore channel. It holds:

$$J_x = J \cos\vartheta = D_0 \left( \frac{\Delta c}{\Delta l} \right) \cos\vartheta = D_0 \left( \frac{\Delta c}{\Delta x} \right) \cos^2\vartheta \,. \tag{7.29}$$

From which it follows that $D_e = D_0 \cos^2\vartheta$ and thus $\Gamma = \cos^2\vartheta$ . If the pore channels are evenly distributed in all directions, an average must be taken over the area of the half unit sphere which provides $\Gamma = 1/3$. This model suggests that the diffusion coefficients of a pollutant in the pore water of a porous material should be about one third of the diffusion coefficients in free water.

The ratio of the effective diffusion coefficient in soils or mineral materials to the diffusion coefficient in free water is, however, also influenced by other effects than only the complexity of the diffusion paths in the pore spaces of a soil. For example, the viscosity of water can decrease in narrow pore spaces, with corresponding effects on the diffusion coefficients of the dissolved substances. "Apparent" tortuosity factors calculated from measured values can therefore be smaller than the value suggested by geometry. It is therefore justifiable to find a conservative estimate of diffusion coefficients for pollutants in soil water, for example when considering mineral landfill liners for which no measured values are available, to use this pure geometrical calculated factor of 1/3 as a typical tortuosity factor.

Mass transport in soils and porous mineral materials is usually calculated based on Eqs. 7.24 to 7.28. Further information about parameters used can be found in the GLR recommendation E1-10 "Mass Transport Models for the Barrier Effect of Liner Layer" (DGGt 1997). Mass transport in clayey barrier systems for waste disposal facilities is discussed in great detail in the book (Rowe et al. 1995a). Programs for the numerical solution of differential equation 7.24 for various boundary conditions and selection of parameters are commercially available (Rowe et al. 1994).

For mass transport in composite liners, which will be dealt with in the next section, further simplifying assumptions can be made. In a perfectly installed composite liner no advective mass transport and associated dispersion effects can take place. The discussion of degradation processes and "natural attenuation" effects is well beyond the scope of this book and therefore neglected. Equation 7.24 is then simplified to the common diffusion equation:

$$R\frac{\partial c}{\partial t} = \Gamma D_0 \frac{\partial^2 c}{\partial x^2}. \tag{7.30}$$

Having a pollutant concentration $c_0$ on the surface of the soil liner and a very low concentration below the liner because steady removal of pollutants from below the liner is assumed, thus $\Delta c = c_0$, the permeation rate $J$ and induction time $t_{ind}$ can be given for a soil liner as:

$$J = \frac{\Theta\,\Gamma D_0\,\Delta c}{d} \qquad t_{ind} = \frac{R d^2}{6\,\Gamma D_0}. \tag{7.31}$$

Sorption at the soil matrix thus causes a delay or retardation in mass transport: it does not affect the permeation rate developing over the long term when pollutant supply is continuous, but it has a substantial influence on "break through" time.

When evaluating data on the diffusion coefficients of pollutant in soils or porous mineral materials some peculiarities in the literature must be considered (Shackelford and Daniel 1991b). Under the simplified assumptions made here, the permeation rates and the permeability and the induction times of pollutant diffusion in the mineral liner are determined by the basic parameters $D_0$, which depend on the pollutant only, by quantities $\Theta$ and $\Gamma$ which depend on the mineral liner material only and by $R$ which depends on the pollutant and the mineral liner material as well. However, diffusion experiments provide no direct values for these basic quantities but only experimental quantities which depend on them. According to the first of Eqs. 7.31, only the permeability $P = \Theta\Gamma D_0$ can be determined directly from a measurement of the permeation rate of steady-state diffusion, i.e. when diffusion has reached a steady state for fixed concentration boundary conditions. The porosity $\Theta$ of the soil material can be determined by special porosity measurements. Under assumption that diffusion affects the entire pore space, which is only approximately true, the effective diffusion coefficient $D_e = \Gamma D_0$ can then be obtained from the permeability measurement. Occasionally, the permeability itself is called effective diffusion coefficient $D_e' = \Theta\Gamma D_0$.

The measurement of induction times (see second of Eqs. 7.31) or the measurements of the temporal change of concentration profiles for transient diffusion, i.e. diffusion before reaching the steady-state condition or with time-dependent concentration boundary conditions, only provide values for experimental quantities in which both $D_e$ and $R$ are included. Usually, only the so-called "apparent" or retarded diffusion coefficient $D_a = D_e/R$ is determined. Due to the typical ranges of values for $\Theta$, $\Gamma$ and $R$ the values for the three diffusion coefficients $D_e$, $D_e'$ and $D_a$ differ by up to 2 orders of magnitude. Since the terminology is sometimes ambiguous, literature data about diffusion coefficients must be scrutinised very carefully to see which coefficient has actually been determined or used.

Since mass transport calculations are in particular used for risk assessment of landfill liners and contaminated sites, some data are compiled in the following tables for mineral liners used in this field. In the literature, however, only a few investigations about the diffusion of organic substances are available for the typical soil or porous mineral materials used in landfills liners. In most cases only $D_a$ was determined. The effective diffusion coefficient can only be calculated from such data when the retardation factor is known for the respective soil material.

The determination of $R$ (see Eq. 7.25) by a measurement of the partition coefficient $k$ in the laboratory, by some type of sorption experiment for the determination of adsorbed solid-phase concentration versus solution-phase concentration in equilibrium (so-called isopleths or isotherm), however, supplies values of $R$ which are not comparable with the actual "in field" retardation factors in a mineral liner, because the specific surface susceptible to the pollutant is very different in both cases. Therefore, based on diffusion coefficients in free aqueous solutions, the effective diffusion coefficients of organic substances for water-saturated mineral liners were estimated by using a tortuosity factor $\Gamma = 1/3$ (Müller et al. 1997a). Table 7.6 shows the results for various soil materials.

**Table 7.6.** Diffusion coefficients of organic substances at 20 °C in dilute aqueous solution and effective diffusion coefficient in soil water of mineral liners calculated with $\Gamma = 1/3$ and diffusion coefficient in the HDPE geomembrane at a small concentration. The units for the diffusion coefficient should be considered (Müller et al. 1997a)

| Group | Substance | Diffusion coefficient ($10^{-10}$ m²/s) | | Diffusion coefficient ($10^{-12}$ m²/s) |
|---|---|---|---|---|
| | | Free water | Soil water | HDPE geomembrane |
| Alcohol | Methanol | 14.5 | 4.8 | 0.8 |
| Ketone | Acetone | 10.2 | 3.4 | 0.6 |
| | Ethylmethylketone | 9.0 | 3.0 | 0.55 |
| Organic acid | Acetic acid | - | - | 0.15 |
| | Propanoic acid | - | - | 0.15 |
| Ester | Acetic acid ethyl ester | 8.4 | 2.8 | 0.15 |
| Aldehyde | Formaldehyde, aqueous solution | 17.8 | 5.9 | 0.8 |
| Chlorinated hydrocarbons | Chloroform | 9.2 | 3.1 | 0.25 |
| | Carbon tetrachloride | 8.7 | 2.9 | 0.25 |
| | Trichloroethylene | 8.4 | 2.9 | 0.25 |
| | 1,2-Dichloroethane | 9.1 | 3.0 | 0.25 |
| | Tetrachloroethylene | 7.6 | 2.5 | 0.25 |
| | 1,2-Dichloropropane | 8.0 | 2.7 | - |
| | Chlorobenzene | 8.1 | 2.7 | 0.25 |
| Aromatic hydrocarbon | Benzene | 9.0 | 3.0 | 0.2 |
| | Ethylbenzene | 6.8 | 2.3 | 0.2 |
| | Xylene | 7.2 | 2.4 | 0.2 |
| | Toluene | 8.0 | 2.7 | 0.2 |
| | Naphthaline | 7.0 | 2.3 | - |
| Aliphatic hydrocarbon | Pentane | 8.0 | 2.7 | 0.2 |
| | Hexane | 7.2 | 2.4 | 0.2 |
| | Heptane | 6.6 | 2.2 | 0.2 |

Numerous measurements of effective diffusion coefficients in mineral liner materials, as they are typically used for landfill liners, are available for cations and anions. In an extensive investigation, H. L Jessberger and collaborators looked into the diffusion coefficients of various test liquids (mono solutions and mixtures of the chlorides, nitrates and sulphates of the metals listed in Table 7.7, organic mono solutions and mixtures and an “artificial” leachate) in different mineral liner materials (natural clayey soils and different kinds of well-graded mineral mixtures with various additives) (Jessberger et al. 1995). A systematic relationship between characteristic geotechnical parameters of the soil material and the diffusion coefficients, however, has not been found.

**Table 7.7.** Diffusion coefficients of cations and anions at 20 °C in dilute aqueous solution and in soil water of mineral liners and tortuosity parameters determined from them

| Cation or anion | Diffusion coefficient ($10^{-10}$ m²/s) | | Tortuosity factor (-) |
|---|---|---|---|
| | Water[d] | Soil water[f] | |
| Lithium | 10.3 | 1.3[c] | 0.16 |
| Potassium | 9.6 | 0.6–2.8[a] | 0.03–0.14 |
| | | 11.7–17.7[b] | 0.6–0.9 |
| Zinc | 7.0 | 0.4–1.6[a] | 0.06–0.23 |
| | | 8.2–10.3[b] | 1 |
| Cadmium | 7.2 | 4.8–7.6[b] | 0.67–1 |
| Lead | 9.5 | 0.4–1.4[a] | 0.04–0.15 |
| Ammonium | 19.6 | 0.6–2.8[a] | 0.03–0.14 |
| | | 2.05; 5.00[a] | 0.10; 0.26 |
| Chloride | 20.3 | 1.4–3.1[a] | 0.07–0.15 |
| | | 4.7–10.6[b] | 0.23–0.52 |
| Bromide | 20.8 | 4.9–9.9[b] | 0.24–0.48 |
| | | 4.8; 6.7[c] | 0.23; 0.32 |
| Iodide | 20.4 | 3.5–14.7[b] | 0.17-0.72 |
| Nitrate | 19.0 | 1.0–3.1[a] | 0.05–0.16 |
| | | 2.27; 1.00[a] | 0.12; 0.05 |
| Sulphate | 10.7 | 0.6–1.5[a] | 0.06–0.14 |
| Acetate-ion | 10.9 | 0.2–2.1[a] | 0.02–0.19 |

a) See reference (Jessberger et al. 1995)
b) Reference (Shackelford and Daniel 1991a) without data from test series 8
c) Reference (Schneider and Göttner 1991)
d) Coefficient of diffusion in infinite dilution of ions
f) Coefficient of salt diffusion

Table 7.7 displays the measured effective diffusion coefficients compared with the values in a free aqueous solution. The last column indicates the ranges of values for the resulting (apparent) tortuosity factor. The values remain almost universally below the value estimated above based on simple geometrical consideration, the smallest values are at most an order of magnitude smaller. It should be noted that the diffusion coefficients of the ion at infinite-dilution aqueous solution are indicated in Table 7.7, Column 2, while permeation measurements usually determine the diffusion coefficients for salt diffusion. In salt diffusion, fast ions are retarded and slow ions accelerated. Strictly speaking, the effective diffusion coefficient in soil water should be compared with the diffusion coefficient of salt diffusion in free aqueous solution. However, such a detailed analysis is beyond the scope of our considerations and it would not substantially change the estimated range of the tortuosity factor.

The overview of retardation coefficients fails to show such a clear picture of the typical range of values for mineral liner materials as was the case with effective diffusion coefficients. Already the discussion of mass transport in plastic geomembranes has shown that the diffusion coefficients lie within a relatively narrow range of values, while partition coefficients for different substances vary by orders of magnitude. The data collected on mass transport in mineral liner materials also indicates that retardation coefficients depend highly on the special material properties of the components and the composition of the soil material and the resulting interaction with the respective pollutant. In addition, the sorption isotherms are far from linear and therefore experimental results, which were obtained at higher concentrations, cannot be reliably used for estimates for the very small actual field concentrations. Thus only a very limited amount of data obtained from diffusion experiments in the field is available here, too. Roughly speaking, the values for highly water-soluble organic materials are fairly small, while in certain cases considerably large values have been obtained for materials with a poor water-solubility.

In (Barone et al. 1992) $R = 1.8$ is given for acetone, $R = 1.7$ for 1,4-dioxan and $R = 7$ for aniline. R. L Johnson and collaborators examined drill cores for pollutant contents of a clay deposit under a hazardous waste landfill which was impacted by leachate (Johnson et al. 1989). The apparent diffusion coefficients were derived from the depth profile of the concentration of the organic pollutants. A tortuosity factor of 0.20–0.33 was found for chloride in this investigation. The following retardation factors can, therefore, be calculated from the measured apparent diffusion coefficients with the effective diffusion coefficients listed in Table 7.5 (see (Müller et al. 1997a), note 16): $R = 15$ (benzene), $R = 56$ (trichloroethylene), $R = 54$ (toluene), $R = 27$ (1,2-dichloropropane), $R = 115$ (ethylbenzene), $R = 115$ (naphthalene). The following values were derived from sorption experiments in the laboratory on unaffected drill core samples: $R = 44$ (benzene), $R = 65$ (trichloroethylene), $R = 82$ (toluene), $R = 269$ (ethylbenzene).

D. Myrand and collaborators (Myrand et al. 1992) determined the apparent diffusion coefficients for benzene, trichloroethylene, toluene and chlorobenzene in diffusion experiments in the laboratory on clay samples which were taken from drill cores of the same natural deposit. Similar apparent diffusion coefficients were determined as in the field experiments. On the assumption of effective diffusion coefficients as per Table 7.6, the following retardation coefficients can be obtained: $R = 27$ (benzene), $R = 47$ (trichloroethylene), $R = 68$ (toluene), $R = 90$ (chlorobenzene).

The retardation coefficients of anions and cations in soil materials are usually about an order of magnitude smaller than for organic substances. They lie within the range of about 1 to 10, while the range 10 to 100 is typical for organic materials, as shown above. A far more detailed compilation of data on effective diffusion coefficients and partition coefficient can be found in (Rowe et al. 1995a).

## 7.4 Mass Transport in Composite Liners

In Germany a composite landfill liner consists of a minimum 2.5 mm thick plastic geomembrane and a mineral liner (Simon and Müller 2004). The mineral liner is constructed from 3 lifts with an overall thickness of at least 0.75 m for conventional municipal waste landfills or class II landfills and from 6 lifts with an overall thickness of at least 1.50 m for hazardous waste landfills. The liner system is supplemented by a protective layer for the geomembrane which is to prevent indentations and deformations from coarse drainage gravels. The surface of the mineral liner must be so smooth and the geomembrane installed with a minimum waviness so that the overburden produces an "intimate contact" between the geomembrane and the mineral layer[8]. A homogeneous load distributing effect within the protective layer will help to achieve intimate contact.

Data about the imperviousness of a faultless liner system under defined boundary conditions for as large a number of pollutants and soil materials as possible form an important component of the characterisation of the efficacy of this liner for instance in comparison to equivalent alternative liner systems. Therefore, in the following, a parameterization will be discussed for the permeation rate (or the permeability) and the induction time for diffusive mass transport in the composite liner consisting of a geomembrane and a compacted clay liner (or more generally: a porous mineral material). Quantities, which refer to the geomembranes, will be denoted with index 1, such as thickness $d_1$ and diffusion coefficient $D_1$, and quantities referring to the mineral liner will have index 2 such as thickness $d_2$ and effective diffusion coefficient $D_2$. The porosity of the water-saturated mineral liner is denoted with $\Theta$ as above.

[8] The intimate contact of the geomembrane and the necessary installation technology is dealt with in Chap. 9 in great detail.

The assumptions for the parameterization made should be emphasized again. The geomembrane has a nonporous internal structure. Advective mass transport processes due to water flow driven by a hydraulic gradient are therefore not possible in an intact geomembrane, neither in the underlying mineral liner. Holes in the geomembrane would lead locally to an advective mass transport in the mineral liner. However, this is excluded since only the faultless liner system is considered here. It is assumed that the pore space of the mineral liner accessible to diffusion is water-saturated. Mass transport processes in the gas phase of a partially saturated mineral liner are neglected. Similarly, material flow driven by temperature gradients is not considered. Thus only diffusion processes described by Fick's law can take place. It is further assumed that in the steady-sate diffusion condition, equilibria for the sorption and solution processes at the boundary surfaces are achieved instantaneously which can be fully described by the simple linear Eq. 7.22 with a constant partition coefficient.

The concentration profile $c(x)$ of the pollutant in the liner is divided into different ranges (Fig. 7.1): $c_0$ is the concentration of the substance in the leachate, which remains constant, $c_1(x)$ is the concentration profile in the geomembrane and $c_2(x)$ in the mineral liner, the concentration beneath the liner is assumed always to be zero. One may imagine a highly permeable aquifer under the liner where a very rapid dilution takes place. These steady-state diffusion conditions are surely an extreme assumption which might, in practice, be the case only at very unfavourable locations. Thus these boundary conditions also represent a worst case scenario.

The concentration gradient, which generates diffusion in the geomembrane, develops through the sorption and desorption of pollutants at the surface of the geomembrane. Concentration boundary conditions at the leachate-geomembrane interface ($x = 0$) are described, as explained above, by the partition coefficient $\sigma_{0,1}$ in accordance with:

$$c_1(0) = \sigma_{0,1} c_0 \,. \tag{7.32}$$

In the composite liner comprising a geomembrane and a mineral liner, the partition coefficient $\sigma_{2,1}$ at the interface ($x = d_1$) of geomembrane and mineral layer is a further parameter for the composite liner:

$$c_1(d_1) = \sigma_{2,1} c_2(d_1) \,. \tag{7.33}$$

In the simplest case $\sigma_{0,1} = \sigma_{2,1}$. It is assumed that a pollutant distributes itself between the geomembrane and the leachate in the same way as between geomembrane and pore water in the mineral liner. Strictly speaking however, $\sigma_{2,1}$ depends on the condition of the interface of the two liner components and on $c_1(d_1)$ and therefore differs from $\sigma_{0,1}$. In addition, the different chemical substances in leachate and soil water can have an influence on the sorption process.

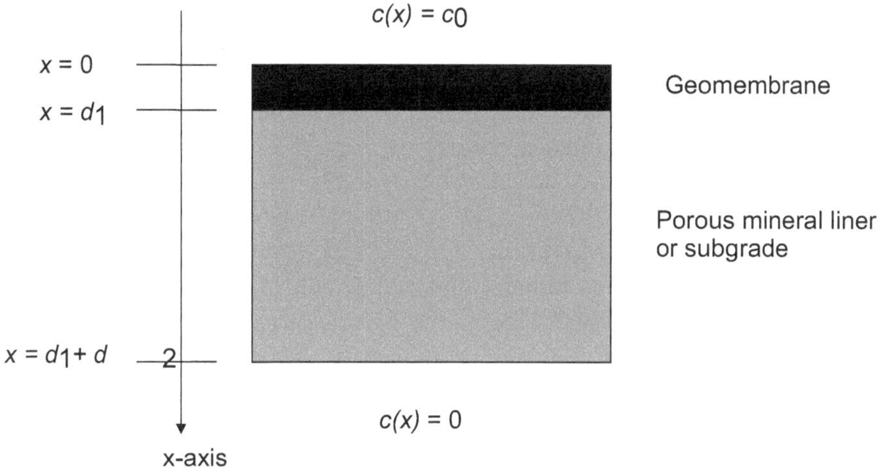


**Fig. 7.1.** Coordinates and boundary conditions for the description of mass transport in composite liners consisting of a geomembrane and a mineral layer

If the permeability is used as a material parameter, then it holds for the geomembrane and/or for the mineral liner according to Eqs. 7.9 and 7.31:

$$P_1 = \sigma_{0,1} D_1 \qquad P_2 = \Theta \Gamma D_0 \,. \tag{7.34}$$

Permeability $P$ and permeation rate $J$ of the composite liner with the thickness $d = d_1 + d_2$ can now simply be calculated for the boundary conditions 7.32 and 7.33 from the conservation of the material flow in a steady-state condition, i.e. $J = J_1 = J_2$. It is obtained:

$$\frac{d}{P} = \frac{d_1}{P_1} + \frac{d_2}{\left(\dfrac{\sigma_{0,1}}{\sigma_{2,1}}\right) P_2} \,. \tag{7.35}$$

The formula for the induction time of the overall system can be calculated by the method of Laplace transformation, see (Carslaw and Jaeger 1959) and (Barrie et al. 1963). The induction time $t$ of the composite liner is composed of induction times $t_1$ and $t_2$ and permeabilities $P_1$ and $P_2$ of the components as follows:

$$t = \frac{t_1\left(\frac{d_1}{P_1} + \frac{3d_2}{(\sigma_{0,1}/\sigma_{2,1})P_2}\right) + t_2\left(\frac{3d_1}{P_1} + \frac{d_2}{(\sigma_{0,1}/\sigma_{2,1})P_2}\right)}{\frac{d_1}{P_1} + \frac{d_2}{(\sigma_{0,1}/\sigma_{2,1})P_2}}. \tag{7.36}$$

The permeation rate for the diffusion of a pollutant through the composite liner is thus determined according to Eqs. 7.35 and 7.36 at specified thicknesses (for example. $d_1 = 2.5$ mm and $d_2 = 0.75$ m and/or 1.50 m for municipal waste landfill and/or the hazardous waste landfill) by diffusion coefficients $D_1$ and $D_0$, partition coefficient $\sigma_{0,1}$ and $\sigma_{2,1}$ as well as by the parameters of the porous mineral material $\Theta$ and $\Gamma$. While the parameters $D_1$ and $D_0$ as well as $\Theta$ and $\Gamma$ vary only by one or two orders of magnitude for different pollutants and mineral materials, the partition coefficients $\sigma$ of different pollutants differ by several orders of magnitude. Therefore, it is the partition coefficient between plastic geomembrane and leachate which characterises the permeation rate of the composite liner for different pollutant classes. The partition coefficient for cations and anions is in practice zero, since they cannot be dissolved in the non-polar medium polyethylene. For the diffusive mass transport therefore only undissociated organic and inorganic molecules have to be taken into consideration.

Since the thickness is included in the induction time as a quadratic form, $t_2 >> t_1$ holds for a composite liner. The retardation of mass transport in the mineral liner can substantially increase the induction time. The induction time $t$ of the overall system is then determined according to Eq. 7.36 by the thickness of the mineral liner alone. In addition to the partition coefficient of the plastic geomembrane, the sorption capacity of the mineral liner is therefore of paramount importance for the imperviousness of the composite liner system. Obviously, this does not apply to very thin mineral liners, such as to bentonite mats.

A perfectly installed, faultless composite liner is extremely impermeable to the large number of organic and inorganic substances in the leachate. However, the permeability calculated by Eq. 7.35 only applies to the described simple boundary conditions. In the field, however, more complex conditions emerge which promote diffusive migration on the one hand but also obstruct it on the other: the partition coefficient of a substance in the multiple-substance leachate mixture only approximately corresponds to that in the "pure" aqueous solution. In particular, the partition coefficients can be different at the interfaces above and beneath the geomembrane. Depositions develop on the geomembrane which affect the sorption processes. The pore space of the mineral liner is only partially saturated. The mineral liner will not be completely homogeneous in its characteristics, and preferential migration pathways can be formed for diffusion. The composition of the leachates and thus the "driving" concentration gradients may change with time. No steady-state diffusion conditions with maximum permeation rates will be able to develop. The actual emissions through a perfect composite liner in a landfill structure cannot therefore be quantified in detail. One may, however, assume that the diffusive permeability indicated here for the defined simple boundary conditions and steady state diffusion considerably overestimates the real permeability in the field.

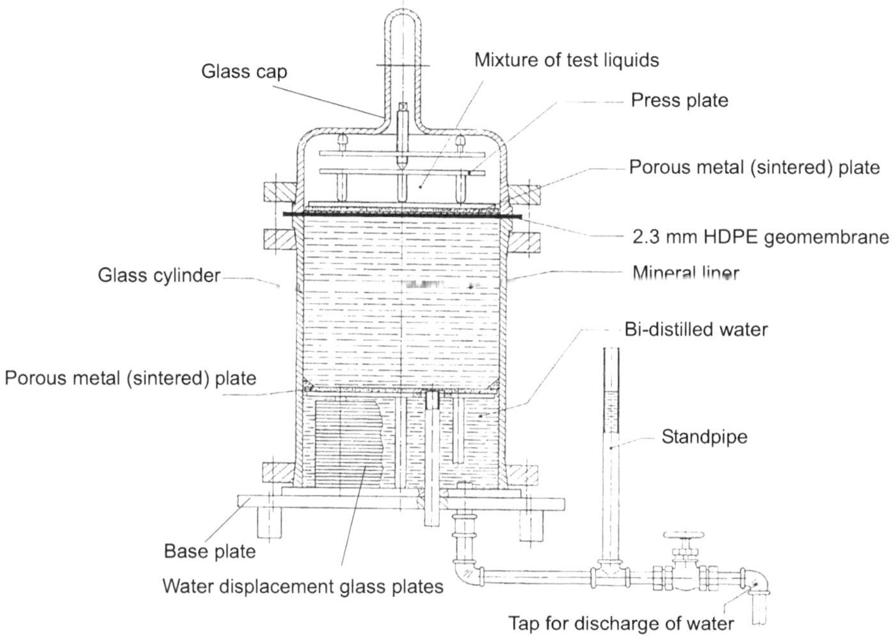


**Fig. 7.2.** Sketch of test cell assembly for permeation tests with a mixture of 9 liquid organic components (methanol, acetone, tetrahydrofurane, isooctane, trichloroethylene, toluene, tetrachloroethylene, chlorobenzene, xylene) (August et al. 1992). The diameter of the cells is 30 cm, the thickness of the mineral liners 7.5 cm, 15 cm, 30 cm and, in two cases, as much as 60 cm

H. August and co-workers carried out extensive investigations into the permeation of liquid hydrocarbons and chlorinated hydrocarbons through plastic geomembranes (August et al. 1984) and through composite liners (August et al. 1992). The results of the investigations on the plastic geomembranes had a substantial influence on the technical discussions about landfill lining in the mid-1980s. On the one hand, the very high resistance of HDPE geomembranes to chemicals was proved in comparison with all other geomembranes used in landfill construction at that time (PVC, ECB, CPE and EPDM)[9]. On the other hand, relatively high permeation rates were measured on the chemically highly resistant HDPE geomembranes under the impact of liquid hydrocarbons and chlorinated hydrocarbons. An effective liner is conceivable under such extreme impacts only when the plastic geomembrane is combined with a polar liner component (mineral liner). This was the starting point for the tests on composite liners.

For this purpose composite liners from a 2.3 mm thick HDPE geomembrane and various mineral liners with a thickness of 7.5 cm, 15 cm and 30 cm were constructed in permeation measuring cells and impacted with a mixture of liquid hydrocarbons and chlorinated hydrocarbons (Fig. 7.2).

The permeation behaviour was observed over a period of approx. 10 years. Afterwards the cells were dismantled and both geomembrane and mineral liner tested in detail (Kalbe et al. 2002). The composite liner did not only prove its chemical resistance without any restriction, but also to be particularly impervious even when the mineral liner was only thin. The combination of a hydrophobic liner material (plastic geomembrane) with a hydrophilic material (mineral liner), comprising the composite liner, operates under a long-term impact of liquid organic materials in the same manner as would be predicted based on the sorption equilibrium at the interfaces. A maximum concentration of the hydrophobic components of the test liquid develops in the soil water under the geomembrane (at the interface to the mineral liner) that corresponds to the (very low) solubility of these components in water, even if the liquid mixture is available above the geomembrane at very high concentration. Mass transport of hydrophilic hydrocarbon components of the test liquid is highly suppressed by the geomembrane alone and the concentration of these components in the soil water of the mineral liner therefore remains very low.

Figures 7.3a and b as well as 7.4a and b show the concentration distribution in the mineral liner and the temporal profile of the pollutant transport through the liner as it can, based on the test results, be assumed for acetone and trichloroethylene when they would accumulate on the composite liner over a long period and in large quantities.

[9] PVC: polyvinyl chloride with low-molecular organic admixtures as softeners, ECB: ethylene copolymer bitumen, a mixture of about 50 % ethylene and 50 % bitumen, CPE: chlorinated polyethylene and/or homogeneous mixture from chlorinated polyethylenes and PVC, EPDM: terpolymer from polyethylene, polypropylene and a small amount of diene monomer. A flexible rubber is obtained by vulcanisation.

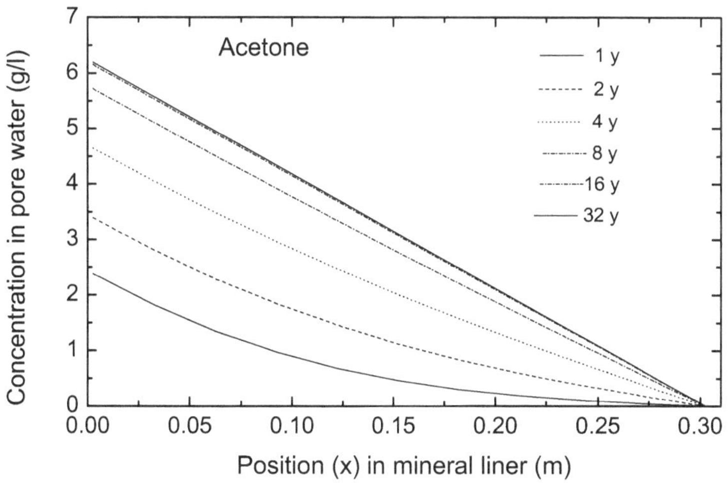


**Fig. 7.3a.** Calculation of the spatial-temporal distribution of the pollutant concentration in the pore water of the 30 cm thick mineral component of a composite liner for pure acetone impact (Kalbe et al. 2000)

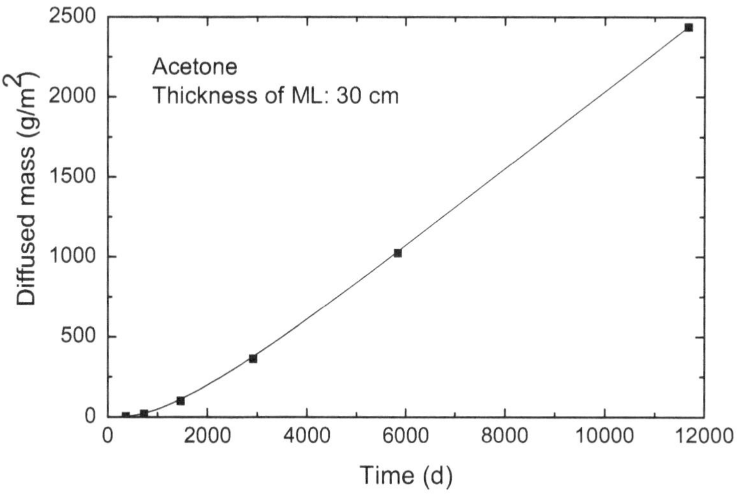


**Fig. 7.3b.** Calculation of the acetone mass diffusing through the composite liner (thickness of the mineral liner (ML): 30 cm) per $m^2$ as a function of time. The slope of the straight line yields the permeation rate and its time axis intercept provides the induction time. The induction time is 1440 days and the permeation rate in a steady-state condition is 0.24 g/($m^2$ d) (Kalbe et al. 2000)

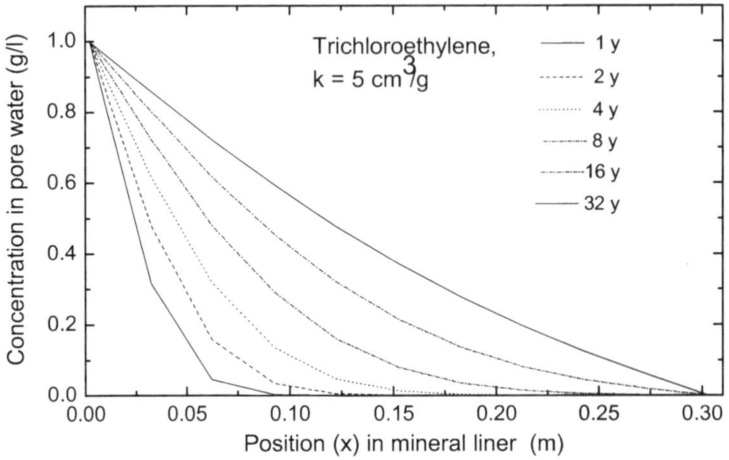


**Fig. 7.4a.** Calculation of the spatial-temporal distribution of the trichloroethylene concentration in the pore water of the 30 cm thick mineral layer (ML) of a composite liner under pure trichloroethylene impact under the assumption of a sorption with $k = 5$ cm$^3$/g (sorption coefficient for the soil matrix) and $R = 20$ (retardation coefficient, see Eq. 7.25) (Kalbe et al. 2000)

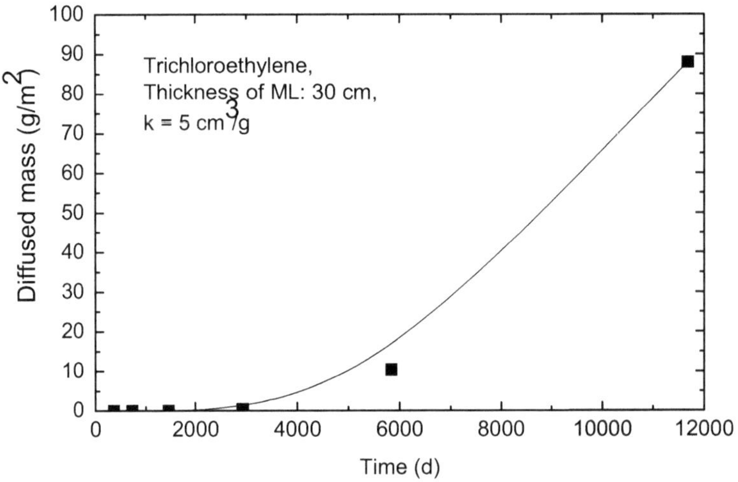


**Fig. 7.4b.** Calculation of the diffusing mass of trichloroethylene through the composite liner (thickness of the mineral layer (ML): 30 cm) per m$^2$ as a function of time (with sorption in the mineral component, see caption of Fig. 7.4a). Induction time is 5000 days and the permeation rate 0.013 g/(m$^2$ d) (Kalbe et al. 2000).

## 7.5 Influence of Holes in Geomembranes

In this section the influence of flaws (cracks or holes) in the geomembrane on the performance of the liner will be quantitatively described. How the flow rate of water can be calculated through the no longer impervious geomembrane liner will be discussed. Such calculations are of interest for the estimation of the field performance of liners, for instance for the evaluation of the composite liner when compared to other alternative liner systems in landfill caps, where a geomembrane is used alone or in combination with a bentonite mat or a capillary barrier. They enable answers to questions as to how the subgrade of the geomembrane should be designed and how carefully the geomembrane must be installed. The quantitative description of the influences of flaws also provides a basis for the evaluation of liner control systems which are currently used in large-area liners. The considerations presented here have already been published (Müller 1999).

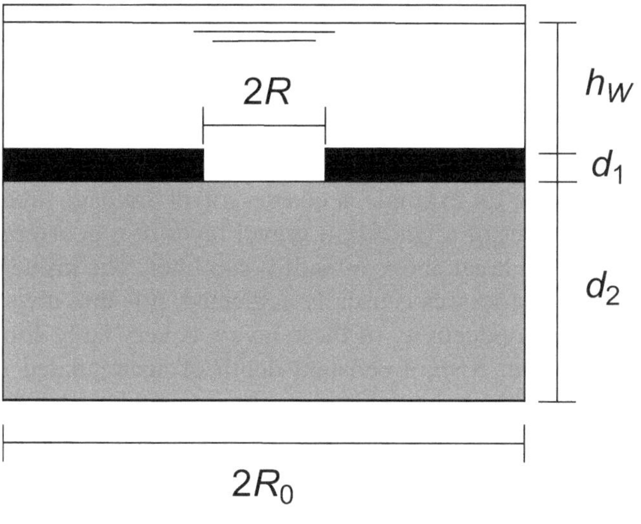


**Fig. 7.5.** Schematic set-up of a permeameter test cell with geomembrane and subgrade for the determination of the flow rate through a hole in the geomembrane. The main geometrical parameters for the description of the experiment are indicated

Theoretical description and calculations of the influence of flaws and the implementation and interpretation of laboratory control experiments are always based on simplifying models. In the following, an overview will be given about the concepts. It should in particular become clear which assumptions and approximations are incorporated in the models and calculations that form the basis of the quantitative statements. Unfortunately, only a limited amount of genuinely reasonable data is available from investigations into the influence of flaws and systematic evaluations of field experience about kind and frequency of flaws and their repercussion on the performance of liners with which such concepts and calculations can be verified (see Sect. 11.4).

Figure 7.5 represents the typical initial situation and the relevant parameters describing this situation. A geomembrane of thickness $d_1$ with a circular hole of radius $R$ overlies a mineral liner, or more generally, a subgrade which has the thickness $d_2$ and whose hydraulic property is characterised by the hydraulic conductivity $k$. The overall thickness of the liner is $d = d_1 + d_2$ and since usually $d_1 << d_2$ holds, $d \approx d_2$ can be assumed for the calculations. If such an assembly is modelled in a cylindrical permeameter for laboratory tests, boundary influences emerge and the radius $R_0$ of the permeameter must be taken into account. Water of depth $h_W$ accumulates on the geomembrane. The value of the hydraulic potential above the geomembrane be $\varphi_0$ and the value directly below the subgrade $\varphi_*$. Water should be able to flow off below the subgrade freely. Therefore $\varphi_* = 0$ is assumed.

In a liner system, further layers are arranged on the geomembrane: usually a protective layer and, depending on the geotechnical construction, more layers such as, for example, a coarse-gravel drainage layer and waste if it is the basal liner of a landfill, a gravel layer or a geocomposite drain and the reclamation layer above a landfill cap liner. The influence of these layers on the flow processes is usually neglected. It is thus always assumed that the hydraulic conductivity of these layers is very large compared with that of the subgrade. Also, a constant depth of accumulated water is assumed. The hydraulic potential within the accumulated water will only change negligibly under these assumptions. $\varphi_0 = h_W + d$ can then be set for the hydraulic potential above the geomembrane. Similarly, the relatively small drop in the hydraulic potential and the small flow resistance in the hole of the geomembrane are neglected. Finally, it is assumed that the subgrade is water-saturated.

The flow processes, however, also depend on the interface between the geomembrane and its subgrade and on how intimate the contact is. The influence of overburden above the geomembrane on the hydraulic potential is neglected, their weight, however, very strongly influences this contact and interface properties. The load on the geomembrane is therefore another very important parameter.

The task is to determine the volumetric flow rate $Q$ for specified parameters, i.e. the volume of water flowing per unit time through the hole in a steady-state condition. $Q$ will simply be called flow rate or leakage rate in the following. A volumetric flow density $v$ is obtained when $Q$ is referred to the liner surface.

If the water can flow off under the geomembrane freely then the flow through the hole in the geomembrane can be calculated using Bernoulli's law:

$$Q = C \cdot \pi R^2 \cdot \sqrt{2 g h_W} \, . \tag{7.37}$$

$C$ is a geometry factor which depends on how sharp edged the boundary region of the hole is. Its value is $C = 0.6$ for a sharp edge and $C = 0.99$ for a strongly rounded-off edge (Richter 1962). As usual, $g$ is the acceleration due to gravity.

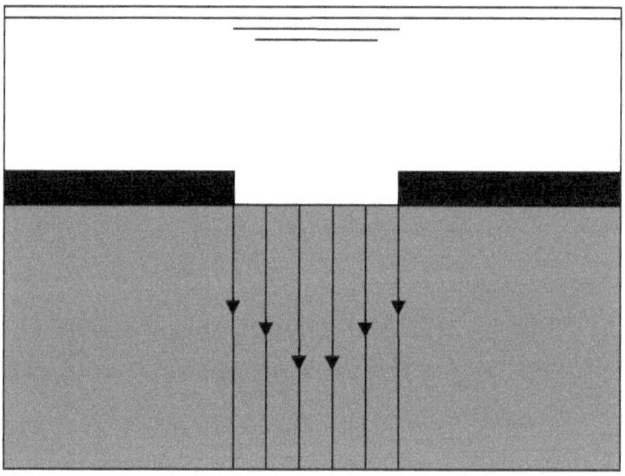

**Fig. 7.6.** Schematic illustration of the streamlines of Model I with the assumption of an only one-dimensional flow in the subgrade below the hole

However, one has to consider the case of a geomembrane installed on a subgrade with some hydraulic resistance. The simplest approach for the solution of this task, called Model I in the following, assumes a one-dimensional, vertical flow through the subgrade and a complete-surface, intimate contact of the geomembrane with the subgrade (Fig. 7.6) as an approximation. Water only flows in a narrow tube below the hole and the hydraulic gradient $i$ is uniformly equal in this approach:

$$i = \frac{\varphi_0 - \varphi_*}{d_2} = \frac{h_W + d}{d_2} \approx 1 + \frac{h_W}{d_2} \, . \tag{7.38}$$

Based on Darcy's law one obtains:

$$Q = \pi R^2 ki = \pi R^2 k \left(1 + \frac{h_W}{d_2}\right). \tag{7.39}$$

This model considerably underestimates the flow since, in reality, horizontal flow components do exist and a diverging flow cone develops in the subgrade.

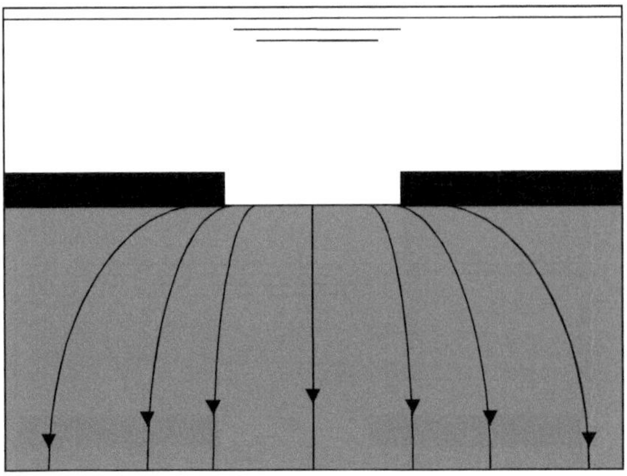

**Fig. 7.7.** Schematic illustration of the streamlines of Model II. Laplace's equation, which correctly describes the water flow in the subgrade, can analytically or numerically be solved

Using the physically correct approach, called Model II, the complete equation of motion is solved analytically or numerically for the hydraulic potential for the boundary conditions described above and the "true" flow rate is then calculated from this taking account of horizontal and vertical flow components (Fig. 7.7). Here, too, an intimate contact is assumed between the geomembrane and the subgrade. The details are described in (Walton and Sagar 1990). The potential and streamline pattern of water flow in the subgrade can formally be described as the potential and streamline pattern of an electric current in a conducting medium. Laplace's equation

$$\nabla^2 \varphi = 0 \tag{7.40}$$

is to be solved with the boundary conditions describing the situation shown in Fig. 7.7, i.e. $\varphi = \varphi_0$ within the range of the hole, $\nabla \varphi \cdot \boldsymbol{n} = 0$ at the lower surface of the geomembrane and at the wall surfaces of the permeameter (**n** is the unit vector in the direction of the surface-normal) and $\varphi = \varphi_*$ at the bottom of the permeameter.

The volumetric flow density can be calculated from solution $\varphi$ of Eq. 7.40:

$$v = -k\nabla\varphi . \tag{7.41}$$

The flow rate through the hole is then given by the integral of the flow density over the hole area:

$$Q = -\int_{\substack{hole \\ area}} k\nabla\varphi\cdot\boldsymbol{n}\, dA = -2\pi\int_0^R k\frac{\partial\varphi}{\partial z} r dr \ . \tag{7.42}$$

The second integral is obtained by applying a cylindrical coordinate system $(r,z)$ with the origin in the centre of the hole.

To deal with the problem independently of the units chosen for the geometric parameters, the dimensionless coordinates $r' = r/R$ and $z' = z/R$ and the dimensionless hydraulic potential $\varphi' = \dfrac{\varphi - \varphi_*}{\varphi_0 - \varphi_*}$ are introduced. The dimensionless flow rate $q$ is then:

$$q = -2\pi\int_0^1 \frac{\partial\varphi'}{\partial z'} r' dr' \ , \tag{7.43}$$

and the relation between $Q$ and $q$ follows from the comparison of Eqs. 7.42 and 7.43:

$$q = \frac{Q}{kR(\varphi_0 - \varphi_*)} . \tag{7.44}$$

In the case that $d_2$ and $R_0$ tend to infinity, i.e. are very large compared to the hole, thus for the infinite half-space, Laplace's equation can analytically be solved and the dimensionless flow rate calculated exactly. One obtains $q = 4$ and therefore the following expression for the leakage rate:

$$Q = 4kR(\varphi_0 - \varphi_*) = 4kR(h_W + d_2) . \tag{7.45}$$

For finite values of $d_2$ and $R_0$ the equation can only be solved numerically, the result is shown in Fig. 7.8. With a sufficiently large $R_0$, as it is typically the case under field conditions, the analytical solution and the relevant flow rate, Eq. 7.45, can be used in all cases where $d_2/R > 50$. Equation 7.45 answers the questions initially asked: how great is the flow rate through a hole and how the extent of the hole and the hydraulic conductivity of the subgrade affects this quantity.

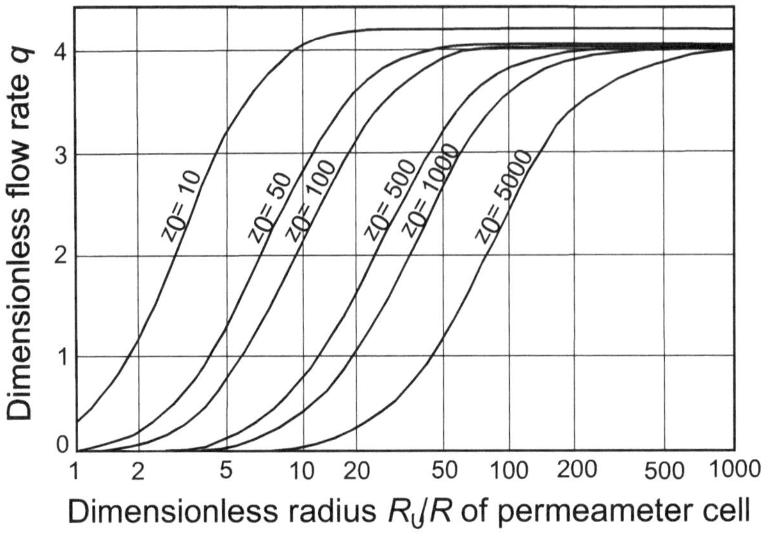


**Fig. 7.8.** Result for the dimensionless flow rate $q$ of the numerical solution of Laplace's equation for the permeameter test with a circular hole, from (Walton et al. 1997; Walton and Sagar 1990). The dimensionless flow rate $q$ through the hole is not only a function of the dimensionless thickness of the mineral subgrade $z_0 = d_2/R$, but also of the dimensionless radius $R_0/R$ of the permeameter cylinder because it sets boundaries to the horizontal flow in the subgrade

Comparing Model I with Model II, it can be seen how the assumption of a one-dimensional flow underestimates the leakage rate. It is:

$$\frac{Q_{\text{model I}}}{Q_{\text{model II}}} \approx \frac{R}{d_2}. \tag{7.46}$$

For a mineral liner with $d_2 = 75$ cm and $R = 0.5$ cm the ratio of the flow rates or leakage rates equals 1/150.

In (Walton and Sagar 1990) Laplace's equation was numerically solved not only for a round hole, but also for an oblong "linear" flaw in the geomembrane with the half width $\Delta$ (Fig. 7.9) assuming an infinite length of the slit for the calculation. Therefore the length $l$ of the slit must be very much greater than $\Delta$ to be able to use the results. Cartesian coordinates are used for the description of the slit: $x$ (along the width of the slot), $y$ (along the length of the slit) and $z$. The dimensionless coordinates are then

$x' = x/\Delta$, $y' = y/\Delta$ and $z' = z/\Delta$. The flow rate per unit length of the slit $dQ/dy$, with the dimension (m²/s), is similar to Eq. 7.42:

$$\frac{dQ}{dy} = -2\int_0^{\Delta} k\frac{\partial \varphi}{\partial z}dx\,. \tag{7.47}$$

The dimensionless flow rate $q$ is then similarly to Eq. 7.43 given by:

$$q = -2\int_0^{1} \frac{\partial \varphi'}{\partial z'}dx'\,, \tag{7.48}$$

and the connection of flow rate $dQ/dy$ per unit length of the slit with the dimensionless flow rate $q$, similarly to Eq. 7.44:

$$q = \frac{\dfrac{dQ}{dy}}{k(\varphi_0 - \varphi_*)}\,. \tag{7.49}$$

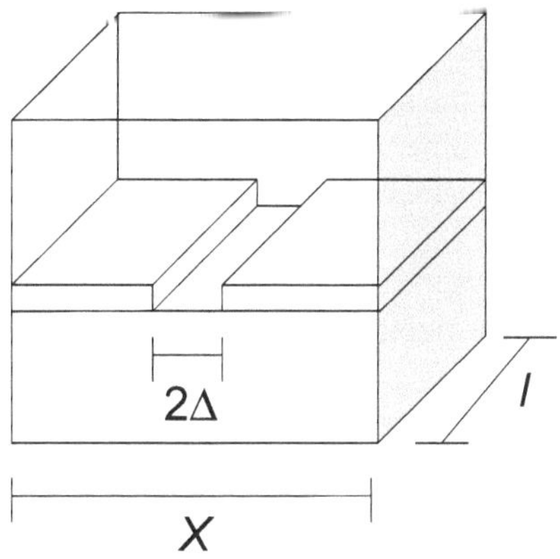


**Fig. 7.9.** Schematic illustration of a permeameter cell for the investigation of the geomembrane with a slit. The main geometrical parameters are indicated: length $l$ and width $2\Delta$ of the slit and width $X$ of the permeameter cell

Figure 7.10 shows the numerically determined values for $q(x_0, z_0)$ as functions of the dimensionless width $x_0 = X/\Delta$ of a permeameter cell with width $X$ and the dimensionless thickness of the subgrade $z_0 = d_2/\Delta$. Finally, the flow rate through the slit over the entire length $l$ can be calculated as:

$$Q = q(x_0, z_0)\cdot k(\varphi_0 - \varphi_*)\cdot l\,. \tag{7.50}$$

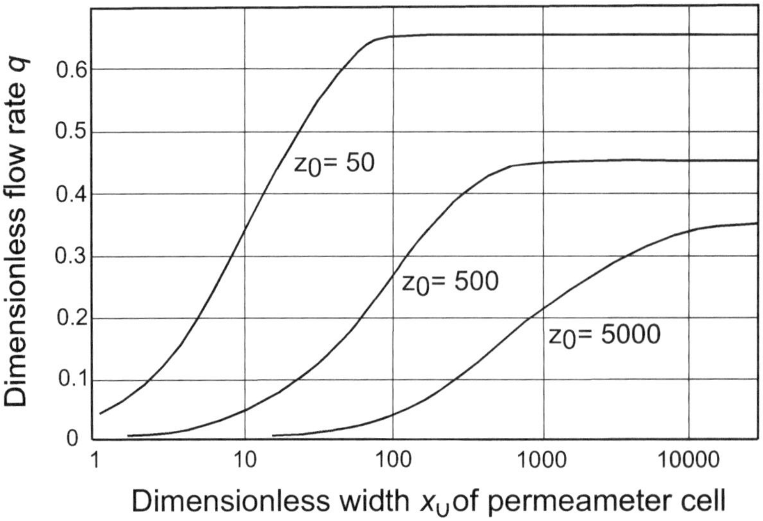


**Fig. 7.10.** Result for the dimensionless flow rate $q$ from the numerical solution of Laplace's equation for the permeameter test with a slit of half width Δ, see Fig. 7.9. Source of diagram: (Walton et al. 1997; Walton and Sagar 1990). The curves apply only to $l >> \Delta$, thus, in principle, for an infinitely long permeameter cell. The dimensionless flow rate $q$ through the slit, see Eq. 7.50, is not only a function of the dimensionless thickness of the mineral subgrade, $z_0 = d_2/\Delta$, but also of the dimensionless width, $x_0 = X/\Delta$, of permeameter cell because of boundary effects

Such very long, slit-shaped flaws, i.e. long tears, hardly ever occur in geomembranes or at least they only occur, when the installation procedure of the backfilling material and welding procedures are very poor. The connection between installation-technical expenditure and kind and frequency of flaws is dealt with in Sect. 11.4. However, geomembranes are quite often installed with waves and wrinkles. Under the load of the overburden, small residual waves are then unavoidably created which form very long narrow tubes (Soong and Koerner 1999). If a flaw is caused in the range of such residual waves which, for example, is likely when the waves cross seams, then the tube is rapidly filled with water and a slit-shaped flaw is formed. Formula 7.50 is therefore presented in the first place since it can be used to estimate the influence of such flaws within the range of a residual wave in the geomembrane, see Model IV discussed further down and Sect. 9.3.1. Such estimates will be of relevance in the controversy about installation procedures for "wave-free" geomembrane liners.

If a geomembrane is installed in a more or less wavy fashion or if the subgrade surface is very rough, the question arises as to what extent a lateral flow between the geomembrane and subgrade determines the flow rate through a hole. To answer this question the following model was developed (Jayawickrama et al. 1988), called here Model III. This model assumes a gap between the geomembrane and the subgrade (Fig. 7.11) with a new important parameter $t$ denoting the height of this gap. The geomembrane is assumed to "float" over the subgrade. Water can spread by flowing freely in the gap until the hydrostatic pressure gradually subsides. Simultaneously it seeps into the underlying subgrade where only a one-dimensional flow is assumed in accordance with Model I. The model differs principally from Model II: the substantial physical aspect is not the spread of water in the porous mineral layer, but the spread in the gap between the geomembrane and subgrade. This model was developed by K. W. Brown and collaborators for the evaluation of permeameter tests (Jayawickrama et al. 1988). It was popularised by J. P. Giroud and R. Bonaparte's papers (Giroud and Bonaparte 1989a, 1989b) and found a wide-spread application in the analysis of the influence of flaws on the efficacy of liners (Rowe et al. 1995a). In the following only a short overview will be given, details can be found in the above references.

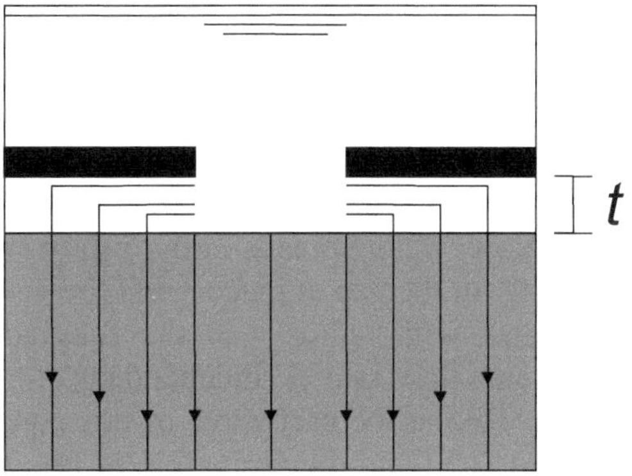


**Fig. 7.11.** Schematic view of the streamlines of Model III. A lateral flow is assumed in the gap between geomembrane and subgrade. The height $t$ of the gap is the main geometrical parameter of the model. Only a one-dimensional flow is assumed in the subgrade

Let $r$ be the cylindrical coordinate which measures the radial distance inside the gap from the centre of the hole to the outward. Within the range of the hole, i.e. $0 < r < R$ (see Fig. 7.5) water pressure is $p = \rho g h_W$ with $\rho$ the density of water. In the water, which spreads in the gap, the pressure drops radially outward and thus the lateral flow rate $Q_r$ also becomes ever smaller in the gap, until both pressure and flow rate become zero at some radius $R_W$. A differential equation can be derived from the equilibrium of pressure and intrinsic shear forces (viscosity) acting on a volume element of water to describe the connection between pressure drop and flow velocity gradient. By integrating, flow rate $Q_r(r)$ at a place $r$ in the gap can be calculated as a function of pressure drop. The calculation corresponds to the well-known derivation of Hagen-Poiseuille's law for a flow in a pipe applied to the cylindrical disk of the gap. It is:

$$Q_r = -\frac{\pi r t^3}{6\eta}\left(\frac{dp}{dr}\right), \tag{7.51}$$

where $\eta$ is the viscosity of water.

However, water does not only spread in the gap, it also seeps simultaneously into the mineral liner due to the still existing local water pressure. To describe this phenomenon the one-dimensional approach of Model I is used. In the circular disk of width $dr$ and radius $r$ the vertical differential flow rate $dQ_m$ through the subgrade is:

$$dQ_m = (2\pi r dr)\, k \left( \frac{\frac{p}{\rho g} + d_2}{d_2} \right). \tag{7.52}$$

One obtains the flow rate $Q_m$ through the circular area of radius $r$ by integrating from $0$ to $r$. Now, water spreads in the gap in such a way that the lateral water flow rate $Q_r$ in the gap at place $r$ and the vertical flow rate $Q_m$ through the circular disk with radius $r$ of the subgrade always exactly equals the specified water flow rate $Q$ (independent of $r$) from above into the hole: $Q = Q_m + Q_r$. Taking the derivative of this equation with respect to $r$ and substituting Eq. 7.52 and the derivative of Eq. 7.51, a differential equation for the pressure as a function of $r$ is obtained:

$$\frac{d}{dr}\left(r\frac{dp}{dr}\right) - \frac{12\eta k}{\rho g d_2 t^3}(pr) = \frac{12\eta k}{t^3} r. \tag{7.53}$$

The solution of this equation is:

$$p(r) = AI_0(\lambda r) + BK_0(\lambda r) - \rho g d_2. \tag{7.54}$$

Here $I_n(z)$ and $K_n(z)$ are the modified Bessel functions of the first and second kind and $\lambda = \sqrt{12\eta k / \rho g d_2 t^3}$.

Having solved Eq. 7.53 for $p(r)$ with the appropriate boundary conditions (see below) the flows $Q_m(r)$ and $Q_r(r)$ as functions of $r$ can be calculated. After all these considerations, flow rate $Q$ through the hole is finally given as a sum of the water flow $Q_m(R)$ that seeps into the subgrade directly within the hole area and the water flow $Q_r(R)$, which flows from the edge of the hole area directly into the gap:

$$Q = \pi R^2 k\left(1+\frac{h_W}{d_2}\right) - \frac{\pi R t^3}{6\eta}\lambda\left(AI_1(\lambda R) - BK_1(\lambda R)\right). \tag{7.55}$$

For a large liner surface with an unlimited laterally extended gap, the water in the gap spreads as far as the pressure subsides to zero at some radius $R_W$ and then the lateral flow succumbs. The boundary conditions, which describe this case, must therefore be formulated as follows:

$$p(R) = \rho g h_W \qquad p(R_W) = 0 \qquad Q_r(R_W) = 0\,. \tag{7.56}$$

From these three boundary conditions the still unknown constants $A$, $B$ and $R_W$ can be calculated by solving the following three equations:

$$A = \rho g \frac{(d_2 + h_W)K_0(\lambda R_W) - d_2 K_0(\lambda R)}{I_0(\lambda R)K_0(\lambda R_W) - I_0(\lambda R_W)K_0(\lambda R)} \tag{7.57a}$$

$$B = \rho g \frac{(d_2 + h_W)I_0(\lambda R_W) - d_2 I_0(\lambda R)}{K_0(\lambda R)I_0(\lambda R_W) - K_0(\lambda R_W)I_0(\lambda R)} \tag{7.57b}$$

$$AI_1(\lambda R_W) - BK_1(\lambda R_W) = 0\,. \tag{7.57c}$$

The boundary conditions are very different in the laboratory experiment. In a permeameter test $R_W = R_0$ is firmly specified: the permeameter wall stops the lateral flow. The unknown constants $A$ and $B$ can then be determined from the two remaining boundary conditions

$$p(R) = \rho g h_W \qquad Q_r(R_0) = 0\,. \tag{7.58}$$

Model III may be considered as an interpolation between Models I and II. In Model I a confined flow cylinder around the hole is considered and the actual flow rate is underestimated. Model II shows the correct, extensive flow cone within the subgrade. Though Model III assumes only the one-dimensional flow cylinder of Model I, it extends it in a lateral direction by allowing the spread of water in a hypothetical gap between geomembrane and mineral layer. If data of a permeameter experiment are interpreted with the help of Model III, a certain value for the gap height $t$ is necessarily obtained, irrespective of whether or not this gap really exists. Therefore gap heights derived from water flow experiments in a permeameter have no independent physical meaning. They are only artefacts of the one-dimensional treatment of water flow within the subgrade. It is particularly important to emphasise this aspect for further discussions.

It will be further discussed below as to what extent Model III is suitable for the description of water flow through holes in a geomembrane overlying a mineral liner or another subgrade. The model approach, however, can be used in connection with Eq. 7.50 without closer inspection to correctly describe the spread of water under a residual wave in a geomembrane when there is a flaw in the range of the wave. As already mentioned a narrow tube is formed by the residual wave in which water can spread in a free flow.

Let $\Delta$ be its half spatial dimension in the cross-section of the tube. Again a Cartesian coordinate system is used for the description where $x$ points in the direction of the width and $y$ does in the direction of the length of the tube. Equation 7.51 is now replaced by Hagen-Poiseuille's law that describes the flow rate $Q_y$ in the tube:

$$Q_y = -\frac{\pi\Delta^4}{8\eta}\frac{dp}{dy} \,. \tag{7.59}$$

Firstly the flow rate $dQ_m$ of water seeping under the wave over the length element $dy$ in the subgrade can be treated as a one-dimensional flow as in Eq. 7.52. Secondly however, in order to consider the flow cone here, one should use Eq. 7.50 as a better approximation. One obtains in the first case:

$$dQ_m = \Delta \cdot k \frac{\frac{p(y)}{\rho g} + d_2}{d_2} dy \tag{7.60a}$$

and in the second case:

$$dQ_m = q(x_0, z_0) k \left( \frac{p(y)}{\rho g} + d_2 \right) dy \,. \tag{7.60b}$$

Using $Q = 2(Q_y + Q_m)$ and by deriving and substituting as above with Model III, one obtains the differential equation for the pressure drop:

$$-\frac{1}{\mu^2}\frac{d^2 p}{dy^2} + p(y) + \rho g d_2 = 0 \,, \tag{7.61}$$

with $\mu = \sqrt{(\Delta / d_2)(8\eta k / \pi\Delta^4 \rho g)}$ in the first case and $\mu = \sqrt{q(x_0, z_0)(8\eta k / \pi\Delta^4 \rho g)}$ in the second one.

$p(y) = -\rho g d_2$ is a particular solution of Eq. 7.61. The general solution of the corresponding homogeneous equation is $A \exp(-\mu y) + B \exp(\mu y)$. One therefore obtains the following general solution, similarly to Eq. 7.54:

$$p(y) = Ae^{-\mu y} + Be^{\mu y} - \rho g d_2 \,. \tag{7.62}$$

The constants $A$ and $B$, as well as length $l$, over which water would spread in an arbitrarily long a tube, are obtained from the three boundary conditions $p(0) = \rho g h_W$, $p(l) = 0$, $Q_y(l) = 0$. For a specified length $l$ one receives the constants $A$ and $B$ for the pressure drop from the two boundary conditions $p(0) = \rho g h_W$ and $Q_y(l) = 0$.

Under common field conditions the parameters will have the following typical values: $\Delta$ some centimetres, $l$ up to some tens of metres. Values > 0.1 mm are assumed for the radius $R$ of the hole. An evaluation of Eq. 7.62 and comparison with Eq. 7.37 with this parameter selection shows that with accumulated water above the geomembrane sufficient water will flow through the hole into the tube formed by the residual wave to completely fill the cavity with water and that the same hydrostatic pressure will develop everywhere. Therefore Eq. 7.50 can be directly applied for the influence of flaws within the range of residual waves, where $\Delta$ characterises the spatial dimension of the cavity cross-section under the wave and $l$ is the length of the wave. Only for perforation sizes of < 0.1 mm is the flow resistance of the hole the decisive quantity, in which case Eq. 7.37 must be used.

After this excursion into "wave theory", the question, which of Models II and III better suit an estimation of the flow through a hole as realistically as possible, has to be discussed. K. W. Brown and collaborators and M. Fukuoka carried out permeameter tests in the mid-80s with the set-up schematically shown in Fig. 7.5 (Fukuoka 1986; Jayawickrama et al. 1988). The $k$ values of the mineral subgrade under the geomembrane were in the range of between $10^{-6}$ and $10^{-9}$ m/s. The overburden was, however, small in nearly all tests. In K. W. Brown's experiments 15 cm gravel was applied. Most of M. Fukuoka's tests were performed without any special cover over the geomembrane. The purpose of these tests was to investigate the lining of ponds, reservoirs and dams with geomembranes. In the few cases where a backfill was placed on the geomembrane (Fukuoka) or a greater overburden was applied by mechanical devices (Brown), a significant flow reduction through the hole was observed. In K. W. Brown's test an overburden of 160 kPa led to a drastic reduction in the flow rate by a factor of 200.

The tests were interpreted from the beginning using Model III. The correct boundary conditions Eq. 7.58 were applied for the permeameter test: the lateral flow in the gap ends at the permeameter wall and the flow rate was calculated by Eq. 7.55 for different values of $t$. From the comparison with the measured flow rates the value required for the height of the gap was determined based on the model calculation. The gap or the associated wetted area around the hole, however, has never been directly observed or measured. Doubtless, interface flow played an important role in the experiments with small overburdens and high water pressures on the geomembrane. The values listed in Table 7.8 for the height of the gap were derived from the analysis of the permeameter results and it was suggested to use these for the calculation of field leakage rates, based on Model III with the relevant boundary conditions Eq. 7.56. An enormous area of water flow in the interface between geomembrane and subgrade can be obtained under these boundary conditions of water spreading in a free interface flow. K. W. Brown and collaborators point out that the overburden has a very large influence on the test results and they come to the restriction that "the model will be limited to shallow waste storage facilities where the overburden due to solid sludge is minimal" (Jayawickrama et al. 1988).

**Table 7.8.** Height $t$ of the gap between geomembrane and subgrade which is formally attributed to the geomembrane liner in Model III (Giroud and Bonaparte 1989b) as a function of hydraulic conductivity $k$ of the subgrade

| Hydraulic conductivity $k$ (m/s) | Gap height $t$ (mm) |
|---|---|
| $10^{-6}$ | 0.15 |
| $10^{-7}$ | 0.08 |
| $10^{-8}$ | 0.04 |
| $10^{-9}$ | 0.02 |

In the mid-90s, J. Walton and collaborators carried out permeameter tests to clarify the importance of overburden on the geomembrane (Walton et al. 1997). Sand with a permeability of $10^{-4}$ m/s was used as the geomembrane subgrade. These tests have shown that the results for overburdens greater than 20 kPa could easily be interpreted with the help of Model II and the associated numerical solution for the boundary condition of the permeameter test. Clear indications of interface flow emerged only below 10 kPa. The authors recommend an earth cover of the geomembrane of at least 0.5 m. Under this condition, Eq. 7.45 can be used for the calculation of leakage rates. It was observed that the hole in thick geomembranes is silted up with soil material, further reducing the flow rate. This effect is also discussed in the paper which should be consulted for further details.

Model III has found a wide-spread application through J. P. Giroud and R. Bonaparte's papers (Giroud and Bonaparte 1989a, 1989b) published in 1989. It must be noted however that the gap which is only an artefact of the special test conditions (no overburden, high water pressure, limited radius of the permeameter) and of the interpretational approach (Modell III with only one-dimensional water flow in the subgrade) and which never has been directly observed, notwithstanding has been declared as a generally valid basic physical characteristic of a liner containing a geomembrane. Conversely, the generally observed experimental fact of the overburden's strong influence on the flow has been interpreted as an artefact of the test procedure and has failed to find any special consideration.

The calculation based on Model III requires considerable numerical calculations. J. P. Giroud and collaborators have therefore developed approximate formulae for "good contact" and "poor contact" between geomembrane and subgrade (Giroud et al. 1989) which are supposed to approximate the functional relations of Model III:

$$Q = 0.21\, h_W^{\;0.9} (\pi R^2)^{0.1} k^{0.74} \quad \text{(good contact)}, \tag{7.63}$$

$$Q = 1.15\, h_W^{\;0.9} (\pi R^2)^{0.1} k^{0.74} \quad \text{(poor contact)}. \tag{7.64}$$

The quantities have to be measured in the following units: $Q$ (m³/s), $R$ (m), $k$ (m/s) and $h_W$ (m). However, only formula 7.63 provides values which more or less correspond to those calculated in Model III with gap parameters of Table 7.8.

The calculation of the flow rate or leakage rate through a hole in the geomembrane leads to very different results depending on whether Model II or III is applied. Two examples should illustrate this. For a uniform comparison of the results from the different approaches, the dimensionless flow rate $q$ is used in each case: flow rates calculated from Eqs. 7.55, 7.63 and 7.64 are converted into dimensionless flow rates using Eq. 7.44.

In the first example, a geomembrane is considered installed on a $d_2 = 0.30$ m thick mineral subgrade with a permeability factor of $k = 10^{-6}$ m/s. In the geomembrane there is a hole with a radius of $R = 5$ mm. A 30 cm drainage gravel layer and a 1 m thick reclamation layer over the geomembrane ensure an overburden greater than 10 kPa.

Based on K. W. Brown's procedurc, the height of the gap is $t = 0.15$ mm. $Q$ must be calculated by Eq. 7.55 of Model III with the boundary conditions 7.56 which apply to the field conditions. The result of such a calculation is presented in (Wu 1994). For water accumulation heights of $h_W = 0.5$ m, 1 m and 5 m the values $q = 490$, 570 and 640 can be obtained for the dimensionless flow rate. However, J. P. Giroud and R. Bonaparte's approximation formulae yield the dimensionless flow rates $q = 400$, 460 and 480 for a good contact (Eq. 7.63) and $q = 2200$, 2500 and 2600 when poor contact is assumed (Eq. 7.64).

For a basal liner which corresponds approximately, for example, to the requirements for German landfill liners, i.e. $d_2 = 0.90$ m, $k = 10^{-9}$ m/s, and for a hole with $R = 5$ mm, a flow rate $q = 1200$ is expected based on Model III and Table 7.8 at an accumulation height of $h_W = 5$ m. An area with a radius of $R_W = 3$ m of lateral flow in the interface is expected under the geomembrane (Wu 1994).

If it is assumed that the overburden ensures an intimate contact and Model II is used, the dimensionless flow rate is generally only $q = 4$ for all accumulation heights and a sufficiently small hole size in comparison to the thickness of the subgrade, i.e. $d_2/R > 50$.

The examples show that the Model III assumption about a gap between the geomembrane and the subgrade with the values of Table 7.8 might overestimate the actual flow rates which would result for an intimate contact by about 2 to 3 orders of magnitude. This is true irrespective of the specific properties of the subgrade, hole sizes and accumulation heights. Such an intimate contact seems to develop, however, with relatively small overburdens.

The basic properties of Model III, i.e. a gap between the geomembrane and the mineral subgrade and a large-area spread of water in this gap, have neither been verified in the laboratory nor in field tests. However, for the application of the model regardless of kind and extent of the overburden on the geomembrane, the following arguments have been stated. Normally, one would not raise any special requirements to the subgrade of the geomembrane. It is usually a soil or sand-gravel layer without being specifically rolled or scraped, or prepared in any other way. Pores, wheel marks, cracks, impressions, outstanding gravels, overlying gravels, foreign bodies would shape the surface. One would attach no great importance to the surface contact of the geomembrane. One would install the geomembranes over large areas without any further protection. Due to temperature differences during the course of a day, the geomembranes would develop a number of waves which would be over-folded and covered during ballasting. One would not place protective layers on the geomembrane, or only such poor protection with thin geotextiles that would possibly prevent perforation. Under these conditions the geomembrane would fail to reach an intimate surface contact even at high levels of overburden and there would always remain space for lateral flow in the interface. Indeed, this hypothetical description (unfortunately) applies world-wide to many geotechnical geomembrane installations in landfills, storage basins etc. However, one may question whether, even under such poor installation conditions, the gap Model III will be applicable. Even with a rough subgrade and a wavy deployment of the geomembranes, however, no large-area gap will probably develop between the geomembrane and the subgrade. Rather, the geomembrane is pressed by the overburden as far as possible onto the subgrade and the "surplus area" of the geomembrane is "stored" in waves (see Sect. 9.3.1, Fig. 9.5), whose effect on the flow would need a special consideration (see Model IV). Model III is even less applicable when geomembranes are installed according to the state of the art, see Chap. 9.

From these considerations we conclude that Model III, and Eqs. 7.63 and 7.64, is not capable of giving even an approximate description of the effect of holes on the performance of geomembrane liners. Instead, Eq. 7.45 of Model II should be used in cases of overburden on the geomembrane. Individual waves can remain in the geomembrane even at very high overburden loads. Flow rates, resulting from flaws in the range of a wave, should however not be described by Model III of a large-area interface flow, instead, the special geometry of the interface flow in tubes of residual waves should be adhered to explicitly in accordance with the Model IV.

A "worst-case" estimation based on Model III with the gap heights in Table 7.8 appears reasonable only in liners with small overburden loads, i.e. less than 20 kPa, in addition to a calculation as per Eq. 7.45 of Model II. Here, for simplicity, the approximate formula 7.63 can be used.

When applying the models, however, the following must also be considered carefully. The indicated models describe the flow when water accumulates over the long term and steady-state conditions develop in the water-saturated subgrade. Such conditions are not, however, available under all conditions. For example, a long-term accumulation of water will only form in exceptional cases on the landfill cap, for example in subsidence troughs. A temporary water accumulation after a heavy rain event will run off within a short period of time. In this case, water is absorbed through holes in the usually only partially saturated mineral subgrade during this time and released only gradually via vapour transport between the waste body and the subgrade of the geomembrane driven by the moisture gradient and temperature gradient. Such effects must be considered in the evaluation of the model calculations presented: From this viewpoint, too, they tend to overestimate the repercussions of geomembrane flaws in composite liners. If, on the other hand, water permanently accumulates on a geomembrane and the geomembrane is installed on a subgrade with low hydraulic conductivity, then only a few holes are necessary to strongly impair the original imperviousness of a geomembrane liner.

## References

August H et al. (1984) Untersuchungen zum Permeationsverhalten von handelsüblichen Kunststoffdichtungsbahnen als Deponieabdichtungen gegenüber Sickerwasser, organischen Lösungsmitteln und deren wässerigen Lösungen. Bundesanstalt für Materialforschung und -prüfung (BAM), Labor Deponietechnik, Berlin

August H et al. (1992) Permeationsverhalten von Kombinationsdichtungen bei Deponien und Altlasten gegenüber wassergefährdenden Stoffen, Bericht zum FuE Vorhaben 10302208 des BMBF. Bundesanstalt für Materialforschung und -prüfung (BAM), Labor Deponietechnik, Berlin

Barone FS et al. (1992) A laboratory estimation of diffusion and adsorption coefficient for several volatile organics in natural clayey soil. J Contam Hydrol 10: 225–250

Barrie JA et al. (1963) Diffusion and Sorption of Gases in Composite Rubber Membranes. Trans Faraday Soc 59: 869

Bear J (1988) Dynamics of fluids in porous media. Dower Publication, New York

Carslaw HS and Jaeger CJ (1959) Conduction of Heat in Solids. Oxford University Press, Oxford

DGGt (ed) (1997) GDA-Empfehlungen. Verlag Ernst & Sohn, Berlin

Einstein A (1905) Über die von der molekularkinetischen Theorie der Wärme geforderte Bewegung von in ruhenden Flüssigkeiten suspendierten Teilchen. Annalen der Physik 17: 549–560

Faure YH et al. (1989) Study of a Water Tightness Test for Geomembranes. In: Christensen T et al. (eds) Proceedings of the 2nd International Landfill Symposium. Environmental Sanitary Engineering Center (CISA), Cagliari, Italien, pp BIII-1–BIII-11

Giroud JP and Bonaparte R (1989a) Leakage through Liners Constructed with Geomembranes-Part I. Geomembrane Liners. Geotextiles and Geomembranes 8: 27–67

Giroud JP and Bonaparte R (1989b) Leakage through Liners Constructed with Geomembranes-Part II. Composite Liners. Geotextiles and Geomembranes 8: 71–111

Giroud JP et al. (1989) Technical Note, Evaluation of the Rate of Leakage Through Composite Liners. Geotextiles and Geomembranes 8: 337–340

Gmehling J and Kolbe B (1992) Thermodynamik. VCH Verlagsgesellschaft mbH, New York

Goydan R et al. (1989) Estimation of the Solubilities of Organic Compounds in Polymers by Group-Contribution Methods. Ind Eng Chem Res 28: 445

Jayawickrama PW et al. (1988) Leakage Rate Through Flaws in Membrane Liners. Journal of Environmental Engineering 114: 1401–1421

Jessberger HL et al. (1995) Versuche und Berechnungen zum Schadstofftransport durch mineralische Abdichtungen und daraus resultierende Materialentwicklung, Abschlußbericht zum Teilvorhaben 25 des BMBF Verbundforschungsvorhabens Weiterentwicklung von Deponieabdichtungssystemen (Förderkennzeichen 1440569). Ruhr-Universität, Institut für Grundbau und Bodenmechanik, Bochum

Johnson RL et al. (1989) Diffusive contaminant transport in natural clay: A field example and implication for clay-lined waste disposal sites. Environ Sci Technol 23: 340

Kalbe U et al. (2000) Mineralogische und chemisch-physikalische Auswirkungen der Permeation von Kohlenwasserstoffen in Kombinationsdichtungen und -dichtwänden, Bericht zum Forschungsvorhaben 1461027 des BMBF. Bundesanstalt für Materialforschung und -prüfung (BAM), Labor Kontaminationsbewertung, Berlin

Kalbe U et al. (2002) Transport of organic contaminants within composite liner systems. Applied Clay Science 21: 67–76

Landau LD and Lifshitz EM (2000) Fluid Mechanics, 2nd Edition. Reed Educational and Professional Publishing Ltd., Oxford

Müller WW (1999) Stofftransport in Deponieabdichtungssystemen, Teil 3: Auswirkungen von Fehlstellen in der Dichtungsbahn, ein Überblick. Bautechnik 76: 757–768

Müller WW et al. (1997a) Stofftransport in Deponieabdichtungssystemen, Teil 2: Permeation in der Kombinationsdichtung. Bautechnik 74: 331–344

Müller WW et al. (1997b) Stofftransport in Deponieabdichtungssystemen, Teil 1: Diffusions- und Verteilungskoeffizienten von Schadstoffen bei der Permeation in PEHD-Dichtungsbahnen. Bautechnik 74: 176–190

Müller WW et al. (1998) Solubilities, Diffusion and Partition Coefficients of Organic Pollutants in HDPE Geomembranes: Experimental Results and Calculations. In: Rowe RK (ed) Proceedings of the Sixth International Conference on Geosynthetics. Industrial Fabrics Association International (IFAI), Roseville, MN, USA, pp 239–248

Myrand D et al. (1992) Diffusion of volatile organic combounds in natural clay deposits: laboratory tests. J Contam Hydrol 10: 159

Oishi T and Prausnitz JM (1978) Estimation of Solvent Activities in Polymer Solutions, Using a Group-Contribution Method. Ind Eng Chem Process Des Dev 17: 333

Piringer OG (1993) Verpackungen für Lebensmittel, Eignung, Wechselwirkungen, Sicherheit. VCH Verlagsgesellschaft mbH, Weinheim

Prasad TV et al. (1994) Diffusion coefficients of organics in high density polyethylene (HDPE). Waste Management & Research 12: 61

Prigogine I (1961) Introduction to Thermodynamics of Irreversible Processes. John Wiley & Sons, New York

Reid RC et al. (1977) The Properties of Gases and Liquids. McGraw-Hill Book Company, New York

Richter H (1962) Rohrhydraulik, Ein Handbuch zur praktischen Strömungsberechnung. Springer Verlag, Berlin

Rowe KR et al. (1995a) Clayey Barrier Systems for Waste Disposal Facilities. E & FN Spon, an imprint of Chapman and Hall, London

Rowe RK et al. (1994) Pollutev6 and Pollute GUI. GAEA Environmental Engineering Ltd., 44 Canadian Oaks Drive, Whitby, Ontario L1N6W8, Canada

Rowe RK et al. (1995b) Diffusion of chloride and dichloromethane through an HDPE geomembrane. Geosynthetics International 2: 507–536

Scheidegger AE (1963) Hydrodynamics in Porous Media. In: Flügge S and Truesdell C (eds) Encyclopedia of Physics Fluid Dynamics II. Springer Verlag, Berlin, pp 625–662

Schneider W and Göttner J (1991) Schadstofftransport in mineralischen Deponieabdichtungen und natürlichen Tonschichten, Geologisches Jahrbuch, Reihe C, Heft 58. E. Schweizerbart'sche Verlagsbuchhandlung, Stuttgart

Shackelford CC and Daniel DE (1991a) Diffusion in saturated soil. II: Results for compacted clay. J of Geotech Eng ASCE 117: 485

Shackelford CD and Daniel DE (1991b) Diffusion in saturated soil. I: Background. J of Geotech Eng ASCE 117: 467

Simon FG and Müller WW (2004) Standard and alternative landfill capping design in Germany. Environmental Science & Policy 7: 277–290

Soong T-Y and Koerner RM (1999) Behavior of waves in high density polyethylene geomembranes: a laboratory study. Geotextiles and Geomembranes 17: 81–104

van Genuchten MT and Wagenet RJ (1989) Two-Site/Two-region Models for Pesticide Transport and Degradation: Theoretical Development and Analytical Solutions. Soil Science Society of America Journal 53: 1303–1310

Walton J et al. (1997) Leakage Through Flaws in Geomembrane Liners. Journal of Geotechnical and Geoenvironmental Engineering 123: 534–539

Walton JC and Sagar B (1990) Aspects of Fluid Flow through Small Flaws in Membrane Liners. Environ Sci Technol 24: 920–924

Wu J-H (1994) Leakage Rates through Composite Liners due to Defects in Geomembranes. Proceedings of the 5th IGS Conference, Singapore, pp 933–936

# 8 Requirements for Protective Layers

## 8.1 Function of Protective Layers

Geomembranes must be protected against damage by coarse objects with sharp edges and points. The geomembrane subgrade or foundation layer and, mainly, the drainage layer or the earthen layers above the geomembranes contain gravel, stones or even foreign bodies of various sizes. Under dynamic and static loadings during the construction phase or in use, these objects may cause unacceptably large indentations and imprints or even holes and tears. Therefore special conditions apply to the subgrade (Sect. 9.3), while a protective layer is placed over the geomembrane as part of the installation. The type and design of the protective layers depend on the characteristics of the neighbouring layers and the loading conditions. In principle two tasks can be distinguished.

On the one hand it may be necessary to protect the geomembrane from perforation by sharp-edged or pointed objects for a short or medium service life (see Table 5.1), and sometimes only during installation. The task of the protective layer is to prevent mechanical effects perforating the geomembrane, i.e. to prevent ductile failure (Sect. 5.3.1) (Koerner et al. 1996).

On the other hand the task may be to ensure that deformations imposed by indentations and imprints do not exceed the permissible limiting strain values for normal or over an especially long service life of the structure. This means that local deformations in the geomembrane which develop through creeping, embrittlement or stress cracking, i.e. through ageing processes in the broadest sense, into holes or tears are excluded. Therefore, the protective layer also has the task of preventing "weak points" which are susceptible in the long run to ageing processes, i.e. to prevent brittle failure (Sect. 5.3.1) (Gallagher et al. 1999; Seeger and Müller 1996b; Seeger and Müller 2003).

Protection of geomembranes in landfill liners provides an excellent example for this second task. A drainage layer is installed above the geomembrane, which in many cases consists of very coarse gravel. Waste deposition exerts a considerable dynamic load on the basal liner and the static load increases gradually with increasing mass of the deposited waste. Pressures up to 1500 kPa result on the basal liners, when waste heights reach 100 m. Under such loads the gravel particles of the drainage layer would exert an unacceptably high loading, which could deform the geomembrane or even perforate it. For the extremely long service life required for landfill liners (> 100 years) the reliability of the geomembrane could not be ensured without a suitably proportioned protective layer with an outstanding long-term performance. The protective layer therefore is an essential element of the liner: geomembrane and protective layer must be considered together as the liner. A quantitative criterion for specifying such protective layers will be developed in Sect. 8.3.1.

One also has to pay the same attention to materials selection, design and installation methods for the protective layer based on the state of the art, as to the geomembrane, the actual sealing component. This especially applies if geotextiles are used as components of the protective layers or other pure geosynthetic protective layers[1]. The protective effect of geosynthetics can be likewise impaired by ageing. In addition, mineral protective layers are also subject to the action of leachate flow on the boundary surface between mineral protective layer and drainage layer, which can destroy the protec-

---

[1] Geosynthetics are "planar products manufactured from polymeric materials used with soil, rock, earth, or other geotechnical engineering related material as an integral part of a man-made project, structure or system". (Forschungsgesellschaft für Straßen- und Verkehrswesen (FGSV) (ed) (1994) Merkblatt für die Anwendung von Geotextilien und Geogittern im Erdbau des Straßenbaus. Forschungsgesellschaft für Straßen und Verkehrswesen (FGSV), Köln

Koerner RM (1990) Designing with Geosynthetics. Prentice Hall, Englewood Cliffs, USA. ASTM D4439-01 *Standard Terminology for Geosynthetic* is referred to for all the following definitions. Geosynthetics are distinguished as:

- geotextiles (permeable nonwoven, woven or knitted fabrics)
- geogrids (a network of connected elements with relatively large apertures), which, depending on the manufacture process, are called woven, stretched (from punched plastic sheet) or placed (and at the crossing points bonded) geogrids, sometimes plastic tapes or rod-shaped plastic elements are included, too
- geonets (a network of stiff rips used for drainage of liquids and gases) are exclusively produced by a special extrusion process
- geomembranes
- geosynthetic clay liners (or bentonite mats, a manufactured hydraulic barrier consisting of clay bonded to a layer or layers of geosynthetic material)
- geocomposites, in which such geosynthetics are combined with each other, such as geocomposite drains or geocontainers

The fields of application of geosynthetics can be characterised by the terms separation, protection, filtration, drainage, reinforcement, containment and sealing.

tive layer by erosion (contact erosion). Therefore long-term behaviour is an issue for mineral protective layers, too.

Testing and design of protective layers that provide protection from perforation from only brief mechanical contact by sharp-edged and pointed objects will be summarised in Sect. 8.3.3. Chapter 8 is concerned mainly with the second task, i.e. that of long-term protection. The presentation follows the papers of S. Seeger (Seeger and Müller 1996b; Seeger et al. 1995; Seeger and Müller 2003). The types of protective layers and their design concepts can best be discussed using the example of protective layers for geomembranes in landfill liners. For this use a special certification procedure for protective layers was established in Germany (Müller 1996, 1995).

The importance of protective layers for geomembranes was recognized in other application fields, especially where high stresses are present, in particular during the installation phase, and the sealing must be as reliable and effective as possible. This in particular applies to tunnel building and especially to tunnels where pressurised water is present. Protective layers for geomembranes in tunnel construction and their testing are reported on in (Brummermann et al. 1999; Brummermann and Schlütter 2004).

## 8.2 Types of Protective Layers

### 8.2.1 Overview

In accordance with the relevant German technical guides and recommendations, the use of coarse gravel (round grain) or stone (double broken stone chips) of grain size 16/32 mm is suggested for the drainage layer in landfill liners and caps. This recommendation does not arise so much from the hydraulic requirements – technical guidelines require a long-term permeability of a minimum $10^{-3}$ m/s –, rather, from the fear that a drainage layer of too fine particles might fail prematurely due to incrustations, i.e. closing up and clogging of the pore volume by the excretions of biochemical metabolic processes of micro-organisms or chemical precipitations in the leachate. This coarse grain size, combined with high static and dynamic loads, poses a particularly high demand on the protective layers. Initially, protective layers composed of mineral materials (earthen materials) were used. However, of late, geotextile protective layers have gained wide application.

Originally, a sand layer (0/2 mm medium sand) was used as a mineral protective layer against the coarse-grained drainage layer. To aid installation, a thin nonwoven geotextile was often placed on the geomembrane. In order to achieve an even installation without damage to the geomembrane, a sand layer of at least 10 cm thickness, placed using special installation equipment, is necessary. However, in the interface between sand and drainage gravel, flowing leachate may cause contact erosion: gravel particles gradually dig themselves into the sand. Therefore a nonwoven geotextile as separator needs to be placed on the sand.

A mineral protection layer of 0/8 mm crushed stone exhibits a filtration stability based on geometrical criteria with respect to a 16/32 mm coarse gravel drainage layer, and can be placed directly under the coarse-grained drainage layer without any risk of erosion. Fig. 8.1 shows grain size distribution range that must be adhered to for filter stability. With high loads, however, this crushed stone can lead to unacceptable deformations in the geomembrane. In order to achieve a safe protection against such deformations, depending on load, a nonwoven geotextile with a mass per unit area of up to 1200 g/m$^2$ or a similar suitable protection layer must be placed between the geomembrane and the crushed stone layer as an additional load-distribution layer. This arrangement has been used in a large number of landfill projects.

Labour and costs as well as installation difficulties with mineral protective layers on slopes has led to the fact that simple geosynthetic protective layers of nonwoven geotextiles with a mass per unit area of up to 3000 g/m$^2$ are increasingly used for 16/32 mm coarse gravel. Initially the question arose as to whether nonwoven geotextiles provided sufficient protective effect and stability. The issues of material suitability and testing methods for geotextile protective layers were therefore examined by three research projects carried out within the "Advanced Landfill Liner Systems Integrated Research Programme", promoted by the German Federal Ministry for Education, Science, Research and Technology (BMBF), at the Franzius Institute for Hydraulic Engineering (Brummermann 1997a), at the State Materials Testing Institute of the Technical University of Hanover (AMPA) (Witte 1997) and at the Federal Institute for Materials Research and Testing (BAM) (August and Lüders 1997). The results can be summarised as follows:

In the case of stringent requirements (i.e. only small indentations in the geomembrane, see Sect. 8.3.1), and extreme static and dynamic loads, current protective layers of nonwoven geotextiles, even when manufactured with a mass per unit area of fibre material of up to 3000 g/m$^2$, do not provide sufficient protection against 16/32 mm coarse gravel. If the drainage layer consists of fine gravel (e.g. 8/16 mm) or wide graded gravel (e.g. 0/32 mm) and the loads are not too high (e.g. 20 m of waste), a nonwoven geotextile made of durable material can be considered a suitable protective layer.

In view of the special loads in landfills some companies have developed composite protective systems, in which sand and geotextiles are combined in such a way that a thin sand layer develops on the geomembrane. This sand layer is the actual long-term protective layer. Geotextiles serve as packing material and installation system for the mineral component and provide erosion stability.

**Fig. 8.1.** The installation of the sand-filled protective mat or geocontainer (MDDS mat) on a geomembrane is shown in Fig. 9.9. This picture illustrates a technical weaving trick in the production of the double fabric with spacers, into which the sand is filled. The double fabric is woven in such a way that it thins out toward the edges in a wedge-shaped way and the rolls are installed in a manner that the two wedges lie on top of one another. Thus a sharp joint edge or a sharp-edged overlap can be avoided and the protective effect can be ensured within the joint range without any restriction

The Gebrüder Friedrich GmbH company in Salzgitter, Germany offers up to 100 m long, approx. 2.20 m wide and approx. 20 mm thick sand-filled protection layers (MDDS layer) (Figs. 8.1 and 9.9). These consist of a double spacer woven fabric filled with sand in the factory and are installed with overlapping joints. The Naue GmbH & Co. KG, company in Espelkamp-Fiestel, Germany has developed the DEPOMAT protective layer system. It consists of a random open array or web of extruded PP monofilaments, 25 mm thick, 2 m wide and generally 35 m long, which is delivered in rolls, and of a separation geotextile. The monofilament array is rolled on the geomembrane, filled with sand up to a thickness of approx. 25 mm on site with a sand blower system and finally covered with the nonwoven geotextile separation layer.

Even a thin lift of sand provides protection against very high static loads: it functions particularly well under very strong dynamic effects. Even with a large number of dynamic load cycles, gravel particles of the 16/32 mm size range dig themselves into the sand only to about 10 mm and then achieve a stable bedding (Brummermann 1997b; Seeger et al. 1995). A mere 2 centimetres thick sand layer is thus a sufficient protective layer for practically all cases. These composite systems therefore provide an excellent protection under almost all conditions in civil engineering and are easy and quick to install.

If one generalizes these examples, in principle three different classes of protective layer can currently be distinguished:

1. Protective layers of a mineral fill, whose grains exhibit filter stability towards the gravel in the area drainage. Usually, an additional geotextile must be placed between the mineral protection layer and geomembrane.
2. Protective layers of pre-packaged sand. The composite protective layers (MDDS layer or DEPOMAT), and the protective layer structure, which originated from a Directive of Lower Saxony, also belong to this group: a 400 g/m$^2$ nonwoven geotextile is placed on the geomembrane, covered with at least a 10 cm thick sand layer on which a separation nonwoven geotextile is placed to prevent erosion.
3. Simple geosynthetic protection layers using nonwoven geotextiles, geogrids etc.

### 8.2.2 Mineral Protective Layers

The problem with mineral protective layers is that the particles must be fine enough so that the protective layer itself does not cause unacceptable indentations and imprints in the geomembrane, but is coarse enough to prevent erosion.

The so-called contact erosion results from the movement of particles from the protective layer by flowing water in the boundary layer into the coarse drainage gravel layer. The erosion may cause the coarse gravel particles to gradually move through the fine-grained protective layer. The process of contact erosion depends on the particle size distributions, the surcharge and the water flow conditions. Erosion stability is guaranteed if migration of finer fractions into the pore space is geometrically impossible due to the particle size proportions irrespective of the strength of leachate flow. K. Terzaghi has formulated the geometrical filter criterion which contains the resulting essentials of particle size distributions of the adjacent layers: the $d_{15}$ size of the mineral particles of the drainage layer (i.e. 15 % of the particles are finer and 85 % are coarser than this diameter) may not be larger than quadruple the diameter of the $d_{85}$ particle size of the protective sand layer to be filtered (Smoltczyk 1980; Terzaghi et al. 1996). A combination of 0/2 mm medium sand and 16/32 mm drainage gravel, obviously, fails to meet this geometrical criterion.

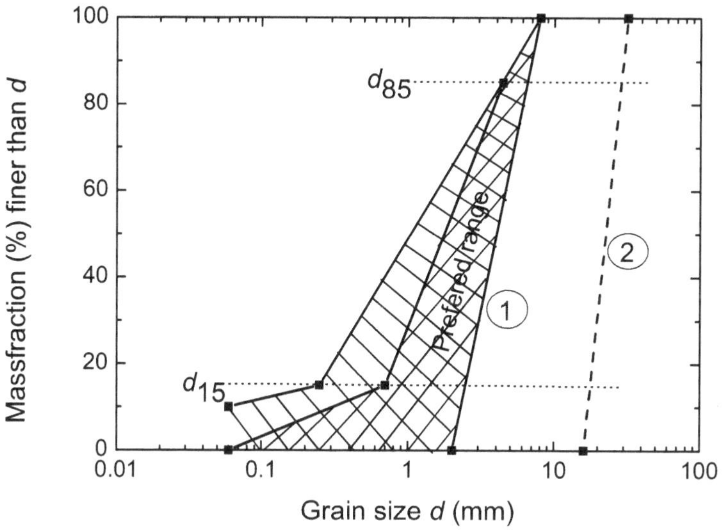


**Fig. 8.2.** Grain size range (1) of a mineral protective layer, which is stable to erosion on the geometrical filter criterion, as opposed to the gravel of the drainage layer (2) with 16/32 mm grain size curve. With high surcharges an additional nonwoven geotextilen with 1200 g/m² mass per unit area must be installed under such a mineral protective layer, in order to avoid unacceptable deformations in the geomembrane. The figure originates from (NLÖ 1994)

The criterion is sufficient but not necessary as the flow rate of the leachate is crucial to the erosion process. Even if the geometrical criterion is not fulfilled, erosion occurs only if the forces of flowing water promote materials transport. Only a few studies are available from which a filter criterion could be derived applicable to the actual boundary condition, i.e. horizontal flow in the boundary of the two layers. For instance, research results of J. Brauns on the erosion behaviour of layered soil with horizontal flow may be considered (Brauns 1985). The filter criterion derived from this leaves more clearance than the pure geometrical approach, but even for a small gradient the above combination of sand and coarse gravel would not exhibit erosion stability according to this criterion.

A 0/8 mm fine gravel with the particle size range given by J. Drescher (Fig. 8.2) exhibits an erosion or filter stability in the geometrical sense in accordance with K. Terzaghi's criterion. However, in order to reach a protective effect for high loads, which meet the requirements discussed in Sect. 8.3.1, an additional protective nonwoven geotextile must usually be placed between the fine gravel and the geomembrane.

These considerations may be summarized as follows: Anywhere where contact erosion due to flowing water plays a role, and a sufficiently fine-grained sand material must be used because of a high surcharge load, simple sand protective layers cannot be used.

Both installation and erosion problems can be solved by enclosing sand in geotextiles as mentioned in Sect. 8.2.1., i.e. using some kind of geocontainer. However, if erosion risk is expected to prevail over the long-term, the geotextile packaging material must also exhibit very good long-term performance. In landfills, the packaging material must exhibit functional integrity at least during the operational phase and during construction of the final cap. BAM-certified sand-filled protection geocontainers commercially available today fulfil this requirement and can provide safe long-term protection for the geomembranes against extreme loads and are quick and easy to install.

## 8.2.3 Geosynthetic Protective Layers

In mineral protective layers, geosynthetics are used as separation layers to avoid erosion and as a fabric for containing mineral materials. If there are not too coarse objects in the adjacent layers and the surcharge is not too high, pure geosynthetic protective layers can also be used, which are easy to handle and more prise-efficient. The plastic material itself takes over the load-distributing protective function. For this purpose, very thick non-woven geotextiles are generally used. However, a fine-meshed geogrid covered on both sides by a thin non-woven fabric can be used as the protective layer, too. Such a special composite system has so far not become generally established, principally due to financial considerations.

A nonwoven geotextile is a fabric, which consists of irregular or pre-arranged fibres, formed as a mat over one-another to build a planar structure. This structure is then reinforced by needle punching or heat bonding. The fibres are approx. 10 cm long (stable fibres). However, there are products, which are manufactured by arranging "endless" fibres (monofilaments) on each other. The fibres are extruded from the resin through a spinneret of many minute orifices (like a showerhead) (Walczak 2002). The fibre bundle composed of individual fibres, which leaves the spinneret, is moistened by a liquid processing or "finishing" agent ("avivage"), stretched on a drawing roller line, rippled/crinkled, cut and finally pressed as stable fibres into bundles, then packed and delivered to the geotextile manufacturer. The plastic deformation and orientation of the crystallites due to stretching results in a considerable increase in fibre strength. The individual fibres are each stretched to a different extent and a more or less broad distribution of values of strength and elongation at break is observed when testing the fibres. The finish is applied to facilitate drawing and/or crimping and to prevent electrostatic charging.

Woven fabrics, as their name says, are woven from yarns or slit film monofilaments using different weaving methods. For slit film monofilament production foils are first extruded and stretched, and then cut into small monofilaments.

The fineness[2] of the fibre is typically, depending on the product, about 6 dtex, 12 dtex or 18 dtex, which correspond to fibre diameters of approx. 30 μm, 40 μm and 50 μm. Likewise the thickness of the slit film monofilaments typically lies within the range of a few 10 μm, the width amounts to a few millimetres. These dimensions and the resulting low surface to volume ratio, may render geotextiles susceptible to the ageing processes.

The polyolefins polypropylene (PP) and polyethylene (PE) are often used as raw materials for geotextiles, in particular when protracted or complex chemical effects are present in the application. In the fibre and foil extrusion process the actual raw material (resin) is mixed in the extruder through a so-called master batch with light stabilizers, antioxidants etc. The carrier substrate of the master batch is also a polyolefin. Additives are of substantial importance for stability and long-term behaviour. Long-term tests and concluding statements on the long-term behaviour, as they are needed for applications in landfill engineering, can only be made where the product is clearly characterized and steady parameters of materials and manufacturing processes are guaranteed.

Ageing (see definition in Sect. 5.1) may limit the life span of geotextiles. Therefore sufficient stability must be available for the special functionality of the geotextile as a protective layer or as the geotextile component in the protective layer against the effects, which may cause internal

[2] The mass per unit length of a fibre is called titer. Its size characterises the fineness of fibres. The unit used to measure titers is the so-called "tex". The unit is defined as follows: 1 tex corresponds to 1 g fibre measured on a fibre length of 1000 m and 1 dtex corresponds to 1 g fibre measured on a fibre length of 10,000 m.

and external ageing processes. First of all oxidative degradation is of importance (see Sect. 5.2) (Mueller et al. 2003), then in all partially crystalline materials, stress crack formation (see Sect. 5.3) must be taken into account as a relevant ageing process (Müller et al. 2004). Even, additional crystallisation and the accompanying embrittlement may come into consideration as internal physical ageing processes. Such a case is reported in the literature (Tisinger et al. 1990).

The extent of the oxidative degradation strongly depends on the $U = F/V$ ratio of the surface $F$, over which the impact is made, to the volume $V$ of the material under impact. For fibres and slit film monofilaments this ratio is substantially greater than for geomembranes; Let $d$ be the thickness of the geomembrane and $r$ the radius of the fibres, typical values for these parameters yield $U_{fibre}/U_{GM} \approx d/r \approx 10^2$. A simple example can illustrate the effects. Antioxidants mixed into the resin gradually deplete, particularly through migration and leaching processes. The time over which most antioxidants deplete (antioxidant depletion time) affects the lifespan of the material. The amount of antioxidant lost per unit of time should be proportional to the surface area $F$. The initially amount available is proportional to the volume $V$. Time $t$, over which antioxidants deplete and after which the material become susceptible to oxidation, would then be inversely proportional to $U$. Therefore, the ratio of these times for geomembrane and fibre can be estimated as:

$$\frac{t_{fibre}}{t_{GM}} \approx \frac{U_{GM}}{U_{fibre}} \approx \frac{r}{d} \approx 10^{-2}. \tag{8.1}$$

This example rather underestimates the differences. If the diffusion path is considered and the time of migration processes for a given volume $V$ is assumed to be proportional to the linear dimension ($r$ or $d$) and inversely proportional to $U$ or for given volumes to the surface area ($F_{fibre}$ and $F_{GM}$ respectively), then Eq. 8.2 applies:

$$\frac{t_{fibre}}{t_{GM}} \approx \frac{r/F_{fibre}}{d/F_{GM}} \approx \frac{r^2}{d^2} \approx 10^{-4}. \tag{8.2}$$

On the other hand, when fibres and slit films are more or less strongly and homogeneous stretched, the orientation generally leads to an additional structural stabilisation (see Sect. 5.2.3). Therefore it is not easy to estimate the long-term performance of geotextiles in comparison to the HDPE geomembranes. Protection of HDPE geomembranes with geotextiles, however, only makes sense if the life spans are to some extent comparable. The long-term behaviour of geosynthetics is therefore an important field of research which is still actively being investigated (Hsuan and Koerner 2002; Mueller et al. 2003; Müller and Jakob 2000; Müller et al. 2004).

The ageing processes are, however, only relevant if they actually impair the protection capacity, i.e. the load-distributing effect. The results of research projects on the stability of nonwoven geotextile show two trends (August and Lüders 1995): if the test conditions (particle size distribution of drainage layer and surcharge) are selected in such a way that there is a strong impact on the nonwoven geotextile and large indentations are formed in the geomembrane, then the effect of chemicals, in particular quelling and oxidizing media, leads to a further substantial degradation of the protective effect. When the test conditions are selected in such a way, or the mass per unit area of the nonwoven geotextile material is substantially designed so that a strong protective effect is present and only small indentations occur in the geomembrane, then a medium impact remains within the test conditions without substantially influencing the protective effect. This effect can also be seen when mechanical protection capacity testing is carried out on a nonwoven geotextile which has been strongly impaired by an oxidative action through submersion in nitric acid (Seeger et al. 1995). The results suggest that, where the thick nonwoven material "fills" the unevenness and voids in the surface of the coarse-grained mineral protection layer by its shear mass, chemical impact and the accompanied ageing exert their influence only after extensive degradation. However, where the tensile strength of the geotextile is taken up due to the load distribution in the geotextile protection layer, ageing leads to a substantial degradation of the protective effect.

Section 8.3.3 deals with the design of protective layers that only have to prevent perforation. The tensile strength of nonwoven geotextile appears to play a substantial role here. Together with the geomembrane, protective nonwoven fabrics are strongly stretched and to a certain extent, take up tensile forces when covering the perforating objects. Ageing processes will have a direct effect under such a load condition. Therefore one cannot assume that protective layers designed along the lines of Sect. 8.3.3 would provide sufficient protection over the extremely long operational time of HDPE geomembranes. If one wants to make use of the extraordinarily high long-term durability of HDPE geomembranes with a high margin of safety, a rather elaborate means of protection must be employed.

## 8.3 Design and Testing of Protective Layers

### 8.3.1 Indentations in the Geomembrane

An object with sharp edges and points initially has only a small contact area on the protective layer. The acting load therefore creates a more or less local compressive stress. In an ideal case, the protective layer has to distribute these perforating compressive stresses in such a manner that the compressive stress load on the geomembrane is homogeneously distributed over the surface without local peaks. In real life, the protective effect of a protection layer is sufficient if the load distribution in the protective layer is dispersed to such an extent that only slight indentations arise in the geomembrane. However, an inadequate protection layer leads to clear indentations and imprints in the geomembrane, and when this happens, biaxial tensile stresses also arise in the geomembrane in addition to the compressive stresses. The load distributing effect of a protective layer must be derived from an estimation of the damaging potential of such deformation and stress states in the geomembrane.

In Sect. 5.3.4 the term critical limiting strain of thermoplastic materials has been defined and discussed. Critical limiting strain means that damage in the microstructure of the partially crystalline material develops when strains exceed this limit, which might then develop into macroscopic stress cracks. Conversely, stress crack formation is impossible when deformations stay below this limiting strain regardless of the stresses imposed. The critical limiting strain of HDPE materials lies within the range of 3–5 %.

Such a limiting value for the permissible deformation can also be derived in another way. Section 5.4 looks into this matter in greater detail. As suggested by R. Koch and his colleagues (Koch et al. 1988), tensile stresses are considered which arise from different deformation events taking into account stress relaxation in the geomembrane. These stresses are then compared with the stress level that the HDPE material can tolerate over the long term without stress crack formation (long-term pipe pressure test). Again one finds permissible strains around 3 % (Seeger and Müller 1996a).

This strain can therefore be used as a criterion for the design of protective layers: protective layers must be designed in such a way that the local strains resulting from indentations by objects with edges and points do not exceed the limiting strain.

Section 4.4 discussed how the local deformation of a geomembrane is calculated if a certain contour line of a subsidence or an indentation is forced upon it. If the strain due an indentation in the geomembrane caused by an object is considered, then two contributions become obvious: the geomembrane experiences an overall lengthening within the range of the indentation, since the indented surface is larger than the original area. This elongation can be quantitatively estimated, in as much that the contour of the indentation is described by a segment of a circle with the smallest extension of the indentation as chord $a$, the greatest depth as height $h$ and the opening angle $2\alpha$. Following the evaluation of the burst test the strain due to the lengthening can be calculated by the arch elongation $\varepsilon_L$ of this segment of a circle, see Sect. 3.2.9 and Fig. 3.11:

$$\varepsilon_L = \frac{\operatorname{arc}\alpha}{\sin\alpha} - 1 \quad \text{with } \sin\alpha = \frac{a\,h}{(a/2)^2 + h^2} \quad \text{and } \operatorname{arc}\alpha = \frac{2\pi\alpha}{360}. \tag{8.3}$$

On the other hand the geomembrane also experiences a bending deformation. The outer part of the bent geomembrane (outer edge fibres) is additionally stretched and the inner part (inner edge fibre) compressed due to the bending. This edge fibre elongation $\varepsilon_B$ is determined by the ratio of thickness $d$ of the geomembrane to the local radius of curvature of the indentation $r_c$ and is approximately:

$$\varepsilon_B = \frac{d}{2r_c}. \tag{8.4}$$

Indentations in the geomembrane, caused by gravel particles when load distribution in the protective layer is not complete, have relatively small radii of curvature ($r_c < 50$ mm)[3]. The accompanied edge fibre elongation then makes up the larger part of the total elongation $\varepsilon = \varepsilon_L + \varepsilon_B$. The relevant time-dependent tensile stresses can be determined e.g. from isochronous stress-strain curves for plane state of stress (Menges and Schmachtenberg 1985).

When evaluating the deformations due to subsidence, the bending deformation can be neglected, since the radii of curvature are always very much larger than the thickness of the geomembrane. The arising deformation of a subsidence trough is therefore well approximated by the lengthening. However, for the indentations due to gravel particles this is not the case for the reasons specified above. To illustrate the orders of magnitude of edge fibre elongation and overall lengthening (arc elongation), S. Seeger has provided an example by carrying out a finite element calculation (FEM) which simulated the deformations by a gravel particle in a simple

[3] With a critical elongation limit of 3 % a permissible radius of curvature of 40 mm can be calculated for a 2.5 mm thick geomembrane from Eq. 8.4. In practice, when handling the geomembrane in the field (e.g. rolling up the geomembrane on a spindle) or on the construction site this value is never exceeded.

way (Fig. 8.3). An indentation of only a small arc elongation (0.25 %) was produced. The table of values of the FEM calculation indicates that the maximum local strain due to the non-homogeneous curvature of the geomembrane can amount to a few per cent (even up to about 4 % in this example).

The example makes clear that even indentations, that appear minor based on the arc elongation, may exhaust locally the permissible range for the strain. If one aims at a high safety factor and a long-term effective protection that is required in landfill construction today, protective layers must be designed in such a way that only slight indentations and imprints are visible in the geomembrane.

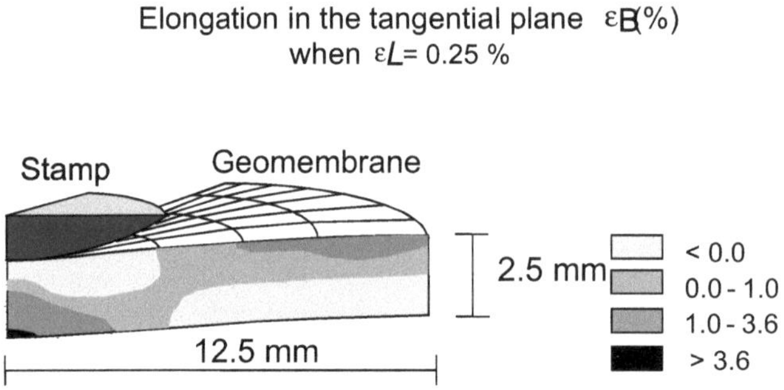


**Fig. 8.3.** Deformation state in a geomembrane, following the impressions with a stamp of given size. The stamp form simulates the contact area of a gravel particle on the geomembrane increased by the protective layer. The lateral dimension of the indentation was approx. 25 mm and the depth approx. 0.77 mm. Such indentations are typically seen, when protection layers are tested with 16/32 mm gravel. The calculated arch elongation equalled 0.25 %. However, local bending strain of the outer fibre is about ten times higher than the calculated arc elongation

The design concept, presented here, is surely very conservative. As discussed in Sect. 5.3.4 (structure particle model), when the critical limiting strain is exceeded, macroscopic cracks do not occur automatically. Macroscopic cracking can only be expected if "dangerous loads" occur as specified there, to which deformations caused by gravel particles usually do not belong. On the other hand, stress crack promoting environments may substantially worsen the load situation. As yet, how far the permissible strain under field condition over the long-term actually may exceed the critical limiting strain from laboratory experiments has not been properly scrutinised: for long-term applications there is no other way than to orient the design at the critical limiting strain.

### 8.3.2 Protective Efficiency Test

The question arises now, as to how this quantitative criterion can practically be converted and how the protective effect can be tested (Gallagher et al. 1999). On a suggestion by G. Heerten, at the beginning of the 90's, the "Quo Vadis Protective Layers" Working Group specified the procedure for the so-called modified load plate test method (or sometimes called cylinder test) initially conceived by F. W. Knipschild as a performance test for protective layers (Witte 1990). Thus a comparative evaluation of the mechanical protection efficiency (without taking into account durability) became possible. The test is very time-consuming. In the meantime, however, sufficient experience has become available on the protective layer classes as specified above, so that for individual systems, e.g. sand mats, an acceptable protective effect may be assumed over a wide range of conditions, without having to test this in each individual case.

The procedure of the mechanical protection efficiency test is described in detail in the GLR-recommendation E3-9 *Suitability Test for Geosynthetics* of the German Geotechnical Society (DGGt 1997). A modified procedure using a structured steel plate as the test body, that simulates the impact of a drainage layer in a standardized way, is also described there. It can particularly be used as an index test for the comparison of protection efficiency of different products (Brummermann 1997a). Meanwhile the European standard EN 13719:2002 *Geotextiles and Geotextile-related Products – Determination of the Long-term Protection Efficiency of Geotextiles in Contact with Geosynthetic Barriers* is available. Following the load plate test method an index test is described, in which a heap of 20 mm diameter steel balls is used as the test body on the protective layer. The mechanical protective effect is tested under this specific test body.

Testing of the mechanical protective efficiency is carried out based on the defaults of the GLR-recommendation E3-9 as follows: in a cylinder of 30–50 cm in diameter are placed (from bottom to top) a steel plate, a support layer, a 0.5–1 mm thick soft metal plate, a part of the geomembrane and the protective layer, the drainage layer or a test body that simulates the impact of the layer above the protective layer, a separation nonwoven geotextile, a load-distributing sand layer and finally another steel plate (load plate). The desired load is then applied with a pressurised ram on the load plate and the force measured using a pressure cell underneath the lower steel plate. The prospective subgrade for the geomembrane, e.g. a sand layer or mineral layer, can be applied as a support layer. However it is usual to use an approx. 2 cm thick elastomer disk (dense rubber pad) of shore A hardness 45–55. In order to keep wall friction small, the test cylinder is lined with geotextiles. Figure 8.4 shows schematically the structure of the load plate or cylinder test equipment. The intended test load is applied quickly but as smoothly as possible. The equipment then remains under load for the test period. The deformations in the geomembrane produce durable, plastic deformations in the soft metal plate. After the test period, the equipment is dismantled and the soft metal plate carefully removed and embedded horizontally in gypsum. Thus the indentations developed in the soft metal plate will be conserved. These indentations are then evaluated and the mechanical protection efficiency is assessed.

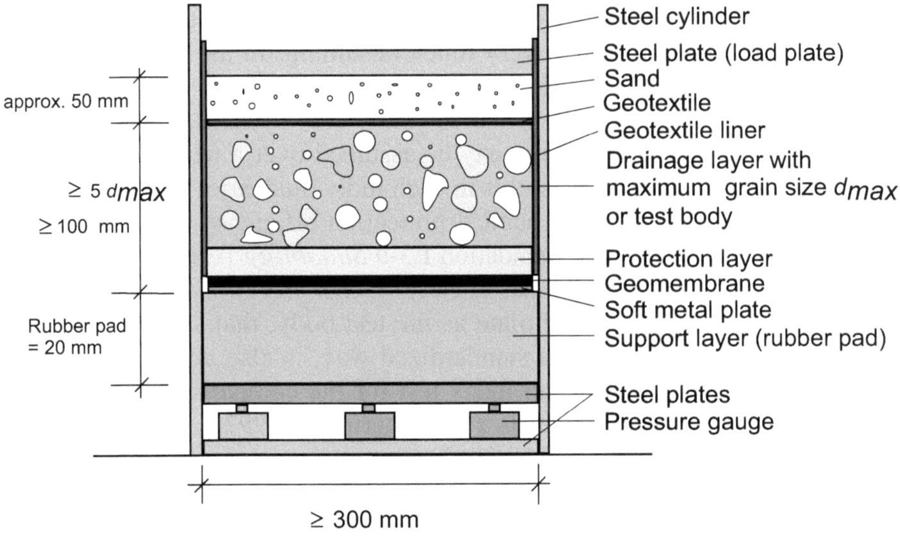


**Fig. 8.4.** Schematic view of the test equipment for testing mechanical protection efficiency, also called modified load plate test or cylinder test. Load is applied hydraulically on the upper steel plate. The drawing has been adopted from (Brummermann 1997b)

**Table 8.1.** Load increase factors for various test conditions

| Test condition | Load increase factor |
|---|---|
| 1000 h, 40 °C | 1.5 |
| 1000 h, room temperature | 2.25 |
| 100 h, room temperature | 2.5 |

For landfill construction, special test conditions have been agreed upon. The test period should be 1000 hours. For testing the protective efficiency in basal liners, a test temperature of 40 °C and a test load have to be set which exceed the maximum load expected in the landfill by a factor of 1.5. Where no reliable data are available, 15 kN/m³ is assumed for the specific gravity of the waste. The increased load and to certain extent the increased temperature also serve to accelerated the test conditions for the deformation behaviour of geotextile protective layers. Due to creep processes in the viscoelastic materials, the indentations will only fully develop after long periods. Since a test at 40 °C requires additional technical expenditure and a test period of 1000 hours is often too long, additional load increase factors (Table 8.1) have been established for simplified and shortened test conditions.

For all other applications in civil and hydraulic engineering the test can be carried out at room temperature with a test period of 100 hours using geotextile protective layers with an increased load factor of 2.

The test method described in the above mentioned European Standard EN 13719:2002 is essentially the same as the load plate test method of the GLR-recommendation E3-9. The standard gives various additional technical details of the components of the test equipment and the test procedure. For example, the thickness of the rubber pad is (25 ± 1) mm. A 1.3 mm thick grade 3 lead to EN 12588 is recommended as soft metal plate. 20 mm diameter steel balls to a minimum depth of 15 cm are used as test body. Protection efficiency was defined in the original standard document as the reciprocal of the slope of a line approximating the curve of arc elongation (see Eq. 8.3) versus registered load. If the arc elongation is measured as dimensionless ration, the unit of this quantity is kN/m². This quantity cannot be used to characterize long-term protection efficiency as the standard's title suggests, since it leaves open whether the protection actually achieved is appropriate with respect to the geomembrane material properties. It may be used as an index for ranking protection layers. According to the correction EN 13719:2002/AC:2005 the part about protection efficiency was recently cancelled.

The GLR-recommendation E3-9 specifies minimum test sample areas for bulk gravel of various particle sizes and the number of individual tests for a test cylinder with a diameter of 30 cm necessary for the assessment of the protective effect: three tests are required for 16/32 mm and two tests for 8/16 mm gravel while one load plate compressive test is sufficient for the 0/8 mm gravel.

K. Brummermann and R. Witte conceived a structured plate as a test body, which enables a simplified form of load plate test (Brummermann 1997b) (Fig. 8.5). Using the structured plate a well-defined index test is available which can be used to compare the protective effect of protective layer systems. When the dimensions of the structured plate are adapted to gravel sizes, bulk gravel can also be simulated. Finally, the long-term load plate test using a structured plate can be used for testing long-term ageing influences on the protective effect, e.g. by oxidation or chemical influence

(Figs. 8.6 and 8.7). Such tests were carried out by G. Lüders and U. Müller (August and Lüders 1997; Lüders et al. 1995).

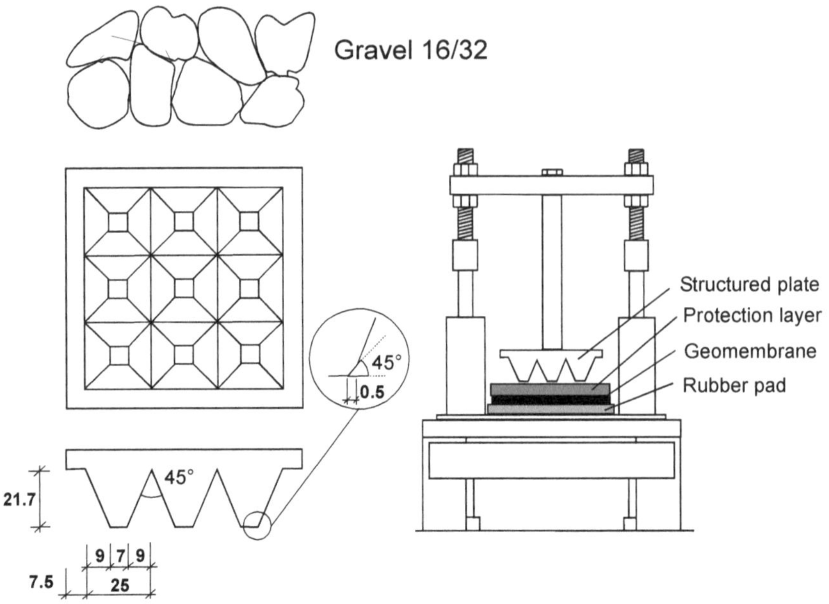


**Fig. 8.5.** Using a standardized structured plate (left) the mechanical protection efficiency test can be carried out in a simplified form as an index test. The test equipment is schematically illustrated on the right. Suitable structured plates may simulate the impact by gravel fills. Dimensions of the structured plate simulating 16/32 mm gravel are indicated on the left. The drawing has been adopted from (Brummermann 1997b)

The reproducible and unambiguous quantitative determination of the deformation of the geomembrane by measuring the indentations in the soft metal plate is confronted with measurement problems that have been studied by various research projects. The status of the evaluation technique, experience gained through performing the tests and the gathered results are summarised in (Sehrbrock 1993) and (Brummermann 1997b). Two methods are possible: using an electronically controlled mechanical sensing or other scanning device, the whole surface of the soft metal plate can be scanned using a sufficiently exact raster. The local radii of curvature and the maximum radii of curvature used for the evaluation can then be determined. This procedure is, however, very complex and expensive. Normally, the most prominent indentations are selected visually and only the contour line of these indentations is scanned along the visually strongest deformation. Strains connected with this contour line are then determined, see Sect. 4.4. Frequently, only the arch elongation is determined: the contour line is approximated by a polygon and its relative change of length in comparison to the baseline is calculated.

**Fig. 8.6.** U. Müller developed test equipment in BAM, which can be used to investigate the long-term behaviour of protective layers (long-term load plate test). A load prop and a structured plate are used to simulate the gravel fill (Fig. 8.5). Elastomer plate, geomembrane and protective layer are installed in a heatable high-grade steel container. This can be filled with water or other liquid chemicals

When designing protective layers, the error of assuming that the arch elongation is the measure for the actual strain in the geomembrane and can be compared with the critical limiting strain is still committed today. Even EN 13719 defines local strain as arc elongation according to Eq. 8.3. The arch elongation or lengthening of the geomembrane in indentations is actually small in relation to the bending strain, as has been shown in Sects.

4.4 and 8.3.1. It is therefore the local radii of curvature that determine the deformation and not the arch elongation. This is because of the fact that the elongation of indentations is in the same order of magnitude as the thickness of the geomembrane. Only for deformations due to subsidence in the subgrade of the geomembrane, whose size is always large in relation to the thickness of the geomembrane, does the arch elongation determine the total deformation value and bending strain can be neglected.

**Fig. 8.7.** View of the test equipment for long-term load plate tests. Load is applied over a lever arm, the steel container is covered with a glass cylinder, temperature is controlled and measured continuously. Settlement in the structure is recorded over a displacement gauge. This test equipment enables running tests of > 10,000 h. Reaction of oxidative degradation, stress crack formation on the protective effect, influence of swelling chemicals or creep behaviour can be investigated

However, S. Seeger and U. Sehrbrock have shown that for a bulk gravel with 16/32 mm particles as a test body, the criterion of permissible arch elongation of maximum 0.25 % can be used as a simplified rule for the design (Seeger 1995; Sehrbrock 1993). Their tests have shown that if the effectiveness of the protective layer allows indentations with an arch elongation of only 0.25 %, then only radii of curvature with edge fibre strains of 3–5 % will occur in the contour line of the typical indentation[4] (Fig. 8.3).

Still, a physically conclusive assessment of the protection efficiency for various conditions requires the determination of the radii of curvature in the contour line of the indentation. From the radii of curvature and the arc elongation the local strain can be calculated according to Eqs. 8.3 and 8.4. Then, this local strain has to be compared with the limiting strain of the plastic material used for the geomembranes (Menges et al. 1975).

Limiting strains of semicrystalline thermoplastics are of the order of magnitude of about $10^{-2}$. Arc elongations are about a factor of 10 lower than (true) local strains, i.e. to be acceptable, they have to be of the order of magnitude of about $10^{-3}$. Therefore, multiplying the protection efficiency of a protection layer, as defined in and measured according to EN 13719:2002, by a factor of 1000 gives a rough estimate of the acceptable load in $kN/m^2$ for this protection layer.

### 8.3.3 Testing for Puncturing of the Geomembrane

In a series of 3 papers R. M. Koerner, D Narejo and R. F. Wilson-Fahmy investigated theoretically and experimentally the question of the protection of the HDPE geomembrane against a perforation (puncture protection) by objects with edges and points, which stand out from the subgrade of the geomembrane or which lie on top of the geomembrane (Koerner et al. 1996; Narejo et al. 1996; Wilson-Fahmy et al. 1996).

If the geomembrane covers an object and is pressed against it by the surcharge, or an object initially resting upon the smooth geomembrane is pressed into it, if the object is large and pointed enough, then the geomembrane can become so strained that it gradually overstretches and finally breaks. These studies are concerned with preventing the geomembrane from breaking by overstretching contrary to the considerations discussed

[4] The limiting value of 0.25 % for the arch elongation has developed from the consideration that a visual inspection finds only minor indentations, but this criterion has to be able to be quantified by a simple measurement. With 16/32 mm gravel this criterion fits coincidentally with the requirement of plastic technology that even local elongations may not exceed the critical limiting strain. Since it is simple to test and already established, it was maintained as a design criterion within the BAM certification for protective layers. However, 0.25 % still remains a mystery, even for some experts.

above that are intended to prevent indentations and accompanied deformations to exceed a specified strain limit. They therefore studied the task of perforation protection and not the long-term protection against the combined impact of deformation and ageing processes, as discussed in the first section of this chapter. This different starting point has to be emphasized, since the authors make a comparison with the protective layers in landfill construction in Germany, without recognizing that two completely different tasks are pursued in the design. Depending upon the tasks, however, the type and extent of the protective layers may be completely different.

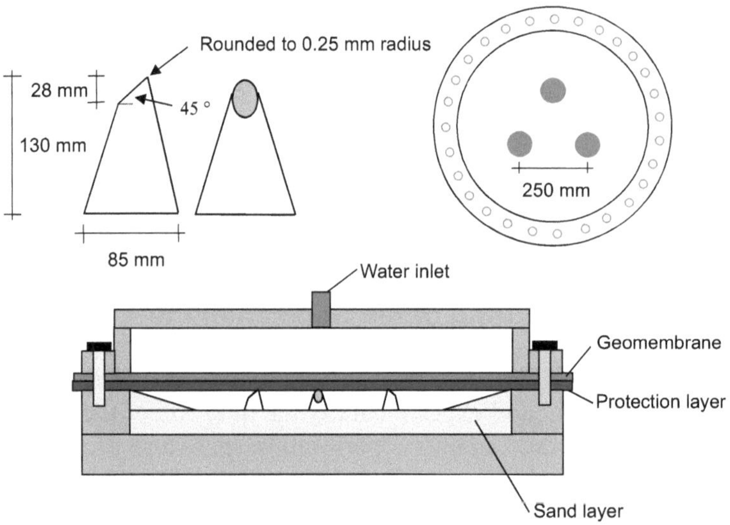


**Fig. 8.8.** Schematic drawing (not to scale) of the puncture test. The tapered truncated cone with its dimensions (left), the arrangement of the truncated cones on the base plate of the pressure vessel (right) and the test equipment with inserted geomembrane and protective nonwoven geotextile as well as the sand bed are indicated. The test procedure is described in the text

The differences also become clear in the test technology. For the puncture test (or so-called hydrostatic pressure truncated cone puncture test) a modified form of ASTM D5514-94 *Test Method for Large Scale Hydrostatic Puncture Testing of Geosynthetics* was used which had been generally conceived for testing the strength of geosynthetics against perforation. The test facility consists of a pressure vessel with three tapered truncated cones installed onto the round base plate. Figure 8.8 shows the dimensions of the truncated cones and their arrangement. A sand layer is then placed on the base plate so that the truncated cones stand out with a variable height $H$ from the sand layer depending on the thickness of the sand layer. By varying the protruding cone height $H$ (protrusion height) different types and sizes of objects can be simulated. The protective layer and the geomembrane are then placed flat on the truncated cones and the flange. Finally the upper section of the pressure vessel is placed on top and screwed on to the flange. Protective layer and geomembrane extend beyond the flange ring and so are clamped when bolting on the cover. Water is lead in above the geomembrane to produce an effective hydrostatic pressure acting in every direction. The pressure is increased continuously by 7 kPa/min. Electrodes connected with the water body are inserted into the points of the truncated cones. At the moment of perforation an electrical short-circuit occurs: a lamp or bell is activated. The effective hydrostatic pressure $p_F$ acting at the instant of this signal is the actual test quantity. The relationship between $p_F$ and $H$ is then investigated.

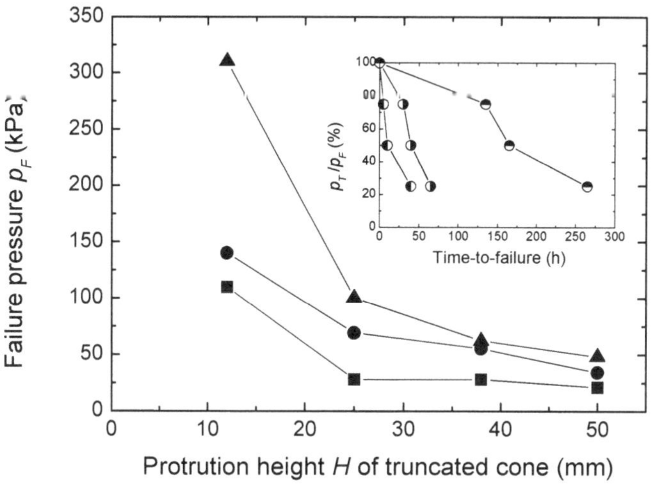


**Fig. 8.9.** Hydrostatic pressure at the instant of puncture (failure pressure $p_F$) as a function of height $H$ with which the tapered truncated cone protrudes from the sand bed. The test was performed on HDPE geomembranes with thicknesses of 1 mm (squares), 1.5 mm (circles) and 2 mm (triangles). Pressure was continuously increased at a speed of 7 kPa/min up to failure. A diagram showing the results of long-term tests is inserted. Here a test pressure $p_T$ was applied, kept constant and the time measured until a hole emerged (time-to-failure). Long-term behaviour was measured on a 1.5 mm thick HDPE geomembrane with different cone heights: $H = 12$ mm (circle, upper half black), $H = 25$ mm (circle, right half black), $H = 38$ mm (circle, left half black)

In the load plate test (Sect. 8.3.2) the indentation is determined at a preset surcharge after a given test period with the help of a soft metal plate. In the puncture test the hydrostatic pressure at which a hole develops in the geomembrane during a dynamic, uniform pressing of the geomembrane onto the truncated cones is determined. This highlights the two completely different approaches.

First the behaviour of the geomembrane was tested without any protective layer. Figure 8.9 shows the connection between failure pressure $p_F$ and protruding cone height $H$ as measured for different geomembrane thicknesses by D. Narejo and colleagues (Narejo et al. 1996). At large cone protrusions and low pressures all thicknesses lead to failure. Small cone heights show greater differences between different geomembrane thicknesses with the failure pressure rising steeply. It appears that a geomembrane of certain thickness is not perforated when the height of the protruding truncated cone is less than a certain value.

However, the measured failure pressures do not mean too much. In the experimental part of the work the failure performance was investigated in a simplified set-up using only one truncated cone. Test pressure $p_T$ was applied below the failure pressure $p_F$, firmly adjusted and observed whether and how long it was before a hole developed. The insert in Fig. 8.9 shows the result for the 1.5 mm thick geomembrane: failure occurred after a certain time at clearly lower pressures than the failure pressures from the short-term test with a continuous, fast pressure increase. Therefore the measured failure pressures, i.e. the pressures along the steeply rising flank of the curves in Fig. 8.9, depend on the speed of pressure increase, and therefore on the imposed deformation rates.

These findings can be interpreted as follows: if uniaxial deformation dominates, the HDPE geomembrane only tears at extremely large deformations (Fig. 3.9). However, with a multi-axial deformation, the HDPE geomembrane tears at substantially smaller strains (Fig. 3.12). In practice, perforating objects always impose multi-axial deformations onto the geomembrane. The failure strain and stress depends on the deformation rate. Failure stress decreases with decreasing deformation rate while failure strain increases. The geomembrane is pressed onto the truncated cone by the hydrostatic pressure from all sides. Tensile stresses develop and the geomembrane experiences a plane state of stress. If the truncated cone protrudes too far at a given deformation rate, the geomembrane will tear before it can completely match the contour of the truncated cone. If the truncated cone is, however, low enough, the geomembrane can stretch sufficiently and adapt to the truncated cone without exceeding the failure strain. Hence even if the deformation rate is taken into account, no failure occurs below a certain cone height, which depends on the geomembrane thickness. Unfortunately, failure pressures were not measured exactly over the range of low protrusion values. It appears, however, that practically no puncture occurs in a 2 mm thick geomembrane if the height of the protruding truncated cone is less than 10 mm.

The aim of the investigations was essentially to establish design rules for protective layers against puncture. Therefore puncture tests were also performed – both as short-term test and long-term test – with protective needle-punched nonwoven polyester and polypropylen geotextiles of different mass per unit area values, but only for a 1.5 mm thick HDPE geomembrane. In the short-term tests it was found that the failure pressure $p_F$ increases linearly with the mass per unit area $M_A$ of the nonwoven geotextile. The failure pressure also depended strongly on the deformation rate. On the basis of the long-term tests it was determined that if the initially rapidly applied test pressure is sufficiently small and the protective nonwoven geotextile has a sufficient large mass per unit area, protection against puncture is ensured independently of the deformation rate. For this purpose long-term tests were carried out over 10,000 hours. Table 8.2 shows the test conditions, which produced such long service livetimes with respect to puncture.

**Table 8.2.** Test results from long-term puncture tests

| Cone height $H$ (mm) | Mass per unit area of the nonwoven geotextile $M_A$ (g/m²) | Test pressure $p_T$ (kPa) or (kN/m²) | Time (h) |
|---|---|---|---|
| 12 | 550 | 1300 | >10,000 |
| 25 | 1080 | 460 | >10,000 |
| 38 | 1080 | 270 | >10,000 |

The tapered truncated cone only represents existing objects in the subgrade of the geomembrane or in the layer above the geomembrane – usually more or less sharp-edged gravel particles – in a strongly idealized way. Furthermore an overburden of earth layers will not exert homogeneous and isotropically distributed pressure as is the case with hydrostatic pressure. In (Narejo et al. 1996) these aspects were also investigated: it was concluded that the test conditions are on the safe side compared with real conditions.

Although the database for practically applied design rules remained small, nevertheless, it was tried to derive unusually extensive and detailed design rules. The required mass per unit area of the protection geotextile can be formally calculated by these rules as a function of maximum particle size and particle shape (angular, semi-rounded, rounded), stone arrangement (isolated or packed), soil arching effects and partial and global safety factors for various other effects. These factors are arbitrary and it is questionable whether the classification fits to the broad variation in actual field conditions. Therefore, one cannot avoid the impression that a design margin for the mass per unit area of the geotextile is thereby suggested by laboratory results, which is actually not present due to the broad variation of field conditions. As far as the details of the design concept are concerned, one should refer to (Narejo et al. 1996), where the design rules were established and to (Koerner et al. 1996), where they were illustrated by examples. The values in Tables 8.3 and 8.4 can serve as reference points for such a design; they were calculated for a 1.5 mm thick HDPE geomembrane for a gravel particle load and/or by a gravel fill under the load of a waste body with the specific weight of 11.8 kN/m³ (Koerner et al. 1996).

**Table 8.3.** Permissible heights of fill (specific weight 11.8 kN/m³), when individual particles with a maximum particle size $d_{max}$ lie on or under a 1.5 mm thick HDPE geomembrane and this is protected by a protective nonwoven needle-punched geotextile of mass per unit area $M_A$ If the mass per unit area is undershot or the surcharge is exceeded, puncture of the geomembrane can be expected

| Maximum particle size $d_{max}$ (mm) | Mass per unit area of the protective nonwoven geotextile $M_A$ (g/m²) | Height of overburden (m) |
|---|---|---|
| 12 | 550 | 108 |
| 24 | 1100 | 39 |
| 50 | 2200 | 12 |
| 76 | 2200 | 4 |

**Table 8.4.** Permissible heights of fill (specific weight 11.8 kN/m³), when a gravel fill with a maximum particle size $d_{max}$ lies on or under a 1.5 mm thick HDPE geomembrane and this is protected by a protective nonwoven needle-punched geotextile of mass per unit area $M_A$ If the mass per unit area is undershot or the surcharge is exceeded, puncture of the geomembrane can be expected

| Maximum particle size $d_{max}$ (mm) | Mass per unit area of the protective geotextile nonwoven $M_A$ (g/m²) | Height of overburden (m) |
|---|---|---|
| 12 | 270 | 84 |
| 24 | 550 | 50 |
| 50 | 2200 | 54 |
| 76 | 2200 | 22 |

It should however be once again emphasized that protection is provided by these design rules only against direct mechanical puncture (ductile failure). The design approach accepts that the geomembrane is strained even into the yield range. Therefore, it is possible for cracks to occur over very long periods of time through stress crack formation (brittle failure) in critical local places, where such deformations close to the yield point occur.

## References

August H and Lüders G (1995) Untersuchung der Langzeitbeständigkeit von Schutzmaterialien für Kunststoffdichtungsbahnen von Deponiebasisabdichtungen, Bericht zum Teilvorhaben 36 des BMBF-Verbundforschungsvorhabens Weiterentwicklung von Deponieabdichtungssystemen. BAM, Labor Deponietechnik, Berlin

August H and Lüders G (1997) Investigation to the long-term integrity of protective materials for geomembranes in landfill basal liners. In: August H et al. (eds) Advanced landfill liner systems. Thomas Telford Publishing, London, pp 261–269

Brauns J (1985) Erosionsverhalten geschichteten Bodens bei horizontaler Durchströmung. Wasserwirtschaft 75: 448

Brummermann K (1997a) Geomembranes under punctiform loads. In: August H et al. (eds) Advanced Landfill Liner Systems. Thomas Telford Publishing, Londen, pp 251–260

Brummermann K (1997b) Schutzschichten für Kunststoffdichtungsbahnen in Deponiebasisabdichtungen - Prüfung und Bewertung ihrer Wirksamkeit. Institut für Grundbau, Bodenmechanik und Energiewasserbau, Universität Hannover, Hannover

Brummermann K et al. (1999) Geotextile Schutzschichten für Kunststoffdichtungsbahnen im Tunnelbau. In: Sonderheft zur K-Geo. DGGt, Essen

Brummermann K and Schlütter A (2004) New German Recommendations for Geomembrane Sealing Systems in Tunnel Constructions. In: Floss R et al. (eds) Geotechnical Engineering with Geosynthetics, Proceedings of the Third European Geosynthetics Conference. Deutsche Gesellschaft für Geotechnik (DGGt) und Technische Universität München, Zentrum für Geotechnik (TUM-ZG), München, pp 211–214

DGGt (1997) GDA-Empfehlungen. Verlag Ernst & Sohn, Berlin

FGSV (1994) Merkblatt für die Anwendung von Geotextilien und Geogittern im Erdbau des Straßenbaus. Forschungsgesellschaft für Straßen und Verkehrswesen (FGSV), Köln

Gallagher EM et al. (1999) Performance Testing of Landfill Geoprotectors: Background, Critique, Developement, and current U. K. Practice. Geosynthetics International 6: 283–302

Hsuan YG and Koerner RM (2002) Durability and lifetime of polymer fibers with respect to reinforced geosynthetic clay barriers; i.e. reinforced GCLs. In: Zanzinger H et al. (eds) Clay Geosynthetic Barriers. A.A. Balkema Publishers, Lisse, The Netherlands, pp 73–86

Koch R et al. (1988) Langzeitfestigkeit von Deponiedichtungsbahnen aus Polyethylen. Müll und Abfall 20: 3–12

Koerner RM (1990) Designing with Geosynthetics. Prentice Hall, Englewood Cliffs, USA

Koerner RM et al. (1996) Puncture Protection of Geomembranes Part III: Examples. Geosynthetics International 3: 655–675

Menges G and Schmachtenberg E (1985) Das Verformungsverhalten von Kunststoffdichtungsbahnen bei mehrachsiger Beanspruchung. In: Knipschild FW (ed) Deponieabdichtungen mit Kunststoffdichtungsbahnen, Müll und Abfall, Beiheft 22. Erich Schmidt Verlag, Berlin, pp 68–74

Menges G et al. (1975) Ermittlung der kritischen Dehnung teilkristalliner Thermoplaste. Kunststoffe 65: 368–371

Mueller WW et al. (2003) Comparison of the oxidative resistance of various polyolefin geotextiles. Geotextiles and Geomembranes 21: 289–315

Müller WW (1996) Anforderungen an die Schutzschicht für die Dichtungsbahnen in der Kombinationsdichtung, Teil 2: Zulassungsanforderungen. Müll und Abfall 28: 90–99

Müller WW (ed) (1995) Anforderungen an die Schutzschicht für die Dichtungsbahnen in der Kombinationsdichtung, Zulassungsrichtlinie für Schutzschichten. BAM, Labor Deponietechnik, Berlin

Müller WW and Jakob I (2000) Comparison of Oxidation Stability of various Geosynthetics. In: Cancelli A et al. (eds) Proceedings of the Second European Geosynthetics Conference. Pàtron Editore, Bologna, pp 449–454

Müller WW et al. (2004) Long-term shear strength of multilayer geosynthetics. In: Floss R et al. (eds) Geotechnical Engineering with Geosynthetics, Proceedings of the Third European Geosynthetics Conference. Deutsche Gesellschaft für Geotechnik (DGGt) und Technische Universität München, Zentrum für Geotechnik (TUM-ZG), München, pp 429–434

Narejo D et al. (1996) Puncture Protection of Geomembranes Part II: Experimental. Geosynthetics International 3: 629–653

NLÖ (ed) (1994) Anforderungen an Siedlungsabfalldeponien in Niedersachsen, Deponiehandbuch. Niedersächsisches Landesamt für Ökologie, Hildesheim

Seeger S (1995) Zulassung von Schutzlagen. In: Knipschild FW (ed) Tagungsband der 11 Fachtagung "Die sichere Deponie, Wirksamer Grundwasserschutz mit Kunststoffen". Süddeutsches Kunststoffzentrum (SKZ), Würzburg, pp 35–48

Seeger S and Müller W (1996a) Limits of stress and strain: Design criteria for protective layers for geomembrane in landfill liner systems. In: DeGroot MB et al. (eds) Geosynthetics: Applications, Design and Construction, Proceedings of the First International European Geosynthetics Conference (Eurogeo 1). A. A. Balkema, Rotterdam, pp 153–157

Seeger S and Müller W (1996b) Requirement and Testing of Protective Layer Systems for Geomembranes. Geotextiles and Geomembranes 14: 365–376

Seeger S et al. (1995) Anforderungen an die Schutzschicht für die Dichtungsbahn in der Kombinationsdichtung, Teil 1: Wirksamkeit (lastverteilende Wirkung und Beständigkeit), Materialien und Prüfverfahren bei Schutzschichten. Müll und Abfall 27: 544–560

Seeger S and Müller WW (2003) Theoretical approach to designing protection: selecting a geomembrane strain criterion. In: Dixon N et al. (eds) Geosynthetics: Protecting the Environment. Thomas Telford, London, pp 137–152

Sehrbrock U (1993) Prüfung von Schutzlagen für Deponieabdichtungen aus Kunststoff. Institut für Grundbau und Bodenmechanik, Technische Universität Braunschweig, Braunschweig

Smoltczyk U (ed) (1980) Grundbau Taschenbuch. Verlag Ernst & Sohn, Berlin, München, Düsseldorf

Terzaghi K et al. (1996) Soil Mechanics in Engineering Practice. John Wiley & Sons, New York

Tisinger LG et al. (1990) Microstructural Analysis of a Polypropylene Geotextile after Long-term Outdoor Exposure. In: Koerner RM (ed) Geosynthetic Testing for Waste Containment Applications, ASTM Special Technical Publication 1081. ASTM, Philadelphia, USA, pp 335–354

Walczak ZK (2002) Processes of Fiber Formation. Elsevier Science Ltd, Amsterdam

Wilson-Fahmy RF et al. (1996) Puncture Protection of Geomembranes Part I: Theory. Geosynthetics International 3: 605–628

Witte R (1990) Kurzbericht: Weiterentwicklung des Schutzwirksamkeitsnachweises für geotextile Schutzsysteme in der Deponiebasisabdichtung. Müll und Abfall 22: 788–789

Witte R (1997) Practice-oriented investigations to improve geotextile protective layer systems for geomembranes with regard to their long-term protective efficacy. In: August H et al. (eds) Advanced landfill liner systems. Thomas Telford Publishing, London, pp 270–277

# 9 Installation of HDPE Geomembranes

## 9.1 Introduction: HDPE Geomembranes in Landfill Engineering

In Germany landfills are considered as final storage systems from which virtually no environmental burden should arise over a period of centuries. The basic principle for landfill design is the so-called multi-barrier concept. It consists of the combination of three independent barriers: the geological barrier (hydrological situation, relevance of the aquifer and properties of the subgrade at the landfill site), the technical barriers (basal liner and capping) and the waste body itself as barrier (immobilisation of hazardous substances by waste pre-treatment and appropriate disposal techniques). Two waste management ordinances were issued in Germany in 2001 and 2002 (Bundesregierung 2001, 2002) which describe in detail the design of the basal liner and capping system for the four landfill classes. Class 0 is for inert waste materials, classes I and II are for low-contamination mineral waste (e.g. construction debris and excavation waste) and treated municipal waste and class III is for wastes with high amounts of hazardous substances (hazardous waste). Along with an increasing hazardous potential of the waste disposed in the various types of landfills the requirements on the basal liner and capping systems are increasing.

The standard basal liner and capping system for landfill classes II and III is the composite liner made of a plastic geomembrane in intimate contact with a compacted clay liner. The composite liner comprises a hydrophobic and a hydrophilic sealing material. The intimate contact enables the compacted clay liner to seal possible faults in the geomembrane, while the geomembrane protects the compacted clay liner. It has been proved that the composite liner is therefore a highly efficient and fault-tolerant system, which is practically impervious to the convective flow of water or gas and the diffusion of all kinds of pollutants (August et al. 1992; Kalbe et al. 2000; Kalbe et al. 2002; Müller et al. 1998), when constructed according to the state of the art. Plastic geomembranes must be certified for use in landfill liners and containment of contaminated land. For over 15 years the Federal Institute for Materials Research and Testing (BAM) in Berlin has been the certification authority in Germany. The technical conditions of geomembranes are described in the "Guidelines for the Certification of Geomembranes as a Component of Composite Liners for Municipal and Hazardous Waste Landfills and for Lining Contaminated Land" (Müller 1999a, 1999b; Müller 2001). Currently only geomembranes made from special HDPE materials are certified. The principle conditions of the mineral layer are compiled and described in the Guideline No. 18 "Mineral Landfill Liners" by North Rhine-Westphalia State Environment Office (N.N. 1993). Certified geomembranes are made of HDPE polyethylene, UV-stabilized with carbon black and have a minimum thickness $d$ of 2.5 mm. The compacted clay liner is made of a fine-grained cohesive soil material with a certain amount of clay minerals. The compacted clay liner is installed and compacted with water content higher than Proctor optimum to achieve very low hydraulic conductivity ($K \leq 5 \cdot 10^{-10}$ m/s at hydraulic gradient $i = 30$).

**Table 9.1.** Standard capping design for German landfill classes I and II

| Position | System component | Landfill Class I | Landfill Class II |
|---|---|---|---|
| Top | Vegetation | necessary | necessary |
| | Restoration layer | > 1 m | > 1 m |
| | Drainage layer[1] | $d \geq 0.3$ m; $K \geq 1 \cdot 10^{-3}$ m/s | $d \geq 0.3$ m; $K \geq 1 \cdot 10^{-3}$ m/s |
| | Protective layer | not necessary | necessary |
| | Geomembrane | not necessary | $d \geq 2.5$ mm |
| | Compacted clay liner | $d \geq 0.5$ m, $K \leq 5 \cdot 10^{-10}$ m/s | $d \geq 0.5$ m, $K \leq 5 \cdot 10^{-10}$ m/s |
| | Gas venting layer | not necessary | necessary |
| Bottom | Regulating layer (foundation) | $d \geq 0.5$ m | $d \geq 0.5$ m |

[1]) The responsible local authority may permit deviations from thickness and hydraulic conductivity of the drainage layer if it is verified that the necessary long-term water flow capacity and long-term slope stability are guaranteed

For Landfill Class I (disposal of mineral or mineralised wastes with low contamination) and for conventional landfills for demolition waste, excavated soil etc. only a compacted clay liner is required. The system components for class I and II cappings, being the most common types nowadays installed in Germany, are listed in Table 9.1. Guidelines with strict requirements for materials selection and testing, construction technique and quality control were published for the standard systems (Müller 1996; Müller 2001; N.N. 1993).

The regulations permit alternative basal liner and capping systems to be used for landfills if it can be proved that their imperviousness and durability is equivalent to the properties of the standard system. As already mentioned in Chap. 1, the HDPE geomembrane is nowadays considered by many experts as the part of the composite liner having equal if not greater importance than the compacted clay liner and there is considerable concern about the compacted clay liner being used as single liner in a capping system[1].

Originally the question of crack formation in compacted clay liners due to desiccation arose for basal liners. Clay liners are compacted with high water content (near to the optimum water content on the wet side of the proctor curve) to achieve low hydraulic conductivity. However, a small decrease in the water content causes a strong increase in soil water tension in the homogenizied clay material. Water tension induces tensile stresses on soil particles, which eventually leads to the aggregation and irreversible cracking of the clay. It was speculated that heat production in the waste body that induces elevated temperatures on top of the composite basal liner will cause water vapour in the compacted clay component to flow to the colder areas in the subgrade below the liner and thereby reduce the water content of the clay which eventually induces cracks. However, it has been found that two processes counteract this desiccation cracking (Holzlöhner 1997): firstly, capillary rise of liquid water from the subgrade compensates for the loss by water vapour transport and a stable equilibrium with high enough water content of the clay liner will arise under appropriate site conditions (capillarity and conductivity of the subgrade and depth of the water table are the relevant parameters). Secondly, and above all, the pressure caused by the high load of the waste body will balance tensile stresses due to soil water tension and prevent the formation of cracks.

In the capping system the desiccation processes are driven by water evaporation into the atmosphere and by the transpiration of the plants. Small deviations of water vapour saturation in the air will cause enormous water tensions thereby effectively reducing the water content of a soil

[1] As a reaction to this critic on compacted clay liners it was suggested that the waste should be buried under an enormously thick soil reclamation layer (> 3 m) with an evergreen wood where the trees' crowns are contiguous. This concept represents the belief that leachate generation could be prevented for ever in a "natural" way, even under climate conditions with annual rainfall of > 500 mm. One may speculated whether such design concepts arise from a psychological bias: The deep-rooting desire to bury garbage and let it disappear not only from the eye, but also entirely from the senses. There are, however, various plausible technical and ecological reasons against this idea.

layer. High water tension in the restoration and drainage layer will trigger the desiccation of the compacted clay liner. Under this condition the compacted clay liner becomes attractive as a water reservoir for the plants thereby accelerating the desiccation process. In the capping system the pressure on the clay liner by the overburden is not relevant and water transport due to capillary rise or vapour flow from the waste body can be ignored at least over the long-term. It is therefore highly questionable whether stable conditions can be achieved in a capping system with high water content in the compacted clay liner and water tension in a stable equilibrium with its environment above and below. At least very thick restoration layers with high storage capacity for plant-available water are necessary to achieve such conditions from above. Detailed information about the state of discussion concerning the long-term performance of compacted clay liners might be obtained from the proceedings of a state of the art workshop (Ramke et al. 2002).

As consequence of these findings various alternative single liner systems, and especially HDPE geomembranes, are used for capping systems of Class I. Figure 9.1 shows schematic diagrams of such single liner systems. HDPE geomembranes can be combined with several of these components to form alternative composite liners (Fig. 9.2).

In Chap. 1 it was shown that there is a preference for HDPE geomembrane in the landfill liner regulations of various countries and similar technical discussion take place in these countries (see Table 1.1). The construction of large-area liners in landfill engineering and containment of contaminated land therefore offers a wide range of applications for HDPE geomembranes and it can be said that the topic of landfill liner construction has been very intensively investigated both scientifically and technically. Knowledge and experience are useful therefore for all other fields of large-area liner construction. Using the example of landfill liner construction, this chapter will deal with the construction of large-area landfill liners using HDPE geomembranes.

Installation or placement or deployment (these terms are often used synonymously) of HDPE geomembranes is composed of the following phases[2]:

1. Installation planning
2. Construction and preparation of the subgrade of the geomembrane

[2] These six phases apply not only to landfill liner construction, but in general to geomembrane installation. In tunnel construction for example the subgrade (called sealing carrier here) consists of the outer tunnel shell (possibly covered with a protective layer) with integrated fastening elements for the geomembrane. Concreting the inner tunnel shell replaces the installation of the drainage layer as phase 6.

3. Placement of the geomembrane, i.e. transport of the geomembrane rolls to the construction site, unreeling, placing and cutting them to size (geomembrane panels).
4. Welding of the placed geomembranes panels, connection to structure penetration systems or buildings
5. Placement of the protective layer
6. Construction of the drainage layer

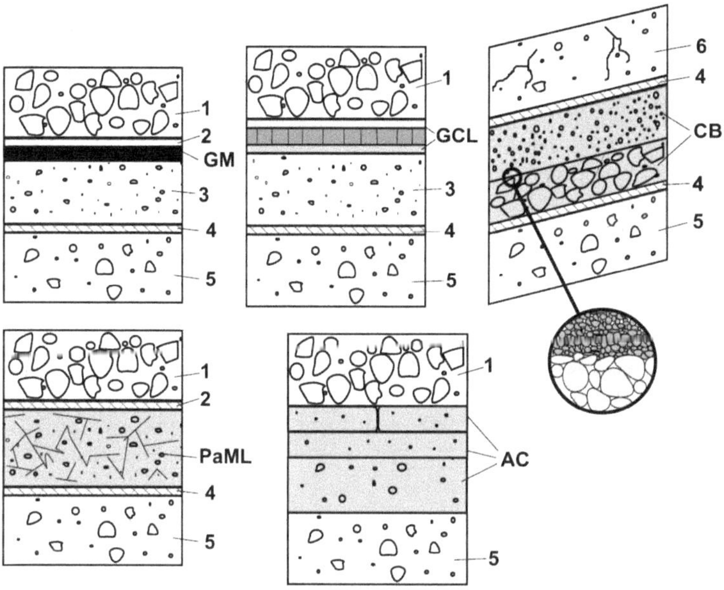


**Fig. 9.1.** Schematic diagrams of lining systems for caps. GM: geomembrane; GCL: geosynthetic clay liner; CB: capillary barrier (capillary layer and capillary block, meanwhile, capillary block mats, only a few centimetre thick, are available, which may be easily installed (von der Hude et al. 2001)); PaML: polymer amended mineral liner; AC: asphaltic concrete (base course layer and two sealing layers), 1: Coarse gravel drainage layer, alternatively a geocomposite drain (GCD) might be used, which has been approved for landfill lining; 2: Protective geotextile, necessary for coarse drainage gravel and high load; 3: Foundation layer (sand), optional, dependent on the grain size distribution of the regulating or gas venting layer; 4: Separation or filter geotextile, optional, dependent on the grain size distribution of neighbouring mineral layers; 5: Regulating or gas venting layer; 6: Restoration layer

All details of the phases of installation cannot be dealt with here. Naturally, the general rules of the geotechnics and the regulations described in Chap. 1 must be considered, particularly those that concern the installation of the geomembrane. Welding of the HDPE geomembranes, machines and devices used for welding, types of weld seams and the quality assurance measures will be specially dealt with in Chap. 10. In this chapter, however, only those aspects of the installation, which have repeatedly proved particularly important will be described. The construction programme can be only theoretically divided into such separate phases. In the actual construction process the phases must be integrated spatially and temporally into a well defined installation method. Only if one succeeds in achieving this, will a liner element emerge that meets all technical state of the art requirements. As an example of a well defined installation method, which has been proved in practice, the so called "anchoring technique" or "fixed modular installation" (both are free translations from the German name "Riegelbauweise") will be described. The photographs in this chapter (with the exception of Fig. 9.9) are courtesy of R. Schicketanz.

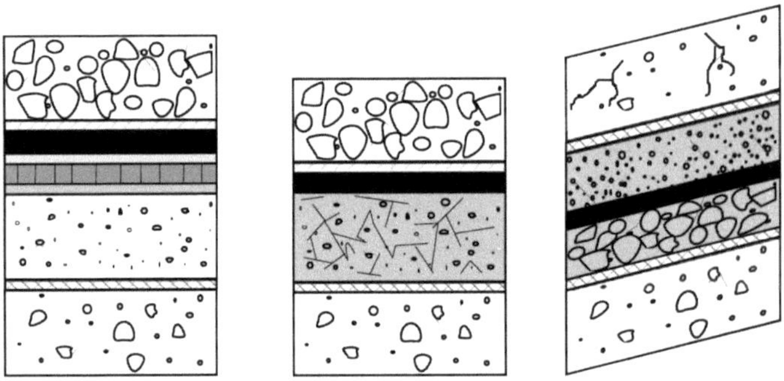

**Fig. 9.2.** HDPE-GM might be combined with various other liner components (see Fig. 9.1) to form alternative composite liners. There are pros and cons for installing the geomembrane above or below the capillary barrier. Recently, it has been suggested that it should be placed between the capillary layer and block (Sehrbrock 2003)

## 9.2 Installation Planning

Each installation of a geomembrane must be preceded by planning phase, resulting in a detailed panel layout drawing or shop drawing, irrespective of the type of application. In landfill construction the installation plan specifies to scale the arrangement of the geomembrane panels in the area to be sealed and the penetrations (e.g. pipe culverts) and connections to buildings (e.g. leachate collection shafts). In addition, how the geomembranes have to be laid out and cut to shape in difficult-to-install areas of corners on short or long slopes and in slope areas where the slope length is gradually changing. The type and extent of on-site cutting, the construction and the necessary lengths of the geomembrane rolls, which must be supplied by the factory, result from the plan. Later on, the division into construction sections is specified. The type of weld seams (extrusion fillet seam or dual hot wedge seam) and technical instructions for welding connections and penetrations are specified. Design details of connections and penetrations, usually pipe penetrations for leachate and landfill gas pipes and connections to shaft construction, supplement the panel layout drawing.

When planning the installation of a geomembrane, some principles have to be adhered to, for instance with regard to the arrangement of the geomembranes panels and therefore of the welding seams. The general rule is that the geomembranes have to be arranged in such a way that as few welding seams as possible are necessary and that dual hot wedge seams, i.e. machine-welded seams, can be used to the largest possible extent. The welding seams of the geomembranes may not cross (no cruciform connections). So-called T-connections, where two seams collide perpendicularly, must exhibit a distance of at least 0.5 m. On slopes, the seams must to a large extent run parallel to the line of maximum slope. Patching geomembrane panels using transverse joints on slopes is not permitted. The connecting seam between geomembranes on the slope and the base should be located in the base at a distance of at least 1.5 m from the slope toe. Radii of curvature in the transition zone between slope and base and generally those of slope changes should be at least 1.0 m. The anchor trench for geomembranes in the berm crown must be made in such a way that a geomembrane can be welded up to its end with the welding machine.

W. Bräcker, A. Schlütter and F. Sänger as well as F. W. Knipschild discussed the design of building components for composite liners (anchoring in the berm crown, connection to already installed sections, pipe culverts, shaft structures, principles of installation planning etc.) in detail (Bräcker et al. 1994; Knipschild 1994). Principles of installation planning as well as alternatives for the geomembrane installation on slope corners and slope areas with continuously changing slope length are further given in the DVS guideline 2225-4:1992 *Welding of Geomembranes from Polyethylene (PE) for the Lining of Landfills and Contaminated Land (Schweißen von Dichtungsbahnen aus Polyethylen (PE) für die Abdichtung von Deponien und Altlasten)*. The document also contains examples and solutions for the design of pipe penetration systems and shaft structures. Also the GLR-recommendation E2-27 *Liner Penetration Systems* dedicates itself to this topic.

Outline installation planning starts simultaneously with construction design and forms one of the basic documents for preparation of the tenders. Specialist knowledge in geotechnical engineering with geosynthetics is therefore required at the beginning of the early design phase. The tender documents for the geomembrane installation, which result from planning, usually include the general contract conditions (in Germany the so called "Allgemeine Technische Vertragsbedingungen (ATV)" and the site specific technical contract conditions (in Germany the so called "Zusätzliche Technische Vertragsbedingungen (ZTV)")). The general contract conditions reflect the national contracting regulation for building work (in Germany the so called "Verdingungsordnung für Bauleistungen (VOB)"). For example, conditions for the material warranty and the installation warranty are specified. The technical contract conditions comprise the descriptions of the construction project, the listing of the individual steps of the installation work and the quality control and the respective requirements, the outline of the layout drawing, the design drawings of individual constructions as well as the analysis of structural stability. They must specify the technical requirements for the installation of the geomembrane with respect to the construction project in such a detailed and differentiated way that a realistic, comprehensible pricing is possible and the suitability of the offered prices can be judged. Management of the tender documents cannot be dealt with here in great detail. However, it is advisable to consult model drafts of technical contract conditions when planning. The German association of geomembrane manufacturers and installers (Arbeitskreis Grundwasserschutz e. V. (AK GWS, www.akgws.de) has compiled such a draft of relevant technical contract conditions for the installation of the geomembranes and geotextile protective layers in landfill liners (AK-GWS 1994, 1995). The International Association of Geosynthetic Installers (IAGI, www.iagi.org) has prepared a detailed document about HDPE geomembrane installation specifications.

The contractor must provide proof that the geomembranes and protective layers provided by him achieve the necessary friction parameters between the various liner elements as planned. Therefore, the designer must indicate the boundary conditions of the construction project (slope angles, surcharge, site specific stresses, safety factors) and resulting test conditions for the shear box test for the determination of the friction parameters. The friction parameters are usually determined in a shear box test according to GLR-recommendation E3-8 *Friction Behaviour of Geosynthetics (Reibungsverhalten von Geokunststoffen)* (see Sect. 3.2.17), ASTM D5321-97 *Standard Test Method for Determining the Coefficient of Soil and Geosynthetic or Geosynthetic and Geosynthetic Friction by the Direct Shear*

*Method* or EN ISO 12957-1:2005 *Geosynthetics – Determination of Friction Characteristics - Part 1: Direct Shear Test.*

The installation plan, which is finally prepared in detail by the contractor, then becomes a part of the quality assurance program of the construction project and finally becomes the as-built plan. In the as-built plan, the construction progress is recorded in detail. An identification code is assigned to the geomembrane panel installed in certain places and can be inferred from the as-built plan. Manufacture date, manufacturing records, test certificates issued by the manufacturer's in-house quality control and third-party control as well as the installation records (daily report and/or logs in a prescribed form) and the inspection results of the third-party control on the construction site have to be clearly and unequivocally assigned to these identification codes. The as-built plan likewise contains the relevant seam identification code for each individual weld seam. Based on these codes, the welding reports and test reports, e.g. under DVS guideline 2225-4, of the in-house quality control of the installation contractor and the inspection results of the third-party control can be assigned to the seam. In addition, the as-built plan contains location and unequivocal references to records of repairs, cuts, connections, anchoring, test and sampling points.

## 9.3 Installation

Before the geomembrane installation can start, the subgrade (or supporting layer), on which the geomembrane will be unrolled and welded, must be produced and its surface prepared. When a composite liner is constructed, it is a mineral layer on which the geomembrane is placed. If an alternative composite liner is constructed, the geomembrane can be placed for example on a bentonite mat or the capillary layer of a capillary barrier. In other fields of geotechnical engineering, however, it can be a supporting layer made of completely different materials. As a rule non-cohesive or only light cohesive soils serve as supporting layer. Since the use of recycled materials is increasingly desired, even such exotic supporting layer material as broken glass can occur.

In principle it mainly holds true that particle shape, particle size and particle size distribution of the supporting layer material must be such that the loads during the course of construction and use do not result in inadmissible deformations by indentations and imprints in the geomembrane. What this means in detail and how the suitability of a supporting layer material is tested, if necessary, were dealt with in detail in Chap. 8. All criteria for the protective layers on and under the geomembrane can be analogously transferred to the supporting layers. If a geomembrane has to be placed on an already existing soil subgrade, which does not fulfil these requirements an additional protection layer has to be installed. The following two paragraphs describe additional requirements for supporting layers as specified in the BAM certification guideline for plastic geomembranes.

The supporting layer must not be damaged by the equipment and machines for transporting the geomembranes. Before the geomembranes are rolled out, the supporting layer must be scraped or rolled so that wheel-tracks or other deep imprints are smoothed out and a flat surface is produced. Imprints and protrusions must not be larger than 2 cm. The surface of the supporting layer must follow the planned radii of curvature and inclinations and the deviations between nominal and actual heights should not exceed ± 3 cm.

Special criteria for the surface apply when the geomembrane is placed on a mineral layer to form a composite liner (Schicketanz 1992). Possible defects in the geomembrane can be sealed by the very low-permeability supporting layer. The more intimate the contact between the geomembrane and the mineral layer, the better this effect is (see Chap. 7). The BAM certification guideline for geomembranes therefore contains particularly high material-technical and geometrical criteria for the surface of the mineral component in the composite liner. The subgrade surface must be stable bearing, homogeneous, fine-grained and free of holes; gravel particles with a diameter > 10 mm and foreign bodies must not be included. All finer gravel components must be embedded in such a way that they are surrounded on all sides by cohesive liner material. Gravel particles and foreign bodies must not lie on the surface. Generally, abrupt changes in height should be smoothed to a large extent. As a reference point a permissible height of 0.5 mm is considered for steps (impression differences). Unevenness when measured beneath a lath (straightedge) 4 m in length resting on the surface may not exceed 2 cm. These are evaluation criteria, which must be interpreted and illustrated on a test site using the same material as used for the final compacted clay liner. The production of such a surface requires a substantial constructional engineering input (Dornbusch et al. 1996).

In geotechnical engineering, geomembrane installation is nearly always at the bidding of the weather. Wet geomembranes cannot be professionally welded. Therefore the geomembranes may not be installed if it rains, hails, snows or if the subgrade is very wet. Even a temperature below the dew point can lead to considerable problems. Strong winds blowing into the welding machine or equipment can likewise drastically reduce the quality of the weld seam. Storm gusts can lift whole geomembranes or even whirl completed areas through the air. Therefore, sand bags or equivalent ballast have to be used to secure in placed geomembrane panels when installation has to be interrupted because of critical weather conditions.

Temperature and humidity conditions contribute substantially to a decision whether or not welding is possible. Therefore the connections between temperature and humidity should be considered in greater detail (Baehr 1966). Moist air is a mixture of dry air[3] and water vapour. The vapour content of air or the humidity is determined by the physical quantity "absolute humidity" $\rho_W$. It is defined as the ratio of water vapour mass $m_W$ contained in a volume $V$ of moist air to this volume:

$$\rho_W = \frac{m_W}{V}. \tag{9.1}$$

A certain mass of water vapour leads to a partial pressure $p_W$ depending on temperature $T$. The connection can approximately be described by the state equation of the ideal gas ($R_W$ is the gas constant of water vapour):

$$p_W = R_W \rho_W T. \tag{9.2}$$

Now, air cannot take up more than a certain mass of water vapour in a given volume at a certain temperature $T$. If this mass is achieved, then air is fully saturated with water vapour. A saturation partial pressure of the water vapour $p_{Ws}(T)$ belongs to the saturation density $\rho_{Ws}(T)$ according to the Eq. 9.2. Humidity is therefore often referred to not only as absolute humidity, but also as "relative humidity" $\varphi$. Quantity $\varphi$ gives the ratio of actual water vapour to saturation density at a given temperature:

$$\varphi_W = \frac{\rho_W}{\rho_{Ws}(T)} \cdot 100 = \frac{p_W(T)}{p_{Ws}(T)} \cdot 100. \tag{9.3}$$

Saturation density of water vapour in air at 20 °C is 17.27 g/m³. A measured absolute humidity of 9.39 g/m³ for example would correspond to a relative humidity of 54 % at this temperature. At 10 °C the saturation content is only 9.39 g/m³, and this absolute humidity would result in a relative humidity of 100 %. Figure 9.3 shows the connection between water vapour partial pressure, relative humidity and temperature.

If non-saturated moist air cools down at constant air pressure, then the water vapour partial pressure also remains constant. Since the saturation value decreases with temperature, the existing water vapour partial pres-

[3] Dry air is a mixture of 78.09 % by volume $N_2$, 20.95 % by volume $O_2$, 0.93 % by volume Ar, 0.03 % by volume $CO_2$ and other gases (Ne, He, Kr, $H_2$, Xe, $O_3$), which occur however only in negligibly small quantities.

sure will finally correspond to the saturation pressure at a certain temperature. If air continues to cool down, the surplus water vapour condenses and settles as dew on surfaces. The state, when this transition occurs, is called the dew point and the corresponding temperature dew point temperature.

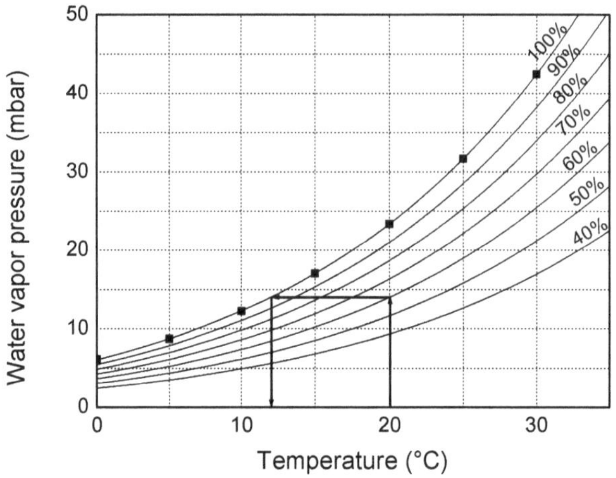


**Fig. 9.3.** Water vapour pressure curves for water vapour in moist air. The curves for the various relative humidities were calculated from the interpolated data of the saturation curve (100 %). Dew point temperature (e.g. 12 °C) for certain ambient temperature (e.g. 20 °C) and certain humidity (e.g. 60 %) can be read from the graph by moving from the respective point on the 60-%-curve to the saturation curve. The arrows indicate the reading steps

Ambient and geomembrane temperature may change due to a shift in the equilibrium between heat irradiation into and out of the soil when the clouds or the sun change their position while air pressure and water vapour partial pressure remain constant. If the temperature drops at or under the dew point temperature, a fine moisture film can form on the surface of the geomembrane. Under such conditions it may not be welded. Inversely, it should only be welded if the ambient temperature has a certain safety margin at the dominant humidity to the dew point temperature. This margin can be determined using Fig. 9.3: if, for instance, a relative humidity of 60 % is measured at 20 °C, the dew point temperature read off from the saturation curve is about 12 °C and thus the margin is 8 °C. At 15 °C and about 82–83 % relative humidity one would be very close to the dew point temperature of 12 °C. Normally, it is required that the margin should be at least 3 °C. Above approximately 83 % relative humidity welding becomes critical, since this is closer than 3 °C to the dew point temperature. Table 9.2 shows an excerpt from a so-called dew point table, set up on the basis of Fig. 9.3. Such tables or appropriate slide rules are used by the welders on the construction site to assess weather conditions.

**Table 9.2.** Dew point table

| Air temperature (°C) | Dew point temperature (°C) at a given relative moisture | | | | | | | | |
|---|---|---|---|---|---|---|---|---|---|
| | 30 % | 35 % | 40 % | 45 % | 50 % | 55 % | 60 % | 65 % | 85 % |
| 30 | 10.5 | 12.9 | 14.9 | 16.8 | 18.4 | 20.0 | 21.4 | 22.7 | 27.2 |
| 29 | 9.7 | 12.0 | 14.0 | 15.9 | 17.5 | 19.0 | 20.4 | 21.7 | 26.2 |
| 28 | 8.8 | 11.1 | 13.1 | 15.0 | 16.6 | 18.1 | 19.5 | 20.8 | 25.2 |
| 27 | 8.0 | 10.2 | 12.2 | 14.1 | 15.7 | 17.2 | 18.6 | 19.9 | 24.3 |
| 26 | 7.1 | 9.4 | 11.4 | 13.2 | 14.8 | 16.3 | 17.6 | 18.9 | 23.3 |
| 25 | 6.2 | 8.5 | 10.5 | 12.2 | 13.9 | 15.3 | 16.7 | 18.0 | 22.3 |
| 24 | 5.4 | 7.6 | 9.6 | 11.3 | 12.9 | 14.4 | 15.8 | 17.0 | 21.3 |
| 23 | 4.5 | 6.7 | 8.7 | 10.4 | 12.0 | 13.5 | 14.8 | 16.1 | 20.3 |
| 22 | 3.6 | 5.9 | 7.8 | 9.5 | 11.1 | 12.5 | 13.9 | 15.1 | 19.4 |
| 21 | 2.8 | 5.0 | 6.9 | 8.6 | 10.2 | 11.6 | 12.9 | 14.2 | 18.4 |
| 20 | 1.9 | 4.1 | 6.0 | 7.7 | 9.3 | 10.7 | 12.0 | 13.2 | 17.4 |
| 19 | 1.0 | 3.2 | 5.1 | 6.8 | 8.3 | 9.8 | 11.1 | 12.3 | 16.4 |
| 18 | 0.2 | 2.3 | 4.2 | 5.9 | 7.4 | 8.8 | 10.1 | 11.3 | 15.4 |
| 17 | -0.6 | 1.4 | 3.3 | 5.0 | 6.5 | 7.9 | 9.2 | 10.4 | 14.5 |
| 16 | -1.4 | 0.5 | 2.4 | 4.1 | 5.6 | 7.0 | 8.2 | 9.4 | 13.5 |
| 15 | -2.2 | -0.3 | 1.5 | 3.2 | 4.7 | 6.1 | 7.3 | 8.5 | 12.5 |
| 10 | -6.0 | -4.2 | -2.6 | -1.2 | 0.1 | 1.4 | 2.6 | 3.7 | 4.8 |

**Table 9.3.** Weather-related limiting conditions for geomembrane installation (Schicketanz 1995)

| Weather-related limiting conditions for geomembrane installation | |
|---|---|
| Ambient temperature $T_U$ | ≥ 5 °C |
| Relative moisture (rel. to saturation water content at $T_U$) | < 83 % |
| Dew point distance (distance between $T_U$ and dew point temperature according to measured relative moisture) | > 3 °C |
| Maximum temperature difference which occurs in the course of the whole day (day and night) | > 10 °C |
| Weather condition | No precipitation, no fog or atmospheric haze, no strong gust or storm |

R. Schicketanz concludes that welding can usually be done (with reference to the DVS guideline 2225-4 and BAM certification guidelines) at the weather conditions specified in Table 9.3 without any preventive measures (Schicketanz 1995). Naturally there is a "grey area" between weather conditions at which welding is possible without any problem, and conditions where welding is just as clearly not possible. Only specialist expertise and extensive experience allows one to make correct decisions in such marginal situations. It might be also necessary to increase control frequency in order to be able to terminate welding early enough before the quality of the seams becomes unsatisfactory.

Geomembrane installation requires a certain amount of time: the welding of a 100 m long seam at a typical welding speed of approximately 1 to 1.2 m/min, with preparation and testing, takes about 2.5–3 hours. Temperature differences in the course of day and night must be used so that geomembranes stretch smoothly and are wave-free due to the thermally induced variations in length after their installation (see Sect. 9.3.2). Therefore longer phases of dry weather are necessary for the production of a large-area geomembrane liner. Such weather conditions usually prevail at Central European latitudes only between April and October.

**Fig. 9.4.** In a large heated tent a composite liner was constructed in the middle of winter. Anchoring technique (Riegelbauweise) can ensure a complete intimate contact between geomembrane and mineral layer even under these conditions (Sect. 9.3.2) (Krath and Schwarz 1993; Schicketanz 1995). Courtesy Schicketanz Consulting Engineers, Aachen

In a heated tent it is possible to work in difficult weather conditions outside this season. Usually, however, one will only want to perform repair and patching or smaller finishing work. But there are also spectacular construction projects where large-area liners have been manufactured in a tent in the middle of winter (Krath and Schwarz 1993; Schicketanz 1995) (Fig. 9.4).

**Table 9.4.** Evaluation of construction and installation days, installation performance and personal and equipment assignment in composite liner construction in 12 projects in 1994 (Knipschild 1995)

| Project | Liner area | Construction period | Installation period | Effectivity | Installation performance | | Deployment of persons and equipment | | |
|---|---|---|---|---|---|---|---|---|---|
| No. | $m^2$ | Construction days | Installation days | | $m^2$ per construction day | $m^2$ per installation day | Number of crew members | Number of HW-WM | Number of WG-E |
| 1 | 120,000 | 120 | 75 | 0.62 | 1.000 | 1600 | 6 | 2 | 2 |
| 2 | 50,000 | 120 | 53 | 0.44 | 420 | 950 | 3 | 1 | 1 |
| 3 | 45,000 | 106 | 58 | 0.55 | 425 | 775 | 4 | 2 | 2 |
| 4 | 45,000 | 132 | 68 | 0.52 | 340 | 660 | 4 | 2 | 2 |
| 5 | 35,000 | 102 | 34 | 0.33 | 340 | 1.030 | 5 | 1 | 2 |
| 6 | 35,000 | 72 | 36 | 0.50 | 470 | 970 | 3 | 1 | 1 |
| 7 | 25,000 | 60 | 26 | 0.44 | 420 | 960 | 4 | 1 | 1 |
| 8 | 22,500 | 48 | 31 | 0.65 | 460 | 710 | 4 | 1 | 1 |
| 9 | 20,000 | 180 | 68 | 0.38 | 105 | 280 | 3 | 1 | 1 |
| 10 | 17,500 | 66 | 30 | 0.45 | 260 | 570 | 3 | 1 | 1 |
| 11 | 15,000 | 60 | 38 | 0.63 | 250 | 400 | 3 | 1 | 1 |
| 12 | 10,000 | 54 | 22 | 0.40 | 100 | 390 | 3 | 2 | 1 |

HW-WM: Hot wedge welding machine, WG-E: warm gas extrusion device

Weather conditions necessary for geomembrane installation hardly differ from the conditions that must prevail for constructing other large-area liners, for instance asphaltic concrete liners (Burkhardt and Egloffstein 1995) or pure mineral liners. The German Association of Construction Industry (Zentralverband des Baugewerbes) generally reckons on 184 working days a year. Experience with landfill construction shows, however, that only about 120 working days are actually available (season April until October, less weekends and holidays plus poor weather) (Averesch and Schicketanz 1998). Weekends must often be used, in order to advance with the construction project when the weather is favourable.

Table 9.4 shows installation performance achieved in Germany on different construction sites with composite liner construction (Knipschild

1995). The installation performance defined as installed square metres of geomembrane relative to the installation days, i.e. those days on which the contractor's installation crew (see Sect. 9.3.2) worked on the construction site, averaged over various projects was about 650 $m^2$ per day. On individual sites an average of approximately 1500 $m^2$ were installed in one installation day. Of the available construction days, i.e. the overall working days available in the construction period, however, an average of only just half and in no case more than two-thirds were used as installation days. These figures suggest that the time requirement of geomembrane installation is seldom the limiting time parameter within the building programme for the production of a composite liner. Rather, it is the progress of the actual earthwork and, above all, co-ordination of the construction steps themselves, which set the time frame. A careful and exact planning of the construction programme and the punctual provision of the necessary equipment are therefore of central importance for rapid progress.

### 9.3.1 Excursion: Development and Effect of Waves in Geomembranes

The HDPE material is incompressible, since the Poisson number ($\nu \approx 0.49$) is very close to 0.5. Waves in the installed geomembrane cannot therefore be simply forced away by the ballast. What happens here can be reasonably simulated with a sheet of paper. One pushes the sheet together in such a way that a flat, broad swinging wave arises (Fig. 9.5a) and fixes the ends. Then one carefully ballasts the area of the paper with the wave in the middle. The original wave will then be compressed to a narrow, low, but very steep wave with high edge fibre strain (Fig. 9.5b). This small wave remains even if very high surcharges prevail. If it is high enough, it folds over and a crease develops.

Already the as-produced geomembrane has a tendency to produce small amplitude waves when unrolled on a flat surface. The BAM certification guideline sets a limit on the acceptable amplitude of such waviness. The maximum clearance between geomembrane and level supporting surface is measured over a length of 10 m when rolled out over a length of 12 m. The measured value must be less or equal to 5 cm. The in-house manufacture quality control (MQC) of the geomembrane manufacturer has to perform this test at every production start-up. Waves (or wrinkles, as is sometimes said) are to a considerable extent unavoidable when the geomembrane panels are placed in the field. The waviness tends to increase when the long geomembrane panels lie uncovered for some time, especially on slopes. If geomembranes, which have lain for a long time in the open and are therefore very wavy, are covered, then different folds can be formed, in particular by pushing waves together.

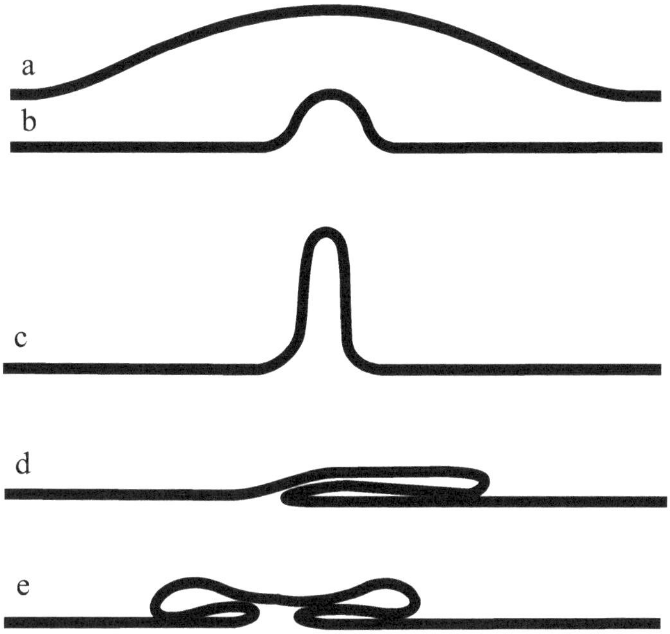


**Fig. 9.5.** Types of waves observed in trial pits (Koerner et al. 1999; Schicketanz 1992). A wave in the geomembrane (a) cannot be smoothed out. In the best case a small residual wave remains (b). Larger waves or a number of waves pushed together produce standing folds (c), laying flat folds (d) or mushroom-shaped waves (e). These waves typically emerge when geomembranes are installed over large areas, remain uncovered over long periods of time and are finally backfilled without ensuring an intimate contact between geomembrane and mineral layer. G. R. Koerner has even suggested a terminology for wave typology: (c) "prayer waves", (d) "S-wave" and (e) "mushroom waves"

One can then find steeply rising folds jammed into the backfilling layer above the geomembrane (Fig. 9.5c). These folds can also be bent over (Fig. 9.5d). A large wave can also be flattened into a mushroom-shaped form, especially if it has been pushed together from a series of smaller waves (Fig. 9.5e). The last three types of waves are typically found in trial pits on all landfill projects where the geomembranes have been placed and covered over large areas without having first established a special installation method to achieve intimate contact between the geomembrane and the mineral layer (Koerner et al. 1999; Schicketanz 1992). The shape of a wave may even change its form from one into another along the wave.

Figure 9.6 shows a trial pit, which clearly indicates that the waves have not been smoothed out by the backfilling. Figure 9.7 shows a steep wave pushed together by the backfilling process.

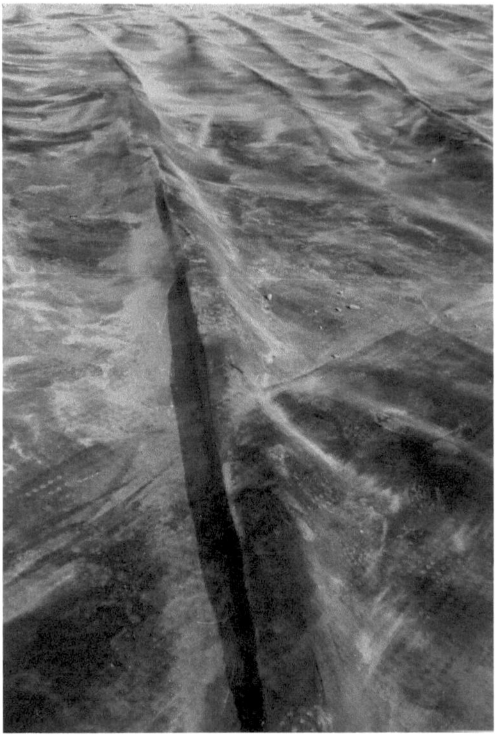

**Fig. 9.6.** Trial pitting of a composite liner after drainage layer installation showing that the geomembrane has been placed and ballasted without ensuring flatness. The waves have not been smoothed out by the load. Courtesy Schicketanz Consulting Engineers, Aachen

The behaviour of a small residual wave, as shown in Fig. 9.5b, and the mechanical load of the geomembrane, which arises in such waves, were systematically tested by T.-Y. Soong and R. M. Koerner in the laboratory (Soong and Koerner 1999). It has been found that even small residual waves persist at high temperatures (50 °C), high loads (1000 kPa) and over long times (1000 h). In none of the cases was it possible to smooth out such waves. The deformations of the geomembrane were in the range of up to 5 %, even in these small waves, i.e. just at the corner of the range of permissible limiting strain (see Sect. 5.3.4). The long-term effective stress calculated, taking into account relaxation, was a maximum 22 % of the yield stress.

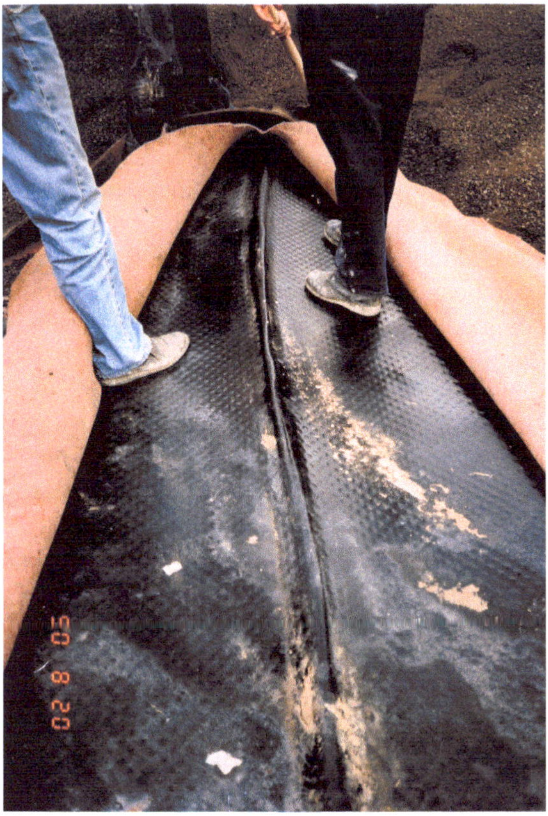

**Fig. 9.7.** Steep wave pushed together during backfilling. Excessive deformations may arise in the range of such waves. Courtesy Schicketanz Consulting Engineers, Aachen

The effect of the residual waves on weld seams across or parallel to the wave was not tested. However, because of the larger thickness within the seam range, substantially larger edge fibre strains might arise, which might lie then outside of the permissible limiting strain. This applies above all to the waves shown in Figs. 9.3c, d and e. In the case of these types of waves, and depending upon surcharge, one will have to contend with large deformations in the geomembrane. This means that there are ranges in which stress crack formation can be expected over the long term. In the weld seams, and in particular in extrusion fillet seams, the load may be large enough to induce weld seam to break or tear in the boundary region of the weld seams. It is, however, difficult to reliably assess short and long-term effects of these waves on the performance of the sealing element in individual cases. There are however indications of the negative effects of waves.

In the USA the following observation was made: the practice exists in landfill engineering that installation is carried out over large areas (daily performances are up to 15,000 $m^2$ (Corbet and Peters 1996)), where the geomembranes are not covered during installation, but only backfilled at the end of the construction without taking care of the waviness of the geomembranes. The formation and covering of waves as described above therefore significantly belong to every-day practice. Based on the regulations of various Federal States, double basal landfill liners were constructed which consist of a composite liner (secondary liner), a control drainage (secondary leachate collection and removal system) and on top of it a geomembrane (primary liner). In connection with these liners it was frequently noted that there were extensive leachate leakage rates seen in the control drainage system (see (Bonarparte and Gross 1990) and the discussion to the contribution by R. M. Koerner in (Corbet and Peters 1996)). It had to be assumed that holes in the geomembrane over the control drainage were a substantial cause for water seeping into the control drainage (Bonarparte and Gross 1990). Therefore, the assumption seems to be rather plausible that – in addition to a thinness of the installed geomembranes, the poor status of welding technology and the general earthworks performance – the strongly wavy installation during construction had to some extent contributed to the pure liner performance in these cases. However, no field studies are available to prove the detrimental effects of waves on the liner performance. They will to a large extent appear in the long run and would therefore only be revealed by long-term follow-up studies of liner performance.

The waves are, however, not only weak points where perforations (tears or holes) may occur in the geomembrane over the long term. There are other detrimental effects. In all cases a pipe-like flow channel of some centimetres wide also develops inside the wave. In Sect. 7.5 it was discussed that if there is a hole within the range of the wave and water piles up on the geomembrane, the cavity under the wave rapidly fills with water. In this cavity, which is formed by the wave contour and the surface of the mineral layer, the entire pressure of the water standing on the geomembrane will exert its hydraulic effect. The hydraulically effective perforation size can thus become substantially larger than the actual one. The flow rate can be calculated for such a configuration, see Sect. 7.5. An evaluation of the equations shows that even small holes and tears within the proximity of a wave can have a substantial influence on the liner systems' permeability.

T.-Y. Soong and G. R. Koerner emphasize other problems (Koerner et al. 1999). Just on the slopes, the water flow above the geomembrane in the drainage layer could be obstructed by "mini dams" of high-standing folds and the hydraulic load concomitantly increased within the range of the weak points. Air could circulate through the flow channels to a certain extent and lead to desiccation of the mineral layer just in this critical area.

From all that is known about repercussions of waviness on geomembrane liner performance and properties it is reasonable to require that geomembranes must be installed with as little waviness as possible. F. W. Knipschild (Knipschild 1992) and R. M. Koerner and colleagues (Koerner et al. 1999) give instructions on the installation technology, which can help to achieve this goal. There is ongoing controversy about which individual steps of the installation method must be followed. Two approaches are discussed for example: complete backfilling carried out on every day of installation in the cool periods of the day after achieving complete flatness (Schicketanz 1992) or applying first only a provisional ballast using a heavy geotextile (> 2000 g/m$^2$) and then finally backfilling the mineral drainage layer at the end of the whole installation period during a cool period of a day (Knipschild 1992). However, other arguments then waviness should be considered: Avoidance of condensation of water or the collection of consolidation water between the geomembrane and the mineral subgrade, flow of such water in the interface and accompanied changes in the surface of the mineral layer also suggest as rapid a surface ballasting of the installed geomembrane as possible.

R. Schicketanz has developed an installation method, the so-called "anchoring technique" ("Riegelbauweise"), which is described in (Averesch and Schicketanz 1998; Dornbusch et al. 1996; Schicketanz 1992). This technique provides a systematic method for ensuring the intimate contact between the mineral layer and the installed geomembrane over a large area and it has in the meantime successfully been applied on numerous landfill sites. All types of production and installation waves can only be systematically smoothed out by using this anchoring technique approach, see Sect. 9.3.2.

### 9.3.2 Anchoring Technique (Riegelbauweise)

A special construction method, which uses the temperature gradient over the day, can guarantee the complete intimate contact of the geomembrane, which is theoretically highly desirable and necessary for the full effectiveness of the composite liner to develop. This technique can be implemented by an experienced and qualified installation contractor. The coefficient of linear thermal expansion of a HDPE material is (15–20)·10$^{-4}$ K$^{-1}$ in the relevant temperature range of 20 °C to 60 °C. A change in temperature of 10 °C over the day can alter the length $L$ of a geomembrane section around 0.015 $L$ to 0.020 $L$, which can amount to 2 m for a 100 m long section.

Therefore if a geomembrane acclimatised to the temperatures of the warmer time of day is aligned and installed with minimum undulation, then welded and anchored by "fixing berms" at the ends, it will pull itself completely smooth as it cools in the course of the day. Also, the waviness of the geomembrane, which is due to the production process, will disappear, if the conditions on dimensional stability of BAM certified geomembranes has been fulfilled to some extent (see Sect. 3.2.5). Strain imposed on the geomembrane in this procedure is in any case far below the permissible limiting strain of 3 % (see Sects. 5.3.4 and 8.3.1). The skill of installation is to also use this effect consistently for difficult geometry (Fig. 9.12b).

The anchoring technique will be explained by describing a typical installation day using the example of a composite liner construction, though the span of events within an installation day can change quite broadly. This will then be a somewhat simplified and idealized representation, but all key aspects will be clearly explained. Fig. 9.8a and b schematically shows the different installation phases, which comprise the anchoring technique.

The working day begins early in the morning with the surface preparation of the compacted clay liner. This had been installed and compacted on the previous day in the construction section. The surface had been pushed away by a bulldozer and compacted and smoothed with a smooth drum roller. Now the remaining unevenness, like imprinted compaction roller marks and ridges and large foot imprints are eliminated using a small (2 to 3 t) tandem roller, hand rammer or shovel (Fig. 9.9). Included particles greater than 10 mm or foreign matter are removed. The on-site third-party inspector responsible for geomembrane QC (see Sect. 9.4.2) inspects the surface and releases it for geomembrane installation. Meanwhile the geomembrane rolls, which have already been identified on delivery and their condition and the shipping documents controlled and randomly taken samples tested (e.g. thickness, surface appearance, dimensional stability) and released by the third-party inspector, are being delivered by a hydraulic excavator from the stock pile to the installation section. Special equipment is necessary to transport and deploy 5 to 9 m broad and 1.5 to 3 t heavy rolls. A steel rod or mandrel is pushed through the roll core and hung from a suitable cross beam. The cross beam is fastened to the bucket of an excavator. The roll is jacked up on the berm crown or just placed down ready for unreeling.

After the surface of the mineral layer has been released, the installation work begins. The geomembrane is unreeled from the jacked-up roll and spread across the surface. The geomembrane roll can also be unrolled in a controlled fashion from the slope berm, the steel rod held by a cable winch. With the rising sun the geomembrane gradually warms up. The unrolled geomembrane is carefully aligned after getting acclimatised (adoption to the ambient temperature needs about 1 hour) and pulled smooth to the extent possible by manual work (Fig. 9.9). Since, for instance, a 100 m long and 5 m wide geomembrane cannot be pushed arbitrarily back and forth by the labourers, the roll had to be positioned exactly right from the beginning. Smoothing and aligning is repeated during welding on each roll.

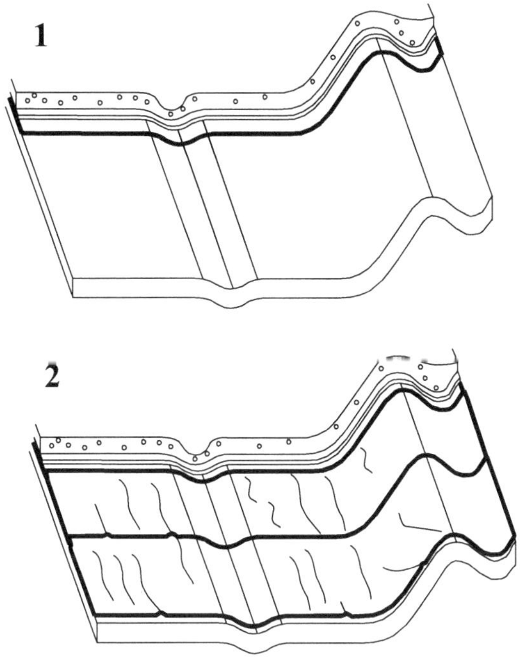


**Fig. 9.8a.** Schematic illustration of construction sequence for anchoring technique (Riegelbauweise). Diagram 1 shows a schematic view on a section of supporting layer, which, in the composite liner, consists of the compacted clay liner. Typical geometrical structures occurring in a landfill liner are drawn: a collecting trench, in which the leachate pipes are placed, a slope with a slope toe, a berm crown and an anchor trench. In the background the edge of a cell with an already fully installed geomembrane is drawn. I.e. a geotextile protective layer lies on the geomembrane and on top of it, the mineral protective layer and the gravel of the drainage layer. A part of the installed geomembrane protrudes for connection. Diagram 2: On the prepared supporting layer further geomembranes are unreeled and welded with each other and along with the geomembrane already installed. Due to production, installation and thermal expansion due to heating by solar radiation, waves are always present in the geomembranes

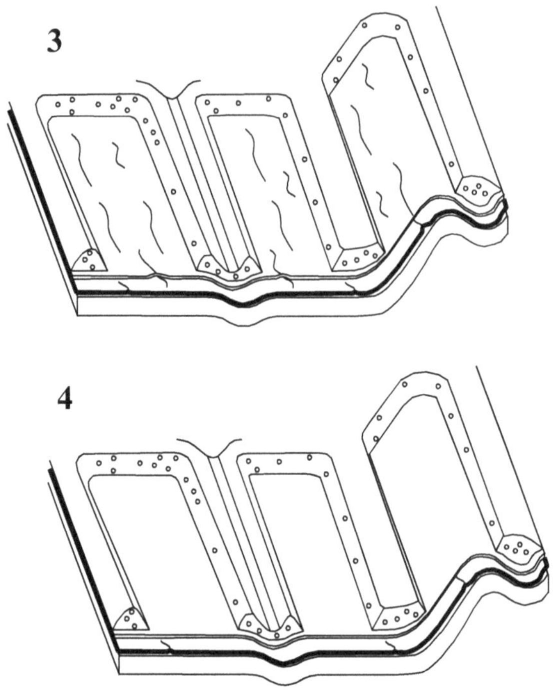


**Fig. 9.8b.** Construction sequence for anchoring technique (continued). Diagram 3 shows the main step. In more or less regular intervals anchoring bars are installed on the geomembrane. This consists of an unrolled geotextile protective layer or sand mat loaded with sand and/or gravel fillings of the mineral protective layer and drainage layer. The dead weight of the rows fixes the geomembranes at regular intervals on the surface and along critical ranges (e.g. ditch, slope toe and berm crown). This action must take place before cooling occurs, i.e. in the late afternoon or early evening.Diagram 4: During cooling the anchored geomembrane contracts and reaches complete flatness. If this state is achieved, flatness is set by the installation of gravel piles on the entire area. For this section the installation is completed and work in the next section can begin on the next day of installation (Diagram 1)

Since one may not drive on the surface of the mineral layer, installation is still manually intensive work. Mostly, the members of the welding crew (foreman, welder, assistant, see Sect. 9.4.1) of the installation contractor obtain help from the construction workers of the earthwork company who are being instructed and supervised by them. Furthermore, constantly present on site is an equipment driver of the construction company and the third-party inspector. There are thus five to seven people who take part directly in the installation of the geomembranes.

After placing the geomembranes the welding begins. The welding machine is controlled, sample seams are welded, site conditions (temperature and humidity) recorded and, finally, the welding parameters specified. The protective strips (or edge tapes) on the edges of the geomembranes are only removed directly before welding, i.e. if possible directly in front of the welding machine. The welders will be busy installing and welding three 100 m long geomembranes for the next 8 hours.

**Fig. 9.9.** This photograph illustrates the different stages of composite liner construction. In the far background the subgrade of the landfill liner system is being prepared and right in the background the compacted clay liner is being built. On the surface of the completed mineral layer a geomembrane has been unrolled. Small defects in the surface are being manually eliminated even after unrolling the geomembrane. The geomembrane is then aligned and pulled flat. In this construction project MDDS sand mats of the Gebrüder Friedrich GmbH company are used as a protective layer, which cover the installed geomembranes and ballasted with drainage gravel. The anchoring technique becomes particularly simple using the sand mat rolls since no fine-gravel mineral protective layer needs to be installed. Courtesy of Gebrüder Friedrich GmbH

Each seam is tested over the entire length by inflating the test channel of the dual hot wedge seam with about 5 bar pressurised air. Peel tests are performed on the seams of the test weld or on samples from the beginning and final ranges of the seam. The geometry of the seams is measured. Site conditions, welding parameters and test results are registered in standard forms of the welding records. The welding records are supplemented by the machine protocol of the welding machine, in which welding parameters are recorded consecutively. The third-party inspector inspects the seams on site. He measures the reduction in seam thickness non-destructively (see Sect. 10.2) using ultrasonic testing along the seams already welded and examines their condition. He takes samples for his laboratory tests from the sample seams or from the beginning and end pieces of each seam and controls the welding and inspection records as well as the machine protocols. By the time welding is finished, it is late afternoon.

**Fig. 9.10.** This photograph shows the anchoring bar or fixing berms. On the front end of the geomembrane a thick nonwoven geotextile was unrolled as a protective layer and on top of it a heap of gravel was applied. Thus the geomembrane is fixed. Courtesy Schicketanz Consulting Engineers, Aachen

After welding and testing are finished the next step is the construction of anchoring bars. This step forms the core of the anchoring technique. At both ends of the welded and cleaned (swept) geomembranes, a roll of sand protection mat or geotextile protective layer is unreeled and then loaded by additional heaps of sand of the protection layer and the gravel of the drainage layer. This forms the so-called anchoring bars or fixing berms. (Figs. 9.10 and 9.11). Such anchoring bars should be constructed across the installed geomembrane at a maximum pitch of 50 m over the geomembrane area. The dead weight of these bars fixes the geomembrane. An anchoring bar must be place along the trenches for the leachate collection pipes. Normally, the foundation for pipes is installed in the trenches and thus an anchoring bar is established. Likewise an anchoring bar has to be arranged along the toe of the slope. Otherwise, as the geomembrane contracts due to evening cooling, it would lift and bridge-over the trench and transition zone between the base and the slope (trampoline effect). The geomembrane is also fixed on the berm crown in an anchor trench by backfilling or earth nails.

**Fig. 9.11.** In the background of the four installed geomembranes the anchor bar is already completed. In the front the gravel filling is being applied with an excavator early in the evening. With evening cooling after sunset the four geomembranes will become completely flat. The nonwoven geotextile rolls are then unrolled over the whole surface and the gravel filling applied. Courtesy Schicketanz Consulting Engineers, Aachen

The preparation and the actual construction work of the anchoring bars has already begun and proceeds during the welding operation. A hydraulic excavator with a long-range telescope arm (12–15 m) or a long-arm articulated excavator is used to place sand and gravel. The anchoring bars can be arranged as small rows on which the material is moved and secured using a small excavator and small caterpillar. When installing sand or gravel close to the edges of the anchoring bars, attention must be paid to the fact that any smaller waves do not become covered, but are moved from the range of the narrow anchor area to the free area. In the early evening all anchoring bars are in place and the geomembranes are fixed. With the gradually increasing evening coolness the geomembranes increasingly contract and become taut. Mostly by late evening, but early next morning at the latest, the clamped geomembranes reach complete flatness (see Figs. 9.12a, 9.12b and 9.12c). Now they can be covered with the protective layer and buried with gravel using a long reach bucket excavator over the entire area. Not only the excavator driver, but also the third-party inspector and the foreman of the installation contractor are – sometimes even before sunrise – on the construction site to supervise the installation of the drainage layer. The new working day begins again in the early morning with the preparation of the surface of the next section of compacted clay liner.

**Fig. 9.12a.** This and the following photographs demonstrate that using the anchoring technique, complete flatness can be achieved for the geomembrane. Courtesy Schicketanz Consulting Engineers, Aachen

## 9.4 Quality Assurance

In the last twenty years intensive discussions and sometimes hectic activities developed around quality and quality management. The development was accompanied by a flood of new terms and standards: Quality Management (QM), Quality Assurance (QA), Quality Control (QC), QM Plan,

Quality Politics, Quality Objectives, Audit, Verification, Accreditation, Certification, DIN EN ISO 9000, 9001, 9002, 9003, 9004, DIN EN ISO 17025, 17020 and so on. Some of the definitions of the terms are difficult to understand. It is however often only “pouring old wine into new wineskins”[4]. For example, simply saying that for a successful construction project, a detailed construction schedule is necessary, which co-ordinates the work of the trades carefully, or that for a standardised test, certain organisational sequences must be adhered to, does not constitute new inventions. Often, in themselves well-known procedures are only formalized and uniformly specified in standards.

**Fig. 9.12b.** This photograph shows that the anchoring technique provides complete flatness even in critical areas (slope toe and pipe penetration constructions). In such places where hand-welded extrusion fillet seams occur in higher frequency and standing leachate can be expected, special care and high technical installation skills are required. Only experienced, particularly qualified and well-equipped specialist installation companies can master this task. Courtesy Schicketanz Consulting Engineers, Aachen

[4]Marc 2:22: “And no one pours new wine into old wineskins. If he does, the wine will burst the skins, and both the wine and the wineskins will be ruined. No, he pours new wine into new wineskins.”

The formalisation and standardization of terms and procedures in the field of quality management should promote the development of a common marketing area and cross boundary goods traffic. The idea of eliminating trade and communication barriers is the driving force behind these activities which, however, often aim at adhering only to certain formal rules. In all this, however, the essential aspects, i.e. which technical conditions have to be actually fulfilled, must not be forgotten.

**Fig. 9.12c.** Complete flatness can be achieved even on long and steep slopes, as this photograph shows. Courtesy Schicketanz Consulting Engineers, Aachen

The complex topics of quality management are beyond the scope of our consideration. The issue as to how a quality management system is established in landfill engineering has been looked into in great detail in (Dornbusch et al. 1996). In the following we try to explain and discuss some of the central terms which play an important role in tendering and placing orders and in the implementation of quality management for geomembrane production and installation on landfill construction sites.

Colloquially the term quality is used in the sense of property or condition of a product, and it is associated with the product being of particularly high value or particularly well fulfilling certain needs in comparison, with for instance, competitive products. This association does not play any role in the concept development in the standards specified above. Quality here simply means the degree to which certain distinguishing features (characteristics) of a product (in the broadest sense) provably fulfil the given requirements of a customer or internal product planning, guidelines or product standards. To exemplify the difference: various features used by geosynthetic manufacturers to define the quality of their products according to standards are actually irrelevant for the performance in a geotechnical application. Therefore a high-quality product might perform really poorly and indeed have poor quality in the colloquial sense.

Each manufacturer of a product has a somewhat different system comprising an organisational structure and sequences of management and production processes, competences and responsibilities. Such a system develops into a Quality Management System (QMS) if it fulfils certain formal conditions specified in the standards. It is essential that all elements of the system are fully identified and specified and thus it is clear, at least for the responsible people, what should happen and that, secondly, monitoring and control tools and processes are available which can test and record the product quality in above mentioned sense (agreement of the characteristics with given requirements). The lastly mentioned part of the management system, which focuses on fulfilling quality requirements of the product is called Quality Control (QC). The Quality Management Manual (QMM), in which the QM System is described, and the Quality Control Plan (QCP), in which the procedures and associated resources, applied by whom and when, to control the quality is described, are therefore not the only ones, but the most palpable components of a quality management system. Tables A1-5 and A1-6 in Appendix 1 describe the test procedures and frequencies for in-house quality control of geomembranes by manufacturers of BAM certified HDPE geomembranes.

In the context of quality assurance in the production of a product or a structure, independent test laboratories or inspection bodies supervise the manufacturing or construction procedure, perform controls and inspections and examine the manufactured products using test methods, which are described in standards or guidelines. Table A1.7 of Appendix 1 describes the test procedures and frequencies for third-party control of resin, carbon black batch and geomembranes as part of the in-house QMS of the manufacturers of BAM certified HDPE geomembranes.

When an engineering structure is about to be constructed, the consulting engineer, construction companies, building material manufacturer etc. unite to carry out a unique manufacturing procedure together. It is usually only for the production of this individual structure and afterwards they separate again. Also such an individual manufacturing procedure must be subject to quality management, which is based on the QM systems of the companies involved but is not fully covered by these.

Appling these considerations to the construction of a liner system there are two aspects: Firstly, every contractor involved should have an in-house QMS and QC documented in his QMM and QCP. Secondly, a site-specific quality management system should be established for the construction of the liner. However, in this respect one often speaks of quality assurance and a Quality Assurance System (QAS) instead of quality management, defined as all means and actions used to assure that the liner materials and the installed liner system fulfil the site specific contractual and regulatory requirements. The quality assurance is documented in the Quality Assurance Manual (QAM). Each construction project must have a quality assurance plan, which integrates the individual plans of the different trades. For each individual construction project, competencies and decision hierarchy must be clearly regulated, the construction schedule using the available equipment planned in detail and the actual quality assurance measures – e.g. control of schedule and quality control testing – must be specified.

The responsibility for the quality assurance is often assigned to third-party inspectors who are independent of owner, designer, manufacturers and installers. For landfill liner construction the German administrative provisions requires third-party inspection in agreement with the responsible authority and monitoring by the responsible authority. Quality control within the quality assurance system is then provided by the manufacturers, the installers and the third party inspector himself. The conditions, which are placed on third-party inspection bodies for the installation of geosynthetics components and plastic construction components in landfill liners will be described in the section after the next.

Using a quality management system based on the conditions of the standards, the manufacturer can prove to the customer in a systematic, standardized and comprehensible way that he, as the standard says, possesses the "capability to produce quality". How this happens is described in the EN ISO 9000 standard family. EN ISO 9000:2005 *Quality Management Systems – Fundamentals and Vocabulary* explains the quality management concept and specifies the terminology. EN ISO 9001:2000 *Quality Management Systems – Requirements* describes the quality management system over the entire range of development and construction and production, assembly and customer service. It serves as a guideline for the implementation of the quality management system. The older standards EN ISO 9001, EN ISO 9002 and EN ISO 9003 have been integrated in the new version of EN ISO 9001. EN ISO 9004:2000 *Quality Management Systems – Guidelines for Performance Improvements* considers the effectiveness in achieving the planned results and the efficiency in use of the resource within the quality management systems. It aims to improve the performance of the quality management system. Finally EN ISO 19011:2002

*Guidelines on Quality and/or Environmental Management Systems – Auditing* describes the auditing process for a quality management system. Further information about the use of these standards might be obtained from the e-brochure "selection and use of the ISO 9000:2000 standard family" developed by the ISO Technical Committee 176 (www.iso.org).

Manufacturers, whose QM system fulfils the requirement of these standards, can get themselves certified by a recognized certification body. In the meantime there are qualified specialist consulting engineering companies, which can help in the development of a QM system and its certification. Capability to produce quality can then be proved to the customer by presenting the certificate. It has to be made clear however that the capability to produce quality proven in such a way refers only to purely formal aspects. It says nothing about which conditions manage the product. A geomembrane manufacturer therefore should be able to present a certificate in accordance with DIN ISO 9001. In addition however, one will have to specify the technical requirements for the geomembrane, for instance by requesting an approval by a BAM certification or a GRI GM13 test certificate (see Chap. 1).

How a quality management for the construction of a composite liner based on the EN ISO 9000 standard family and the technical guidelines should look like, has been described by J. Dornbusch, U. Averesch and M. El Khafif (Dornbusch et al. 1996). Quality management for landfill liners has been reported in (Anemüller and Schicketanz 1998). One component of the total quality assurance plan is a section, which regulates the installation of the geomembrane, protective layers and plastic penetration and building elements. The German association of geomembrane manufacturers and installers (AK GWS) has compiled a prototype of a QCP for this purpose[5].

However, the formal proof of "capability to produce quality" of the companies involved and as precise a description as possible of the requirements and conditions of the construction project is not sufficient to succeed with the construction project. Those involved must also not only have the formal qualifications but also the capability, i.e. specialist knowledge, experience and availability of specialists, machines and equipment, to meet the project requirements Accreditation procedures and voluntarily certification associations offer the possibility that both testing and inspection bodies, which are involved in the context of quality assurance, and the companies involved in construction work can prove their specialist competence.

---

[5] The document can be viewed at AK GWS's website: www.akgws.de.

Accreditation as a test laboratory must be distinguished from accreditation as an inspection body. The latter can have the first as a prerequisite. Accreditation of a test laboratory and/or an inspection body means to possess the recognition by a nationally recognized accreditation body that the test laboratory and/or the inspection body possess the competence to carry out certain tests or test types and/or inspections. The general requirements for test bodies are specified in EN ISO/IEC 17025:2000 *General Requirements for the Competence of Testing and Calibration Laboratories*, the requirements on inspection bodies in EN ISO/IEC 17020:2004 *General Criteria for the Operation of Various Types of Bodies Performing Inspections*. EN 45002:1990 *General Criteria for Expert Opinions on Testing Laboratories* and EN 45003:1995 *Calibrating and Testing Laboratories Accreditation System – General Requirements for Operation and Recognition,* from the EN 45000 standard family stipulate the execution of the accreditation procedure. An accreditation thus certifies a specific intrinsic competence. It does not make any sense to say in itself that a test laboratory and/or an inspection body is accredited, it must always be stated what it is accredited for. The accreditation certificate precisely sets out those test methods or types of test methods that the accreditation applies to, and/or the inspection measures for which the accreditation as an inspection body applies. Special experts, so-called auditors, designated by the accreditation body from the specialist field to which the respective standards or regulations belong, examine the test bodies and/or inspection bodies. These audits are regularly repeated. In view of the status of accreditation available Europe-wide today inspection or test bodies should only be commissioned when they are accredited for the actual measures.

The desire to be able to prove their highly specialist qualifications and suitability easily and convincingly has led the German building industry to offer voluntary quality control to their associations' member. In 1989 the companies working in the filed of landfill engineering and contaminated land remediation came together and founded the certification association "Building for Environmental Protection" (Bauen für den Umweltschutz, (BU), www.ueberwachungsgemeinschaft-BU.de) (BU 1995; Görg 1997). In different fields of performance requirements were made for specialist knowledge, efficiency and reliability, which are regularly monitored by independent experts. Members of the association, who fulfil the conditions, obtain a certification emblem. This should enable a check on the suitability of the contractor in connection with bidding for and assignment of construction work.

Following the example of BU, the German association of geomembrane manufacturers and installers (AK GWS), founded a certification program for specialist installer companies in 1997. This was based on the recommendation of BAM "Installation Contractor for the Installation of Plastic Components in Landfill Liner Systems, BAM's Recommendation for Requirements on the Qualification and Tasks of a Installation contractor" (AK-GWS 1997; Albers 1997; Albers and Preuschmann 1998). BAM audits and supervises the specialist installation companies within the framework of this certification program. A list of certified installers may be taken from AK GWS's homepage: www.akgws.de.

**Fig. 9.13.** The certification emblem of the German professional association of geomembrane manufacturers and installers (AK GWS). An installation contractor may carry this emblem if it belongs to the association, fulfils the certification conditions and proves this in regular audits

**Fig. 9.14.** The certification emblem of the certification program of the International Association of Geosynthetic Installers

The BAM certification of geomembranes demands, with reference to the general administrative provisions, that geomembranes are installed by an installation contractor that is provably equipped with experienced and qualified personnel and the necessary machines and equipment. The proof can be provided by the certification program of AK GWS. The certified installers are decorated by a certification emblem (Fig. 9.13). A comparable certification program was established by the International Association of Geosynthetic Installers (IAGI). Figure 9.14 shows the certification emblem. Details and a list of certified members may be viewed at the IAGI's homepage: www.iagi.org.

## 9.4.1 Conditions placed on Installation Companies

The following excerpt from the above mentioned recommendation of BAM, which forms the basis of the certification program for geomembrane installers of the German association of geomembrane manufacturers and installers (AK GWS), reveals what experience, qualifications, equipment and general organisational conditions must be available in an installation contractor within the field of landfill engineering in Germany (Albers and Preuschmann 1998). In a similar way, this not only applies to landfill construction but also to other fields of application of geotechnical engineering.

### *General Conditions*

The installer must fulfil the personnel, technical, equipment-related and organisational conditions listed below for work preparation and the installation of geomembranes, geotextiles and protective layers, and quality control in landfill liner systems. Even in the event of illness and holidays and/or equipment failure sufficient personnel and equipment must be available in order to ensure a continuous installation progress. The company must be legally identifiably, financially sound (certificates of impeccability), and must have an operating liability insurance with a covering sum of 1 million euros and an appropriate product liability insurance (at least 2.5 million euros in individual cases). The installer must possess the means and qualifications to produce easily readable shop and design drawings and as-built drawings with all relevant details.

The installer has to work under a quality assurance management, described in a quality management manual (QMM), which corresponds to the conditions of DIN EN ISO 9000 et sequ. Here all work and quality control procedures, the quality management measures and the form of their implementation have to be described; detailed information has to be given about equipment (number and types) and personal (number and qualification).

Acceptance, monitoring and calibration of working and test equipment relevant to installation and quality control must be regulated in the QMM, for example: the test parameters of the hot wedge welding machine such as roller pressure, hot wedge temperature, speed. Maintenance and inspection plans must be available and the actions taken must be recorded. The necessary calibration and inspection devices must be provided. For the documentation of the installation work and quality control in a project the installer must hold an archive of written documents (design documents, shop and as-built drawings, welding and testing protocols, acceptance documents). The duration of archiving is decided by the authorizing body, but must be at least 6 years.

### *Personnel Conditions*

The installer must have trained specialists with expert knowledge and experience in plastics and civil engineering, in plastics quality control and in the installation methods of large-area geosynthetic liner systems. The installer's personnel must be able to compile a technically sound proposal on the basis of tender documents and provide a full service in accordance with the conditions. They must be able to assess the effect of deficiencies detected on the liner system's performance in order to be able to perform a practical and technically competent repair[6].

Qualification and further training of the personnel must be ensured and recorded. The records should cover the fields of

- basic training in accordance with the specialist activity of the employees,
- probation time in practice,
- independent employment in the respective field,
- measures for further education.

The company must have at least one installation manager (superintendent, supervisor), one QA officer and five geomembrane welders (seamers), including a foreman (master seamer). These personnel must fulfil the following conditions:

#### *Installation Manager*[7]

Qualification

The responsible installation manager has to prove he has at least a full engineering degree (polytechnics or university) and an at least three years practical activity in planning or execution in construction of landfill liner systems using geosynthetic components or plastic building elements[7]. The responsible installation manager must have the authority to issue orders to the employees of the installation contractor on site. He must be able to prove his qualification by having completed training as a plastics welder in accordance with DVS 2212 Part 3 *Testing of Plastics Welders, Test Group III, Geomembranes in Earthworks and Hydraulic Engineering*, at least Subgroup III-1, III-2, III-3 (landfill welder).

[6] Concerning experience, the IAGI's "HDPE geomembrane installation specification" requires that the installer "shall have installed at least 10 projects involving a total of 500,000 m² of the specified type of geomembrane or similar during the last years". The same requirement holds for the field installation supervisor. The master seamer should have seamed a minimum of 300,000 m² of geomembrane.

[7] The installation manager is the responsible leader of the geomembrane installation for the installation contractor within the building project.

Tasks
The installation manager must be able to produce a technical assessment of a design plan. He has to produce the necessary installation plans and manage its practical execution. He must present objections on recognizable planning faults in time and make suggestions for improvements to ensure a proper installation. He has to lead the negotiations with the other parties (owner, designer, geomembrane manufacturer, earthwork contractor, third party inspector) during the construction phase and attends the pre-construction and progress meetings. He has to supervise the schedules and the co-ordination with the preceding and follow-up trades. He has to oversee in particular the organisational conditions for the acceptance of a proper geomembrane subgrade (mineral layer) from the contractor who produces it and the speedy installation of the geomembrane. He must ensure the complete flatness of the properly welded geomembrane before applying the protective layer so that the overburden can produce an intimate contact. In the event of differences that cannot immediately be clarified with the installation team during installation, he is the partner for the construction management and the third-party inspector.

### *Geomembrane Welder*

Qualification
The geomembrane welder (or seamer) must have acquired his qualification through, as a minimum, a completed training as a plastics welder in accordance with DVS 2212 Part 3 *Testing of Plastics Welders, Test Group III, Geomembranes in Earthworks and Hydraulic Engineering*, at least Subgroup III-1, III-2, III-3 (landfill welder). He must successfully have participated in the basic and preparatory courses for the acquisition of the qualification of a plastics welder in accordance with DVS 2212. Education and further training are to be proved by presenting the test certificates in accordance with DVS 2212, the annual tests by an entry in the record sheets of the welder ID. The documents or copies must be available on the construction site.

Tasks
In addition to proper placement and welding, his responsibility covers the control of devices, machines and materials for serviceability. If necessary, he has to commission calibration, repair, replacement or other measures on the devices. Compliance with the conditions of work safety and the conditions of the quality assurance is also part of his responsibilities. In exceptional cases he keeps the welding records.

### *Foreman*

Qualification
The qualification of the foreman (master seamer) corresponds to that of the geomembrane welder. An employment as a foreman may only take place after a minimum of three years successful work as a geomembrane welder[7].

Tasks
He is constantly on the construction site and leads the installation crew guided by the instructions of the installation manager and, as a rule, keeps the records, in

particular those of the in-house monitoring (as-built plans, welding protocols, test records). He is also responsible for the proper condition of the equipment on the construction site.

He performs the tasks of the quality control officer:

- execution of quality control tests in accordance with the quality assurance plan,
- recording the quality control results and keeping the as-built plan and the construction logs.

*Helper*

The handling of geomembranes and building components on the construction site, transport and proper installation has to be guaranteed by a sufficient number of instructed helpers. The helpers are to be instructed in their tasks referring to the characteristics of the individual construction project. The extent of the instruction must be documented (participant, type of instruction, duration and date).

### Conditions Placed on Equipment and Devices

The installer must be equipped with equipment, machinery and devices for the installation and welding, in accordance with the relevant welding methods, for the construction site QC tests and for transporting his equipment on the construction site. For transport and unreeling of the geomembranes and protective layers suitable handling and transport devices must be available. The available test equipment, machinery and devices must enable a professional installation or construction and the execution of the tests in compliance with the guidelines, regulations and state of the art technology. The test equipment and machinery must be maintained and calibrated in accordance with documented instructions and records must be kept. The amount of equipment, machinery and devices must be chosen in such a way that the building schedule is not unduly delayed during installation. In case of failure, a replacement must be made available immediately. It is advisable to establish a contractor's company-owned store or an equipment pool accessible to cooperating contractors or arrange a co-operation with the geomembrane manufacturers.

For the installation and welding of geomembranes and geotextile protection layers as well as for their welding against plastic penetration and building components the following at least must be available:

*For Seam Preparation*

Portable grinding machines, drawing knife, suitable equipment for the removal of surface unevenness (e. g. milling cutter), cleaning appliances: buckets, clothes, sponges, water containers, brooms.

*For Welding*

3 hot wedge welding machines with data logger for dual seams with a test channel in accordance with the BAM certification guidelines and DVS 2225, Part 4; at least one computer unit for the evaluation of the electronic recordings depending

on the type of machine; 3 warm gas extrusion welding apparatus for extrusion fillet seams in accordance with the BAM certification guideline and DVS 2225, including devices for the supply of the welding extrudate; 3 warm gas welding apparatus and/or devices for temporarily tacking geomembrane panels.

*For Record Keeping*

Thermometers for melt and hot wedge temperatures; thermometers for air temperature; hygrometers; dew point tables, graphs or slide rules; length and thickness measuring instruments (tape measure, dead weight thickness gauge, calliper gauges).

*For Seam Testing*

Pressure gauge up to 10 bar with data logger (accuracy class 1 – drum-chart recorder or electronic recording) in accordance with DIN EN 837[8], pressure hoses, fittings and connections; compressor; vacuum box test equipment with boxes for standard seams and also for inner and outer edges at seams on buildings or penetration systems and for repair patches; foaming surfactant (released by resin and geomembrane manufacturers); tensiometer for peel tests with a motor-driven winder, capable of peeling with a constant test speed of 50 mm/min and with maximum force indicator; mechanical peel test devices (simple version of a tensiometer suitable for quick operation on construction site).

*Other Appliances*

Cutting device, chalk string; utilities for marking and labelling; equipment for transport and handling of rolls, unreeling equipment; equipment (sand bags) for wind protection. All work instructions, standards, guidelines, instruction sheets or reference data that concern the activity of the installation contractor must be held on the latest status and be easily available on the construction site.

### 9.4.2 Conditions for Third-Party Inspectors

In the three-tier quality assurance system composed of quality control measures of the installer, third-party inspection and administrative control and supervision, special importance is attributed to third-party inspection (Knipschild 1996). The third-party inspector is the independent third person, who ensures that the installer has provided correct and complete documentation of his construction work and has proved it by his quality

[8] DIN EN 837-1:1997-02 *Pressure gauges – Part 1: Bourdon gauges; Dimensions, Measurement Methods, Requirements and Testing*;
DIN EN 837-2:1997-05 *Pressure gauges – Part 2: Selection and Installation Suggestions for Pressure Gauges*;
DIN EN 837-3:1997-02 *Pressure gauges – Part 3: Diaphragm Gauges and Capsule Manometers; Dimensions, Measurement Method, Requirements and Testing.*

control measures. The third-party inspector also checks this through his own inspections and issues a final construction quality assurance report so enabling supervising and acceptance for the authority. Rapid and smooth construction progress and reliable and thorough quality assurance measures inevitably conflict to some extent. Only a third-party inspection body that has the necessary machine, personnel and technical capacity to rapidly perform the necessary tests and who possesses sufficient specialist knowledge and experience to rapidly assess the type and extent of deficiencies and remedial measures can solve this contradiction. A third-party inspector lacking technical knowledge and experience tends not to use his area of discretion for a technical sound judgement but makes only formal judgement by purely insisting on the letter of the rules: construction progress comes to a halt, or the uncertainty of the third-party inspector leads to "laisser faire": the quality of the product becomes poor and the construction progress gets entangled in the developing mess (Müller 1994).

Since the supervising authority relies on the third-party inspection, it must therefore exert an influence on the selection of third-party inspectors for which selection criteria are needed. BAM has published a guideline, which provides detailed requirements for third-party inspection bodies (Müller 1998). This guideline should provide a selection criterion for the authority and the owner concerning third-party inspectors.

The guideline contains a list of conditions on which basis the suitability of a third-party inspection body can be relatively and simply assessed. The inspection body must fulfil at least the following conditions:

### Organizational/Firm-specific Conditions

1. Legal identifiability
2. Occupation liability insurance over in each case 0.5 million euros for personal and other damage
3. Accreditation in accordance with EN ISO/IEC 17020:2004 *General Criteria for the Operation of various Types of Bodies Performing Inspections* for inspection measures necessary within the framework of third-party control for geomembrane liner systems.
4. Reference list of the last 3 years with clients certificates

### Personnel Conditions

Proof of trained specialists with expert knowledge and experience in plastics engineering and quality management as well as landfill-specific geotechnical engineering procedures by documented reference and training course certificates. In detail the following must be provided:

1. A responsible third-party inspector (quality assurance managing engineer) with a completed engineering training (engineering degree from technical high school or university) and relevant training course certificates and at least three years proven activity of third-party inspector for geosynthetic components and plastic building and penetration elements in connection with landfill liners.
2. An on-site third-party inspector (quality assurance monitor) with completed training as a certified technician or technical assistant with a materials engineering background or engineer training and a minimum one year's construction site test activity under the supervision of an experienced third-party inspector. The successful participation in the basic and preparatory course for the acquisition of the qualification of a plastic welder in accordance with the DVS Guideline 2212-3:1994-10 *Testing of Plastics Welders, Test Group III, Geomembranes in the Earthworks and Hydraulic Engineering.*

#### *Equipment Conditions*

1. Own test laboratory; valid accreditation of the test laboratory according to EN ISO/IEC 17025:2000 *General Requirements for the Competence of Testing and Calibration Laboratories* for the necessary tests in third-party control. These are:

   DVS 2203-5:1999 *Testing of Welded Joints of Thermoplastics Plates and Tubes – Technological Bend Test*, or EN 12814-1:1999 *Testing of Welded Joints of Thermoplastics Semi-finished Products – Part 1: Bend Test*

   DVS 2226-2:1997 *Testing of Fused Joints on Liners Made of Polymer Materials – Lap Shear Test*, or EN 12814-2:2000 *Testing of Welded Joints of Thermoplastics Semi-finished Products – Part 2: Tensile Test*

   DVS 2226-3:1997 *Test of Fusion on PE Liners – Peeling Test*, or EN 12814-4:2001 *Testing of Welded Joints of Thermoplastics Semi-finished products – Part 4: Peel Test*

   EN ISO 527-1:1996 *Plastics – Determination of Tensile Properties – Part 1: General Principles*

   ISO 1133:2005 *Plastics – Determination of the Melt Mass-flow Rate (MFR) and the Melt Volume-flow Rate (MVR) of Thermoplastics*

BAM test method for determination of dimensional stability (Sect. 3.2.5)

EN ISO 60:1999 *Plastics – Determination of Apparent Density of Material that can be Poured from a Specified Funnel*, or EN ISO 1183-1:2004, *Plastics – Methods for Determining the Density of Non-cellular Plastics – Part 1: Immersion Method, Liquid Pyknometer Method and Titration Method*

EN ISO 9863-1:2003 *Geosynthetics – Determination of Thickness at Specified Pressures – Part 1: Single Layers*

EN ISO 9864:2005 *Geosynthetics – Test Method for the Determination of Mass per Unit Area of Geotextiles and Geotextile-related Products* or EN 29073-1:1992: *Textiles – Test Methods for Nonwovens – Part 1: Determination of Mass per Unit Area*

EN ISO 10319:1996 *Geotextiles – Wide-width Tensile Test* or EN 29073-3:1992, *Textiles – Test Methods for Nonwovens – Part 3: Determination of Tensile Strength and Elongation*

2. Personal test equipment (measurement and inspection device) for the on-site tests as per the Guideline DVS 2225-4:1996 *Welding of Polyethylene Geomembranes (PE) for Lining of Landfills and Contaminate Land.*

Third-party inspection bodies can easily provide the proof that they fulfil these conditions by a valid document, which proves an accreditation as an inspection body in accordance with EN ISO/IEC17020 on the basis of the BAM guideline for third-party inspection bodies.

## References

AK-GWS (1994) Dichtungssysteme mit Geokunststoffen, Zusätzliche technische Vertragsbedingungen für das Gewerk Geokunststoffe bei Kombinationsdichtungssytemen im Deponiebau, Teil 1: Kunststoffdichtungsbahnen. Müll und Abfall 26: 162–170

AK-GWS (1995) Dichtungssysteme mit Geokunststoffen, Zusätzliche technische Vertragsbedingungen für das Gewerk Geokunststoffe bei Kombinationsdichtungssytemen im Deponiebau, Teil 2: Geotextilien zum Schützen, Trennen, Filtern und Bewehren. Müll und Abfall 27: 702–706

AK-GWS (1997) Ordnung über die Durchführung des Überwachungsverfahrens. Arbeitskreis Grundwasserschutz e.V. (www.akgws.de), Berlin

Albers K-H (1997) Wirksamer Grundwasserschutz durch Oberflächenabdichtungen mit Kunststoff-Flächenabdichtungen. In: Egloffstein T and Burkhardt G (eds) Oberflächenabdichtungen von Deponien und Altlasten, Planung – Bau – Kosten. Erich Schmidt Verlag, Berlin, pp 41–48

Albers K-H and Preuschmann R (1998) Die Überwachungsordnung des AK GWS e.V. als Instrument zur Qualitätssicherung bei der Verlegung von Kunststoffdichtungsbahnen. In: Knipschild FW (ed) Tagungsband der 14 Fachtagung "Die sichere Deponie, Wirksamer Grundwasserschutz mit Kunststoffen". Süddeutsches Kunststoffzentrum Würzburg (SKZ), Würzburg, pp F1–F25

Anemüller M and Schicketanz R (1998) Das Qualitätsmanagement bei Deponieabdichtungsarbeiten – derzeitiger Stand. In: Egloffstein T et al. (eds) Oberflächenabdichtungen von Deponien und Altlasten ′98, Wirksame und kostengünstige Systeme – Reststoffe als alternative Dichtungsmaterialien. Erich Schmidt Verlag, Berlin, pp 91–120

August H et al. (1992) Permeationsverhalten von Kombinationsdichtungen bei Deponien und Altlasten gegenüber wassergefährdenden Stoffen, Bericht zum FuE Vorhaben 10302208 des BMBF. Bundesanstalt für Materialforschung und -prüfung (BAM), Labor Deponietechnik, Berlin

Averesch UB and Schicketanz RT (1998) Installation procedure and welding of geomembranes in the construction of composite landfill liner systems – focus on "Riegelbauweise". In: Rowe RK (ed) Proceedings of the Sixth International Conference on Geosynthetics. Industrial Fabrics Association International (IFAI), Roseville, MN, USA, pp 307–313

Baehr HD (1966) Thermodynamik, Eine Einführung in die Grundlagen und ihre technischen Anwendungen. Springer Verlag, Berlin, Heidelberg, New York

Bonarparte R and Gross BA (1990) Field Behavior of Double-Liner Systems. In: Waste Containment Systems: Construction, Regulations, and Performance Geotechnical Special Publication No 26. ASCE, New York pp 52–83

Bräcker W et al. (1994) Deponiebauteile für Kombinationsabdichtungssysteme. Bautechnik 71: 293–299

BU (1995) Leistungskatalog der anerkannten Fachbetriebe der Überwachungsgemeinschaft "Bauen für den Umweltschutz". Überwachungsgemeinschaft "Bauen für den Umweltschutz" e. V., Wiesbaden

Bundesregierung (2001) Abfallablagerungsverordnung. Bundesgesetzblatt I: 305

Bundesregierung (2002) Verordnung über Deponien und Langzeitlager. Bundesgesetzblatt I: 2807

Burkhardt G and Egloffstein T (eds) (1995) Asphaltdichtungen im Deponiebau (Asphalt concrete barriers for landfill lining). expert-Verlag, Renningen-Malmsheim

Corbet SP and Peters M (1996) First Germany/USA Geomembrane Workshop. Geotextiles and Geomembranes 14: 647–726

Dornbusch J et al. (1996) Bauverfahrenstechnik und Qualitätsmanagement bei der Herstellung von Kombinationsdichtungen für Deponien. Shaker Verlag, Aachen

Görg H (1997) Qualitätsmanagement für Bauleistungen beim Bau von Oberflächenabdichtungen. In: Egloffstein T and Burkhardt G (eds) Oberflächenabdichtungen von Deponien und Altlasten, Planung – Bau – Kosten. Erich Schmidt Verlag, Berlin, pp 71–86

Holzlöhner U (1997) Water balance, the risk of desiccation in earthen liners. In: August H et al. (eds) Advanced landfill liner systems. Thomas Telford Publishing, London, pp 19–30

Kalbe U et al. (2000) Mineralogische und chemisch-physikalische Auswirkungen der Permeation von Kohlenwasserstoffen in Kombinationsdichtungen und -dichtwänden, Bericht zum Forschungsvorhaben 1461027 des BMBF. Bundesanstalt für Materialforschung und -prüfung (BAM), Labor Kontaminationsbewertung, Berlin

Kalbe U et al. (2002) Transport of organic contaminants within composite liner systems. Applied Clay Science 21: 67–76

Knipschild FW (1992) Qualitätssicherung beim Bau von Deponieabdichtungen – Einbau der Kunststoffdichtungsbahn. In: Fehlau K-P and Stief K (eds) Fortschritte der Deponietechnik 1992, Qualitätssicherung für Deponieabdichtungssysteme und Eigenkontrollen beim Aufbau der Deponie. Erich Schmidt Verlag, Berlin, pp 41–58

Knipschild FW (1994) Konstruktive Einzelheiten von Kombinationsdichtungen. In: Knipschild FW (ed) Tagungsband der 10 Fachtagung "Die sichere Deponie, Wirksamer Grundwasserschutz mit Kunststoffen". Süddeutsches Kunststoffzentrum Würzburg (SKZ), Würzburg, pp 103–119

Knipschild FW (1995) Erfahrungen beim Bau von Kombinationsdichtungen. In: Knipschild FW (ed) Tagungsband der 11 Fachtagung "Die sichere Deponie, Wirksamer Grundwasserschutz mit Kunststoffen". Süddeutsches Kunststoffzentrum (SKZ), Würzburg, pp 89–105

Knipschild FW (1996) Kunststofftechnische Überwachung beim Bau von Deponieabdichtungssystemen. In: Knipschild FW (ed) Tagungsband der 12 Fachtagung "Die sichere Deponie, Wirksamer Grundwasserschutz mit Kunststoffen". Süddeutsches Kunststoffzentrum (SKZ), Würzburg, pp E1–E10

Koerner GR et al. (1999) Properties of exhumed HDPE field waves. Geotextiles and Geomembranes 17: 247–261

Krath U and Schwarz T (1993) Kombinationsdichtungsbau im Winter am Beispiel der Zentralen Mülldeponie Eiterköpfe. Müll und Abfall 25: 366–374

Müller WW (1994) Die Anwendung der BAM-Zulassung beim Bau von Kombinationsdichtungen. Müll und Abfall 26: 601–606

Müller WW (1996) Anforderungen an die Schutzschicht für die Dichtungsbahnen in der Kombinationsdichtung, Teil 2: Zulassungsanforderungen. Müll und Abfall 28: 90–99

Müller WW (1999a) Die neue BAM-Richtlinie für die Zulassung von Kunststoffdichtungsbahnen für die Abdichtung von Deponien und die Sicherung von Altlasten. In: Egloffstein T et al. (eds) Oberflächenabdichtung von Deponien und Altlasten 1999, Zeitgemäße Oberflächenabdichtungssysteme – ist die Regelabdichtung nach TA-Si noch zeitgemäß? Erich-Schmidt Verlag, Berlin, pp 41–50

Müller WW (ed) (1998) Fremdprüfung beim Einbau von Kunststoffkomponenten und -bauteilen in Deponieabdichtungssystemen – Richtlinie der Bundesanstalt für Materialforschung und -prüfung (BAM) für die Anforderungen an die

Qualifikation und die Aufgaben einer fremdprüfenden Stelle. BAM, Labor IV.32, Deponietechnik, Berlin

Müller WW (ed) (1999b) Richtlinie für die Zulassung von Kunststoffdichtungsbahnen für die Abdichtung von Deponien und Altlasten. Wirtschaftsverlag NW, Verlag für neue Wissenschaften GmbH, Bremerhaven

Müller WW (ed) (2001) Certification Guidelines for Plastic Geomembranes Used to Line Landfills and Contaminated Sites. Laboratory of Landfill Engineering, BAM, Berlin

Müller WW et al. (1998) Solubilities, Diffusion and Partition Coefficients of Organic Pollutants in HDPE Geomembranes: Experimental Results and Calculations. In: Rowe RK (ed) Proceedings of the Sixth International Conference on Geosynthetics. Industrial Fabrics Association International (IFAI), Roseville, MN, USA, pp 239–248

N.N. (1993) Mineralische Deponieabdichtung – Richtlinie Nr. 18. Landesumweltamt Nordrhein-Westfalen, Essen

Ramke H-G et al. (2002) Ergebnisse des Status-Workshops "Austrocknungsverhalten von mineralischen Abdichtungsschichten in Deponie-Oberflächenabdichtungssystemen". In: Egloffstein TA et al. (eds) Oberflächenabdichtung von Deponien und Altlasten 2002. Erich Schmidt Verlag, Berlin, pp 167–182

Schicketanz R (1992) Wirkungsweise der Kombinationsdichtung und Anforderungen an die mineralische Oberfläche. Müll und Abfall 24: 287–295

Schicketanz R (1995) Bau von Kombinationsabdichtungen unter einem Zeltschutz. In: Knipschild FW (ed) Tagungsband der 11 Fachtagung "Die sichere Deponie, Wirksamer Grundwasserschutz mit Kunststoffen". Süddeutsches Kunststoffzentrum (SKZ), Würzburg, pp147–167

Sehrbrock U (2003) Kombi-Kapillardichtung, kostengünstige Dichtungsvariante für ein TASi konformes Oberflächenabdichtungssystem. In: Knipschild FW (ed) Tagungsband der 19 Fachtagung "Die sichere Deponie, Sicherung von Deponien und Altlasten mit Kunststoffen". Süddeutsches Kunststoffzentrum (SKZ), Würzburg, pp F1–F10

Soong T-Y and Koerner RM (1999) Behavior of waves in high density polyethylene geomembranes: a laboratory study. Geotextiles and Geomembranes 17: 81–104

von der Hude N et al. (2001) Testfeldergebnisse der konventionellen Kapillarsperre und der Kapillarblockbahn im Oberflächenabdichtungssystem der Deponie Breinermoor. In: Egloffstein TA et al. (eds) Oberflächenabdichtung von Deponien und Altlasten 2001, Neue Erkenntnisse aus Wissenschaft und Praxis. Erich Schmidt Verlag, Berlin, pp 295–316

# 10 Welding of HDPE Geomembranes

## 10.1 Welding Machines, Devices and Weld Seams

HDPE geomembranes are manufactured from a nonpolar thermoplastic polymer which is chemically very stable but forms a melt above 140 °C, and so can be extruded at approximately 200 °C. Therefore in practice, HDPE geomembranes cannot be glued by chemical processes and welding by a thermal process is the jointing technology of choice.

The Plastics Handbook (Saechtling et al. 1998) defines welding of thermoplastic polymers as "connecting … with the use of heat and pressure and with or without the use of aid materials. The surface is heated to a temperature above melting point and joined under pressure in such a way that a connection as uniform as possible develops". The welding process therefore consists of a thermal process (melting of the material, the technical literature often speaks of plasticising the material) and a rheological process (melt flow and mixing of the melted material areas). It is beyond the scope of this book to deal with the "mixing" of polymeric materials, i.e. the behaviour of polymer-polymer interfaces and their dissolution by molecular interdiffusion in detail (Potente 1977; Wool et al. 1989). However, for the following it is important only that the thermal process is triggered and controlled by a heat supply and the rheological process of the melt flow by the application of an external force. The two processes can occur simultaneously or proceed consecutively depending on the way the welding process is performed in terms of process engineering.

In principle, different welding methods (extrusion fillet welding, extrusion flat welding, hot wedge welding and hot air welding) and corresponding types of seams (extrusion fillet seams, extrusion flat seams, hot wedge seams and hot air seams) are possible. Welding of large-area HDPE geomembranes is, however, predominantly carried out using hot wedge welding or extrusion fillet welding. Hot wedge welding usually produces two parallel seams with a channel between them, while extrusion fillet welding provides single extrusion fillet seams. In the following, these two procedures and the two relevant seam types will be dealt with in more detail.

The relevant guidelines of the German Society for Welding Technology and Associated Methods (Deutscher Verband für Schweißen und verwandte Verfahren e. V. (DVS), www.dvs-ev.de), which describes the state of the art of welding geomembranes in geotechnics in Germany, will be considered in detail. The operating conditions for welding machines and devices and the weld seams are described in the guidelines DVS 2225-3:1997 *Joining of Lining Membranes Made of Polymer Materials (Geomembranes) in Geotechnical and Hydraulic Applications – Requirements for Welding Machines and Welding Devices*, and DVS 2225-1:1991 *Joining of Lining Membranes Made of Polymer Materials (Geomembranes) in Geotechnical and Hydraulic Applications – Welding, Adhesive Bonding and Vulcanisation.* There is a special guideline available for the use of HDPE geomembranes for lining landfills and contaminated land: DVS 2225-4:1992 *Welding of Geomembranes from Polyethylene (PE) for Lining Landfills and contaminated Land.*

The EPA/600/2-88/052:1991-05 *Technical Guidance Document: Inspection Techniques for the Fabrication of Geomembrane Field Seams* of the US Environmental Protection Agency (US-EPA) describes in detail welding methods, machines and devices as well as weld seams for HDPE geomembranes (Landreth 1991). However, US American and German guidelines are not completely similar in all details. Comparable to the German DVS, in the United Kingdom, The Welting Institute (TWI, www.twi.co.uk), offers information and advice on welding of plastic geomembranes.

Hot wedge welding is carried out by a hot wedge (Fig. 10.1), which is heated to a temperature of 300–400 °C and is pulled between the overlapping lower and upper geomembranes. A system of guide rollers provides a complete surface contact between geomembrane and the two separate tracks of the dual hot wedge. The surface layers of the geomembranes are melted and the two melt layers are pressed together by a squeeze roller system immediately behind the wedge. Figure 10.2 gives a schematic representation of the procedure. In the hot wedge welding machine[1] the three substantial functional elements are integrated into one basic unit: the heating system, i.e. the heatable wedge with its guide rollers, the pressure system, i.e. the squeeze roller and the pressure device, and the driver system, i.e. the drive rollers and the drive motor. The driver system ensures that the welding machine proceeds at a constant speed along the overlap joint and the two geomembranes are continuously welded. Usually the squeeze roll-

[1] The equipment is called welding machine if it is self-propelled and the pressure is produced by the machine. It is called a welding device or apparatus, if it is moved by hand and the pressure results from the welder's muscular force.

ers have a knurled surface and serve as drive rollers as well. Figure 10.3 shows a commercial hot wedge welding machine.

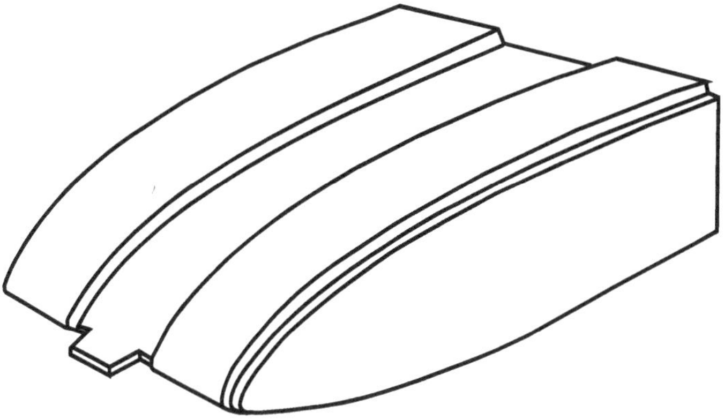

**Fig. 10.1.** Sketch of a hot wedge. The two tracks, which glide along the geomembrane surface and heat up a stripe of material, can be recognised. The geomembrane surfaces are united by the wedge shaped arrangement and pressed together by the squeeze rollers immediately behind the wedge nose. The groove between the tracks results in a test channel between the seams. The nipple is to prevent the welding bead or squeeze-out, i.e. melt, which is squeezed out when the melted areas are pressed together, from clogging the test channel. Hot wedges are manufactured in different forms and lengths as well as with grooved or smooth tracks

The key welding parameters that determine the hot wedge welding process are connected with these three functional elements: hot wedge temperature and hot wedge track length, which determine how the geomembrane surface is melted by the hot wedge, the roller pressure which squeezes the melt layers together (more exactly: the force which is applied over the contact surface of the squeeze rollers) and finally the welding speed or seam velocity, which is adjusted with the driver system so determining the contact time of the geomembranes on the hot wedge and under pressure. When the squeeze roller system simultaneously serves as driver system, it is assumed that the permissible pressure necessary for the welding is always larger than the pressure necessary for advance and transport. This is usually, but not necessarily always, the case. One has to consider the case of a heavy machine, which has to weld up a steep slope.

Hot wedge temperature, roller force and welding speed, as process engineering welding parameters, must allow the independent regulation and adjustment to the nominal value. The hot wedge track length is a given machine constant. The actual roller pressure induced by a certain force is machine dependent, too. The hot wedge welding process itself and the choice of the welding parameters necessary for a high seam quality will be dealt with in detail in Sect. 10.3.

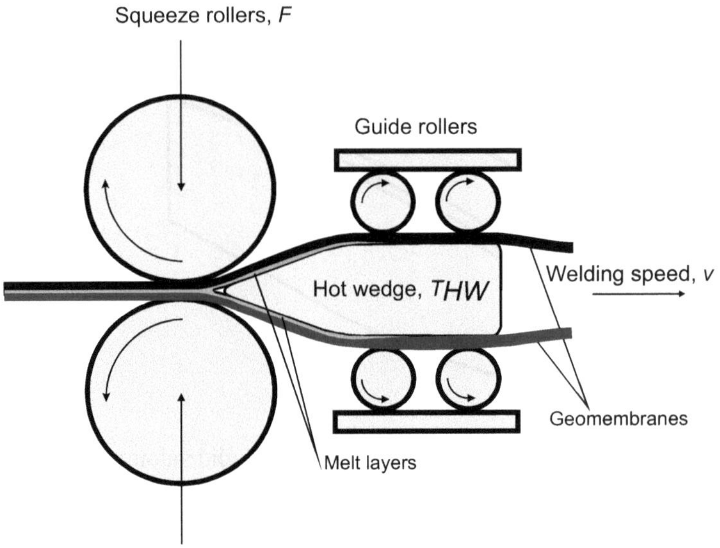


**Fig. 10.2.** Schematic diagram of a hot wedge welding machine with the three substantial functional elements: hot wedge (hot wedge temperature $T_{HW}$) with guide rollers and squeeze rollers (roller force $F$), which serve here simultaneously as driving rollers (Gehde 1999) (see also Fig. 4 in DVS 2225-1). The welding machine travels at a welding speed $v$ (in the picture to the right) on its travel rollers pushed by the squeeze rollers. If the subgrade of the geomembrane is too soft, the travel roller may "bulldoze" into the ground. In this case a base has to be provided, e.g. a "drag strip" (from a piece of geomembrane), which is pulled along step by step

In dual hot wedge welding the wedge (Fig. 10.1) and the squeeze roller system are built in such a way that in the overlap, two seams, called front and rear seam or first and second seam, are produced separated by a small gap, the so-called test channel, which is used for the non-destructive testing of the tightness of the seam (Fig. 10.4). This weld seam is called dual hot wedge seam or hot wedge seam with air channel. The shape of the seam must meet certain geometrical requirements. Figure 10.4 shows the requirements in accordance with DVS 2225-4 for a hot wedge seam with air channel on 2.5 mm thick HDPE geomembranes, used in Germany for lining of landfills and contaminated land. Of special importance for the assessment of seam quality is the so-called thickness reduction due to joining $s_r$ defined as:

$$s_r = (d_t + d_b) - d_S . \tag{10.1}$$

The symbols are explained in Fig. 10.4. The importance of this quantity will be dealt with in more detail in the section after next.

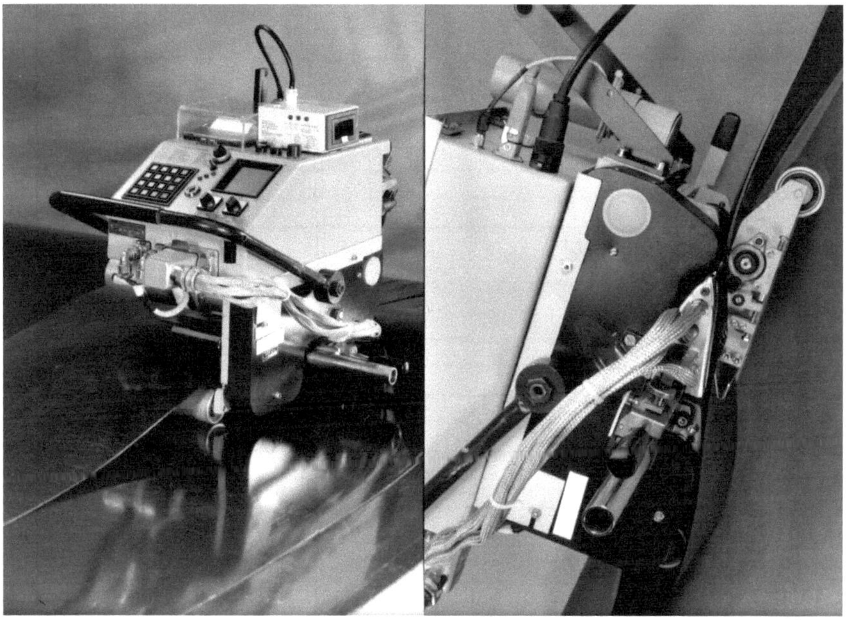

**Fig. 10.3.** Photographs of a hot wedge welding machine. Compact, electronically controlled machines with display, control desk and data logger are offered by several manufacturers. On the right, the travel rollers are seen underneath and on the lower geomembrane as well as the hot wedge over which the two geomembranes are guided

Today robust and easy-to-use hot wedge welding machines are available which fulfil the high technical requirements. Impairment by mechanical influences during operation and transport, and by dirt and moisture cannot be avoided on civil and hydraulic engineering construction sites. The machines must function faultlessly under these conditions. The conclusion is that the electrical and electronic components particularly must be protected against corrosion and contamination and that the functional elements, in particular the hot wedge, must be easily accessible, easy to clean and maintain under these conditions. The basic frame must be light and easy to handle. On the other hand, it must be so stable and torsion-resistant that the forces developed when applying the pressure can be absorbed with little deformation. The mechanics of the guide rollers, the pressure element and the hot wedge must allow a limited mobility, so that the hot wedge has proper access to the surfaces of the geomembranes under operating conditions.

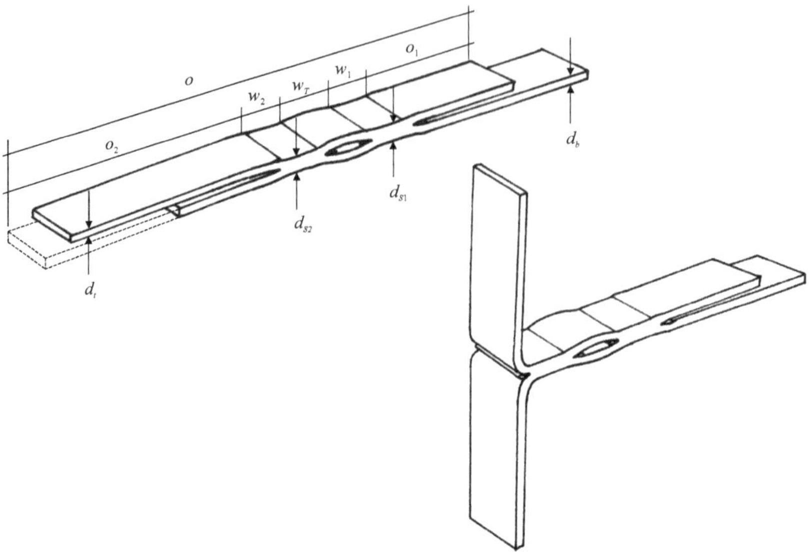


**Fig. 10.4.** Dual hot wedge seam with test channel: schematic view of the test specimen for the tensile test (top left), cut from a sample, which is taken from the weld seam. The upper geomembrane extends to the left, the lower to the right. The rear overlap shortened for the tensile shear test is indicated with dotted lines. The dimensions, characteristic for this seam shape, are indicated: $d_t$ (thickness of the top geomembrane), $d_b$ (thickness of the bottom geomembrane), $d_{S1}$, $d_{S2}$ (thickness of the front and rear seam), $w_{S1}$, $w_{S2}$ (width of the front and rear seam), $w_T$ (width of the test channel), $o_1$ and $o_2$ (overlap in front and in the back). According to DVS 2225-4 the following requirements apply to dual hot wedge seams of HDPE geomembranes for landfill lining: $d_t$ and $d_b \geq 2.5$ mm, 5 mm $\leq o_1 \leq 15$ mm, $o_2 \geq 40$ mm, $w_{S1}$ and $w_{S2} \geq 15$ mm, as well as $w_T \geq 10$ mm. Further, requirements are made on the seam thickness (and thickness reduction), see text. DVS 2226-2 stipulates the following dimensions for the test specimen for tensile shear test: width ≥ 15 mm and at least 5 times geomembrane thickness, gauge length (clamp distance) = 100 mm + seam width. Overall length ≥ gauge length + 50 mm. Bottom right: schematic view of the test specimen as inserted in the short-term or long-term peel test device. Its dimensions in accordance with DVS 2226-3: width ≥ 15 mm and at least 5 times geomembrane thickness, length of the two legs = at least 10 times geomembrane thickness in each case, gauge length = 40 times geomembrane thickness, overall length ≥ clamping length + 50 mm. The squeeze-out, which forms at the seam edges, is indicated

Some of the current requirements of DVS 2225-4 on the operation and control of the functional elements are set out here. The hot wedge temperature must allow a continuous temperature adjustment up to 400 °C and regulated to between ± 5 °C. The temperature measured on the surface of the hot wedge point near the place where the geomembrane leaves the hot wedge can be used as a control parameter. The roller force must allow it to be regulated with a maximum tolerance of ± 100 N. With sudden changes in geomembrane thickness, such as at T-junctions, the roller pressure usually exceeds the permissible tolerance. The guideline requires that this excess does not exceed 30 % of the adjusted value. The pressure system must apply the roller pressure evenly on front and rear seams. Permissible difference in thickness reduction $s_r$ of the two seams may not exceed 0.1 mm. Welding speed must allow adjustment and regulation with an accuracy of ± 5 cm/min.

Modern machines regulate the welding parameters (hot wedge temperature, roller force and welding speed) to the adjusted nominal values, indicate the actual values on a display, announce inadmissible deviations with an audio signal and print an error log. They are additionally equipped with data acquisition, which stores the welding parameter values electronically at regular intervals (e.g. every 2.5 cm seam length). The data can be read out after welding the seam, graphically represented on a laptop and analysed. Ranges of abrupt thickness changes are indicated where the roller force ran out of the permissible parameter window. Such places can be more thoroughly looked at, evaluated and examined if necessary. Apart from the welding parameters, surface temperature of the geomembrane and air humidity should be included in the data acquisition. Using such an automatic welding machine is often referred to as "smart welding".

Using hot wedge welding, geomembranes can be welded by the machine continuously over large seam lengths. Seams can be produced which have been welded with welding parameters clearly specified in terms of process engineering and regulated to the required nominal value. The dual seam provides additional safety. The tightness can be evaluated nondestructively using the test channel over a large seam length. Therefore, as far as technically possible, the geomembrane should be welded using this welding method. There are however difficult-to-access weld areas, connections to buildings and penetration systems, repairs and rework, which cannot be machine-welded. In such cases extrusion fillet welding using an extruding fillet welding device should be selected as the joining technology (Struve 1990).

In extrusion fillet welding of geomembranes a melt strand (extrudate) of the same or very similar HDPE resin material as the geomembrane is applied along the edge of the overlap (Fig. 10.5). The extruded material strand has to be merged with both geomembranes. Therefore the geomembrane surface is heated by hot air directly before applying the strand. Heat transfer between the strand and the unprepared surface of the geomembrane is, however, poor. Therefore the range, which has to be over-welded, must be additionally prepared. Seam preparation requires special attention with the extrusion fillet welding. A thin layer of wax, which can form from low-molecular polyethylene molecules diffusing to the surface of the geomembrane and any oxidation layer must be ground off, the leading edge of the upper geomembrane must be tapered to a 45° bevel and the two geomembranes must be tacked using simple portable hot air welding devices so that an overlap develops with a rigid fixed contact. Small hand held electric rotary grinders and sufficiently fine sandpaper are used for surface preparation. When tapering the edge of the upper geomembrane, the welding zone of the lower geomembrane must not be damaged by deep scoring or grooving. The end result must be an even, finely grained surface, with the grinding marks running predominantly perpendicularly to the seam. All of the material ground off must be wiped or blown away from the welding zone. The worked welding zone must later be completely covered with the extrudate. Careful attention must be paid not only to the surface properties of the worked welding zone, but also its width. Seam preparation in extrusion fillet welding is thus laborious and lengthy and requires patience and manual skill. This is one of the reasons why hot wedge machine welding should be used as far as technically possible.

The extrudate used with extrusion fillet welding should consist of the same resin as that of the geomembrane to be welded, or at least of an HDPE material with very similar flow properties[2]. A seam of sufficient quality to meet the required standards can only be reached under such conditions. In accordance with DVS 2207:1995-08 *Welding of Thermoplastics – Heated Tool Welding of Pipes, Pipeline Components and Sheets Made of PE-HD* the rule applies that HDPE resins with a melt mass-flow rate (MFR 190/5, see Sect. 3.2.4) within the range between 0.3 and 1.7 g/10min may be welded with each other. In the outdated standard DIN 16776-1:1984 *Polyethylene and Ethylene Copolymer Thermoplastics; Classification and Designation* the resins had been classified according to melt flow rate classes defined in Table 3.4. The above-mentioned range covers half of the melt mass-flow rate class T003 and, in addition, the melt mass-flow rate classes T006 and T012 and then a small part of the melt mass-flow rate class T022. This rule was slightly extended for HDPE geomembranes so that geomembranes themselves and geomembrane and ex-

---

[2] Of course, stress crack resistance and oxidation stability must be comparable too.

trudate can be welded if they are in the same melt flow rate class, or if the resins belong to the neighbouring classes T006 and T012 (Müller 2001).

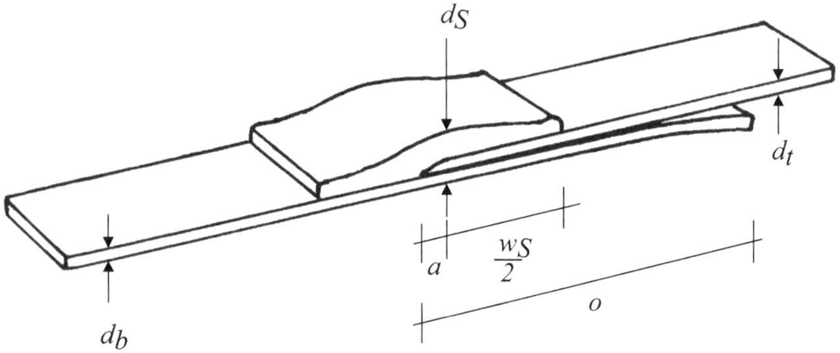


**Fig. 10.5.** Schematic view of a test specimen from an extrusion fillet seam. The top geomembrane extends to the right, the lower to the left. The dimensions characteristic for this weld shape are indicated: $d_t$ (thickness of the top geomembrane), $d_b$ (thickness of the bottom geomembrane), $d_S$ (thickness of the seam), $w_S$ (width of the seam), $a$ (off-setting, disalignment), $o$ (overlap). DVS 2225-4 stipulates the following requirements on extrusion fillet seams of HDPE geomembranes for landfill lining: $d_t$ and $d_b \geq 2.5$ mm, $o \geq 40$ mm, $w_S \geq 30$ mm, $a \leq 5$ mm. Further requirements are made on the factor of seam thickness, see the text

The technical terms of delivery for extrudates are described in the DVS 2211:1979 *Filler Materials for Thermoplastics – Scope, Designation, Requirements and Tests*. If there is a deviation from the rule specified above, which is unnecessary with newly installed geomembranes but may especially occur when new geomembranes are joined to geomembranes already installed in a former construction section, then the guidelines demand that a suitability test is performed on a case-by-case basis. What a suitability test should consist of for an extrusion fillet seam or a dual hot wedge seam is so far, however, not clearly regulated. Most often, the results of short time tensile shear tests and short time peel tests are used to evaluate the seam. The assessment procedure described in Sect. 10.3, however, offers the possibility of performing a systematic quantitative evaluation of the seam quality for dual hot wedge seams and should therefore be used as suitability test.

The hand-held portable extrusion welding device or apparatus also consist of three functional elements: preheat system for the hot air, which is used to preheat the welding zone; the plasticising system with temperature control, a small extruder, in which the extrudate pellets or an extrudate rod are melted, homogenised and brought to a controlled mass temperature and from where the melt strand is discharged at a certain discharge speed through the Teflon welding die which is the third functional element. Figure 10.6 shows a schematic of the device. The extrudate is smoothed, formed and transported to the weld area over the welding die. The die has two lateral support surfaces with which the device rests on the geomembrane.

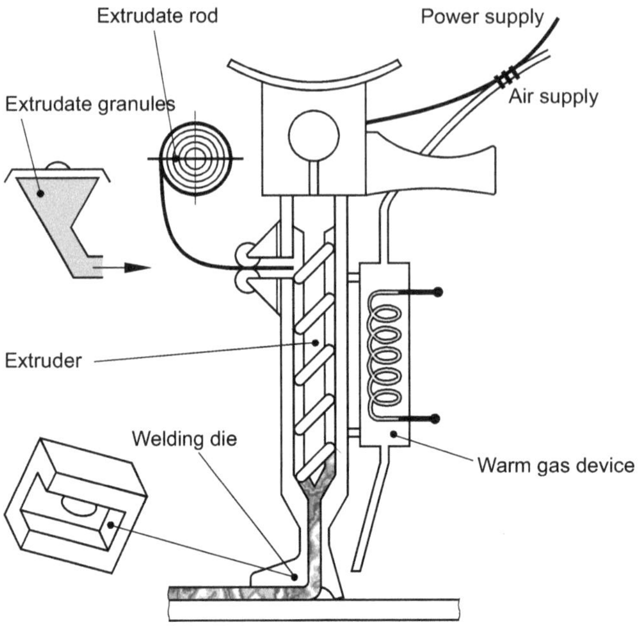


**Fig. 10.6.** Schematic view of extrusion welding equipment (the figure is based on Fig. 2 of DVS 2209-1). The resin of the extrudate is filled as granules or as a rod in a small extruder. The melt flows over a Teflon welding die. The Teflon welding die consists of a horseshoe-shaped skid (schematic diagram, view from below), which slides along the geomembrane. The extrudate is discharged between the legs and is advanced onto the geomembrane. Immediately before the front end of the Teflon welding die the geomembrane is warmed up by a hot air unit, which is connected to the mainframe

Extrusion welding produces a so-called extrusion fillet seam. Figure 10.5 shows the seam shape. Certain geometrical conditions are required of this seam shape as well. Figure 10.5 shows the dimensions of the extrusion fillet seam in accordance with DVS 2225-4 for minimum 2.5 mm thick HDPE geomembranes used in Germany for lining landfills and contaminated land. In addition to the actual dimensions, the so-called factor of seam thickness $f_{SA}$ is used to assess seam quality:

$$f_{SA} = \frac{d_S}{d_t + d_b}. \quad (10.2)$$

The symbols are defined in Fig. 10.5. Subject to repeated discussion is the issue of whether or not beads may be squeezed out on the right and left under the support surfaces of the welding die. In this respect one should pay attention to two aspects: first, large squeeze-out are an indication that proper craftsmanship has not been applied, i.e. the lateral supports of the extrusion die were not directly seated on the geomembrane, the extrudate temperature was improper for adequate flow or the seaming velocity was too slow. Second, the squeeze-out may not under any circumstances have damaging consequences, such as obstructing vacuum testing or encouraging the whole seam to peel off when the squeeze-out is pulled up.

Not all welding parameters are defined exactly or regulated in terms of process engineering within extrusion welding procedure. In accordance with the guidelines the mass temperature of the extrudate must be regulated and controlled to ± 10 °C and the hot air temperature continuously adjusted to 350 °C and regulated up to ± 20 °C. However, the temperature of the hot air preheated geomembrane is not yet explicitly determined by adjusting these parameters. Welding pressure and welding speed depend on the capacity of the extruder, geometry of the Teflon welding die and the welder's skill. None of the process parameters are quantitatively recorded in current welding engineering practice. The quality of the extrusion fillet seam chiefly depends on the welder's experience, knowledge, manual skills and physical ability.

In those critical places of a geomembrane liner, such as penetration points and connections to structures in the deepest point of the liner system, extrusion fillet seams are often necessary. Therefore, the importance of a technically qualified and experienced installation contractor to produce seals of high quality with geomembranes can hardly be overestimated. On the other hand, thermoplastic geomembranes are the only sealing materials that enable the manufacture of a really watertight and homogenous seam without any bonding agent especially in such critical places.

## 10.2 Testing Seams

The dual hot wedge seams and extrusion fillet seams must be tested on the construction site. The construction-site tests are supplemented to a certain extent with tests under laboratory conditions. In the laboratory, long-term tests on weld seams can be performed. Long-term tests are in principle aimed at investigating the influence of the welding process parameters on seam quality, long-term behaviour of the seams and welding properties of various resins.

Construction site tests are described in the guideline DVS 2225-2:1992 *Joining of Lining Membranes Made of Polymer Material in Geotechnical and Hydraulic Engineering, Site testing* and in the DVS 2225-4. The standard ASTM D4437-84 *Practice for Determining the Integrity of Field Seams Used in Joining Flexible Polymeric Sheet Geomembranes* also contains a brief overview. Pressurised-air testing and vacuum testing are also covered by ASTM standards: ASTM D5820-95 *Practice for Pressurized Air Channel Evaluation of Dual Seamed Geomembranes* and ASTM D5641-94 *Practice for Geomembrane Seam Evaluation by Vacuum Chamber*. The tests are to help adjust the process parameters by trial and error. (A systematic adjustment of the process parameters with regard to the optimum seam quality can be performed using the process model described below.) The tests mainly enable the determination of the quality of the seams with respect to specifications developed from field experience and non-destructive proof of its water tightness. The following seam properties are investigated: external appearance, the dimensions, strength and, of course, water tightness.

The external appearance is controlled by visual inspection. The naked eye of the examiner assesses whether the seam exhibits perfect craftsmanship. The assistance of a blunt instrument (e.g. a screwdriver) is needed, which is moved along the seam edge and some pressure applied. Even this simple, so called mechanical point stressing test may reveal individual defects or unbonded areas, in particular in the case of extrusion fillet seams. Usually, the instrument only leads the eye of the examiner. Only qualitative instructions can be given concerning the requirements of the visual inspection. The examiner must observe shape and appearance, central position and uniform boundary regions of the seam. He must evaluate the squeeze-out on the edge of dual hot wedge seams and extrusion fillet seams and the smooth and streak-free texture of the surface of the fillet seams and look for inadmissible notches and scoring from the preparation of the joining zones with extrusion fillet seams and of T-junctions and structured geomembranes with the hot wedge welding. Obviously, this test can only be performed by an experienced specialist with extensive training in welding technology, no matter if he acts for the in-house monitoring by the installation contractor or for third-party inspection by an independent body. When performing such an important, but only qualitatively described test, which is so crucial for extrusion fillet seams whose quality depends to a large extent on the welder's manual skills, all involved should agree on the quality criteria for perfect welding zone preparation, for appearance and surface texture prior to the commencement of welding by inspecting a longer stripe of the test seams.

The dimensions of the seams are determined on specimens from test samples, which are usually taken at the beginning and end of the seam. The relevant dimensions are shown in Figs. 10.4 and 10.5. The width of the rear overlap should be at least 40 mm to allow easy handling in tensile tests, such as peel tests and tensile shear tests. For HDPE geomembranes used in landfills and for containment of contaminated land, there are minimum requirements on other dimensions in the DVS 2225-4 guideline. In addition to the actual dimensions, the previously mentioned reduction in thickness $s_r$ (Eq. 10.1) for dual hot wedge seams and the factor of seam thickness $f_{SA}$ (Eq. 10.2) for extrusion fillet seam are important parameters. There are defaults given in the guidelines. It holds for the thickness reduction:

$$0.2 \leq s_r \leq 0.8\,. \tag{10.3}$$

The factor of seam thickness for extrusion fillet seam must meet the following requirement:

$$1.25 \leq f_{SA} \leq 1.75\,. \tag{10.4}$$

In the case of extrusion fillet seams, off-setting of the centre of the seam from the edge of the upper geomembrane may not exceed 5 mm.

The dimensions and, to certain extent, homogeneity of the seam can be tested non-destructively using ultrasonic measurements. A small ultrasonic measuring head is placed on a clean, even part of the seam, impulses at an ultrasonic frequency of 4 to 6 MHz are sent through it and the delay time of the echoes from the back of the seam and the geomembrane, or a defect in the seam, is measured (pulse-echo testing). The thickness can be determined from the delay time of the geomembrane back echo. Coupling of the probe is made over water or special pastes. The test should not be started within one hour of welding the seam. On the construction site small hand devices are used which, after adjustment and calibration on planar reference plates, directly indicate the thickness. Such devices cannot only be used for random sampling but also for systematic measurement of the thickness reduction along the seam. Shorter echo delay times indicate defects in the seam. Enclosed dirt, pores and air gaps in the seam also generate ultrasonic echoes. Ultrasonic measurements can therefore indicate seam inhomogeneity. Welded areas, which are only superficially attached or not sufficiently melted, cannot be recognised.

The strength of the seam is tested in the peel test. The test is described in the guideline DVS 2226:1997-07 *Testing of Fusion on PE-Liner – Peel Testing* and in the Standard ASTM D6392-99 *Standard Test Method for Determining the Integrity of Nonreinforced Geomembrane Seams Produced Using Thermo-Fusion Methods.* It is now state of the art that peel tests can be carried out on the construction site with a tensile test device, which allows a peel test of the required constant test speed. Times are long forgotten when converted car jacks with a crank handle were used. However, the test has to be performed in the laboratory in strict accordance with the requirements of the guidelines or standards by the third party inspector as well. A minimum of 15 mm but usually a 20 mm broad strip is cut off transversely to the seam as a test specimen. Figure 10.4 shows the test specimen. The overlapping ends are unfolded and then clamped in such a way in a tensile test device that the joining plane or welded area, i.e. the imaginary plane in the centre of the seam, lies in the middle between the clamps. The specimen is pulled at a test speed of 50 mm/min and the deformation and failure behaviour is observed. The test result is the description of the deformation and failure behaviour, which is, however, only qualitatively assessed, and the quantitative evaluation of peel strength (maximum force at break) and peel separation (ratio of the area of separation to the original bonded area, both areas estimated by "visual approximation").

Standards for peel testing are described in the above-mentioned guidelines DVS 2225-1 to 2225-4 and in the GRI Standard GM19 *Seam Strength and Related Properties of Thermally Bonded Polyolefin Geomembranes.* A weld seam between HDPE geomembranes has good seam strength, if it does not peel off and the basic material in the test strip strains and breaks outside the seam. With the dual hot wedge seam, straining and peel-off in the boundary region of the seam (i.e. peel separation) is still permissible by some regulations, if the residual width of the seam is larger than the minimum width required for the respective application, 15 mm according to DVS 2225-4 for instance, or if it is less than 10 % of the intended seam width. However, as was stated by I. Peggs: zero peel separation is regularly achieved by capable operators, therefore it should be recommended (Peggs 2005). When extrusion fillet seams are tested they can still be passed if the extrudate strains and tears in certain cases and the maximum tensile force reached is "within the order of magnitude" (DVS 2225-4) of the comparable maximum tensile force of the tensile shear test (see below) on the extrusion fillet seams. A seam (dual hot wedge seam or extrusion fillet seam) does not have undoubtedly sufficient strength when it either peels off or the test specimen breaks in the boundary region or outside the seam in a brittle way without any clear elongation.

The peel test in accordance with DVS 2226-3 or other standards generally serves to test the jointing connections in polymeric material geomembranes. Seams of soft PVC geomembranes, ECB geomembranes and elastomer geomembranes, such as EPDM geomembranes[3] are also tested using this test. Usually, these seams peel off. The medium force, which must be applied when peeling, related to the width of the test strip is called peel strength. The unit of this quantity is therefore N/mm. The seam strength exhibited when peeling seams is then assessed by the size of the peel strength. Although a perfect seam between HDPE geomembranes may not peel off, the DVS guideline defines the maximum tensile force related to the specimen width that can be measured, even when elongating and tearing of the test strip outside the seam, as peel strength. The DVS 2226-1:2000 *Testing of Fused Joints on Liners Made of Polymer Materials - Testing Procedure, Requirements* specifies a thickness-dependent minimum value for this "peel strength" for HDPE geomembranes: $15 \cdot d$ N/mm ($d$: value of thickness of the geomembrane measured in mm). For a 2.5 mm geomembrane one obtains 37.5 N/mm. Considerable lower values are specified in the GM19 standard for smooth and structured HDPE geomembranes, e.g. 26 N/mm for hot wedge seams of 2.5 mm geomembranes.

The use of the same term for two different fracture modes – namely peeling off a seam and elongating and tearing the test strip outside the seam – can easily lead to misunderstandings. The so-called "peel strength" for HDPE geomembranes should therefore be used very carefully. Since a clear elongation is required before the test strip tears, the maximum tensile force related to the initial cross section of the strip should be at least in the range of the yield stress, which formally corresponds to the required minimum value of the peel strength. On the other hand, it is doubtful whether one may compare the maximum tensile forces on specimens deformed by shear flow near or at the edge of the seam from different HDPE materials and with different seam geometry. The maximum tensile force in the tensile test with HDPE geomembranes is highly sensitive to the test specimen characteristics.

---

[3] Soft PVC: polyvinyl chloride with low-molecular organic mixtures as softeners; ECB: ethylene copolymer bitumen; EPDM: terpolymer from polyethylene, polypropylene and dien-monomers. This terpolymer is a rubber, which can be interlaced with sulphur to an elastomer by the double bonds, brought in over the dien-monomers (vulcanisation).

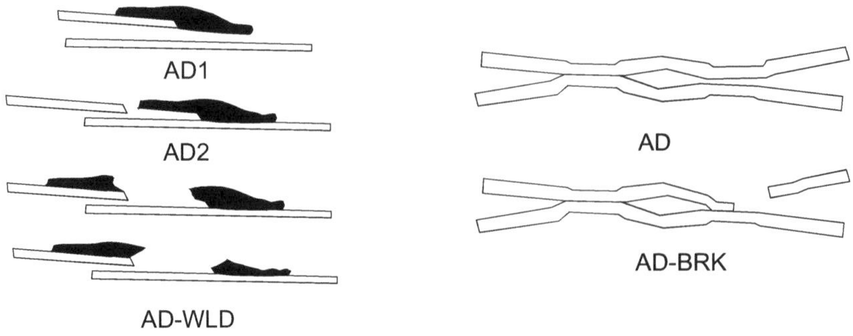


**Fig. 10.7.** Examples for the locus-of-break code of ASTM D6392 and GM19. AD is described as failure in adhesion, AD-BRK as break in a seam of either the top or bottom sheet after some adhesion failure, AD-WLD as a break in the fillet either in centre or off centre

A quantitative evaluation of the peel test, i.e. the determination of the peel strength or peel separation, is usually only possible in the laboratory. Here a tensile shear test, another short-time tensile test, can be carried out for the quantitative assessment of the seam, too. The test method is described in the DVS 2226-2:1997 *Testing of Fused Joints on Liners Made of Polymer Materials – Lap Shear Test* and in ASTM D6392. The tensile shear test is similar to the peel test with respect to the test parameters. However in this test, the specimen is clamped at the strip end of the lower geomembrane on one side of the seam and at the strip end of the upper geomembrane on the other side of the seam. Figure 10.4 shows the test specimen. The seam lies in the middle and across the tensile direction. If the test strip elongates and tears in this test outside the seam, then good seam strength is assigned to the seam. The ratio of the maximum tensile force in the tensile shear test on the test strip with a seam to the maximum tensile force determined on a test strip of the geomembrane without a seam, is called the short-time welding factor or short-time seam strength factor. Experience-supported typical values and minimum values can be derived for this welding factor. However, similar reservations apply, which were made in connection with the peel strength above. DVS 2226-1 requires a factor $\geq 0.9$. GM19 specifies thickness dependent values for the tensile strength measured in N/25mm. To make the important qualitative criteria of clear and unambiguous yielding of the specimen material outside of the seam more quantitative, the shear elongation is defined in ASTM D6392 as percent ratio of extension at test end (after breaking) to original gauge length. Usually a value of greater than 100 % is required. However, allowable shear elongation for structured geomembranes should be determined on a case-by-case basis. Since some manufacturing processes have strong influence on the elongation at break point of the structured geomembrane in the tensile test.

It was repeatedly mentioned that the qualitative assessment of the deformation behaviour and rupture mode of the specimens in the peel test as well as in the shear test are of paramount importance. It is getting tedious to say that these tests can only be performed and evaluated by trained and experienced specialists. To offer some help in communicating test result GM19 and ASTM D6392 offer a classification scheme for rupture modes or "a locus-of-break code" (Fig. 10.7). Using this code it may be said that adhesion failures (AD, AD1 and AD2) or break in the seam (AD-BRK) are clearly unacceptable, break through the fillet (AD-WLD) is acceptable only when certain minimum specification values for strength and elongation at break are met.

Finally, on the construction site, the tightness of the seams must be tested. Dual hot wedge seams with a test channel are tested using a pressurised air test, the extrusion fillet seams with a vacuum box or high voltage test. The pressurised air test is described in the guideline DVS 2225-4 and with slight differences in parameters and procedures in the standard ASTM D5820-95 *Standard Practice for Pressurized Air Channel Evaluation of Dual Seamed Geomembranes*. Dual hot wedge seams with a test channel can thus be non-destructively tested over the entire seam length, which can be as long as 300 metres. A compressed air test can begin about 1 hour after welding at the earliest. At one end of the seam an HDPE quick coupling hose connector nipple is welded to the test channel, to which a compressor with a pressure gauge and a pressure recorder is attached. The test channel is blown through and welded shut or clamped hermetically at the other end of the seam. Compressed air is then applied. First, pressure is adjusted above the actual test pressure for approx. 1 minute. The test channel must open and bulge out first. After this pre-loading the proper test pressure is adjusted. The selected test pressure is, to certain extent, based on the geomembrane temperature and test channel width. It is usually about 3–5 bar (300–500 kPa). After the test pressure has been adjusted, the actual test begins. The pressure is recorded over a test period of 10 minutes continuously with the pressure recorder. The pressure gauge must correspond to the test device class 1.0 in accordance with EN 837[4]. The meas-

[4] EN 837-1:1996 *Pressure Gauges – Part 1: Bourdon Tube Pressure Gauges - Dimensions, Metrology, Requirements and Testing*
EN 837-1:1996 *Pressure Gauges – Part 2: Selection and Installation Recommendations for Pressure Gauges*
EN 837-1:1996 *Pressure Gauges – Part 3: Diaphragm and Capsule Pressure Gauges – Dimensions, Metrology, Requirements and Testing*

urement range of the pressure gauge and recorder should be no larger than double the test pressure and the scale not courser than 0.1 bar (10 kPa). During the test time the pressure must not drop more than 10 % of the initial value. At the end of the test time the test channel is opened at the clamped or welded end. Air must suddenly escape and the pressure gauge reading drop rapidly. With this inspection results the tested seam section qualifies as tight. If the inspection result deviates from these requirements, then error tracing begins. Sometimes the seam section must be further tested piece by piece, until the error sources are identified.

Testing tightness of the extrusion fillet seams is more laborious. They must be tested piece by piece by applying a vacuum box or chamber using the vacuum box method. The vacuum box test is described in the guideline DVS 2225-4 and with slight differences in parameters and procedures in the standard ASTM D5641-94 *Standard Practice for Geomembrane Seam Evaluation by Vacuum Chamber*. The testing device consists typically of a 10 cm long and about 10–15 cm wide transparent test box, whose edge is provided with a flexible sealing ring, so that the box can be pressed hermetically on the weld seam section. A small pump and a pressure gauge are attached to the box. The measurement range of the pressure gauge and recorder should be no larger than double the test pressure and the scale not courser than 0.1 bar (10 kPa). For transition zones between slope and base, edges and corners there are specially formed test boxes available.

This test should also start 1 hour after welding at the earliest. The seam section, which is to be tested, is covered or sprayed with a bubble-forming liquid. The vacuum box is placed on the geomembrane and then a vacuum applied. During the test a vacuum of at least 0.5 bar (50 kPa) must be kept constant for at least 10 seconds. In a leaky place the liquid will make bubbles. If the vacuum can be built up "rapidly", the pressure maintained for the duration of the test and no bubbles are observed, then the tested section is considered as tight. The test box is ventilated. Places, in which bubbles had formed, are marked and repaired later. The test box is then placed on the next section brushed-in or sprayed with test liquid. The test sections must overlap by at least 10 cm.

Extrusion fillet seams can be non-destructively tested using another testing method, namely a high voltage electrical test. This procedure is used instead of the vacuum test above all on places, which are difficult to access with the vacuum box. The description of the procedure again follows the guideline DVS 2225-1. Further details may be obtained from the standard ASTM D6365-99 *Standard practice for the Nondestructive Testing of Geomembrane Seams using the Spark Test*. Use is made of the fact that a gas discharge occurs between two electrodes when high voltage is applied. The spark discharge is visible and audible: it sparks and cracks. The test equipment consists of a high voltage source to which a brush electrode or a ball electrode is attached. At the back of the extrusion fillet seam, along the overlapping edge of the upper geomembrane, an electrode, such as a wire of good electrical conductivity, is placed and over-welded. The welded-in electrode is earthed. The testing voltage applied to the brush electrode must not exceed the breakdown voltage of the HDPE geomembrane. The height of the voltage on the other hand determines the possible length of the discharge distance. The permissible test voltage is 60 kV for HDPE geomembranes with a thickness of 2.5 mm. Thus a sparking distance of approx. 20 mm can be reached. A central electrode lying in the seam thus enables 30 to 40 mm wide extrusion fillet seams to be tested. The brush travels at a speed of approx. 10 m/min along the seam edge. However, only such defects can be detected where a sufficiently short discharge distance develops over a sufficiently large air duct running almost perpendicularly to the seam. Such defects trigger a spark discharge. The relevant places are marked and repaired. What "sufficiently" really means is not clear and this test fails to recognise closed but poorly attached seam zones anyway. The effectiveness and reliability of this test method is therefore disputed.

So far construction site tests have been discussed which are supplemented by tests in the laboratory (peel and tensile shear test), for example by an independent inspection body that carries out the third-party control. In addition, there are laboratory tests on weld seams where the aim is to clarify fundamental questions, i.e. the dependence of quality of the weld seam on the welding parameters and the long-term behaviour. These are the long-term tensile creep test (more exactly: the long-term tensile shear creep test), the long-term slow tensile test, the long-term relaxation test and the long-term peel creep test. The tests are characterised as "long-term" since they are performed at elevated temperature to accelerate processes that might lead to failure. In addition a water bath or a water-surfactant solution is used to accelerate brittle failure.

The long-term tensile creep test was discussed in great detail in Sect. 3.2.16, for testing stress crack resistance of structured geomembranes. The same test can be performed on weld seams. The guideline DVS 2226-4:2000 *Testing of Joints on Liners Made of Polymer Materials – Tensile Creep Test on PE* describes the test method. The specimens are prepared in accordance with the DVS 2226-2 guideline (see Fig. 10.4) for the short-term tensile shear test and clamped in the long-term tensile creep test apparatus. The long-term peel creep test has not been standardised yet but is carried out analogous to as the long-term tensile creep test. The test specimen is prepared according to the instructions of the peel test guideline DVS 2226-3 (see Fig. 10.4) and clamped in the long-term tensile creep test apparatus. Usually, not the applied stress (unit: $N/mm^2$) is used as a test parameter but a line test force defined as applied tensile force related to the width of the test specimen (unit: N/mm). The long-term peel creep test is dealt with in greater detail in the next section.

The question of the long-term behaviour of geomembrane seams was originally tackled by using concepts from the well-established field of polyethylene pipe seam testing and evaluation. The so called long-term welding factor or long-term seam strength factor was introduced by G. Diedrich and E. Gaube for the description of the long-term behaviour and quality of weld seams in polyethylene pipes and plates compared to the base material (Diedrich and Gaube 1970, 1973). A short discussion of the derivation of this factor might be helpful in understanding the approach to characterise the long-term behaviour of geomembrane seams.

For testing the long-term behaviour of pipes the pipe pressure test is used (Sects. 5.4 and 3.2.13). This test can also be performed on a pipe section welded together from two pipe parts. To evaluate the long-term behaviour, the brittle branch of the hoop stress versus time-to-failure curve is determined i.e. the range of the curve where the fracture mode can be clearly characterised as brittle failure due to stress crack formation. Results from G. Diedrich and E. Gaube as well as other working groups showed that the location of the branch and therefore the medium service lives did not differ in unwelded and correctly butt welded pipes in the brittle range at all test temperatures. Failure usually arose in the base material of the pipe. The result shows that the butt-welded seam is not a weak point, which would dramatically impair the maximum strength in the pipe.

From this, however, no conclusion can be made on the quality of the seam and on its long-term strength in comparison to the base material. In the pipe pressure test the longitudinal component of stress, i.e. the stress perpendicular to the plane of the seam of the butt-welded seam, is only half of the hoop stress. The weld seam is thus exposed to a much smaller tensile stress than the base material in planes outside the seam.

Therefore, G. Diedrich and E. Gaube cut out tensile test bars (parallel test bars or strip specimens and shoulder test bars) from the pipe walls, once without a weld seam and once with a central weld seam and tested them in long-term tensile tests. The plane of the seam is now essentially perpendicular to the tensile stress and the weld seam area and the base material is similarly loaded. By measuring the times-to-failure at different tensile stresses, a brittle branch of the tensile stress versus time-to-failure curve was found both for the base material and for the seam. Depending on the quality of the seam and type of welding method (e.g. hot element butt welding and extrusion welding) different positions in the hoop stress versus time-to-failure diagram were found for the seam's brittle branch in comparison to the branch of the base material. These differences are described by the long-term seam strength factor. The long-term seam strength factor is defined as the ratio of the two stresses, one on the brittle branch of the tensile stress versus time-to-failure curve of the seam sample, the other one on the brittle branch of the base material, which leads to the same time-to-failure. Or expressed in a different way: the long-term seam strength factor indicates, how the tensile stress must be reduced so that the same medium time-to-failure can be reached for the pipe with a weld seam as for a pipe without seam. In an ideal case the long-term seam strength factor should be unity: weld seam and base material behave in the same way. However, the branches of base material and weld seam in the pipe pressure test diagram rarely run in such a way that the same long-term seam strength factor can be calculated for each stress. Usually the long-term seam strength factor depends rather strongly on the testing stress. The test method and the determination of the long-term seam strength factor were standardised in the guideline DVS 2203-4:1997 *Testing Welded Joints on Thermoplastic Plates and Pipes – Long-term Tensile Test.* A long-term seam strength factor can; however; also be defined as the ratio of the time-to-failure of a seam to that of the base material at a pre-set test stress. One then obtains a factor with which the service life of the pipes must be reduced to take account of pipe seams.

In the late eighties it was suggested that this procedure could be used for geomembrane seams as well. The long-term tensile creep test (or more exactly the long-term tensile shear creep test) and the determination of the long-term seam strength factor derived from it was used by J. Hessel and P. John for the characterisation of the long-term behaviour of weld seams (dual hot wedge seam and extrusion fillet seam) with geomembranes (Hessel and John 1987). The test and procedure for the determination of the "long-term seam strength factor" of geomembrane seams was described in the DVS 2226-4 specified above. However; there are three unresolved methodical problems in the application to geomembranes for this procedure.

1. The time-to-failure of a specimen from an unwelded geomembrane and thus the reference value for the time-to-failure of the weld seam depends highly on the type of specimen preparation. The specimens from the base material always fail by stress crack formation, which begins at machining defects located at the edge of the specimen. Punched tensile test bars fail rapidly; specimens sawed with a high-speed tungsten carbide saw or milled with high-speed milling endure longer. The time-to-failure increases if the edges are smoothed with fine sandpaper. Only if completely smooth cut edges are produced by

cutting by a microtome blade or very fine plane will a range of times-to-failure be reached where, in individual cases, stress cracks initiated from the geomembrane surface result in a failure. The time-to-failure obtained in this case is very long for HDPE geomembranes of acceptable stress crack resistance (> 10,000 h). It follows from this observation that the long-term seam strength factor is larger or smaller depending on the quality of the specimen preparation, i.e. it depends on the preparation technique. In accordance with the DVS 2226-4 guideline the tensile test bars must not be punched. Rather they may "... be manufactured by sawing, milling or cutting (e.g. using a water jet). In order to attain a notch-free cut section, these have to be repeated if necessary by smoothing in a longitudinal direction." But even such a detailed description of specimen preparation fails to provide a clear definition of a reference value for the base material.
To avoid this ambiguity in the DVS 2226-1a minimum time-to-failure of the basic material for the determination of the long-term seam strength factor was required: for HDPE geomembranes a time-to failure of 500 h at a test stress of 4 N/mm² in an 80 °C hot surfactant solution with Arkopal N100® surfactant. The seam strength factor for dual hot wedge seams (extrusion fillet seams) should then be at least 0.5 (0.4). This approach seems to be rather "long-winded". Obviously, one can directly specify a minimum value for the time-to-failure of the weld seam instead of defining the minimum time-to-failure for the basic material and requiring a minimum value for the seam strength factor. Under the test conditions mentioned that would simply be 250 or 200 hours respectively.

2. The two branches in the tensile stress versus time-to-failure curve, which describe the range of brittle failure of a weld seam and the base material, do not usually run parallel, but diverge. Therefore the long-term seam strength factor depends not only on specimen preparation but also on the selected test tensile stress. With decreasing test stress the seam strength factor would become lower and lower.

3. In the long-term tensile test on tensile test bars with dual hot wedge seams and extrusion fillet seams the plane of the seam is not perpendicular to the tensile direction, as in butt welded pipe seams, but parallel. The force flow in the loaded specimen is directed over the weld seam from one geomembrane into the other in an offset direction, therefore force flow lines get more dense in the boundary region of the weld seam and a stress concentration develops. In the boundary region there are also notch effects induced by the transition from the welded material area to the base material. A crack that develops there

grows perpendicularly to the force flow lines toward the strongest downward gradient in the line density and therefore leads into the base material. Indeed: the observed stress cracks actually start in the boundary region of the weld seam and then run perpendicular to the seam plane through the base material. The stress crack in dual hot wedge seams therefore practically never runs within the seam (Gehde 1992; Viertel 1997). Therefore the time-to-failure is essentially a function of the specific geometry of the weld seams in the geomembranes.

For these three reasons it is not possible to consider the time-to-failure in the long-term tensile creep shear test as a criterion for the quality of the weld seam and the welding method in geomembranes at all.

However, from the results of long-term tensile test on dual hot wedge and extrusion fillet seams in geomembranes one clearly sees that these types of seam in particular should not be put under long-term tensile stress in the field[5]: the time-to-failure is always substantially smaller than in the non-welded geomembrane itself when put under permanent tensile stress. Therefore the requirement for the HDPE geomembranes be installed in such a way that no long-term effective tensile stresses develop, applies above all to the weld areas. The DVS 2225-4 guideline describes appropriate precautionary measures to avoid permanent tensile stress in installed geomembranes (see Chap. 9). For example, on slopes, the seams must run parallel to the line of maximum slope to a large extent. Patching geomembrane panels using transverse joints on slopes is not permitted. The connecting seam between geomembranes on the slope and the base should be located in the base at a distance of at least 1.5 m from the slope toe, and so on.

The test condition for seams in the long-term relaxation test is closer to actual mechanical effects occurring under field condition of an installed geomembrane. The relaxation test was described in Sect. 3.2.10. This test can also be performed as a long-term test. The test apparatus is modified in such a way that the specimen can be kept in a test liquid at an elevated test temperature while a specified constant strain is applied and the consecutive stress relaxation is measured. There are still no standards or guidelines available, which describe the long-term relaxation test on seams. The specimens are prepared as for the long-term tensile creep test and the test

---

[5] This frequently used formulation is somewhat lax. It is meant that the geomembranes may not be intended to be used to transfer and absorb permanent loads in a structure. Naturally, tensile stresses that are strictly zero can not be required. A tensile stress below 2 $N/mm^2$ is considered harmless even for seams (Heitz and Henkhaus 1992). See also Sect. 3.2.10.

procedure is analogous to the one in the long-term tensile creep test: instead of the rapid application of a constant tensile stress, a strain is swiftly applied and held constant. Not only is the relaxation curve of interest, but also the time-to-failure and the failure modes. The fracture modes resemble those seen in the long-term tensile test. The times-to-failure measured on seams in this test are very long even for a strain close to the yield strength (> 1000 h). Therefore measuring critical limiting strain of seams using this test would be very time-consuming. So far only a few test results have been published (Knipschild 1992).

E. Heitz and R. Henkhaus carried out slow tensile tests on dual hot wedge seams, also called constant strain rate tests by the authors (Heitz and Henkhaus 1992). In this test the tensile test specimen is subjected to a slow constant strain rate (0.2 % a day up to 2.5 % an hour) instead of a constant load in the tensile creep test or a constant strain in the relaxation test. From the measurement of times-to-failure at different strain rates and test temperatures, an attempt was made to determine a critical limiting strain for the seams (defined here as the permissible strain for a theoretical service life of 100 years at 20 °C) using the time-temperature superposition law (see Sect. 4.2). Extrapolation uncertainties are naturally very large. A conservative estimation was tried and limiting strains of 1.7 to 2.7 % were determined for HDPE resins. The tests were performed in a 2 % surfactant solution. The critical limiting strain in this medium is substantially smaller than in water or air (see Sect. 5.3.4). On the other hand the load was only uniaxial. For a planar state of stress the limiting strain should be lower compared to the limiting strain in the state of uniaxial applied force. All in all, the limiting strain of 3 % used for HDPE geomembranes (Sect. 5.3.4) seems to be comparable to the permissible limiting strain for properly manufactured seams as well.

In addition to these mostly “customary” laboratory and construction site tests, further tests have been and will be conceived on geomembrane seams - usually derived from test procedures on the geomembrane itself. However, all these tests only provide necessary, but insufficient conditions for a “perfect” seam. The burst test described in Sect. 3.2.9, used to test the multiaxial deformation behaviour, can be performed on a geomembrane disk with a centrically placed seam. For this purpose, sufficiently thick elastomer rings must be used in order to provide a watertight clamping. With properly manufactured seams the arch-height versus pressure diagram measured in the burst test does not show any difference to the diagram of the geomembrane without a seam (Hutten 1991). The seam does not therefore impair the multiaxial deformation behaviour. From a methodical point of view, long-term relaxation tests on seams, in which a constant planar strain is applied using a test device analogous to the burst test apparatus would probably be the most informative test for the long-term behaviour of seams. These tests are, however, expansive, difficult to design and very lengthy. So far, to the author's knowledge, such tests have not been tackled.

Burst tests were performed on T-junction dual hot wedge seams and also destructive pressure tests with water in the test channel on dual hot wedge seams, analogous to the pressurised air test. Marked differences can be observed in the deformation behaviour of T-junctions. The arch elongation of a good seam T-junction is above 6 %, but it does not in principle reach the values of the geomembrane (at least 15 %) because of the reinforcing effect of the T-junction of dual seams and the change in thickness at the seam edge (Müller and Preuschmann 1992). Instead of using pressurised air, the test channel can be pressurised with water, in similar fashion to the hydrostatic burst test with water. The pressure is increased in steps of 2 bars, and the pressure is maintained at each step for 2 minutes. The increase in pressure is carried on until the seam breaks. Properly manufactured seams reach pressures of 20–40 bar (Müller and Preuschmann 1992). The different failure modes are noteworthy. There are seams where the test channel bulges, elongates and exhibits a ductile failure similar to the ductile failure of a pipe in the pipe pressure test. Other seams peel off in localised limited areas. There are, however, seams where the material shows sudden and sharp-edged brittle fracture in the boundary region of the test channel over a distance of a few centimetres. How these markedly different failure modes depend on material, seam characteristics and welding parameters, has so far not been examined.

Infrared thermography of seams is as yet a somewhat little used test method, which is only occasionally applied. The seam is photographed immediately after welding using an infrared camera. Cavities and zones of poor adhesion in the seam can be seen to some extent as anomalies in the infrared picture (Peggs 1995).

To finish this section a few remarks about the amount of destructive testing as part of construction quality control measures will be added. The tradition has long prevailed in the construction quality control of third party inspectors to use extensive destructive testing to document installation integrity. Coupons are regularly cut out of the dual hot wedge seams with high frequency. The holes are than repaired by patches, which are extrusion fillet welded. As mentioned above the extrusion fillet seam is a much less reliable welding method than a hot wedge machine welding. Therefore, extensive destructive testing and repair work will reduce the overall reliability and performance of the geomembrane liner system. Quality control measures can check the quality of the geomembrane installation, i.e. the extent to which the specifications are met. However, subsequently quality control measures and associated repair work can only "heal" to a very limit extent an already inferior quality of a finished installation work. Against this background the approach to achieve a high quality output from the very beginning was emphasized in this book: to commission only certified installers, to use smart welding, to use non-destructive test methods, like ultrasonic testing, to use geomembranes with taped edges, to choose welding parameters not by trial and error but by systematic application of a process model. A comparable approach was recently suggested by a white paper of the International Association of Geomembrane Installers (IAGI), which summarised the results of a panel discussion at the Geosynthetics 2003 conference in Atlanta. It is recommended that the installer, who has shown that he installs with low failure rates and who uses the various technical improvements in geomembrane installation, should be "rewarded" by systematically reducing the frequency of destructive sampling.

## 10.3 Process Model for Quality Assessment of Dual Hot Wedge Seams

When is a weld seam good? The guidelines, like those of DVS or GRI GM 19, specify criteria based on the experience of specialists over many years and various aspects: seam geometry and appearance must be correct, failure behaviour in the peel and tensile shear tests must correspond to the qualitative description in the guidelines, the seams must be tight in compressed air or vacuum tests, the thickness reduction and the factor of seam thickness must be within given tolerances (see Eqs. 10.3 and 10.4). How are welding parameters selected so that a good seam develops? The guidelines consolidate practical experience into a range of parameters. The exact parameter set of choice is then defined on the construction site by performing and testing a test seam. What is the long-term behaviour of a good seam? Long-term behaviour of weld seams has so far been primarily investigated under tensile shear stress. Service lifetimes achieved here are always substantially smaller than those of the geomembrane due to seam geometry, which produces a stress concentration in the area of the seam edge. The service lifetime achieved is then primarily determined by the stress crack resistance of the material and the quality of the seam itself is of rather secondary importance. From these investigations it follows that, regardless of their quality, seams must not be subjected to continuous tensile stress (see Footnote 5). However, a quality criterion for the welding process and the resulting seam, which would go beyond the short-term tests, has not been established.

These answers are not yet satisfactory from a weld engineering point of view in comparison to the technical level achieved in other fields of welding (Michel 1995). It would be desirable to have a test method, which relates directly to the strength and the characteristics of the welded material area and which differentiates sufficiently strongly between different choices of welding parameters. From such a test method a manageable and systematic relationship between the choice of welding parameters and the characteristic of the seam should be obtained. Geomembrane welding specialists have dealt with this problem for a long time. Attention has so far been predominantly focussed on dual hot wedge seams with test a channel produced by hot wedge welding. Extrusion fillet seams have been rather neglected. Based on the work in the SKZ (Süddeutsches Kunststoffzentrum, www.skz.de) in Würzburg (Bielefeldt et al. 1991) and by F. W. Knipschild and P. Michel (Michel 1995), G. Lüders (Lüders 1998; Lüders 2000, 2002) has developed such a relationship and a associated welding process model for the assessment of the quality of dual hot wedge seams created by hot wedge welding, which can be used for a systematic derivation of acceptable ranges of the welding parameters from long-term tests.

K. Bielefeldt and E. Schmachtenberg first pointed out that the long-term peel test is suitable to characterise the quality of dual hot wedge seams (Bielefeldt and Schmachtenberg 1990). The test is usually performed at 80 °C in water-surfactant solution with a line force of 4–6 N/mm and 2 % by mass of surfactant content. The specimens fail predominantly by slowly peeling off the seam. Here stress crack formation occurs within the actual area of welded material. There are not only clear differences between good and bad seams (in the above sense of short-term testing), but also so-called good seams achieve different times-to-failure. Appearance and condition of the fracture mirror permits a qualitative assessment of the melt flow, which had taken place during the welding procedure (Lüders 1998). Frozen orientation, melt turbulence and melt displacement from the seam centre, which indicate an unsatisfactory welding procedure, only become visible when the seam is opened by the peel test.

G. Lüders pointed out how the results of the long-term peel test depend in a systematic way on welding parameters. This relationship will be explained in the following based on the analysis in (Lüders 1998; Lüders 2000, 2002). It has already been emphasised that the welding process can be divided, at least theoretically, into two steps: producing two layers of melted material on the geomembrane parts to be welded together (thermal process), and mixing the melts (rheological process). In extrusion welding, thermal and rheological processes coincide. In hot wedge welding, however, these processes are separate. The geomembranes are melted on the hot wedge, the squeeze rollers press and mix the melted areas together and a melt flow is produced within and out of the seam area.

The melt depth $L_0$, i.e. the thickness of the two melted layers, can be assigned to the thermal process as a characteristic value and the thickness reduction $s_r$, defined above in Eq. 10.1, at a given melt depth to the rheological process[6]. The assignment is naturally understandable: the larger the heat input, the thicker the melt layer. The larger and longer the squeezing force acts at a given melt depth, the greater the reduction in thickness. The ratio of these two quantities, the so-called thickness reduction ratio $s_r/L_0$, altogether characterises the welding process: said simply, a sufficiently thick melt layer is useful only if it is mixed sufficiently and if there is sufficient contact surface of melted material available for the interdiffusion of the polymer chains. If the ratio is too large, a too large squeezing force and a strong melt flow act on the seam to produce a welded area that is strongly oriented and may contain parts where no melt is available at all: the seam is unsatisfactory. Conversely, if the ratio is too small, the melt layer is not sufficiently mixed, and then a seam develops which has only a pure adhesion. Such seams peel off in the long-term peel test with short times-to-failure. The fracture mirror, as previously mentioned, shows the consequences of a non optimal thickness reduction ratio: whole or partial total displacement of the melt from the seam centre, flaky melt flow and oriented planar formations within the welded area (Lüders 1998). A perfect seam with high service lifetime however exhibits an even and bright fracture surface characterised by finely structured crazes as it emerges for example in fracture surfaces seen in the NCTL test on samples of the base material, see Fig. 5.8.

An optimum thickness reduction ratio is a necessary but not a sufficient criterion for a perfect seam. One must recall that in any case a precondition for a good seam is that in the thermal process a sufficiently thick melt layer is made available: therefore for $L_0$ itself there also is a range of optimum values that must be achieved in order to manufacture a perfect seam. A simple consideration makes this clear: if $L_0$ is for instance very small,

[6] In addition to thickness reduction, the rheological process obviously is characterized by the size of the squeeze-out of the seam, which is produced during melt flow. Type and form of the squeeze-out at the same time influence the notch effect in the welded area and thus the time-to-failure in the long-term tensile test. Measurements of form and area of the squeeze-out were also used for the description of the quality of dual hot wedge seams, see (Bielefeldt et al. 1991; Corbet and Peters 1996). This quantity alone, however, cannot completely and unambiguously characterise the welding process.

then a very small squeezing force can produce an optimum thickness reduction ratio, yet such a seam, where the geomembranes only cling superficially together, will not meet the test requirements.

In DVS guideline 2225-4, as discussed in the previous section, thickness reduction $s_r$ alone is used as a criterion for seam quality. In view of our previous considerations it is obvious that this criterion cannot guarantee a perfect seam. When the melt depth is small, a thickness reduction conforming to the guideline can still be achieved by extensively driving the melt out from the seam. However, one will then most likely produce an unsatisfactory seam and an experienced welder will recognise this. Only the ratio of $s_r$ to the melt depth $L_0$ of a certain range of values represents a quantitative criterion for the seam quality.

Melt depth $L_0$ and thickness reduction ratio $s_r/L_0$ are thus the fundamental characteristics or process parameters which characterise the welding process for hot wedge welding. The times-to-failure in the long-term peel test should therefore strongly depend on these two quantities. Conversely, it should be possible to limit the range of permissible values on the basis of times-to-failure measured in the long-term peel test.

Before the results of the long-term peel tests by G. Lüders are considered, a preliminary comment is necessary. The time-to-failure of a seam in the long-term peel test depends on the quality of the welding process, thus on the production of a sufficiently homogeneously mixed area of melted material. The achievable time-to-failure, however, is also determined by the intrinsic resistance of the HDPE resin material to stress crack formation. Even if the welding process was perfect and a homogeneous, fine-structured fracture mirror was formed during the peeling off, the times-to-failure of such seams might nevertheless be very different due to differences in the stress crack resistances of the basic HDPE resin. An unsatisfactory seam obtained by welding a geomembrane with very high stress crack resistance will still achieve relatively high times-to-failure. On the converse, a highly stress crack sensitive material will rapidly fail even though it may have optimum seam quality with respect to the welding process: usually a fracture in the base material outside the seam will occur. Therefore, if the dependence of the process parameters on seam quality should be extracted from the times-to-failure in the long-term peel test, the dependence on the properties of the HDPE resins must be suppressed. For this purpose the distribution of times-to-failure reached on the seams of geomembranes from a certain resin material with certain values of process parameters $s_r/L_0$ and $L_0$ is normalised to the maximum time-to-failure obtained for these seams. A seam with certain values of $s_r/L_0$ and $L_0$ is not assessed on the basis of the absolute value of the mean time-to-failure but on its ratio to the maximum possible time-to-failure achieved for seams with the material. This maximum time-to-failure of a seam with an optimised choice of process parameters can, however, serve as a criterion for the weldability of a HDPE resin.

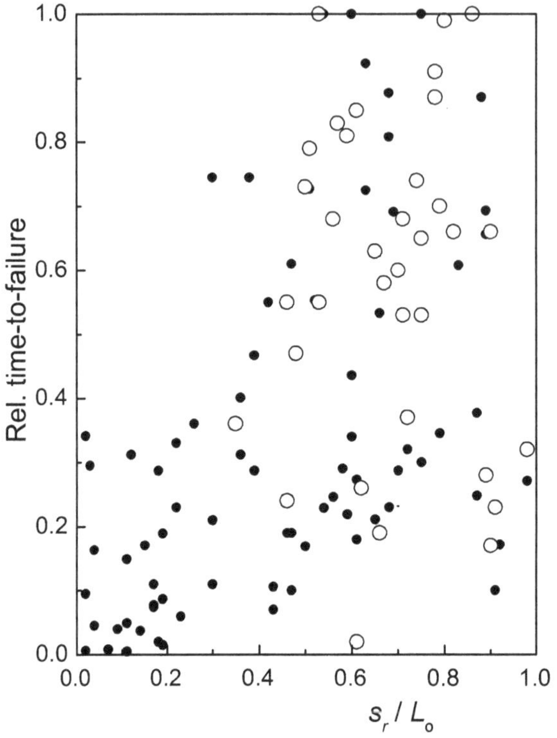


**Fig. 10.8.** Relative times-to-failure of dual hot wedge seams in the long-term peel test, i.e. the times-to-failure related to the maximum time-to-failure of the seams that can be achieved with the actual HDPE resin of the geomembrane, as a function of its thickness reduction ratio ($s_r/L_0$). The figure has been taken from reference (Lüders 2000). The thickness reduction ratio was determined for each seam by calculating $L_0$ from the manufacture conditions of the seam by Eq. 10.6 and by calculating $s_r$ from the seam dimensions by Eq. 10.1. The weld seams were manufactured in the laboratory (black circles) and on site (white circles)

Figure 10.8 shows the relative times-to-failure determined in such a way as a function of the thickness reduction ratio of the respective seams. The data originate from a number of tests on seams from geomembranes which were made of 4 different resins and which were welded with 7 different welding machines with different choices of the welding parameters. Not only seams made in the laboratory tested (black circles, Fig. 10.8), but also seams produced directly on landfill construction sites by different installation contractors (white circles, Fig. 10.8) had been included in the long-term peel tests. A large relative time-to-failure will obviously be achieved when the thickness reduction ratio is in a relatively narrow range of about 0.5 to 0.9.

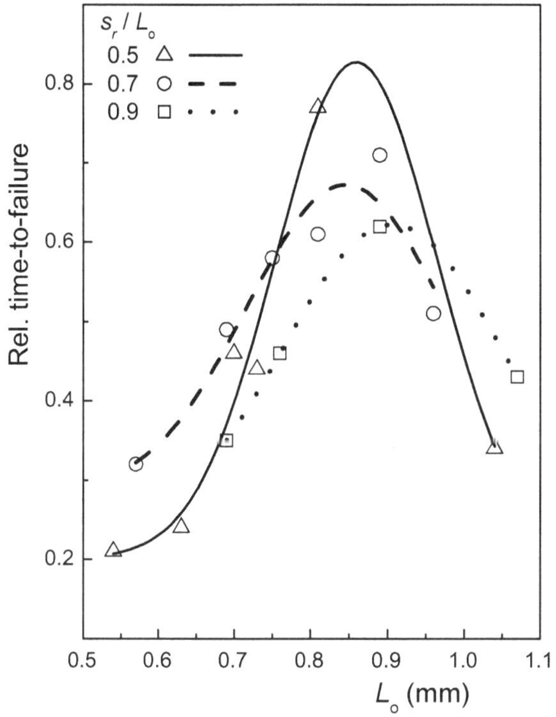


**Fig. 10.9.** Relative time-to-failure of a seam in the long-term peel test as a function of the melt depth $L_0$ for a given thickness reduction ratio ($s_r/L_0$). The figure has been taken from reference (Lüders 2000). Seams with $s_r/L_0$ values of 0.5 (triangles), 0.7 (circles) and 0.9 (squares), thus those within the optimum range of this process parameter, were selected. Each data point represents the average value of 18 to 24 individual seams. The lines of the bell shaped curves lead the eye. Obviously, a very long time-to-failure (about 60 % of the maximum time-to-failure achievable) is achieved only if the $L_0$ values of the seams are within the range of 0.75 mm to 0.95 mm. Based on a more extensive investigation G. Lüders suggests a permissible range from 0.75 mm to 0.9 mm

As expected, this is not yet sufficient. In this range there are also a lot of seams, which achieve only short relative times-to-failure. However, if one looks at the melt depth $L_0$ of these seams, it can be seen that the long relative times-to-failure are reached whenever $L_0$ is also in a relatively narrow range (Fig. 10.9). The relative times-to-failures of the long-term peel test thus clearly show a strong dependence on the characteristic parameters of the welding process, as assumed from theoretical considerations. The criterion[7] is now established that a seam is acceptable or a good seam, when it reaches a minimum of at least approximately 60 % of the possible maximum time-to-failure in the long-term peel test. With this criterion the permissible range for the process parameters can be identified (Table 10.1).

So far, how the quality of seam depends on the process parameters has been discussed. The next step is to correlate the process parameter with the welding parameters, i.e. to explain how the process parameters $L_0$ and $s_r/L_0$ can be determined and correlated with the welding parameters, i.e. hot wedge temperature, roller force and welding velocity, and the welding conditions. For thickness reduction the determination is obvious: it is calculated from the measured thicknesses of the seam and the welded geomembranes in accordance with Eq. 10.1. The determination of the melt depth is, however, more difficult. The melted layer is produced as the geomembrane slides over the hot wedge surface. Its greatest extent is reached directly behind the hot wedge before the melted material is mixed by the squeeze rollers. The thickness of the melt layer therefore is in practice not accessible to direct measurement. However, for the process model such a measurement is not necessarily needed. Rather, it is sufficient when the functional dependence of this process parameter on the welding parameters and the machine-related characteristics are so precisely known that a quantity $L_0$ proportional to the actual melt depth can unambiguously be calculated for each seam (and machine)[8]. The exact knowledge of the propor-

---

[7] The definition of this criterion includes certain arbitrariness. The range should not be too far, since the process model then bears too little fruit. The range may not be narrower either than the scattering of the data permits. For practical reasons a too narrow range would limit the work of the welder too much in the consequence of the process model.

[8] Here discussions often evolve about the exact meaning of $L_0$. From a physical consideration G. Lüders differentiates two melt layer thicknesses. From the thermodynamics' point of view the depth $L_0^{th}$ (measured from the surface of the geomembrane lying on the hot wedge), at which the temperature dropped to the melting temperature of the polyethylene may be defined as melt depth. Within this range a temperature gradient and a corresponding gradient in the viscosity prevail. Only a part of the layer is fluid enough to contribute to the rheological process of melt flow. Thus there is a rheologically relevant melt layer thickness $L_0^{rheo}$, considerably smaller than $L_0^{th}$, but in essence proportional to it. None of these quantities can be directly measured practically at the welding machine during the welding process. $L_0$ is a parameter derived from the modelling of the melt process, which acts as a "representative" of these quantities.

tionality constant is redundant since the permissible parameter range for $L_0$ is empirically defined from the long-term peel tests on the seams (Fig. 10.9).

**Table 10.1.** Process parameters for hot wedge welding of geomembranes and permissible range of parameter values according to (Lüders 2000)

| Process parameters | Permissible range |
|---|---|
| Melt depth $L_0$ | 0.75–0.9 mm |
| Thickness reduction ratio $s_r/L_0$ | 0.5–0.9 |

Naturally, the crucial importance of the melt depth as a process parameter was not only recognised in connection with the hot wedge welding of geomembranes, but also in other fields of plastics welding, and in particular with the hot element butt welding of pipes where this quantity is used for various theoretical and practical investigations. H. Potente set up a theory for hot element butt welding, within which $L_0$ was calculated from the equation of thermal conduction as a function of the boundary surface temperature, the melt temperature of the material and the warm-up time (Potente 1977). Heat conductivity or effective thermal conductivity respectively was also included as further important material characteristics in this theory.

**Table 10.2.** Overview of the used symbols and their meaning

| | | |
|---|---|---|
| Process parameters | $s_r/L_0$ | Thickness reduction ratio |
| | $L_0$ | Melt depth |
| Welding parameters | $F$ | Roller force |
| | $T_{HW}$ | Hot wedge temperature |
| | $v$ | Welding speed |
| Material parameters | $T_M$ | Crystalline melting temperature of the HDPE resin |
| | $a$ | Effective thermal conductivity |
| Machine parameters | $L_{HW}$ | Effective hot wedge track length |
| | $t_W$ | Warm-up time or contact time |
| Environmental conditions | $T_{GM}$ | Surface temperature of the geomembrane |

Starting with H. Potente's theory, P. Michel transferred this concept to hot wedge welding and determined analogously the connection between the melt depth $L_0$ and the various quantities involved, see Table 10.2 (Michel 1995). These are the welding parameters relevant for the melting process, i.e. hot wedge temperature $T_{HW}$ and welding speed $v$. Additionally, machine-specific characteristics will play a role: the design of the hot wedge and the resulting effective hot wedge track length $L_{HW}$, i.e. the length over which there is sufficient contact between geomembrane and hot wedge, which can be identified by the blackening of the hot wedge track due to material residues. The warm-up time $t_W$ of a surface element of the geomembrane sliding over the hot wedge can be calculated from the hot wedge track length and the welding speed:

$$t_W = \frac{L_{HW}}{v}. \tag{10.5}$$

The ambient conditions also affect the welding process, since from them a certain temperature $T_{GM}$ of the geomembrane results. However, the influence of this parameter can be neglected in hot wedge welding. A further relevant variable is naturally the melt temperature $T_M$ of the HDPE materials. The details of the derivation cannot be explained here. They can be found in the literature. Eventually the following equation is obtained:

$$L_0 = 1.905 \cdot \left(1 - \frac{T_M - T_{GM}}{T_{HW} - T_{GM}}\right) \cdot \sqrt{a \cdot t_W}. \tag{10.6}$$

$t_W$ can thereby be attributed as in Eq. 10.5 to the welding speed $v$ and the machine parameter $L_{HW}$.

Equation 10.6 in connection with 10.5 contains the solution of the task to determine the process parameter $L_0$ as a function of the welding parameters, namely hot wedge temperature and welding speed, as well as the parameters of the material and machine. In the evaluation of the long-term peel tests (Figs. 10.8 and 10.9) the values of $L_0$ calculated via Eqs. 10.5 and 10.6 were used for the individual seams. Therefore, to achieve perfect seams at specified $T_M$ and $a$ as well as $L_{HW}$, the welding parameters $T_{HW}$ and $v$ must be selected in such a way that the value of $L_0$ calculated by Eqs. 10.5 and 10.6 falls within the range of between 0.75 and 0.9 (Table 10.1).

Now, in order to obtain a process model, which enables a complete evaluation of the dual hot wedge seams, the relationship between the squeezing roller force[9] and the thickness reduction must be added. If this

[9] Actually the pressure (and the pressure gradient towards the edge of the seam) that acts in the melt layer is the relevant physical quantity. This quantity, however, is clearly determined at given width and radius of the squeeze roller by the squeeze roller pressure or roller force. The following considerations therefore apply only to certain squeeze rollers. Typical squeeze rollers are 15–19 mm wide and have a radius of approx. 50 mm. Strictly speaking, the Table 10.3 of values of functions $A(L_0)$ and $B(L_0)$ must be empirically determined for each squeeze roller system and thus for each machine. It has already been mentioned that the determi-

relationship is known, both process parameters ($L_0$ and $s_r/L_0$) of a dual hot wedge seam, that has been welded on a certain HDPE geomembrane ($T_M$, $a$) under certain conditions ($T_{GM}$) with a certain machine ($L_{HW}$) and specified welding parameters ($F$, $T_{HW}$, $v$), can unambiguously be determined. Therefore, the welding parameters can be systematically selected at given conditions in such a way that a seam of high quality develops, i.e. that the process parameters will fall within the permissible range given by Table 10.1.

The investigation of a multiplicity of seams by G. Lüders showed that, at given melt depth $L_0$, the thickness reduction $s_r$ depends linearly on the squeezing roller force:

$$F = A\, s_r + F_0\,. \tag{10.7}$$

This connection is intuitively plausible: $F_0$ corresponds to a certain minimum force that must be applied so that the actual joining procedure can be set in motion against the rheological resistance of the melt. If this minimum value, which depends on $L_0$, is exceeded, the increase in thickness reduction $\Delta s_r$ is proportional to the increase in the squeezing roller force $\Delta F$, where the proportionality constant $A$ also depends on $L_0$. Dividing Eq. 10.7 by $L_0$, a relationship between the squeezing force ratio and thickness reduction ratio is obtained:

$$\frac{F}{L_0} = A(L_0)\,\frac{s_r}{L_0} + B(L_0)\,. \tag{10.8}$$

Figure 10.10 shows the experimentally derived relationship between $F/L_0$ and $s_r/L_0$. Here the thickness reduction was measured on a number of weld seams and the melt depth calculated with Eq. 10.6. Seams with approximately the same melt depth were combined into a group and the relationship between $F/L_0$ and $s_r/L_0$ was determined for this group. This was done by collecting seams with approximately the same thickness reduction ratio into a subgroup. The average value $F/L_0$ was determined for this subgroup. Each data point in the figure represents such a subgroup of seams with given $L_0$ as well as $s_r/L_0$ and the associated thickness reduction ratio

nation of $L_0$ is also machine-dependent. Strictly speaking the hot wedge temperature indicated by the machine must be distinguished from the melt temperature at the hot wedge point required for the calculation. However, a clear machine-specific relationship also exists here. These details are beyond the scope of our current considerations, since they only disturb the understanding of the principle relations. G. Lüders has the design-specific aspects of the assessment model for nearly all commercial welding machines established and made them available to machine manufacturers and installation contractors (Lüders 2002).

$F/L_0$. Thus each subgroup consists of 18 to 24 evaluated seams. Apparently, the data points for a given $L_0$ are approximately on a straight line and Eq. 10.8 is confirmed. Fitting of straight lines to the data in Fig. 10.10 and the Table 10.3 of values of functions $A(L_0)$ and $B(L_0)$ can now be determined.

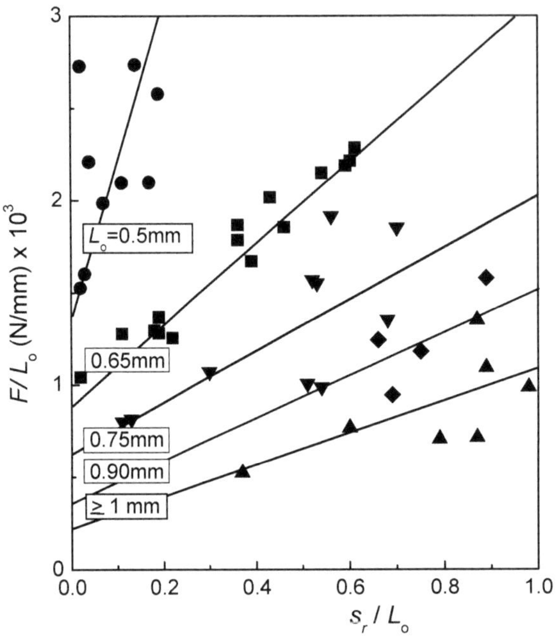


**Fig. 10.10.** Relationship between squeeze roller force and thickness reduction and between squeeze roller force ratio $F/L_0$ and thickness reduction ratio $s_r/L_0$ for a given melt depth. Each data point represents the average value of the squeeze roller force of a group of 18 to 24 seams, which are all close to a certain value of $s_r/L_0$ and $L_0$. The axis intercept $B(L_0)$ and the slope $A(L_0)$ of the straight lines are functions of $L_0$. The figure has been taken from (Lüders 2000)

Thus the task to establish a connection between the process parameters specified in Table 10.2, from which the quality of the weld seam can be read off, and the welding parameters, the material parameters, the machine parameters and the ambient influences has been completely resolved. Equation 10.6 in connection with Eqs. 10.5 and 10.8 using the values of Table 10.3 as input permit the calculation of the process parameters for an actual welded or to-be-welded seam. Handling of these formulas, however, is obviously a little bit tricky and at least laborious. Therefore, machine-related tables and graphs were compiled in close cooperation with manufacturers of hot wedge welding machines, which enable the simple illustration of these relationships and the direct reading-off of the necessary welding parameters. More information can be obtained from machine manufacturers involved in the project. Examples of such machine-related tables and graphs and a detailed explanation are given in (Lüders 2002).

**Table 10.3.** Table of values of the $A(L_0)$ and $B(L_0)$ functions, see Fig. 10.10 (Lüders 2000)

| $L_0$ (mm) | $A(L_0)$ (N/mm) | $B(L_0)$ (N/mm) |
|---|---|---|
| 0.50 | 8800 | 1335 |
| 0.65 | 2200 | 880 |
| 0.75 | 1400 | 625 |
| 0.90 | 1150 | 360 |
| ≥1,00 | ≤870 | ≤225 |

The results presented here were developed in a research project headed by G. Lüders and supported by the German Association of Geomembrane Manufacturers and Installation (AK GWS). The AK GWS had simultaneously developed a certification system for installation contractors who install geomembranes in accordance with German regulations for landfill liners (see Sect. 9.4.1). BAM acts as an independent auditor in the context of the certification procedure. Numerous visits by installation contractors on their premises and landfill construction sites offered a unique opportunity to produce sample seams with arbitrarily selected parameters using various machines and to sample and assess actually manufactured seams of geomembrane landfill liners. This enabled a representative overview of the quality of dual hot wedge-welded seams produced by certified installation contractors. In addition seams with a broad variation of welding parameters were produced in the laboratory. Welding parameters, material properties, machine parameters and ambient influences were determined and seam geometry was measured for each individual seam. The process parameters were then determined on the seams and long-term peel tests were carried out. The data flood is included in Fig. 10.8 and in a hidden way in Fig. 10.9.

Now, the question arises of the quality of the seams of the actually installed geomembranes in the light of the described process model. On the basis of Fig. 10.11 a systematic overview of the results can be obtained. Long times-to-failure are reached in the long-term peel test if the process parameters of the seam fulfil the requirements of Table 10.1. Such a seam is considered to be a good or high-quality seam. A seam is completely characterised by the parameters $F/L_0$ and $s_r/L_0$. To be a high-quality seam the data point ($F/L_0$, $s_r/L_0$), when plotted into a diagram with $s_r/L_0$ as abscissa and $F/L_0$ as ordinate like in Fig. 10.11, should be within a working field that extends between the values 0.5 and 0.9 on the abscissa and is limited by the two straight lines designated with $L_0 = 0.75$ and 0.9. It should be recalled that, as Fig. 10.9 indicates, seams with $L_0$ values of 0.95 are still close to the maximum time-to-failure, thus a working field extending slightly beyond the line designated by $L_0 = 0.9$ still provides good seams.

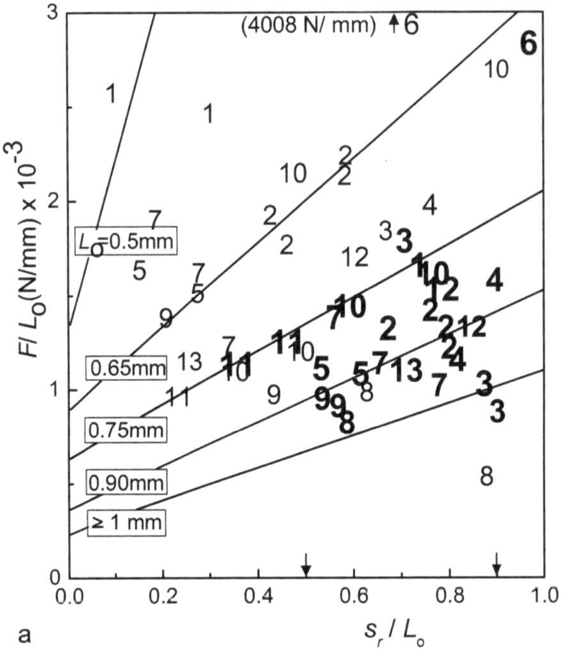


**Fig. 10.11a.** A number of seams, which were welded using arbitrarily selected welding parameters (non-bold figures), as well as seams, which were welded in accordance with the requirements of the DVS guidelines on landfills (bold figures) were analysed in detail, and melt depth $L_0$, squeeze force ratio $F/L_0$ and thickness reduction ratio $s_r/L_0$ were determined from the manufacture conditions and the data points ($F/L_0$, $s_r/L_0$) displayed in this graph. The data points are encoded according to the 13 installation contractors that had manufactured the seams on site. Seams produced on landfill construction sites are almost without exception within the parameter range $0.5 \le s_r/L_0 \le 0.9$ and $0.75 \text{ mm} \le L_0 \le 0.9 \text{ mm}$ to 0.95 mm permissible according to the process model

Figures 10.11a and b displays the data points ($F/L_0$, $s_r/L_0$) of various seams. The seams had been produced by 13 installers with altogether 7 different machines. In diagram (a) digits 1 to 13 are used as data point symbols to indicate the different installers. In diagram (b) letters a to g (lower and upper case letters) are used as data point symbols to indicate the 7 different machines. All normal font digits and lower case letters designate seams whose parameters were arbitrarily selected and in deed these data points are randomly distributed in the diagrams. Bold font digits and bold font upper case letters designate those seams which were from seams of actually installed geomembranes manufactured by the welders as per the state of the art as described in the DVS guideline. Nearly all these seams are within the working field of high-quality seams. In a very few cases, when these seams were outside the working field, errors in craftsmanship or in the machine could be detected in each individual case with the help of the process model which had not been discovered by the conventional methods of quality control.

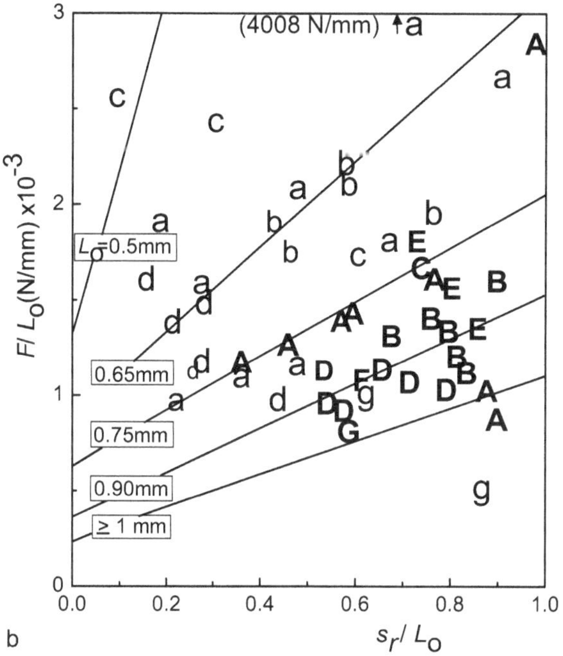


**Fig. 10.11b.** Same date points as in Fig. 10.11a., however, encoded with letters according to the 7 different welding machines used. Seams welded using arbitrarily selected welding parameters (lower case letters), as well as seams, which were welded in accordance with the requirements of the DVS guidelines on landfills (bold capital letters) are included

Dual hot wedge seams produced on landfill construction sites by certified installation contractors applying the quality control methods described in Sect. 10.2 in most cases met the quality requirements, which were set up on the basis of the statistical and morphologic investigations of the failure behaviour of seams in long-term peel tests. However, errors could be detected more reliably by analysing seam quality with the process model and avoided using this process model for the determination of welding parameters right from the start. Most importantly, however, with this process model the basis has been established for which an automatic process control of hot wedge welding machines can be developed.

This section has only dealt with dual hot wedge seams. What is the situation with extrusion fillet seams? Extrusion fillet seams produced by extrusion welding can hardly be avoided in large-area geomembrane installations. Quantitatively describing the quality of extrusion fillet seams over permissible ranges of process parameters and establishing a correlation between process parameters and various influencing welding parameters is an important and rewarding welding engineering challenge – with work towards a solution still being in its initial phase.

## References

Bielefeldt K et al. (1991) Forschungsbericht FV 187, Beurteilung des Langzeitverhaltens der Fügenähte von Kunststoffdichtungsbahnen zur Basis- und Oberflächenabdichtung bei Deponien. Süddeutsches Kunststoffzentrum (SKZ), Würzburg

Bielefeldt K and Schmachtenberg E (1990) Development of a Short Term Testing Method for Welded Liners. In: Koerner RM (ed) Geosynthetic Testing for Waste Containment Applications, ASTM Special Technical Publication 1081. ASTM, Philadelphia, USA, pp 143–154

Corbet SP and Peters M (1996) First Germany/USA Geomembrane Workshop. Geotextiles and Geomembranes 14: 647–726

Diedrich G and Gaube E (1970) Schweißverfahren für Rohre und Platten aus Hart-Polyäthylen, Zeitstandfestigkeit und Langzeitschweißfaktoren. Kunststoffe 60: 74–80

Diedrich G and Gaube E (1973) Zeitstandfestigkeit und Langzeit-Schweißfaktoren von geschweißten Rohren und Platten aus Hart-Polyäthylen und Polypropylen. Kunststoffe 63: 793–797

Gehde M (1992) Schweißverbindungen an Deponiedichtungsbahnen – Prüfung und Versagensverhalten. In: Knipschild FW (ed) Tagungsband der 8 Fachtagung "Die sichere Deponie, wirksamer Grundwasserschutz mit Kunststoffen". Süddeutsches Kunststoffzentrum (SKZ), Würzburg, pp 59–88

Gehde M (1999) Maschinen und Geräte zum Schweißen von Dichtungsbahnen. In: Knipschild FW (ed) Tagungsband der 15 Fachtagung "Die sichere Deponie, Wirksamer Grundwasserschutz mit Kunststoffen". Süddeutsches Kunststoffzentrum (SKZ), Würzburg, pp N1–N20

Heitz E and Henkhaus R (1992) Abschlußbericht des Foschungsvorhabens 7577: Langzeitverhalten von Schweißverbindungen an Deponiedichtungsbahnen aus Polyethylen. DECHEMA, Frankfurt

Hessel J and John P (1987) Langzeitfestigkeit von Schweißverbindungen an Dichtungsbahnen aus Polyethylen. Werkstofftechnik 18: 228–231

Hutten A (1991) Anwendungsspezifische Eigenschaften von PE-HD-Dichtungsbahnen unter besonderer Berücksichtigung des Relaxationsverhaltens. In: Knipschild FW (ed) 7 Tagungsband der Fachtagung "Die sichere Deponie, wirksamer Grundwasserschutz mit Kunststoffen". Süddeutsches Kunststoffzentrum (SKZ), Würzburg, pp

Knipschild FW (1992) Qualitätssicherung beim Bau von Deponieabdichtungen – Einbau der Kunststoffdichtungsbahn. In: Fehlau K-P and Stief K (eds) Fortschritteder Deponietechnik 1992, Qualitätssicherung für Deponieabdichtungssysteme und Eigenkontrollen beim Aufbau der Deponie. Erich Schmidt Verlag, Berlin, pp 141–158

Landreth RE (1991) EPA/530/SW-91/051, Technical Guidance Document: Inspection Techniques for the Fabrication of Geomembrane Field Seams. U. S. Environmental Protection Agency, Washington, D. C.

Lüders G (1998) Assessment of Seam Quality and Qptimization of the Welding Process in HDPE Geomembranes. In: Rowe RK (ed) Sixth International Conference on Geosynthetics, Conference Proceedings. Industrial Fabrics Association International, Atlanta Georgia USA, pp 337–352

Lüders G (2000) Quality assurance in hot wedge welding of HDPE geomembranes. In: Cancelli A et al. (eds) Proceedings of the Second European Geosynthetics Conference. Pàtron Editore, Bologna, pp 591–597

Lüders G (2002) Processmodel assisted control for hot wedge welding of landfill HDPE geomembranes. In: Delmas P and Gourc JP (eds) Geosynthetics, State of the Art, Recent Developement, Proceedings of the Seventh International Conference on Geosynthetics. A.A. Balkema Publishers, Nisse, The Netherlands, pp 1491–1494

Michel P (1995) Qualitätssicherung beim Schweißen von Kunststoffdichtungsbahnen. In: Knipschild FW (ed) Tagungsband der 11 Fachtagung "Die sichere Deponie, wirksamer Grundwasserschutz mit Kunststoffen". Süddeutsches Kunststoffzentrum (SKZ), Würzburg, pp 167–206

Müller WW (ed) (2001) Certification Guidelines for Plastic Geomembranes Used to Line Landfills and Contaminated Sites. Laboratory of Landfill Engineering, BAM, Berlin

Müller WW and Preuschmann R (1992) Zulassung von Kunststoffdichtungsbahnen in Kombinationsdichtungen – Anforderungen an Material, Herstellung und Einbau. AbfallwirtschaftsJournal 4: 61–68

Peggs I (1995) Assessing Geomembrane Seam Strength – Nondestructively. GFR 13: 16–20

Peggs I (2005) Geomembrane Seam Specification HDPE, PVC, PP. www.geosynthetica.net.

Potente H (1977) Zur Theorie des Heizelement-Stumpfschweißens. Kunststoffe 67: 98–102

Saechtling H et al. (1998) Kunststoff-Taschenbuch. Carl Hanser Verlag, München, Wien

Struve F (1990) Extrusion Fillet Welding of Geomembranes. Goetextiles and Geomembranes 9: 281–293

Viertel A (1997) Untersuchung zur Lebensdauer von Überlappschweißverbindungen an Deponiedichtungsbahnen aus PE-HD. In: Knipschild FW (ed) Tagungsband der 13 Fachtagung "Die sichere Deponie". Süddeutsches Kunststoffzentrum (SKZ), Würzburg, pp E1–E21

Wool RP et al. (1989) Welding of Polymer Interfaces. Polymer Engineering and Science 29: 1340–1367

# 11 Leak Detection and Monitoring Systems

## 11.1 Methods for Monitoring Geomembrane Liners

HDPE geomembranes manufactured and installed to high standards (Averesch and Schicketanz 1998; Müller 2001) are impervious, i.e. there are no actual pore channels or faults which would allow water to seep through driven by a hydraulic gradient (hydrostatic pressure, gravitation or capillary force). Consequently, the flow of water and advective transfer (see Sect. 7.3) of dissolved contaminants is prevented. As long as HDPE geomembranes are not damaged mechanically, they remain impervious under any column height of standing water, which is practically relevant in geotechnical applications: barodiffusion (see Sect. 7.2) can be ignored in nearly all cases. HDPE geomembranes are also almost completely insensitive to large-area compression (hydrostatic pressure or a uniformly distributed load). On the other hand, poor installation procedures such as bad craftsmanship in handling construction tools and machines or inappropriate subgrade or cover material (see Chap. 8) may cause damage even to a 2.5 mm HDPE geomembrane, in spite of its considerable mechanical robustness. Under extremely unfavourable conditions (severe damage, highly permeable subgrade and steep hydraulic gradient) large volumes of water may seep through holes and cracks. One of the major advantages of this liner element, i.e. total imperviousness to liquids, would then be compromised.

There are three options to prevent this from happening:

1. Preventing damage from occurring in the first place. For example, by placing strict requirements on technical expertise and the experience of specialist installation companies and implementing multi-level construction quality assurance (CQA) (Müller 2001).
2. Minimising the effects of damage. For example, by combining geomembranes with a low-permeability subgrade (e.g. geomembrane plus a compacted clay liner or other mineral liner or geomembrane plus geosynthetic clay liner (Simon and Müller 2004)), which can considerably reduce the effect of holes, or with a capillary barrier,

which can laterally divert water penetrating through a hole (see Sects. 7.4 and 7.5).

3. Identifying any damage which does occur. For example, by implementing leak monitoring systems which can reliably detect and localise holes and cracks in a geomembrane during construction and operation of the lining system (Hix 1998).

The first option should always be a matter of course. However, in many projects CQA requirements are either not comprehensively met or not sufficiently robust. The second option points out the advantages of composite liners, which help to prevent damage and to minimise other potential defects. In addition, a combination of geomembrane and mineral liner is necessary wherever the prevention of diffusive transport of organic contaminants is crucial. In many countries some composite liners for basal liners and also for caps of landfills were established as technical standards and extensive quality assurance measures and strict standards for certification procedures were determined (Holzlöhner et al. 1995). All in all, the combination of the first two options is regarded as preventive so-called "passive" concept of safety, aiming at a reliable, efficient, failsafe and fault-tolerant liner. The third option, that of long-term monitoring of liner systems, is certainly not available in cases where holes detected during operation or aftercare can only be repaired at extremely high cost if at all (e.g. in landfill basal liners). However, in landfill caps and many other geotechnical applications the actual sealing components are relatively easily accessible for repair. In Germany the "Technical Instructions on Hazardous Wastes" require for solid hazardous waste landfills, that "...landfill capping systems (must be) constructed such that any leak may be located and repaired during the period of aftercare." Therefore a combination of the first and third options represents an equivalent, reasonable and possibly economic alternative, at least in terms of the protection goals that must be achieved. This solution, which makes a multi-component, error-tolerating liner structure superfluous, can be considered an "active" safety concept.

Which combination of the above three options best meets the technical and economic conditions and the requirements on fault tolerance and imperviousness can only be decided in the context of a given geotechnical sealing project. The considerations below will, however, show that an active safety concept makes great demands on materials, specialist companies and quality assurance for the leak monitoring system and only by fulfilling these demands can the active safety concept replace the passive concept of safety. Even if applying leak monitoring systems, no concessions should therefore be made to the preventive quality control requirements.

In the following, leak monitoring systems for geomembranes will be considered (Müller and Seeger 2002). There are various possible options, but three main types of control methods can be distinguished:

1. Leak monitoring systems that measure the effect of water seeping through a hole on the state of the subgrade (moisture, temperature or contaminant content of the subgrade soil air). The hole is only detected indirectly through changes in those characteristic properties of the subgrade which are sensitive to water infiltration.
2. Double geomembrane liners with closed vacuum chambers. In this system the liner property itself is changed through the formation of the hole and damage is indicated by a pressure increase in the relevant chamber segment or by the pump power required to maintain a pre-set vacuum.
3. Leak monitoring systems that enable measurement of electrical anomalies due to a hole. The idea here is that geomembranes are impermeable not only for liquids but also for electric current. It could be said that "virtual electrical holes" or "electrical leak paths" are measured which only form "potential leak paths" for water flow.

Considering these different types of leak monitoring systems in more detail it will become clear that in most cases only electrical systems are practically capable of providing adequately accurate control for large-area geomembrane liners. Therefore only the requirements on these systems will be analysed in more detail in later sections of this chapter.

The first type of leak monitoring system requires sensors applied beneath the geomembrane which exhibit specific reactions to changes in the subgrade due to water input through a hole. This concept has been used by various manufacturers of which three examples will be briefly discussed.

Measurement of the soil's dielectric constant, at sufficiently high spatial resolution, can determine changes in the moisture content. For this purpose a pair of insulated cables is buried in the ground. Propagation of electromagnetic waves in this pair of cables is influenced by the complex dielectric constant of the surrounding soil and this, in turn, by the moisture content. This enables the detection of changes in water content due to water infiltration. The method can achieve an absolute accuracy of ± 2 % for water content determination in typical mineral liners (Brandelik et al. 1996).

In addition to changing moisture content in the ground, water seeping through a hole can alter the temperature locally. Thus if an unexpected local temperature change is observed under a geomembrane, this can be interpreted as an indication of a hole. For this purpose, however, very accurate temperature measurements with a high spatial resolution are required. A leak monitoring system based on fibre-optic temperature measurement

(Hartog and Gamble G. 1991) has been introduced onto the market (Hurtig et al. 1995). The technique is based on laser impulses sent through a meandering glass cable buried in the subgrade under the geomembrane. From the photon bundle running through the cable, photons are continuously backscattered. The intensity of a part of the spectrum of the backscattered photons depends on the temperature around the area of the scattering in a characteristic way. The point within the cable (and thus the spot beneath the geomembrane) from where photons just analysed were scattered, can be calculated from the time between the rise of the laser pulse and the measurement of the spectrum of the scattered light. The intensity as a function of time then provides the spatial temperature distribution with an accuracy of temperature measurement of a few tenths of a degree Celsius.

Finally, the LEOS leak detecting and locating system of Siemens KWU AG has been used for some time for monitoring gas and oil pipelines to measure contaminant contents in the air or soil with high spatial resolution. This system has also been proposed for use in monitoring of landfill liners (Egoffstein and Burkhardt 1993) by burying meandering narrow gas hoses beneath the geomembrane. The hoses exhibit good diffusive permeability for certain contaminants (as a rule, highly volatile organic substances) and a flush gas (air) is pumped through them. Contaminant content in the flush gas at the end of the pipe is continuously measured in a detector. The location of the point of contaminant intrusion into the hose can then be calculated from the time of contaminant detection and flow velocity of the flush gas.

Those systems discussed briefly are very similar conceptually, in that the indirect effects of holes are being detected. This, however, requires the hole to have been hydraulically effective, i.e. a considerable amount of water must have penetrated into the ground through the hole. "Considerable" here means that the moisture content, temperature or contaminant content must have changed in such a large area around the hole that the surroundings of the detection elements (electromagnetic waves, laser pulses or flush gas) have been affected by the change. The damage to which a sensor would respond needs to be quite extensive compared to the extent of damage one would expect from small holes in the geomembrane within the short term: i.e. the "response threshold" is too high. Therefore in those terms, these systems are too insensitive to be used for geomembrane monitoring. They appear to be more suitable for monitoring liner systems, such as mineral liners or capillary barriers, which are permeable to a certain extent anyway.

Double liners comprise two geomembranes – one overlain by another. The top side of the bottom geomembrane is formed with ridges which serve as supports for the overlying geomembrane. The top and bottom geomembranes are welded together to form flat horizontal chambers between the geomembranes. A connection piece is welded to each chamber which can be used to remove air from the chamber. If there is a hole, a vacuum cannot be generated or can only be achieved at the expense of a high pumping power. Using this method, in principle even very small holes can be found. Such double liners have been used for years for lining the so-called LAU plants (Anlagen zum Lagern, Abfüllen und Umschlagen wassergefährdender Stoffe – plants for storage, filling and trans-shipment of substances dangerous to water) and meanwhile their use has been extended to tunnel construction (DGGt 1997). In the latter case it is planned that impervious geomembrane areas, which are no longer accessible, will subsequently be contained by injecting sealants into the chambers.

It is obvious that double liners require extensive welding and the seams are as a rule not produced by machines but manually as extrusion fillet seams. This type of seam is, however, susceptible to errors, difficult to test and therefore controversial among experts. It is sometimes said, with some exaggeration, that the system itself generates the error-susceptibility to justify its implementation. Indeed, it has been difficult to establish its operational efficiency in several cases or to unambiguously assess when the system has reached its designed operational state. For this reason double liners have been only partially successful in achieving acceptance in the field of sealing technology in the building industry, while experience is just now being gathered in tunnel construction. Hence, on the basis of available experience so far, large-area applications, such as in landfill basal liners or capping systems, cannot be wholeheartedly recommended.

The third type of leak monitoring system makes use of one of the plastic geomembrane's secondary properties, specifically that it is usually a very good insulator. The specific electric bulk resistance (resistivity) of HDPE geomembranes exceeds $10^{15}$ $\Omega$ cm and thus lies several orders of magnitude above the wet soil specific resistance which is a few $10^{3}$ $\Omega$ cm. Another advantage is that in the ground electric current and potential are easy to measure with high accuracy. If the conditions are not particularly unfavourable, a hole can be detected much earlier than it can by monitoring any hydraulic effect (Darilek and Laine 1999).

The two schemes in Figs. 11.1a and 11.1b illustrate the functional principle of this type of leak monitoring system. The configuration consists of a regular array of point sensor electrodes under the geomembrane, at least one source electrode for current input above it and a remote earth electrode placed at the side of the lined area, as shown in the diagram. Large areas require more than one source electrode. If a voltage is applied between the source electrodes and the remote electrode, the electric potential or its equipotential lines beneath the geomembrane may be measured with the point electrode array. Beneath an intact geomembrane (Fig. 11.1a) a fairly homogeneous electrical potential develops. If the soil above and beneath the geomembrane is connected around the liner's edge, some current can flow in the peripheral area which impairs monitoring.

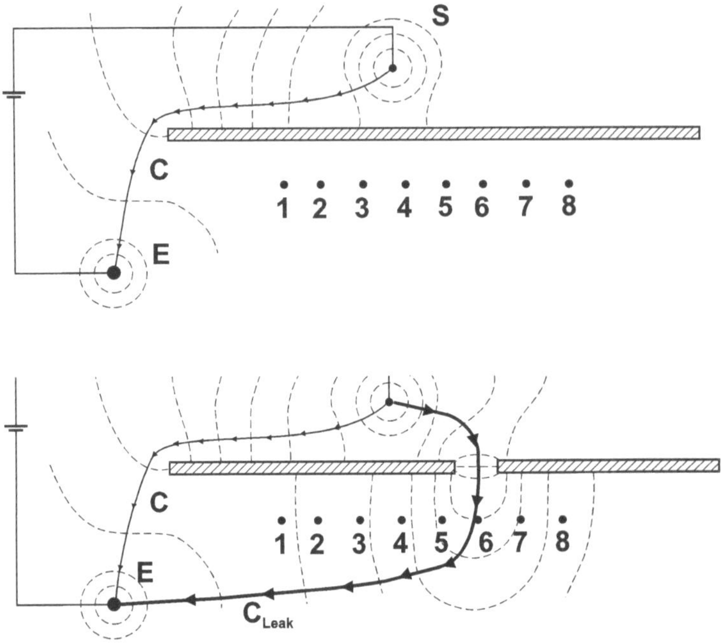


**Fig. 11.1a and b.** Schematic illustration of the function of electrical monitoring systems. A source electrode S is installed above the geomembrane (shaded stripe), and an array of point sensor electrodes (1, 2, 3 ... 8) is arranged beneath it. A remote earth electrode E is positioned at the side. Applying a voltage between the source electrode and the earth electrode generates a potential field (dotted lines). If the geomembrane is intact, only a weak current flows around the periphery of the landfill cap as indicated by C. The electric potential measured between pairs of point sensors is shallow. However, a hole establishes a low impedance transition between the soil layers above and beneath the geomembrane, allowing a relatively strong current flow $C_{Leak}$ through it. This leads to a drastic local anomaly in the electrical potential, as indicated in Fig. 11.1b, that can now be probed (strength and position) by the point electrode array

Potential lines change drastically near a hole in the geomembrane (Fig. 11.1b), provided there is a low-impedance transition across the hole so that a considerably strong current can flow through it. This is the case in particular when the hole is filled with low-resistivity material, such as water or wet soil. However, a flow of water through the hole is not a necessary prerequisite for a low resistance transition. If the transition across the hole has high resistance, e.g. when the hole is dry, filled with air, or the geomembrane around the hole has no intimate contact with the mineral material, only a small anomaly in the electric potential lines occurs and it becomes difficult to prove the existence of that hole. Detailed calculations of potential distribution under the influence of a hole in a geomembrane are reported by J. O. Parra (Parra 1988; Parra and Owen 1988).

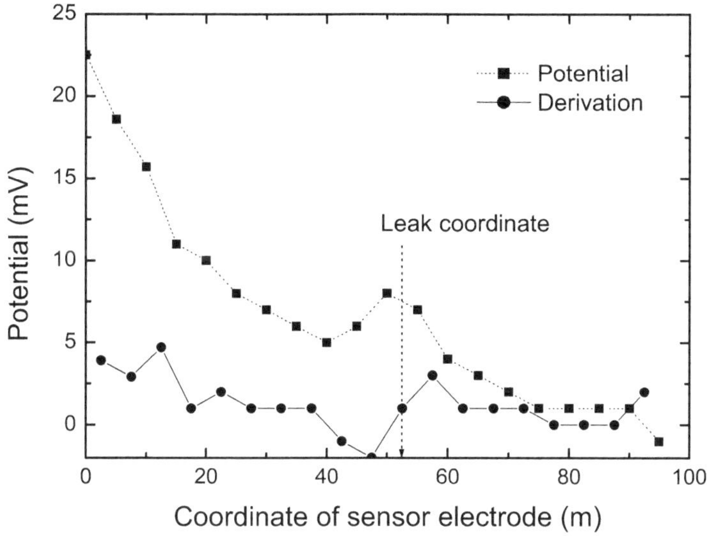


**Fig. 11.2.** Data from a leak location measurement showing the potentials determined between sensor electrodes along a grid line (abscissa) and a reference sensor electrode when voltage is applied between source and earth electrode. The location of the anomaly in the derivation gives the leak coordinate on the grid line axes

Figure 11.2 shows an example of a measurement. For this, 20 sensor electrodes positioned regularly at a distance of 5 m along a straight line were chosen from a grid of permanently installed electrodes. One of the other sensor electrodes located approximately centrally to the inspected area was chosen as reference electrode. The potential between each sensor electrode and the reference electrode was measured for a given voltage between source electrode and the earth electrode. Then the voltage was reversed and the potential was measured again. The absolute value of the difference of both measurements was calculated and differentiated as a function of electrode location. The sinusoid-like part of the derivation is characteristic for an electrical leak, whose coordinate along the considered line of electrodes is given by the coordinate of the reversal point of the sinusoid (Fig. 11.2). Finally, additional measurements on parallel lines of electrodes will reveal the location of the leak.

**Fig. 11.3.** A groove is milled into the surface of a compacted clay liner into which the sensor cable is laid. The extent of preparatory work on the subgrade of the geomembrane depends highly on the subgrade properties. In any case one has to meet the requirement, that indentations imprinted on the geomembrane by the cable and electrodes have to fall short of the acceptable limits (see Sect. 8.3.1)

Various methods are available for high spatial resolution measurement of the electrical anomaly caused by a hole (Nosko 1998; Rödel 1998). The methods rely on long established geoelectrical survey and measurement methods. The sensitivity of detecting holes depends to a large extent on the lattice distance of the point electrode pattern and may be impaired by several disturbing influences. Detecting holes may be complicated in the periphery because electrical "sidewall leakages" may occur as depicted in Fig. 11.1a. This problem can easily be avoided by electrically insulating the lined area, i.e. leaving a small strip uncovered at the periphery of the geomembrane liner outside the area. Provided the geomembrane is covered by at least one subsequent layer of geosynthetic or mineral material with high hydraulic permeability (geotextile protection layer, geocomposite or mineral drain layer) the moisture condition necessary for an electrical measurement may easily be obtained from natural or artificial irrigation.

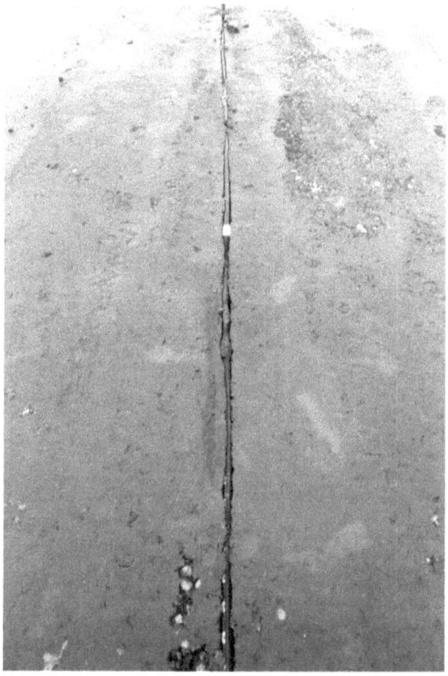

**Fig. 11.4.** A telecommunication cable with the sensor electrodes is laid into the groove. The electrodes are formed by a piece of carbon fibre fabrics wrapped around and connected to the cable about every 5 m

The set of Figs. 11.3 to 11.7 is to give an impression of the installation of two versions of leak monitoring systems. The first version consists of a multiwire communication cable. About every 5 m a carbon fibre fabric is connected to one of the wires. The connection is carefully isolated and the fabric wrapped around the cable acts as a sensor. Figure 11.3 shows how a groove is milled into the surface of a compacted clay liner into which the cable is laid (Fig. 11.4). The grooves are at a distance of about 5 m, so a rectangular electrode grid of 5 m grid length is formed (Fig. 11.5). Figure 11.6 shows another version of a leak monitoring system. Each electrode formed by a rod of electrically conducting polyethylene has its own connecting cable and the electrodes are inserted into the subgrade in a regular grid. All connecting cables of the sensor electrodes, the source electrode and the earth electrode lead into a control cabinet which, in addition, might contain the power supply and measurement electronics for a permanent monitoring and a heater (Fig. 11.7).

**Fig. 11.5.** View of the subgrade prior to the installation of the geomembrane with the cables installed in a regular distance of about 5 m

**Fig. 11.6.** This picture of a prepared subgrade shows another version of a leak monitoring system. The electrodes are distributed regularly over the subgrade and each electrode has its own connecting cable

**Fig. 11.7.** View of the control station of the system. All connecting cables of the sensor electrodes, the source electrode and the earth electrode lead to the switch boards in an electrical cabinet. The electrical cabinet may contain a power supply and all the electronics for data processing. All these components are protected against bad weather by the an appropriately constructed cabinet

A measurement taken immediately after finishing construction work may provide a quality proof for the intactness of the completed liner. The proven functional readiness of the leak monitoring system itself of course is prerequisite. Such a measurement after the completion of the whole construction work is a basic requirement for geomembrane leak monitoring systems as the overwhelming majority of damage is caused by poor subgrade or by stresses during construction work after geomembrane installation (rather than during installation of the geomembrane itself, see Sect. 11.4). In this respect mineral liners are quite different from geomembranes: damage here is hardly ever caused by the construction procedure, but more commonly by long-term stresses such as desiccation, root penetration and settlement.

## 11.2 Types of Electrical Leak Detection Systems for CQA

All leak monitoring systems discussed so far are permanently mounted and therefore capable of reliable long-term liner monitoring if their components are manufactured with the necessary long-term stability. However, electrical systems, in relying upon the electrical insulating characteristic of HDPE geomembranes, can be used as a temporary leak detection and location survey as part of CQA measurements during or immediately after installation and before installing the protection layer. Even if the geomembrane liner is covered with a layer of soil material or flooded with water, such not permanently installed electrical leak detection systems can be used to some extent to detect and localise leaks under appropriate conditions. This section gives an overview about such temporarily used electrical leak detection systems, before the next section concentrates on the requirements for permanently installed geomembrane monitoring systems in landfill caps. The types of leak detection systems are described in ASTM D6747-04 *Standard Guide for Selection of Techniques for Electrical Detection of Potential Leak Path in Geomembranes*, to which in the following is referred.

The guide defines potential leak path as any unintended opening, perforation, breach, slit, tear, puncture, crack, or seam breach. Excluded are scratches, gouges, dents or other aberrations that do not completely penetrate the geomembrane. According to the guide, faults found in leak detection surveys can be typically categorised into five groups: holes, i.e. round shaped voids with downward or upward protruding rims; tears, i.e. linear or areal voids with irregular edge borders; linear cuts, i.e. linear voids with neat closed edges; seam defects, i.e. area of partial or total separation between sheets; and burned through zones, i.e. areas where the geomembrane has been melted over the whole thickness during the welding process. However, the category of seam defect described is only observed as the final stage of a rather poor welding process and more frequently an intermediate stage of a superficial adhesion between the sheets will occur. Also, burned zones will usually not penetrate the geomembrane fully. These types of faults do not constitute a potential leak path for the present but will form potential leak paths with high probability in the future.

Common to all the electrical techniques is that the HDPE geomembrane acts as an insulator and a voltage is applied across this insulator. In all cases where high voltage is used, these techniques are dangerous and can lead to personal injury or death by electrocution. Therefore, the staff have to be qualified for and experienced in the safe handling of the system and safety measures have to be taken to protect the operator or other personnel on the site.

The so called water puddle system or water lance system is applied to survey the geomembranes during installation on an electrically conductive subgrade (usually some layer of moist soil material). The cathode electrode is installed within the subgrade and the anode electrode is placed in a water puddle or connected to the water stream of a lance and a 12 or 24 volts dc supply is connected to the electrodes. The water puddle is systematically pushed over the geomembrane area by a squeegee. If the water finds its way through a hole an increase in the steady background current signal occurs, which could be made audible (increase in pitch and volume). The technique is not applicable where wrinkles and waves and other forms of poor contact between the geomembranes and the subgrade inhibit the electrical contact by the flow of water through a potential leak or where steep slopes prevent the controlled distribution of the water. Under appropriate conditions, however, a survey rate of 500 $m^2/h$ may be achieved. Detection of leaks in seams and especially in repair patches can be tricky, because potential leak paths are often long and small in size and infiltration of water needs a certain time. Details of the procedure are described in the ASTM D7002-03 *Standard Practice for Leak Location on Exposed Geomembranes Using the Water Puddle System.* It is stated there that the leak detection sensitivity – defined as the diameter of the smallest leak that the leak detection equipment and survey methodology are capable of detecting under a specified set of conditions – can be very good and that leaks even smaller than 1 mm in diameter can routinely be found. It is recommended that a realistic test of the leak detection sensitivity with an actual or artificial leak of 1 mm in diameter should be part of each survey and the integrity of the circuit should be checked every 15 to 20 min during the survey.

Another concept to locate leaks uses the so called electrically conductive geomembrane (Cadwallader and Barker 1994). For this purpose an HDPE geomembrane is co-extruded with a double layer die. The thicker layer consists mainly of HDPE resin with a carbon black content of 2 % by mass suitable for UV protection. A thin layer is integrally fused to its base during the extrusion process with a carbon black content of approx. 8 to 14 % by mass. This high carbon black content enables a sufficiently high number of carbon black particles to have contact with each other, producing continuous conducting paths, and the polyethylene then exhibits the characteristics of an electrical conductor (Gilg 1989; Weßling 1986; Wright and Woodham 1989). An electrically conductive pad is placed on top of the geomembrane and a high voltage source is connected to the pad and to a metal broom, which is then swept gradually over the whole surface of the geomembrane (Fig. 11.8). The conductive pad and the co-extruded conducting PE layer form a capacitor, which is charged by the voltage source as soon as an electrical connection is established between the broom and the conducting PE-layer. If the broom reaches a point where the top layer is damaged down to the bottom layer or where a hole or a crack in the geomembrane is located, an electric flashover occurs which can be detected optically or acoustically.

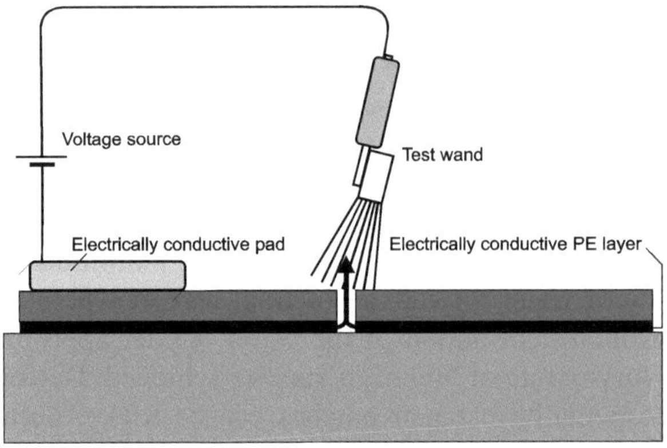


**Fig. 11.8.** Schematic drawing of the electrically conductive geomembrane system. A thin co-extruded layer of electrically conductive PE and an electrically conductive (e.g. neoprene) pad, placed on top of the geomembrane, form a capacitor. If the test wand has electrical contact to the conductive PE layer, the capacitor will be charged by a high voltage source. Therefore, when the test wand hits a hole, a spark is produced closing the circuit

If the electrically conducting bottom side of the geomembrane has good overall electrical contact to an electrically conducting support layer, for instance the drainage layer in a double liner system, the anode might be alternatively connected to the support layer. By applying 15,000 to 35,000 V to the broom, holes of approx. 1 mm in diameter can be detected in 1.5 to 2 mm HDPE geomembranes according to the manufacturer. After installation, an inspector can check an area of approx. 500 to 1500 m² of installed geomembrane by foot in an hour, or up to a hectare using a vehicle-mounted broom. To a certain extent, this detection system also enables detection of faults in the critical extrusion fillet seams, but not in machine-welded dual hot wedge seams with air channels. Here, one has to be satisfied that the dual seams and the possibility of continuous testing through the air channel provide sufficient safety against failures. However, even for extrusion fillet seams the method can only supplement the standard quality assurance measures (see Chap. 9).

The approach described in Sect. 11.1 for permanently installed electrical leak monitoring system can be used to undertake a leak detection survey as a CQA measurement on geomembranes already covered with a layer of soil (the soil-covered geomembrane system). Typically, the layer thickness has to be less than 1 metre to obtain the appropriate efficacy and spatial resolution (Laine and Darilek 1993; Laine and Miklas 1989). However, in some cases leaks might be detected even under a much greater thickness of cover (Colucci et al. 1999; Laine et al. 1997). Instead of using an electrode grid below the geomembrane, the potential distribution, produced by the source electrode and the earth electrode (Fig. 11.1), is measured on top of the soil layer point-by-point using a movable pair of electrodes. The pair of electrodes has a typically a spacing of 0.5 to 5 m and the measurement is typically made along parallel survey lines or on a rectangular pattern. Leak location sensitivity depends highly on the conductivity of the material below, above and within the leak, the electrical homogeneity of the layer above the geomembrane, the distance of the sensor electrodes, the survey grid and the arrangement of the source and earth electrodes, the output level of the power supply, producing the electrical potential and current flow, and the sensitivity of the detector electronics. Since there are large variations in these conditions from site to site a reliable survey is only possible when artificial leaks to test the sensitivity of the system are part of the earth-covered geomembrane liners (Darilek and Laine 1999). Details of the procedure are given in ASTM D7007-03, *Standard Practices for Electrical Methods for Locating Leaks in Geomembranes Covered with Water or Earth Materials*. Annex A3 of the standard describes the use of artificial leaks for leak detection sensitivity tests.

A similar approach can be used to test the geomembrane while it is completely covered with water: the water-covered geomembrane system. A cathode ground is established and an anode is placed in the contained water. A high voltage dc or ac power supply is connected, which produces a low current flow. A hand-held probe or a probe on a long cable is scanned through the water signalling anomalies in the potential or current distribution due to leaks. This method can be used to test all kinds of in-service impoundments. Again a sufficient electrically conductive subgrade is necessary. It is applicable to other installations to some extent, when flooding and draining of the test area (approximately 0.15 to 0.75 m water depth) is practicable.

## 11.3 Requirements on Leak Monitoring Systems

This section deals with the technical performance criteria on leak monitoring systems for geomembranes in landfill caps which are permanently installed and are expected to enable long-term monitoring. However, these criteria are similarly applicable to various others geotechnical applications where long-term monitoring is required. The overview will concentrate on electrical leak monitoring systems, since they are the only type capable of detecting and locating holes with high accuracy and reliability before leakage of water actually occurs. Resolution accuracy, i.e. the minimum area to which the position of a hole can be attributed, is as a rule a few square metres. It does not need to be less than the size of the area usually required for trial pits anyway. In addition to the efficacy and spatial resolution requirements, there are of course requirements on long-term durability, above all concerning the buried components which can only be repaired at very great expense by interfering with the liner system itself. Furthermore, there are performance criteria for operational measures, care and maintenance, and operational safety. For electrical equipment, lightning prevention is of special importance. Compatibility with the liner system must be guaranteed: construction and operation must not interfere with the actual liner system. Further requirements concern quality management in the manufacture of the components of the monitoring system, plus guidelines and quality assurance measures for its installation, including specifications for functional testing. Finally, there are requirements on the technical documentation that should enable a technically qualified user to independently operate and maintain the monitoring system.

In order to respond to these objectives, a recommendation document was published by the "Landfill Liner Monitoring Systems Working Group" (Arbeitskreis Dichtungskontrollsysteme für Deponieabdichtungen – AK DKS) (Seeger and Müller 2001). The Working Group was organised by the Federal Institute for Materials Research and Testing (BAM) and included experts from testing institutes, third party controller, landfill authorities, engineering consultants and manufacturers of leak monitoring systems. The requirements are identified in accordance with permanent leak monitoring in landfill caps aimed at application and a projected long service life.

### 11.3.1 Efficacy and Assessment of Leak Monitoring Systems

Firstly, the question arises as to how the efficacy of a leak monitoring system can be assessed. Efficacy is dependent on the minimum leak size which can be detected and with what resolution accuracy. The detection limit of a leak monitoring system is the minimum size of a hole in the geomembrane, assumed to be circular, that can be detected by the leak monitoring system with specified probability under normal conditions. By stating this, it is assumed that the geomembrane has a good contact to the adjacent layers with a sufficiently wet protection or drainage layer, such as shortly after an average rainfall event.

Efficacy of a system cannot be derived once and for all from the technical features and characteristics of the system components themselves as the real resolution accuracy and the detection limit in a given construction project are highly influenced by the actual conditions of the liner, the mineral materials used and indeed the environmental conditions (e.g. dry period or precipitation) (Nosko and Touze-Foltz 2000). Conversely the decision makers must clarify in advance and independent of a specified system the requirements on efficacy for leak monitoring of a landfill cap. From this the following two tasks emerge: firstly, a list of general needs for detection systems has to be specified, regardless of the individual systems, in order to enable a proper decision about whether a detection and monitoring system should be applied at all and which system seems to be the most appropriate. Secondly, to provide detailed CQA procedures must be provided for each construction project to ensure that the efficacy requirements are met under the prevailing circumstances and which therefore include a precisely defined functional test of the installed leak monitoring system and third party control measures.

The guideline recommends checking the monitoring system's compliance with the requirements in two consecutive key steps:

1. First Step: Single check of general requirements
2. Second Step: Check of applicability and CQA requirements for each construction project.

The general requirements encompass requirements on long-term durability of the buried components, on performance criteria for operational measures, care and maintenance, and operational safety, on lightning prevention, on quality management in the manufacture of the components of the monitoring system, on guidelines and recommendations for quality assurance measures during installation, and, finally, on the technical documentation. The following requirements should be checked once on a test site according to step 1 as well as on each construction site according to step 2. Detection limit: the leak monitoring system must be able to detect a circular hole of 5 mm in diameter (corresponding to an area of 20 $mm^2$ per hole) with high probability. Resolution accuracy: the position of a detected leak must be located within a circle of at most 2.5 m in radius (equivalent to an area of approx. 20 $m^2$) around the real position of the hole. In order to pass the obligatory functional test at least three artificial holes (each with 5 mm in diameter) randomly spread over the construction site must be detected individually, possibly finding a test hole only after having repaired holes already detected. The above requirement sets an arbitrary but reasonable leak size limit (see below) as from a practical point of view it neither makes sense to require detection of extremely small holes nor seems technically possible to confirm them with the required high reliability. This stipulation implies that smaller holes may possibly not be found. To restrict the water flow theoretically attributable to such holes, requirements have been formulated on the permeability and thickness of the subgrade. The permeability coefficient must be $\leq 1 \cdot 10^{-6}$ m/s and the thickness minimum 0.15 m.

### 11.3.2 Permeability of Liners with Leak Monitoring Systems

A liner systems hydraulic permeability, $k_S$, due to undetected holes in the liner, can be calculated using Eq. 7.45, repeated here, for the leakage rate $Q$ (volume of water leaking per unit of time) through a hole,

$$Q = 4kR(\varphi_0 - \varphi_*) = 4kR(h_W + d_2) \tag{11.1}$$

and considering the total area of the liner. The equation is explained in detail in Sect. 7.5. $\varphi_0$ is the hydraulic potentials immediately above the liner and $\varphi_*$ immediately beneath the subgrade. $R$ denotes the radius of the hole, $d_2$ the thickness and $k$ the hydraulic permeability of the subgrade. Putting $\varphi_* = 0$, the second part of the equation is obtained with $h_W$ for the water head above the liner, see Fig. 7.5. For calculations, however, assumptions must be made about the frequency of the holes (see following Sect. 11.4). If, for example five holes per hectare of 3 mm in diameter remain in the liner in spite of leak monitoring, and the water head above the total area is 30 cm, a system permeability of $k_S = 1.4 \cdot 10^{-12}$ m/s is obtained. Under the most unfavourable assumption that the infiltrating water finds large-area cavities between the geomembrane and subgrade, the above equation must be modified and yields a system permeability as high as $4 \cdot 10^{-10}$ m/s. It has to be pointed out, however, that these unfavourable assumptions are contradictory and therefore hardly met in practice for the following reason: it does not make any sense to simultaneously assume a water head over the liner, i.e. a large uniform load acting on the geomembrane *and* the existence of extensive cavities under the geomembrane occur simultaneously. All in all, based on these considerations, the requirements of the AK DKS Committee recommendations on efficacy of leak detection systems appear reasonable.

### 11.3.3 Long-term Behaviour and Handling of Leak Monitoring Systems

Even the schematic diagrams of Figs. 11.1a and b indicate that a leak monitoring system comprises components for which various durability requirements are valid. Buried cables and sensors under the geomembrane can only be repaired at very great expense if at all. Cables and sensors above the geomembrane, plus the components of the control and assessment units, are generally easy to access, repair or replace. The same requirements would apply to these components as to common electrical equipment and installations. However, special durability proof has to be demanded for buried components beneath the geomembrane. The recommendation document assumes a typical service lifetime of the monitoring system of 30 years. To provide a durability proof over this period of time, one can usually rely on available regulations or standards for testing and assessing buried cables in other fields of application. The relevant standards are listed in the recommendation document (the requirements for long-term tests refer to EN 60811-4-2:1999 *Insulating and Sheathing Materials of Electric and Optical Fibre Cables – Common Test Methods – Part 4: Methods Specific to Polyethylene and Polypropylene Compounds – Section 2: Tensile Strength and Elongation at Break after Pre-conditioning – Wrapping Test after Thermal Aging in Air – Measurement of Mass Increase – Long-term Stability Test – Test Method for Copper-catalysed Oxidative Degradation*). Production of such components must be performed under a quality management that extends from incoming inspection of the materials and semi-finished products to the final inspection of the completed systems components.

Notwithstanding that requirement the leak monitoring system should also exhibit a high degree of redundancy with regard to the sensors: failure of a few sensors should not be allowed to lead to the collapse of the whole system.

A few basic requirements also apply to the handling of the leak monitoring system. The system must be able to be operated and, if required, repaired independently of the manufacturer. It is a prerequisite for this that complete technical documentation of the monitoring system (including function descriptions, lists of parts, circuit and construction plans) should be available. The system must exhibit high operational safety. The relevant requirements for the operation of electrical equipment are valid. Staff must be able to quickly recognise any malfunction of the leak monitoring system. The leak monitoring system must therefore possess a self-test device that unambiguously indicates the operability of the system or any malfunction. The control and assessment software should automatically report and record malfunctions, and should help the operator in analysing possible causes of failure.

Lightning prevention is a special problem in leak monitoring systems for large-area liners that are operated over long periods of time. In Germany, up to one million lightning strokes are registered annually. Frequency decreases from north to south and is particularly high in the mountains (Schwarzwald, Alps). Especially high lightning frequencies can be observed in summer (Fig. 11.9).

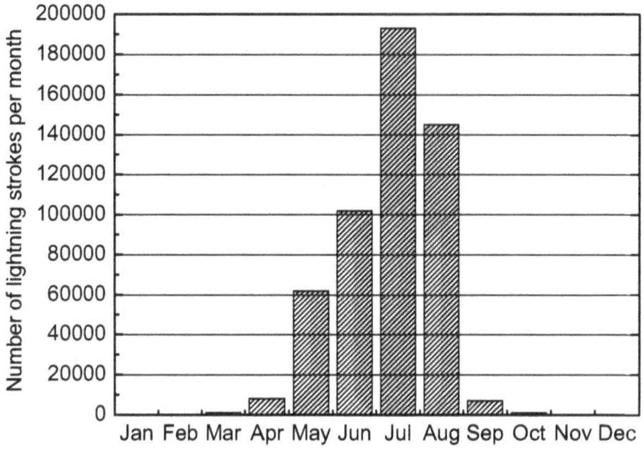


**Fig. 11.9.** Monthly number of lightning strokes observed during a year in Germany. Data taken from (Fabich 1999). Data about world-wide distribution and rate of lightning strokes may be obtained from satellite monitoring, see www.science.nasa.gov

The leak monitoring system must therefore be supplied with equipment which can divert voltage surge on the buried components and protect people and the control and assessment system. An independent expert inspector must have examined and reported on the lightning prevention equipment.

The manufacturer of the monitoring system must also take out a suitable liability insurance policy (operational liability: 2.5m Euro cover, product liability: a minimum of 2.5m Euro cover). This should explicitly include liability for follow-up costs that emerge from false leak alarms in intact liners during acceptance measurement or the period of guarantee. The basic rule is that a leak alarm requires compulsory trial pitting, inspection and repair.

Installation of the leak monitoring system must be integrated into the liner construction technology. The manufacturer's installation procedure must be adapted to these conditions. In the appendix of the recommendation document, the AK DKS Working Committee has published sample drafts for the additional technical contract conditions and the performance list for an invitation to tender for leak monitoring systems plus a quality assurance plan for their installation. It is first advisable to pilot the leak monitoring system in a test field of a minimum area of 1000 $m^2$, which may be part of the liner itself. The structure of the test field should be identical to the structure of the planned liner, as far as type and materials are concerned. The Working Committee strongly advises that installation of the leak monitoring system be first tested on a test field if construction is to proceed on steep slopes.

## 11.4 Types and Frequency of Faults

Finally, data and estimates should be compiled from the literature with regard to the type and frequency of faults in geomembranes. However, the historic data pool is rather limited. To the Author's knowledge, no systematic empirical investigations have been carried out so far in this field in Germany. The reason is that serious faults are rare and experience with geomembranes has been generally positive.

The somewhat dated papers by J.P. Giroud and R. Bonaparte (Giroud and Bonaparte 1989a, 1989b) have been used as a common source of information for types and frequency of faults. Their findings on types and frequencies of faults have been derived from quality assurance data and inspections and reports on failures. They investigated liners comprising all geomembrane types for various fields of application such as reservoirs, landfills etc. They mainly found faults on weld seams. The data collected refer primarily to extrusion fillet seams made by using hand held, portable extruders. The quality of such seams depends greatly on the quality of seam preparation and the welder's skill and experience. Besides, seams can only be checked section by section using a vacuum chamber.

Construction of large-area geomembrane liners using state-of-the-art technology is widely performed by applying machine-made dual hot wedge seams to weld together the overlapping geomembrane sheets. Such joints, produced by hot wedge welding machines under regulated and documented process control, comprise two parallel machine-made seams with an air channel lying in between. They can be checked along their entire lengths by compressed air test through the air channel. Failure frequency of these seams should be considerably lower than that for extrusion fillet seams (see Chap. 10).

J. P. Giroud and R. Bonaparte assumed that there was a fault at every 10 m in a (extrusion fillet) weld seam made on site without any special quality assurance. However, if quality assurance is applied according to what was the then state of the art, faults should only be expected at every 300 m. Assuming this frequency, the authors arrived at 3–5 faults per hectare. Failures may additionally occur due to perforating the geomembranes when the protection or drainage layers or additional soil layers are placed.

Further perforations or cracks are conceivable due to impact by sharp-edged objects under the overburden, due to too high deformation caused by settlement, by lightning strokes etc. The bottom line was that the authors estimated a frequency of one fault per 4000 m². This estimate, as already mentioned, was based on what was, for its time, an intensive and expert quality assurance. However, up to 25 holes per 4000 m² were to be expected if no such measures had been applied.

**Table 11.1.** Survey of the occurrence of frequencies of faults found with electric leak monitoring systems on 94 installation sites totalled one million square meters (Nosko 1998)

| Fault frequency | Occurrence | |
|---|---|---|
| | Number of sites | Percentage of area |
| None | 23 | 17 |
| Up to 5 per hectare | 48 | 49 |
| Up to 20 per hectare | 11 | 7 |
| More than 20 per hectare | 12 | 27 |
| Total | 94 | 100 |

**Table 11.2.** Size of faults and the corresponding percentage of occurrence of fault causes. Data were taken from (Nosko and Touze-Foltz 2000) surveying 300 sites with about 3 million square metre and 4000 detected faults

| Fault size | Occurrence (%) | | | | |
|---|---|---|---|---|---|
| (cm²) | Stones | Heavy equipment | Welds | Cuts | Worker directly |
| < 0.5 | 11 | - | 43 | 5 | - |
| 0.5–2.0 | 58 | 6 | 40 | 36 | 84 |
| 2.0–10 | 28 | 18 | 11 | 18 | 16 |
| > 10 | 3 | 76 | 6 | - | - |

Estimation for the size of faults was made by evaluating interviews with third-party inspectors. This survey concluded that 1 to 3 mm radii could be typically assumed for faults in weld seams, with up to 5 mm in exceptional cases. A radius of 10 mm was presumed for the typical dimension of a perforation.

There is limited experience available on leak monitoring systems. One of the manufacturers has published a report on their experience world-wide with regard to types and sizes of faults (Nosko 1998; Nosko and Touze-Foltz 2000). All types of lining tasks have been investigated, using geomembranes of different materials, and applying a wide range of installation techniques and quality assurance methods. However, the majority of cases covered were landfill liners comprising thin HDPE geomembranes. The examples did not include state-of-the-art composite liners with 2.5 mm thick HDPE geomembranes.

The vast majority of faults (73 %) occurred while subsequent layers were being placed, i.e. during construction but after geomembrane installation, and only 24 % were caused during geomembrane installation and welding. The frequency of faults occurring after the end of construction (3 %) was negligible. Out of the faults during geomembrane installation and welding, the majority (61 %) were faults in weld seams. Table 11.1 gives the frequency of faults. According to the company data 94 construction sites were investigated, totalling nearly one million square metres of installed area. In performing this survey, holes, cracks and faulty weld seams were all counted as faults. Table 11.2 shows data on the sizes of faults. Since holes and cracks are mainly caused during geotechnical construction after geomembrane installation and by an inappropriate protection layer, considerable fault sizes (> 5 mm radius) may be assumed.

G. T. Darilek, D. L. Laine and co-workers report data about leak detection surveys of water filled liners, i.e. directly after liner installation and before any additional construction work of covering the liner (Laine and Darilek 1993; Laine and Miklas 1989). They found an average of 2.4 and 3.2 leaks per 1000 $m^2$, most of them being located in weld seams. For example, from a total of 58 surveys with 363,684 $m^2$ inspected area they found a total of 817 leaks and from these 734 leaks, i.e. 90 %, were located in seams.

These findings indicate that fault frequency depends to a large extent firstly on the quality of craftsmanship in construction work after geomembrane installation and secondly on the quality of craftsmanship in geomembrane welding. Therefore these results cannot be extrapolated to state-of-the-art composite landfill liners constructed in Germany in the 1990s without further interpretation. State of the art includes, for example, especially strict requirements on protection layers (Seeger and Müller 1996; Seeger and Müller 2003). A research project looking into risk assessment for waste disposal was seeking data for such landfill liners. Since there was no such data then available, in 1993 G. Heibrock and H. L. Jessberger, in co-operation with E. Gartung, carried out extensive interviews with experts in the field of landfill lining, especially third party inspectors on geomembranes installation. The results were compiled and evaluated in (Heibrock and Jessberger 1977) and reference therein. Some of the questions were concerned with types, causes and frequency of faults. The results were classified as *typical*, *probable*, *possible*, *less probable* and *not probable*. A piece of information from the experts was classified as *typical,* when the middle of its estimated confidence interval for safety lay above 60 %; *probable*, when it lay between 40–60 %; *possible*, when it lay between 20–40 %; *less probable*, when it lay between 10–20 % and finally *not probable*, when it lay below 10 %.

Geomembrane experts consider it *probable* that geomembranes are placed free of faults. Among those experts expecting failures – and these are mainly non-geomembrane experts – the opinion is quite unanimous about types, causes and frequency. In view of the level of quality assurance and welding technology, faults are not primarily assumed in the weld seam area. Rather they are anticipated in deficiencies of construction technology for protection layers, and in the stresses caused by further stages of construction procedure after geomembrane installation. However, all these fault causes are also considered avoidable. Manually welded extrusion fillet seams in edges connecting geomembranes to constructions penetrating the liner are considered particularly critical as possible sources of faults. 3 to 5 faults per hectare are considered *probable*, a higher number of faults *less probable*. The following statements are made on the sizes of faults. *Typical* lengths of cracks and weld seam faults are in the range of 0–10 mm, larger values are only *less probable*. *Typical* hole areas lie under 100 mm$^2$, while larger values are considered *not probable*.

The gist of the statements is more optimistic than those of J.P. Giroud and R. Bonaparte or the findings of the company report cited earlier (Nosko 1998). Indeed, in view of the dramatic difference in the levels of construction and welding technology, between the fields of experience considered by J. P. Giroud and R. Bonaparte and the company reports, and the field of experience dealt with in the interview of experts, one might have expected considerably greater differences in their statements on frequency and size of faults. It cannot be excluded that those experts less extensively familiar with geomembranes may have gathered their knowledge by reading J. P. Giroud's and R. Bonarparte's papers.

As mentioned, there is a lack of specialist investigations and dependable data acquisition about types and frequency of faults in large-area geomembrane liners as function of geomembrane thickness, subgrade properties, protection layer design, craftsmanship of welding and construction work after installation. In this field an increasing use of leak monitoring systems will begin to provide reliable information.

## 11.5 Leak Monitoring and CQA of Geomembrane Liners

Even the most extensive CQA procedures cannot completely compensate for severe deficiencies in the performance of the geomembrane liner due to poor quality of the geomembrane (i.e. too small thickness, inappropriate resin), poor installation procedures such as bad craftsmanship in welding or handling construction tools and machines, wrong protection layer design or inappropriate subgrade or cover material. Extensive destructive testing might even cause more harm than benefit (Thiel et al. 2003). Faults, caused by sub-standard craftsmanship in geomembrane installation, lie mainly in the area of weld seams. They are rather small and they might be difficult to detect even by electrical leak monitoring systems. Therefore the need for high-standard craftsmanship in geomembrane installation cannot be avoided under any circumstances. Obviously, it is difficult to fully exclude the possibility that geomembranes are subsequently damaged during follow-up earth works, placement of drainage layers, waste, soil materials, capping structures, etc. If there are faults, they are most likely to have been caused by this part of the construction procedure. Faults caused in such a way are easier to detect by electrical leak monitoring systems, since the size of such faults should normally lie above the detection limit of the electrical systems. However, the above mentioned weld seam deficiencies have a considerable potential to cause faults in the future under various service stresses. Therefore, to add the described leak detection procedures to the traditional CQA procedures, as is often suggested (Darilek and Laine 2001; Thiel et al. 2003), is not the "the final defense weapon" to win the battle against all "enemies which would destroy the integrity of the geomembrane" (Thiel et al. 2003).

On the contrary, in the first place the focus should be on appropriate materials choice and geomembrane manufacture and on state-of-the-art design and installation, welding and construction technology. According to the experience in Germany with leak monitoring systems for HDPE geomembranes in landfill caps, experienced specialist installation companies, who are specifically qualified for this task, are capable of constructing fault-free liners when appropriate HDPE geomembranes are used and well defined state-of-the-art installation technology and quality assurance are applied. In this book an effort was made to describe the state-of-the-art and the relevant requirements in detail. Wherever one fails seriously to meet these requirements, for example due to special site conditions, local traditions or economical constraints, leak detection technologies will certainly be of less help in winning the battle but will only prevent the worst case scenario from happening, namely that the incidence of faults becomes so high that the geomembrane liner will become virtually superfluous.

If very high safety standards are mandatory, e. g. for liners of hazardous waste landfills, and a highly reliable and efficient as well as failsafe and fault-tolerant liner is required the approach discussed in the first paragraphs of Sect. 11.1 should be taken: either composite liners or, if repair work is possible, permanent leak monitoring systems.

This consideration leads to the question as to how a liner acceptance test using a leak monitoring system can be integrated into the acceptance procedures of the various trades. First, the leak monitoring system itself must be accepted and its proper function proved. This can be performed in a large scale test field which later becomes a part of the liner. The actual acceptance measurement, using the leak monitoring system, should, however, only proceed after finishing the complete construction procedure of building the lining system. However, it will cause problems to make the acceptance of a geomembrane installation dependent upon an acceptance measurement which occurs much later, e.g. in the case of landfill caps or basal liners usually after the placement of the restoration layer or the first layer of waste; even so, as it is very unlikely that it will be possible to blame the specialist installation company for faults found at that stage.

There are two possibilities: firstly, in the case of a geomembrane installation by a certified installer and accompanied by extensive quality assurance, one should perform the acceptance procedure on the basis of the quality assurance reports alone, without any special check-up leak monitoring measurement. Secondly, a process of repeated measurements, using the leak monitoring system, may accompany the different stages of the building procedure. The parties involved on site have to agree the course to follow.

In 2004 the United Kingdom Environmental Agency published a technical report about the likely medium to long-term generation of defects in geomembrane liners (Needham et al. 2004). The authors reviewed the state of knowledge about durability of polyethylene geomembranes and physical damage mechanisms. A six-stage model was developed for the estimation of defects generation in HDPE geomembrane liners and the service lifetime and frequency of holes calculated for different landfills scenarios. Details can be found in the report. The following key actions to minimise geomembrane hole generations in landfill liners were identified (Author's comments are added in brackets):

1. Select a geomembrane that exceeds the GRI GM13 specification (see Chaps. 1 and 3).
2. Specify a thicker geomembrane to attain greater survivability and resistance to oxidative degradation.
3. Design the containment system to ensure that the geomembrane liner is not exposed to avoidable stresses and that large-scale failure do not occur (this includes an appropriate design of the protection layer, see Sect. 8.3.1).

4. Have design quality assurance undertaken by a suitable experienced third party engineer.
5. Ensure that the durability of geomembrane protection materials is at least comparable to the desired service lifetime of the geomembrane (see Sect. 8.2).
6. Prepare a well-crafted, site-specific technical specification and CQA plan (see Sect. 9.4).
7. Install the liner in fair weather conditions using installers, certified to an appropriate scheme (see Sect. 9.4.1) and monitored by well experienced CQA inspectors with the objective of achieving a wrinkle-free and undamaged liner (see Sect. 9.3.1 and 9.3.2).
8. Full-time CQA monitoring of placement of protection and drainage materials, and the first layer of waste.
9. Install a fixed electric leak location system and monitor quarterly until the completion of waste disposal and annually thereafter. (A protocol has to be established in advance about how to proceed when a leak is indicated. This monitoring concept seems to be relevant only for single geomembrane liner, the alternative approach is the use of composite liner systems, see Chap. 9).
10. The working practices at the site should hold the maintenance of the integrity of the geomembrane liner as a priority (see Sect. 9.3.2).

## References


Averesch UB and Schicketanz RT (1998) Installation procedure and welding of geomembranes in the construction of composite landfill liner systems – focus on "Riegelbauweise". In: Rowe RK (ed) Proceedings of the Sixth International Conference on Geosynthetics. Industrial Fabrics Association International (IFAI), Roseville, MN, USA, pp 307–313

Brandelik A et al. (1996) Zerstörungsfreie in-situ Messung der Feuchte und Dichteänderung von mineralischen Deponieabdichtungen. Müll und Abfall 28: 263–268

Cadwallader MW and Barker PW (1994) Post Installation Leak testing of Geomembranes. In: Proceedings of the Fifth International Conference on Geotextiles, Geomembranes and Related Products. Singapore, pp 919–922

Colucci P et al. (1999) Locating landfill leaks covered with waste. In: Proceedings Sardinia 99, International Waste Management and Landfill Symposium. Environmental Sanitary Engineering Center (CISA), Cagliari, Italy, pp 137–140

Darilek G, T. and Laine DL (1999) Performance-based specifications of electrical leak location surveys for geomembrane liners. In: Proceedings of the Geosynthetics Conference 1999. Industrial Fabric Association International (IFAI), www.ifai.com, Roseville, USA, pp 65–75

Darilek G, T. and Laine DL (2001) Costs and Benefits of Geomembrane Liner Installation CQA. In: Geosynthetics Conference 2001, Conference Proceedings. Industrial Fabric Association International (IFAI), www.ifai.com, Roseville USA, pp 65–75

DGGt (1997) Empfehlung Doppeldichtung Tunnel-EDT. Verlag Ernst & Sohn, Berlin

Egoffstein T and Burkhardt G (1993) Kontrollierbar Abdichtungssysteme für Deponien. In: Tagungsunterlagen des VDI-Seminars Nr. 43-14-01, Deponiedichtungssysteme nach TA Abfall. Verein Deutscher Ingenieure (VDI), VDI-Bildungswerk, Düsseldorf

Fabich C (1999) Himmlische Attacke, Blitze und Überspannungen. c't 17: 130

Gilg R-G (1989) Ruß für leitfähige Kunststoffe. In: Mair HJ and Roth S (eds) Elektrisch leitende Kunststoffe. Carl Hanser Verlag, München, pp 55–76

Giroud JP and Bonaparte R (1989a) Leakage through Liners Constructed with Geomembranes-Part I. Geomembrane Liners. Geotextiles and Geomembranes 8: 27–67

Giroud JP and Bonaparte R (1989b) Leakage through Liners Constructed with Geomembranes-Part II. Composite Liners. Geotextiles and Geomembranes 8: 71–111

Hartog A and Gamble G (1991) Photonic distributed sensing. Physics World 3: 45–49

Heibrock G and Jessberger HL (1977) Development of a safety concept for landfill liner systems. In: August H et al. (eds) Advanced Landfill Liner Systems. Thomas Telford, London, pp 101–109

Hix K (1998) Leak Detection for Landfill Liners, Overview of Tools for Vadose Zone Monitoring, Technical Status Report, EPA-542-R-98-019. U. S. Environmental Protection Agency (EPA), Washington D. C.

Holzlöhner U et al. (eds) (1995) Landfill liner systems, a state of the art report. Penshaw Press, Cleadon, U.K.

Hurtig E et al. (1995) Überwachung von Deponiebasis und Deponiebasisabdichtung mit faseroptischen Temperaturmessungen. In: August H et al. (eds) Tagungsband der 3 Arbeitstagung des BMBF-Forschungsvorhabens Weiterentwicklung von Deponieabdichtungssystemen. Fachgruppe IV.3, Bundesanstalt für Materialforschung und -prüfung (BAM), Berlin, pp 341–350

Laine DL et al. (1997) Locating geomembrane liner leaks under waste in a landfill. In: Geosynthetics '97, Conference Proceedings. Industrial Fabric Association International (IFAI), www.ifai.com, Roseville USA, pp 407–411

Laine DL and Darilek G, T. (1993) Locating leaks in geomembrane liners of landfills covered with a protective soil. In: Geosynthetics '93, Conference Proceedings. Industrial Fabric Association International (IFAI), www.ifai.com, Roseville USA, pp 1403–1411

Laine DL and Miklas M, P. (1989) Detection and location of leaks in geomembrane liners using an electrical method: case history. In: Proceedings of the 10 th National Conference, Superfund '89. Hazardous Materials Control Research Institute, Washington, pp 35–40

Müller WW (ed) (2001) Certification Guidelines for Plastic Geomembranes Used to Line Landfills and Contaminated Sites. Laboratory of Landfill Engineering, BAM, Berlin

Müller WW and Seeger S (2002) Requirements for leak monitoring systems for HDPE geomembrane landfill liners. Land Contamination & Reclamation 10: 91–99

Needham A et al. (2004) Likely medium to long-term generation of defects in geomembranes, R&D technical report P1-500/1/TR. UK-Environmental Agency, Bristol, UK

Nosko V (1998) SENSOR DDS technology – modern and high effective way of testing integrity of geomembranes. In: Knipschild FW (ed) Tagungsband der 14 Fachtagung "Die sichere Deponie". Süddeutsches Kunststoffzentrum (SKZ), Würzburg, pp N1–N11

Nosko V and Touze-Foltz N (2000) Geomembrane Liner Failure: Modelling of its Influence on Contaminant Transfer. In: Cancelli A et al. (eds) Proceedings of the Second European Geosynthetic Conference. Pàtron Editore, Bologna, pp 557–560

Parra JO (1988) Electrical response of a leak in a geomembrane liner. Geophysics 53: 1445–1452

Parra JO and Owen TE (1988) Model studies of electrical leak detection surveys in geomembrane-lined impoundments. Geophysics 53: 1453–1458

Rödel A (1998) Langlebige Überwachungssysteme für die Langzeitüberwachung von Deponie-Oberflächenabdichtungen. In: Knipschild FW (ed) Tagungsband der 14 Fachtagung "Die sichere Deponie". Süddeutsches Kunststoffzentrum (SKZ), Würzburg, pp M1–M19

Seeger S and Müller W (1996) Requirement and Testing of Protective Layer Systems for Geomembranes. Geotextiles and Geomembranes 14: 365–376

Seeger S and Müller WW (2003) Theoretical approach to designing protection: selecting a geomembrane strain criterion. In: Dixon N et al. (eds) Geosynthetics: Protecting the Environment. Thomas Telford, London, pp 137–152

Seeger S and Müller WW (eds) (2001) Anforderungen an Dichtungskontrollsysteme in Oberflächenabdichtungen von Deponien, Amts- und Mitteilungsblatt der BAM, Sonderheft 1/2001. Wirtschaftsverlag NW, Verlag für neue Wissenschaft GmbH, Bremerhaven

Simon FG and Müller WW (2004) Standard and alternative landfill capping design in Germany. Environmental Science & Policy 7: 277–290

Thiel R et al. (2003) Cutting holes for testing vs. testing for holes. Geotechnical Fabric Report: 20–23

Weßling B (1986) Elektrisch leitfähige Kunststoffe, Vergleich leitrußgefüllter Compounds mit intrinsisch leitfähigen Polymeren. Kunststoffe 76: 930–936

Wright WM and Woodham GW (1989) Conductive Plastics. In: Margolis M (ed) Conductive Polymers and Plastics. Chapman and Hall, New York, pp 119–174

# Appendix 1

## Requirement Tables

The following tables are from the "Certification Guidelines for Plastic Geomembranes Used to Line Landfills and Contaminated Sites" of the Federal Institute for Materials Research and Testing (BAM):

Table A1.1. General Physical Requirements
Table A1.2. Mechanical Requirements
Table A1.3. Requirements for Physical Resistance and Long-Term Behaviour
Table A1.4. Additional Requirements for Textured Geomembranes
Table A1.5. Test Procedures and Frequencies for Incoming Quality Control of Resin and Carbon Black Batch
Table A1.6. Test Procedures and Frequencies for In-House Quality Control of Geomembranes by Manufacturer
Table A1.7. Test Procedures and Frequencies for Third-Party Inspection of Resin, Carbon Black Batch and Geomembranes

Some small errors in the original version have been corrected.

**Table A1.1, Part 1.** General Physical Requirements

| No. | Properties | Test Attributes | Requirements | Tests/Test Conditions |
|---|---|---|---|---|
| 1.1 | Surface condition | Appearance | Smooth surface, free of fissures, craters and pores, no damage | Visual inspection in accordance with DIN 16726, Section 5.1 |
| 1.2 | Homogeneity | Appearance of cross section | Free of pores, voids, foreign inclusions | |
| 1.3 | Carbon black content | Weight percentage | Range for nominal value specified in certification report: 1.8–2.6 wt %, individual values determined may not vary from nominal by more than ±10 % | Thermogravimetric analysis based on DIN EN ISO 11358, Section B1 in Supplemental Information on Tests[1], or determination in accordance with ASTM 1603-94 |
| 1.4 | Carbon black distribution | Supplemental Information onTests[1] | Supplemental Information on Tests[1] | ASTM D5596-94, Section B2 in Supplemental Information on Tests[1] |
| 1.5 | Skew | Maximum distance between edge of geomembrane and straight line over a length of 10 m when rolled out over a length of 12 m | All individual measuremcnts must be ≤ 50 mm | Determination of skew and waviness in accordance with DIN 16726, Section 5.2 |
| 1.6 | Waviness | Greatest clearance between geomembrane and level supporting surface over a length of 10 m when rolled out over a length of 12 m | All individual measurements must be ≤ 50 mm | |

[1]) Available from Laboratory IV.32, Landfill Engineering, BAM/Berlin.

**Table A1.1, Part 1, continued.** General Physical Requirements

| No. | Properties | Test Attributes | Requirements | Tests/Test Conditions |
|---|---|---|---|---|
| 1.7 | Thickness | Nominal thickness and average thickness | Arithmetic average of thickness measurements must be ≥ nominal thickness | As called for in DIN 16726 Section 5.3, measure thickness in accordance with DIN 53370 every 0.2 m over the entire width of the geomembrane |
| | | Individual thickness measurements | The minimum permissible thickness is 2.50 mm. Therefore, for geomembranes with a nominal thickness of 2.50 mm all individual measurements must be ≥ 2.50 mm and the maximum permissible deviation of any individual value from average is ± 0.15 mm. For other nominal thicknesses, maximum permissible deviation of any individual value from average is ± 0.20 mm | |

**Table A1.1, Part 2.** General Physical Requirements

| No. | Properties | Test Attributes | Requirements | Tests/Test Conditions |
|---|---|---|---|---|
| 1.8 | Melt mass-flow rate | MFR of resin, MFR of geomembrane | \|MFR\| ≤ 15 %, \|MFR\|: absolute value of difference in MFR between resin and geomembrane | DIN ISO 1133 |
| 1.9 | Dimensional stability | Absolute value of shrinkage \|δL\| of sides of square sample | \|δL\| ≤ 1.0 % for all individual values | Oven curing and shrinkage determination based on DIN 16726, Section 5.13.1, DIN 53377 and draft DIN EN 495-1, Oven exposure at (120±2) °C for 1 hour, shrinkage determination on 10 cm square specimens taken every 1.0 m over the width of the geomembrane. Round off shrinkage value to nearest 0.1 %, see Section B4 of Supplemental Information on Tests[1] |
| 1.10 | Permeability to hydro-carbons | Permeation rate of trichloroethylene at 23 °C | < 80 g/m²d, determined from the steady-state line | Steady-state determination at 23 °C on specimens with 80 mm effective diameter and 2.5 mm thickness based on DIN 53532 |
| | | Permeation rate of acetone at 23 °C | < 0.5 g/m²d, determined from the steady-state line | |
| 1.11 | Oxidative stability[2] | Oxidation Induction Time (OIT) | At 210 °C: 20 min | Determination based on DIN EN 728 |

[1]) Available from Laboratory IV.32, Landfill Engineering, BAM/Berlin.
[2]) The efficiency of antioxidants depends on temperature. An antioxidant which is highly effective at normal ambient temperatures may not be effective at the high temperatures used in the OIT procedure. This has been taken into account in the requirements given under points 1.11 and 3.3. Depending on the stabilization formulation used as disclosed to BAM, in certain cases other analytic procedures are required to determine the change in stabilization caused by immersion testing.

**Table A1.2.** Mechanical Requirements

| No. | Properties | Test Attributes | Requirements | Tests/Test Conditions |
|---|---|---|---|---|
| 2.1 | Behaviour under multiaxial tension | Bulge elongation $\varepsilon_b$ in multiaxial tension test | $\varepsilon_b \geq 15$ %, without yielding | Based on DIN 53861-1 and ASTM D5617 |
| 2.2 | Tensile properties | Yield strength $\sigma_Y$ Elongation at yield $\varepsilon_Y$, Elongation at break $\varepsilon_B$ in both extrusion and transverse directions | $\sigma_Y \geq 15$ N/mm$^2$ $\varepsilon_Y \geq 10$ %, $\varepsilon_B \geq 400$ %, | Tensile test in accordance with DIN EN ISO 527-3, test specimen type 5, (Specimen thickness need not be in conformance with this standard), 23 °C/50 % R.H., crosshead speed 50 mm/min up to 20 % elongation, thereafter 200 mm/min, 5 test specimens for each direction (extrusion and transverse) taken over the entire width of the geomembrane |
| 2.3 | Tear resistance | Tear force | Tear force ≥ 500 N,<br><br>Tear force ≥ 300 N | Tear test in accordance with DIN 53356-A, Tear test (Graves-type crescent specimen) DIN 53515, on specimens taken in extrusion and transverse directions |
| 2.4 | Resistance to static puncture | Puncture force | ≥ 6000 N | Static puncture test DIN EN ISO 12236, crosshead speed 50 mm/min |

**Table A1.2, continued.** Mechanical Requirements

| No. | Properties | Test Attributes | Requirements | Tests/Test Conditions |
|---|---|---|---|---|
| 2.5 | Resistance to dynamic puncture | Water tightness at spot of puncture | No loss of water tightness | Test in accordance with DIN 16726, Section 5.12, drop height 2000 mm |
| 2.6 | Cold temperature brittleness (flexural) | Visual appearance of flex edge | No cracks | Flexural testing in accordance with DIN EN 1876-1, flex edge in extrusion and transverse directions |
| 2.7 | Relaxation behaviour | Stress as a function of time at constant strain (relaxation curve) | Stress at 1000 hours must be ≤ 50 % of stress at 1 minute | Stress relaxation test in accordance with DIN 53441, 3 % strain, 23 °C, 50 % R.H., specimens taken in extrusion and transverse directions |
| 2.8 | Seam quality | Deformation and failure behaviour under shear stress | No shear and no failure of seam, pronounced necking in parent material | Shear test according to DVS R 2226-2, crosshead speed 50 mm/min |
| | | Deformation and failure behaviour under peel stress | No peeling of seam, pronounced necking in parent material | Peel test according to DVS R 2226-2, crosshead speed 50 mm/min |

**Table A1.3, Part 1.** Requirements for Physical Resistance and Long-Term Behaviour

| No. | Properties | Test Attributes | Requirements | Tests/Test Conditions |
|---|---|---|---|---|
| 3.1 | Resistance to chemicals (concentrated liquid solutions)[1] | Change in weight, change in yield strength and elongation at yield | Change in weight after re-drying ≤ 10 % Change in yield strength and elongation at yield ≤ 10 % | Immersion testing on the basis of DIN ISO 175. Immersion temperature: 23 °C. Immersion must be over a period of 90 days or to attainment of steady-state weight, whichever occurs first. Tensile tests to be conducted on re-dried specimens[1]. |
| 3.2 | Resistance to stress cracking | Time to failure in stress crack test | ≥ 200 h at test stress equal to 30 % of the yield strength (determined at 23 °C/50 % R.H.) | Notched constant tensile load test in accordance with ASTM D 5397 |
| 3.3 | Resistance to thermal oxidative degradation in air | Change in appearance; | No change, see Table A1.1.1 | Oven ageing in forced convection oven at 80 °C for 1 year, |
| | | relative change in elongation at break $\varepsilon_B$ | no change beyond experimental uncertainty | |
| | | OIT time after 1/2 year ageing: OIT(0.5 y); | Average time ≥ 10 min | OIT determination based on DIN EN 728 and ASTM D 3895 at 210 °C in aluminium pan |
| | | Relative change in OIT time: (OIT(0.5y)–OIT(1y))/OIT(0.5y) | ≤ 0.3 calculated from average values | |

1) Test liquids, see Table 3.7 in Chap. 3 of this book. Substances 2, 3, 6, 7, 10 and 11 need not be tested for HDPE geomembranes

**Table A1.3, Part 1, Continued.** Requirements for Physical Resistance and Long-Term Behaviour

| No. | Properties | Test Attributes | Requirements | Tests/Test Conditions |
|---|---|---|---|---|
| 3.4 | Long-term behaviour under combined stressing[1] | Graph of time-to-failure vs. hoop stress in long-term pipe burst testing | Extrapolation (in accordance with the extrapolation time limitations given in DIN 16887) of time to failure vs. stress curves generated at high temperatures (e.g. 80 °C, 60 °C) must predict no failure after 50 years at a stress of 4 $N/mm^2$ and a temperature of 40 °C. | Long-term burst testing in accordance with DIN 16887 on pipe specimens extruded or otherwise fabricated from the resin used to make the geomembrane |

[1]) Service life prediction of HDPE geomembranes is based on extrapolation of long-term pipe burst test curves in accordance with DIN 16887 or ISO TR 9080. The validity of this procedure is backed up by scientific studies and decades of experience. Pipe burst test curves are not required for resins for geomembranes if their stress-crack resistance has been shown to fulfil the requirements of point 3.2 above and their oxidative stability has been shown to fulfil the requirements of point 3.3 above in a test analogous to DIN 16887 pipe burst testing – i.e. involving immersion of geomembrane specimens in water at 80 °C for a period of at least $10^4$ h (e.g. long-term tensile test).

**Table A1.3, Part 2.** Requirements for Physical Resistance and Long-Term Behaviour

| No. | Properties | Test Attributes | Requirements | Tests/Test Conditions |
|---|---|---|---|---|
| 3.6 | Weathering resistance | Change in mechanical properties | See Sect. 3.2.14 of this book | See Sect. 3.2.14 of this book |
| 3.7 | Microbe resistance | Visual inspection, change in weight, change in mechanical properties | No substantial change in average values i.e., $\Delta m \leq 5$ %, $\Delta\varepsilon, \Delta\sigma \leq 10$ % | DIN EN ISO 846, Procedure D, soil burial testing in microbe-active soil for 1 year, tensile testing in accordance with DIN ISO 527, see Table A1.2, 2.2 |
| 3.8 | Root penetration resistance | Visual inspection | No penetration | FLL root penetration resistance procedure[1], from FLL Guidelines for Landscaped Roofs, tests on material with and without welding seams |
| 3.9 | Welding properties of resins[2] | Time to failure in long-term peel test | Geometric mean $\geq 35$ h | Long-term peel testing based on DVS R 2203-4 and DVS R 2226-3 of overlap welds with test channels made with hot-wedge welding machine using welding parameters in optimum field, see Chap. 10 of this book |

[1]) Landscape Research, Development & Construction Society (FLL, www.f-l-l.de). The FLL test procedure is currently being revised by CEN Technical Committee TC 254.

[2]) This time-to-failure requirement characterizes the welding properties of different resins. It is not suitable for evaluation of the quality of a welding seam per se. Welding seams not fabricated in the optimum parameter field which do not show times to failure typical of that resin are substandard even if their times to failure exceed 35 h.

**Table A1.4.** Additional Requirements for Textured Geomembranes

| No. | Properties | Test Attributes | Requirements | Tests/Test Conditions |
|---|---|---|---|---|
| 4.1 | Thickness in textured areas | Thickness | The thinnest point in the textured geomembrane must not be below the specified minimum (2.50 mm, or for thicker geomembranes, nominal minus 0.20 mm) | Measurement e.g. in accordance with ASTM D 5994-96 |
| 4.2 | Dimensional stability | Absolute value of shrinkage $\lvert\delta L\rvert$ of sides of square sample | $\lvert\delta L\rvert \leq 1.50$ % for individual values of embossed texturing<br>$\lvert\delta L\rvert \leq 1.00$ % for individual values of applied texturing | Table A1.1, 1.9 |
| 4.3 | Mechanical properties of textured areas | Elongation at break $\varepsilon_B$ | Requirement set for each case individually | Table A1.2, 2.2 |
| 4.4 | Texturing homogeneity of applied texturing | Nominal amount and variation of texturing area weight | Requirement set for each case individually | Determination on a specific specimen area (typically 100 cm²) |
| | | Variation in friction parameters | No substantial additional variation in friction parameters beyond test repeatability | Shear-box testing of 30 cm·30 cm specimens in accordance with GDA E 3-8 |
| | | Homogeneity of texturing appearance | Comparison with reference samples provided to BAM | Visual inspection |

**Table A1.4, continued.** Additional Requirements for Textured Geomembranes

| No. | Properties | Test Attributes | Requirements | Tests/Test Conditions |
|---|---|---|---|---|
| 4.5 | Texturing adhesion for applied texturing | Behaviour in shear-box test | No peel-off or removal of texturing | Shear-box testing under compressive loads typical of landfills in accordance with GDA E 3-8 |
| | | Scrape strength | Requirement set for each case individually | Test procedure of MPA Darmstadt |
| | | Time to failure in long-term shear test | Evaluation of time to failure based on DIN 16887 | Long-term shear test, see Sect. 3.2.18 of this book |
| 4.6 | Chemical resistance of bonding of texturing applied after production | Change in scrape strength | Change in average value ≤ 10 % | Immersion in substances 5 and 9, see Table A1.3, No. 3.1 |
| 4.7 | Stress-crack resistance (long-term tensile test) | Time to failure | Geometric mean of times to failure 700 h | Long-term tensile test based on DVS R2203-4 on at least 5 test specimens at 80 °C und 4 N/mm² tensile stress in a 2 % surfactant solution |

**Table A1.5.** Test Procedures and Frequencies for Incoming Quality Control of Resin and Carbon Black Batch

| No. | Test Attributes | Tests / Test Specimens | Frequency | Requirements and Tolerances |
|---|---|---|---|---|
| 5.1 | Density | DIN 53479, Procedure A, extrudate from MFR determination of resin or base polymer pellets | Random samples from every shipment | Requirement defined in certification report |
| 5.2 | Melt index MFR 190/5 or MFR 190/21.6 | DIN ISO 1133 / resin or base polymer pellets | Random samples from every shipment | Requirement defined in certification report |
| 5.3 | Weight percent carbon black | Thermogravimetric analysis based on DIN EN ISO 11358 (see Table A1.1, No. 1.3) or determination by ASTM 1603-76 / batch pellets | Random samples from every shipment | Requirement defined in certification report |
| 5.4 | Weight percent volatiles, moisture | Weight loss in oven (DIN EN 12099) or in infrared oven / resin pellets or base polymer pellets and batch pellets | Random samples from every shipment and before every production start-up, or at least once per production week | < 0.10 wt % in pre-compounded resin or base polymer<br>< 0.25 wt % in batch |
| 5.5 | Bulk density[1] | DIN EN ISO 60 and DIN 53466/ base polymer pellets and batch pellets | Random samples from every shipment and before every production start-up, or at least once per production week | Metering procedure and process set in quality management manual |

[1]) Required only in the event volumetric batch metering is used.

**TableA1.6, Part 1.** Test Procedures and Frequencies for In-House Quality Control of Geomembranes by Manufacturer

| No. | Test Attribute | Tests / Test Specimens | Frequency | Requirements and Tolerances |
|---|---|---|---|---|
| 6.1 | Thickness | Table A1.1, 1.7 / measured at a minimum of 10 different points across the width of the geomembrane | Continuous automatic monitoring[1] with mechanical control measurement every 300 m | Requirement defined in certification report; roll certificate must show at least the minimum und maximum values of control measurement |
| 6.2 | Visual appearance | Table A1.1, 1.1 | Continuous | Table A1.1, 1.1; roll certificate must confirm satisfactory appearance |
| 6.3 | Skew and waviness | Table A1.1, 1.5 and 1.6 | Per production start up[2] | Table A1.1, 1.5 and 1.6; roll certificate must confirm in-spec skew and waviness |
| 6.4 | Weight percent carbon black[5] | Table A1.1, 1.3 | Per production start-up and change in batch lot[3] and at least every 900 m | Requirement defined in certification report; roll certificate must give test procedure and individual values obtained |

[1]) In production lines without continuous automatic thickness monitoring, thickness must be measured over the entire production width every 10 m using an ultrasonic gauge.

[2]) Definition of production start-up: restart after machine stoppage or change of resin or thickness.

[3]) A higher test frequency may be specified in individual cases after a production start-up or change of batch lot.

**Table A1.6, Part 1, continued.** Test Procedures and Frequencies for In-House Quality Control of Geomembranes by Manufacturer

| No. | Test Attribute | Tests / Test Specimens | Frequency | Requirements and Tolerances |
|---|---|---|---|---|
| 6.5 | Homogeneity of carbon black distribution[5] | Table A1.1, 1.4 | Per production start-up and change in batch lot[3] and at least every 900 m | Table A1.1, 1.4; roll certificate must confirm homogeneous distribution |
| 6.6 | Yield strength, Elongation at yield, Elongation at break | Table A1.2, 2.2; Crosshead speed: 50 mm/min up to 20 % elongation, thereafter 200 mm/min / One test specimen each in extrusion and transverse directions from the edge and the middle of smooth geomembranes or from the land area of textured geomembranes[4] | Per production start-up and at least every 300 m | Requirement defined in certification report; roll certificate must show the minimum und maximum values determined in extrusion and transverse directions |

[1]) In production lines without continuous automatic thickness monitoring, thickness must be measured over the entire production width every 10 m using an ultrasonic gauge.

[2]) Definition of production start-up: restart after machine stoppage or change of resin or thickness.

[3]) A higher test frequency may be specified in individual cases after a production start-up or change of batch lot.

[4]) The certification report will specify the sampling procedure for test specimens from textured areas and the criteria to be used in their evaluation.

[5]) Only required when carbon black (batch) is added by the geomembrane manufacturer.

**Table A1.6, Part 2.** Test Procedures and Frequencies for In-House Quality Control of Geomembranes by Manufacturer

| No. | Test Attributes | Tests / Test Specimens | Frequency | Requirements and Tolerances |
|---|---|---|---|---|
| 6.7 | Melt index and change in melt index | Table A1.1, 1.8 / samples from the geomembrane and the texturing material | Per production start-up and at least every 900 m | Requirement defined in certification report; roll certificate must show the MFR determined in 5.2 of Table A1.5 and difference to MFR of the resin |
| 6.8 | Dimensional stability | Table A1.1, 1.9 / specimens taken from the edges, the middle of the geomembrane and from other critical locations (e.g. transition between smooth and textured areas) | Per production start-up and at least every 900 m | Table A1.1, 1.9; roll certificate must show the individual values determined and their respective sampling points |
| 6.9 | Area weight of texturing for applied texturing | In-house procedure | Per production start-up and at least every 300 m | Requirement set defined in certification report; roll certificate must show the minimum und maximum values determined |
| 6.10 | Texturing adhesion for applied texturing | In-house procedure | Per production start-up and at least every 300 m | Requirement defined in certification report; roll certificate must confirm satisfactory adhesion |

**Table A1.7.** Test Procedures and Frequencies for Third-Party Inspection of Resin, Carbon Black Batch and Geomembranes

| No. | Test Attributes | Tests / Test Specimens | Requirements and Tolerances[1] |
|---|---|---|---|
| 7.1 | Density | Table A1.5, 5.1 | Set in certification report |
| 7.2 | Melt index MFR 190/5 or MFR 190/21.6 | Table A1.5, 5.1 | Set in certification report |
| 7.3 | Change in melt index | Table A1.6, 6.7 | Set in certification report |
| 7.4 | Thickness | Table A1.6, 6.1 | Set in certification report |
| 7.5 | Visual appearance of surface and cross-section | Table A1.1, 1.1 and 1.2 | Table A1.1, 1.1 and 1.2 |
| 7.6 | Visual appearance of identification markings | Visual inspection | Section 2.5 and set in certification report |
| 7.7 | Weight percent carbon black | Table A1.1, 1.3 / geomembrane | Set in certification report |
| 7.8 | Homogeneity of carbon black distribution | Table A1.1, 1.4 / geomembrane | Table A1.1,1.4 |
| 7.9 | Dimensional stability | Table A1.6, 6.8 | Table A1.1, 1.9 and Table A1.4, 4.2 |
| 7.10 | Multiaxial bulge elongation | Table A1.2, 2.1 | Table A1.2, 2.1 |
| 7.11 | Yield strength<br>Elongation at yield<br>Elongation at break | Table A1.6, 6.6 | Set in certification report |
| 7.12 | Static puncture resistance | Table A1.2, 2.4 | Table A1.2, 2.4 |
| 7.13 | Area weight of texturing (applied texturing only) | Table A1.6, 6.9 | Table A1.6, 6.9 |
| 7.14 | Adhesion of texturing (applied texturing only) | Table A1.4, 4.5 and Table A1.6, 6.10 | Table A1.4, 4.5 and Table A1.6, 6.10 |
| 7.15 | Oxidation induction time (OIT) | Based on DIN EN ISO 728 | Table A1.1, 1.11; statistically significant correlation[2] with specification for resin |
| 7.16 | Type and content of tracer | Set on case-by-case basis, test carried out by BAM / geomembrane | Confidential, disclosed to BAM |

[1]) In general, the requirements specified in the requirement tables must be fulfilled. Additional requirements and tolerances characterizing the particular properties of the specific geomembrane certified are given in Appendix 1 of the certification document.

# Appendix 2

## Index of Standards, Guidelines and Recommendations

The following tables compile the standards, guidelines and recommendations which have been mentioned, quoted or discussed in this book. In the first table A2.1 the European and international Standard of the European Committee for Standardization (CEN, www.cenorm.be) and of the International Organization for Standardization (ISO, www.iso.org) are listed. The date refers to the issue of the German edition of the standard. The date of issue in other European Countries might be different. In table A2.2 standards of the American Society for Testing and Materials (ASTM, www.astm.org) are listed. A table A2.3 follows which contains German language standards mostly of the German Institute for Standardisation (Deutsches Institut für Normung (DIN), www.din.de or www.beuth.de). Normally these standards are only available in German. The last table A2.4 compiles guidelines and recommendations from various organisations, above all from the German Society for Welding Technology and Associated Methods (Deutschen Verband für Schweißen und verwandte Verfahren e. V. (DVS), www.dvs-ev.de) and from the German Society for Geotechnical Engineering (Deutsche Gesellschaft für Geotechnik e. V. (DGGt), www.dggt.de). However, most of these documents are available in English.

**Table A2.1.** CEN and ISO standards

| | | |
|---|---|---|
| EN ISO 60 | 1999 | Plastics – Determination of Apparent Density of Material that can be Poured from a Specified Funnel |
| EN 117 | 2005 | Wood Preservatives – Determination of Toxic Values against Reticulitermes Species (European Termites) (Laboratory Method) |
| EN ISO 175 | 2000 | Plastics – Methods of Test for the Determination of the Effects of Immersion in Liquid Chemicals |
| EN ISO 291 | 2005 | Plastics –Standard Atmospheres for Conditioning and Testing |
| EN 495-1 (Draft) | 1991 | Thermoplastic and Elastomeric Roofing and Sealing Sheets; Determination of High Temperature Dimensional Stability |
| EN 495-5 | 2000 | Flexible Sheets for Waterproofing – Determination of Foldability at Low Temperature – Part 5: Plastic and Rubber Sheets for Roof Waterproofing |
| EN ISO 527-1 | 1996 | Plastics – Determination of Tensile Properties – Part 1: General Principles |
| EN ISO 527-3 | 2002 | Plastics – Determination of Tensile Properties – Part 3: Test Conditions for Films and Sheets |
| EN 728 | 1997 | Plastics Piping and Ducting systems – Polyolefin Pipes and Fittings – Determination of Oxidation Induction Time |
| EN 837-1 | 1996 | Pressure gauges – Part 1: Bourdon Tube Pressure Gauges – Dimensions, Metrology, Requirements and Testing |
| EN 837-2 | 1996 | Pressure Gauges – Part 2: Selection and Installation Recommendations for Pressure Gauges |
| EN 837-3 | 1996 | Pressure Gauges – Part 3: Diaphragm and Capsule Pressure Gauges – Dimensions, Metrology, Requirements and Testing |
| EN ISO 846 | 1997 | Plastics – Evaluation of the Action of Microorganisms |
| EN ISO 877 | 1997 | Plastics – Methods of Exposure to Direct Weathering, to Weathering Using Glass-filtered Daylight and to Intensified Weathering by Daylight Using Fresnel Mmirrors |
| EN 1107-2 | 2001 | Flexible sheets for waterproofing – Determination of Dimensional Stability – Part 2: Plastic and Rubber Sheets for Roof Waterproofing |
| EN ISO 1133 | 2005 | Plastics – Determination of the Melt Mass-flow Rrate (MFR) and the Melt Volume-flow Rate (MVR) of Thermoplastics |
| EN ISO 1183-1 | 2004 | Plastics – Methods for Determining the Density of Non-cellular Plastics – Part 1: Immersion Method, Liquid Pyknometer Method and Titration Method |

**Table A2.1, Part 2.** CEN and ISO standards

| | | |
|---|---|---|
| EN ISO 1183-2 | 2004 | Plastics – Methods for Determining the Density of Non-cellular Plastics – Part 2: Density Gradient Column Method |
| EN ISO 1628-1 | 1998 | Plastics - Determination of the Viscosity of Polymers in Dilute Solution Using Capillary Viscometers – Part 1: General Principles |
| EN ISO 1628-3 | 2003 | Plastics – Determination of Viscosity of Polymers in Dilute Solution Using Capillary Viscometers – Part 3: Polyethylenes and Polypropylenes |
| EN 1847 | 2001 | Flexible sheets for Waterproofing – Plastic and Rubber Sheets for Roof Waterproofing – Methods for Exposure to Liquid Chemicals, Including Water |
| EN 1848-2 | 2001 | Flexible Sheets for Waterproofing – Determination of Length, Width, Straightness and Flatness – Part 2: Plastic and Rubber Sheets for Roof Waterproofing |
| EN 1849-2 | 2001 | Flexible Sheets for Waterproofing – Determination of Thickness and Mass per Unit Area – Part 2: Plastic and Rubber Sheets for Roof Waterproofing |
| EN 1850-2 | 2001 | Flexible Sheets for Waterproofing – Determination of Visible Defects – Part 2: Plastic and Rubber Sheets for Roof Waterproofing |
| EN ISO 1872-1 | 1999 | Plastics - Polyethylene (PE) Moulding and Extrusion Materials - Part 1: Designation System and Basis for Specifications |
| EN 1876-1 | 1998 | Rubber or Plastics Coated Fabrics – Low Temperatures Tests – Part 1: Bending Test |
| EN ISO 2286-3 | 1998 | Rubber– or Plastic-coated Fabrics - Determination of Roll Characteristics – Part 3: Method for Determination of Thickness |
| EN ISO 3146 | 2002 | Plastics – Determination of Melting Behaviour (Melting Temperature or Melting Range) of Semi-crystalline Polymers by Capillary Tube and Polarizing-microscope Methods |
| EN ISO 4599 | 1997 | Plastics – Determination of Resistance to Environmental Stress Cracking (ESC) – Bent Strip Method |
| EN ISO 4600 | 1998 | Plastics – Determination of Environmental Stress Cracking (ESC) – Ball or Pin Impression Method |
| EN ISO 6252 | 1998 | Plastics – Determination of Environmental Stress Cracking (ESC) – Constant-Tensile-Stress Method |
| EN ISO 9000 | 2005 | Quality Management Systems – Fundamentals and Vocabulary |
| EN ISO 9001 | 2000 | Quality Management Systems – Requirements |
| EN ISO 9004 | 2000 | Quality Management Systems – Guidelines for Performance Improvements |

**Table A2.1, Part 3.** CEN and ISO standards

| | | |
|---|---|---|
| EN ISO 9863-1 | 2003 | Geosynthetics – Determination of Thickness at Specified Pressures – Part 1: Single Layers |
| EN ISO 9864 | 2005 | Geosynthetics – Test Method for the Determination of Mass per Unit Area of Geotextiles and Geotextile-related Products |
| EN 10204 | 2004 | Metallic products – Types of inspection documents |
| EN ISO 10319 | 1996 | Geotextiles – Wide-width Tensile Test |
| EN ISO 11357-1 | 1997 | Plastics – Differential scanning calorimetry (DSC) – Part 1: General principles |
| EN ISO 11358 | 1997 | Plastics – Thermogravimetry (TG) of Polymers – General Principles |
| EN ISO 11403 | 2001 | Plastics – Acquisition and Presentation of Comparable Multipoint Data – Part 3: Environmental Influences on Properties |
| EN 12099 | 1997 | Plastics Piping Systems – Polyethylene Piping Materials and Components – Determination of Volatile Content |
| EN 12224 | 2000 | Geotextiles and Geotextile-related Products – Determination of the Resistance to Weathering |
| EN 12225 | 2000 | Geotextiles and Geotextile-related Products – Method for Determining the Microbiological Resistance by a Soil Burial Test |
| EN ISO 12236 | 1996 | Geotextiles and Geotextile-related products – Static Puncture Test (CBR-Test) |
| EN 12814-1 | 1999 | Testing of Welded Joints of Thermoplastics Semi-finished Products – Part 1: Bend Test |
| EN 12814-2 | 2000 | Testing of Welded Joints of Thermoplastics Semi-finished Products – Part 2: Tensile Test |
| EN 12814-4 | 2001 | Testing of Welded Joints of Thermoplastics Semi-finished products – Part 4: Peel Test |
| EN ISO 12957-1 | 2005 | Geosynthetics – Determination of Friction Characteristics – Part 1: Direct Shear Test. |
| EN 13067 | 2003 | Plastics Welding Personnel – Approval Testing of Welders – Thermoplastics Welded Assemblies |
| EN ISO 13438 | 2004 | Geotextiles and geotextile-related products – Screening Test Method for Determining the Resistance to Oxidation |
| EN 13719 | 2002 | Geotextiles and Geotextile-Related Products – Determination of the Long-term Protection Efficiency of Geotextiles in Contact with Geosynthetic Barriers |
| prEN 13948 | 2000 | Flexible sheets for waterproofing – Bitumen, Pplastic and Rubber Sheets for Roof Waterproofing – Determination of Resistance to Root Penetration |
| EN 14151 | 2001 | Geoynthetics – Determination of Burst Strength |

**Table A2.1, Part 4.** CEN and ISO standards

| | | |
|---|---|---|
| EN 14576 | 2005 | Geosynthetics – Test Method for Determining the Resistance of Polymeric Geosynthetic Barriers to Environmental Stress Cracking |
| EN ISO/IEC 17020 | 2004 | General Criteria for the Operation of Various Types of Bodies Performing Inspections |
| EN ISO/IEC 17025 | 2005 | General requirements for the competence of testing and calibration laboratories |
| EN ISO 19011 | 2002 | Guidelines on Quality and/or Environmental Management Systems – Auditing |
| EN 29073-1 | 1992 | Textiles – Test Methods for Nonwovens – Part 1: Determination of Mass per Unit Area |
| EN 29073-3 | 1992 | Textiles – Test Methods for Nonwovens – Part 3: Determination of Tensile Strength and Elongation |
| EN 45002 | 1990 | General Criteria for Expert Opinions on Testing Laboratories |
| EN 45003 | 1995 | Calibrating and Testing Laboratories Accreditation System – General Requirements for Operation and Recognition |
| ISO 9080 | 2003 | Plastic Piping and Ducting Systems – Determination of the Long-term Hydrostatic Strength of Thermoplastics Materials in Pipe by Extrapolation |
| ISO 11357-3 | 1999 | Plastics – Differential Scanning Calorimetry (DSC) – Part 3: Determination of Temperature and Enthalpy of Melting and Crystallization |
| ISO/TR 13434 | 1998 | Geotextiles and Geotextile-related Products – Guidelines on Durability |
| ISO 18553 DAM 1 | 2006 | Method for the Assessment of the Degree of Pigment or Carbon Black Dispersion in Polyolefin Pipes, Fittings and Compounds |

**Table A2.2.** ASTM standards

| | | |
|---|---|---|
| ASTM E145 | 1994 | Standard Specification for Gravity-Convection And Forced-Ventilation Ovens |
| ASTM E473 | 2000 | Standard Terminology Relating to Thermal Analysis |
| ASTM D638 | 1999 | Standard Test Method for Tensile Properties of Plastics |
| ASTM F739a | 1999 | Standard Test Method for Resistance of Protective Clothing Materials to Permeation by Liquids or Gases Under Conditions of Continuous Contact |
| ASTM D792 | 1998 | Standard Test Methods for Density and Specific Gravity (Relative Density) of Plastics by Displacement |
| ASTM D883 | 1996 | Standard Terminology Relating to Plastics |
| ASTM D1004 | 1994 | Standard Test Method for Initial Tear Resistance of Plastic Film and Sheeting |
| ASTM D1204 | 1994 | Standard Test Method for Linear Dimensional Changes of Nonrigid Thermoplastic Sheeting or Film at Elevated Temperature |
| ASTM D1238 | 1999 | Standard Test Method for Flow Rates of Thermoplastics by Extrusion Plastometer |
| ASTM D1248 | 1984 | Standard Specification for Polyethylen Plastics Molding and Extrusion Materials |
| ASTM D1505 | 1998 | Standard Test Method for Density of Plastics by the Density-Gradient Technique |
| ASTM D1598 | 2000 | Standard Test Method for Time-to Failure of Plastic Pipe Under Constant Internal Pressure |
| ASTM D1603 | 1994 | Standard Test Method for Carbon Black In Olefin Plastics |
| ASTM D1693 | 1997a | Standard Test Method for Environmental Stress-Cracking of Ethylene Plastics |
| ASTMD2837 | 2001a | Standard Test Method for Obtaining Hydrostatic Design Basis for Thermoplastic Pipe Materials |
| ASTM D2839 | 1987 | Standard Practice for Use of a Melt Index Strand for Determining Density of Polyethylene |
| ASTM D3895 | 1997 | Standard Test Method for Oxidative-Induction Time of Polyolefins by Differential Scanning Calorimetry |
| ASTM D4218 | 1996 | Standard Test Method for Determination of Carbon Black Content in Polyethylene Compounds by the Muffle-Furnace Technique. |
| ASTM 4355 | 2002 | Standard Test Method for Deterioration of Geotextiles by Exposure to Light, Moisture and Heat in a Xenon Arc Type Apparatus |
| ASTM D4437 | 1988 | Practice for Determining the Integrity of Field Seams Used in Joining Flexible Polymeric Sheet Geomembranes |
| ASTM D4833 | 2000 | Standard Test Method for Index Puncture Resistance of Geotextiles, Geomembranes, and Related Products |

**Table A2.2, Part 2.** ASTM standards

| | | |
|---|---|---|
| ASTM D4883 | 1999 | Standard Test Method for Density of Polyethylene by Ultrasound Technique |
| ASTM D5199 | 1995 | Standard Test Method for Measuring Nominal Thickness of Geotextiles and Geomembranes |
| ASTM D5321 | 1997 | Standard Test Method for Determining the Coefficient of Soil and Geosynthetic or Geosynthetic and Geosynthetic Friction by the Direct Shear Method |
| ASTM D5322 | 1992 | Standard Practice for Immersion Procedures for Evaluating the Chemical Resistance of Geosynthetics to Liquids |
| ASTM D5323 | 1999 | Standard Practice for Determination of 2 % Secant Modulus for Polyethylene Geomembranes |
| ASTM D5397 | 1995 | Standard Test Method for Evaluation of Stress Crack Resistance of Polyolefin Geomembranes Using Notched Constant Tensile Load Test |
| ASTM D 5514 | 1994 | Test Method for Large Scale Hydrostatic Puncture Testing of Geosynthetics |
| ASTM D5596 | 1994 | Standard Test Method for Microscopic Evaluation of the Dispersion of Carbon Black in Polyolefin Geosynthetics |
| ASTM D5617 | 1994 | Standard Test Method for Multi-Axial Tension Test of Geosynthetics |
| ASTM D5641 | 1994 | Practice for Geomembrane Seam Evaluation by Vacuum Chamber |
| ASTM D5721 | 1995 | Standard Practice for Air-Oven Aging of Polyolefin Geomembranes |
| ASTM D5747 | 1995a | Standard Practice for Tests to Evaluate the Chemical Resistance of Geomembranes to Liquids |
| ASTM D5820 | 1995 | Practice for Pressurized Air Channel Evaluation of Dual Seamed Geomembranes |
| ASTM D5885 | 1995 | Standard Test Method for Oxidative-Induction Time of Polyolefin Geosynthetics by High-Pressure Differential Scanning Calorimetry |
| ASTM D5886 | 1995 | Standard Guide for Selection of Test Methods to Determine Rate of Fluid Permeation Through Geomembranes for Specific Application |
| ASTM D5994 | 1996 | Standard Test Method for Measuring Core Thickness of Textured Geomembrane |
| ASTM D6365 | 1995 | Standard practice for the Nondestructive Testing of Geomembrane Seams using the Spark Test |
| ASTM D6392 | 1999 | Standard Test Method for Determining the Integrity of Nonreinforced Geomembrane Seams Produced Using Thermo-Fusion Methods |
| ASTM D6747 | 2004 | Standard Guide for Selection of Techniques for Electrical Detection of Potential Leak Path in Geomembranes |

**Table A2.2, Part 3.** ASTM standards

| | | |
|---|---|---|
| ASTM D6693 | 2001 | Standard Test Method for Determining Tensile Properties of Nonreinforced Polyethylene and Nonreinforced Flexible Polypropylene Geomembranes |
| ASTM D7002 | 2003 | Standard Practice for Leak Location on Exposed Geomembranes Using the Water Puddle System |
| ASTM D7007 | 2003 | Standard Practices for Electrical Methods for Locating Leaks in Geomembranes Covered with water or Earth Materials |

**Table A2.3.** National German-language standards

| | | |
|---|---|---|
| DIN 8075 | 1999 | Polyethylene (PE) Pipes - PE 63, PE 80, PE 100, PE-HD - General Quality Requirements, Testing |
| DIN 8075 attachment 1 | 1984 | High Density Polyethylene (HDPE) Pipes; Chemical Resistance of Pipes and Fittings |
| DIN 16726 | 1986 | Plastic Roofing Felt and Waterproofing Sheet; Testing |
| DIN 16739 | 1994 | Geomembranes from Polyethylene (PE) for Landfill Liners; Requirements, Tests |
| DIN 16776-1 | 1984 | Polyethylene and Ethylene Copolymer Thermoplastics; Classification and Designation |
| DIN 16887 | 1990 | Determination of the Long-term Hydrostatic Pressure Resistance of Thermoplastics Pipes |
| DIN 18200 cancelled | 1986 | Control (Quality Control) of Construction Materials, Construction Components and Construction Design, General Principles |
| DIN 18200 | 2000 | Assessment of Conformity for Construction Products – Certification of Construction Products by Certification Body |
| DIN 50011-1 cancelled | 1978 | Testing of Materials, Structural Components and Equipment, Warming Cabinet, Definitions, Requirements |
| DIN 50011-2 | 1960 | Testing of Materials, Structural Components and Equipment; Hot Cabinets; Directions for the Storage of Specimens |
| DIN 50011-11 | 1982 | Climates and their Ttechnical Application; Controlled-atmosphere Test Installations; General Terminology and Requirements |
| DIN 50011-12 | 1987 | Artificial Climates in Technical Applications; Air Temperature as a Climatological Quantity in Controlled-atmosphere Test Installations |
| DIN 50011-13 | 1991 | Technical Climatology; Climate-test-devices; Climatic Parameter: Air-humidity and Air-temperature |
| DIN 50014 | 1985 | Climates and their Technical Application; Standard Atmospheres |
| DIN 50035-1 | 1989 | Terms and definitions used on ageing of materials; basic terms and definitions |

**Table A2.3, Part 2.** National German-language standards

| | | |
|---|---|---|
| DIN 50035-2 | 1989 | Terms and definitions used on ageing of materials; examples concerning polymeric materials |
| DIN 51005 | 1993 | Thermal analysis (TA) - Terms |
| DIN 53370 | 1976 | Testing of Plastic Films; Determination of theThickness by Mechanical Feeling |
| DIN 53377 | 1969 | Testing of Plastic Films; Determination of Dimensional Stability |
| DIN 53353 cancelled | 1971 | Testing of Artificial Leather and Similar Sheet Materials, Determination of Thickness with Mechanical Feelers |
| DIN 53356 | 1982 | Testing of Artificial Leather and Similar Sheet Materials; Tear Growth Test |
| DIN 53 441 cancelled | 1984 | Testing of Plastics, Stress Relaxation Test |
| DIN 53479 cancelled | 1976 | Testing of Plastics and Elastomers, Determination of Density |
| DIN 53521 | 1987 | Testing of Rubber and Elastomers, Determination of the Resistance to Liquids, Vapour and Gases |
| DIN 53532 cancelled | 1989 | Testing of Elastomers, Determination of Permeability of Elastomer-Sheetings to Liquids |
| DIN 53861-1 cancelled | 1992 | Testing of textiles, Vaulting Test and Bursting Test |
| DIN 55350-11 | 1995 | Concepts on Quality and Statistics – Part 11: Concepts of the Quality Management |
| DIN 53515 | 1990 | Testing of Rubber and of Plastic Films; Tear Test Using the Graves Angle Test Piece with Incision |
| ÖNORM S2073 | 2004 | Landfills – Polymer Membrane liners – Requirements and Test Methods – Marking of Conformity |
| ÖNORM S2076-1 | 1999 | Waste Disposal Facilities – Polymer Membrane Liners – Layers Installation |

**Table A2.4.** Guidelines and recommendation concerning plastic geomembranes in geotechnical engineering

| | | |
|---|---|---|
| GRI Standard GM 11 | 1997 | GRI standard GM 11 Accelerated Weathering of Geomembranes using s Fluorescent UVA-Condensation Exposure Devices |
| GRI Standard GM13 | 2006 | Test Properties, Testing Frequency and Recommended Warrant for High Density Polyethylene (HDPE) Smooth and Textured Geomembranes |
| GRI Standard GM19 | 2003 | Seam Strength and Related Properties of Thermally Bonded Polyolefin Geomembranes |
| BAM certification guideline | 2001 | Certification Guidelines for Plastic Geomembranes Used to Line Landfills and Contaminated Sites |
| BAM certification guidelines (protective layer) | 1995 | Requirements of Protective Layers for Geomembranes in Composite Liner Systems, Certification Guidelines |
| EPA/600/2-88/052 | 1991 | Technical Guidance Document: Inspection Techniques for the Fabrication of Geomembrane Field Seams |
| EPA Method 9090 | 1986 | Compatibility Tests for Wastes and Membrane Liners |
| DVS 2203-4 | 1997 | Testing Welded Joints on Thermoplastic Plates and Pipes – Long-term Tensile Test |
| DVS 2203-5 | 1999 | Testing of Welded Joints of Thermoplastics Plates and Tubes – Technological Bend Test |
| DVS M 2211 | 1979 | Filler Materials for Thermoplastics – Scope, Designation, Requirements and Tests |
| DVS 2225-1 | 1991 | Joining of Lining Membranes Made of Polymer Materials (Geomembranes) in Geotechnical and Hydraulic Applications – Welding, Adhesive Bonding and Vulcanisation. |
| DVS 2225-2 | 1992 | Joining of Lining Membranes Made of Polymer Material in Geotechnical and Hydraulic Engineering, Site Testing |
| DVS 2225-3 | 1997 | Joining of Lining Membranes Made of Polymer Materials (Geomembranes) in Geotechnical and Hydraulic Applications – Requirements for Welding Machines and Welding Devices |
| DVS 2225-4 | 1996 | Welding of Geomembranes from Polyethylene (PE) for Lining Landfills and Contaminated Land. |
| DVS 2226-1 | 2000 | Testing of Fused Joints on Liners Made of Polymer Materials – Testing Procedure, Requirements |
| DVS 2226-2 | 1997 | Testing of Fused Joints on Liners Made of Polymer Materials – Lap Shear Test |
| DVS 2226-3 | 1997 | Testing of Fusion on PE-Liner – Peel Testing |
| DVS 2226-4 | 2000 | Testing of Joints on Liners Made of Polymer Materials – Tensile Creep Test |

**Table A2.4, Part 2.** Guidelines and recommendation concerning plastic geomembranes in geotechnical engineering

| | | |
|---|---|---|
| GDA E 2-7 | 1998 | Resistance to Sliding of the Liner System |
| GDA E 2-21 | 1997 | Proof of Spreading stability and Estimation of Deformations for the Landfill Base |
| GDA E 2-27 | 1999 | Liner Penetration Systems |
| GDA E 3-8 | 1997 | Friction Behaviour of Geosynthetics |
| GDA E 3-9 | 1997 | Suitability Tests for Geosynthetics |
| GDA E 7-1 | 1997 | Stability Proof for Geosynthetic reinforcements |
| DIBT Geomembrane Certification Principles | 2000 | Certification Principles for Geomembranes used in Plants for Storage, Filling and Trans-shipment of Substances Dangerous to Water |
| BAM Test Recommendations | 1999 | Recommendation for Testing within the BAM Certification for Plastic Geomembranes, BAM Berlin, laboratory IV.32 |
| NRW-Guideline | 1985 | Landfill Liner System with Plastic Geomembranes, Environmental Protection Office, North Rhine-Westphalia, Essen |
| FLL-Guideline | 2002 | Green-Roof-Guideline, release 2002, www.f-l-l.de |
| Guideline 853 of Deutsche Bahn AG | 1999 | Design, Construction and Maintenance of Railway Tunnels, Part 10 Liners and Drainage |
| SIA Standard V280 | 1996 | Plastic Geomembranes (Polymer Geomembranes) – Threshold Values and Materials Testing |

# Index

**Y**

MIX
Papier aus verantwortungsvollen Quellen
Paper from responsible sources
FSC® C105338

If you have any concerns about our products,
you can contact us on
**ProductSafety@springernature.com**

In case Publisher is established outside the EU,
the EU authorized representative is:
**Springer Nature Customer Service Center GmbH**
**Europaplatz 3, 69115 Heidelberg, Germany**

Printed by Libri Plureos GmbH
in Hamburg, Germany